# Predicting
## Chemical Toxicity
# and Fate

# Predicting
## Chemical Toxicity
# and Fate

## Mark T.D. Cronin
Liverpool John Moores University
Liverpool, England

## David J. Livingstone
University of Portsmouth
Portsmouth, England

## CRC PRESS

Boca Raton   London   New York   Washington, D.C.

## Library of Congress Cataloging-in-Publication Data

Predicting chemical toxicity and fate / edited by Mark T.D. Cronin and David J. Livingstone.
    p.   cm.
  Includes bibliographical references and index.
  ISBN 0-415-27180-0 (alk. paper)
  1. Molecular toxicology. 2. Toxicological chemistry. 3. QSAR (Biochemistry).
I. Cronin, Mark T.D. II. Livingstone, D. (David). III. Title.

RA1220.3.P74 2004
615.9—dc22                                                          2004043999

### Visit the CRC Press Web site at www.crcpress.com

# Dedications

*From MC*

*To AMC and CCFC, for the pleasure and the pain.*

# Table of Contents

⮡ See Pg 161 – 162

# Contributors

**Romualdo Benigni**
Istituto Superiore di Sanità
Rome, Italy

**Jacob G. Bundy**
School of Biological Sciences
University of Aberdeen
Aberdeen, Scotland

**Colin D. Campbell**
Macaulay Institute
Aberdeen, Scotland

**Robert D. Combes**
Fund for the Replacement of Animals in
    Medical Experiments
Russell & Burch House
Nottingham, England

**Mark T.D. Cronin**
School of Pharmacy and Chemistry
Liverpool John Moores University
Liverpool, England

**John C. Dearden**
School of Pharmacy and Chemistry
Liverpool John Moores University
Liverpool, England

**Judith C. Duffy**
School of Pharmacy and Chemistry
Liverpool John Moores University
Liverpool, England

**Hong Fang**
Logicon ROW Sciences
Jefferson, AR, U.S.A.

**Huixiao Hong**
Logicon ROW Sciences
Jefferson, AR, U.S.A.

**Peter R. Fisk**
Peter Fisk Associates
Whitstable, England

**Klaus L.E. Kaiser**
TerraBase, Inc.
Hamilton, Ontario, Canada

**David J. Livingstone**
ChemQuest, Isle of Wight, England
Centre for Molecular Design, University of
    Portsmouth, Portsmouth, U.K.

**Helena Maciel**
School of Biological Sciences
University of Aberdeen
Aberdeen, U.K.

**Louise McLaughlin**
Peter Fisk Associates
Whitstable, England

**Monika Nendza**
Analytisches Laboratorium
Luhnstedt, Germany

**Tatiana I. Netzeva**
School of Pharmacy and Chemistry
Liverpool John Moores University
Liverpool, U.K.

**Graeme I. Paton**
School of Biological Sciences
University of Aberdeen
Aberdeen, Scotland

**Martin P. Payne**
LHASA Ltd.
Department of Chemistry
University of Leeds
Leeds, England

**Roger Perkins**
Logicon ROW Sciences
Jefferson, AR, U.S.A.

**Rosemary A. Rodford**
SoloSTAR Ltd.
Bedford, England

**T. Wayne Schultz**
College of Veterinary Medicine
University of Tennessee
Knoxville, TN, U.S.A.

**Gerrit Schüürmann**
Department of Chemical Ecotoxicology
UFZ Centre for Environmental Research
Leipzig, Germany

**Daniel M. Sheehan**
Food and Drug Administration's National
  Center for Toxicological Research,
  Jefferson, AR, U.S.A.
Daniel M. Sheehan and Associates,
  Little Rock, AR, U.S.A.

**Weida Tong**
Food and Drug Administration's National
  Center for Toxicological Research
Jefferson, AR, U.S.A.

**Cornelius J. Van Leeuwen**
Institute for Health and Consumer Protection,
  Joint Research Centre
European Commission
Ispra, Italy

**Gilman D. Veith**
International QSAR Foundation to Reduce
  Animal Testing
Duluth, MN, U.S.A.

**Rosalind J. Wildey**
Peter Fisk Associates
Whitstable, England

**Andrew P. Worth**
European Center for the Validation of
  Alternative Methods
Institute for Health and Consumer Protection,
  Joint Research Center
European Commission
Ispra, Italy

**Qian Xie**
Logicon ROW Sciences
Jefferson, AR, U.S.A.

# Foreword

When Corwin Hansch and Al Leo encouraged me in applying quantitative structure-activity relationships (QSARs) to the screening of environmental hazards, the U.S. Toxic Substances Control Act was still only a concept, and most QSAR calculations were still being made with a pencil. Their encouragement included two principles for QSAR along with a word of caution. The principles were that QSAR ought to be based on well-defined endpoints of intrinsic chemical activities as well as on molecular descriptors that could be interpreted mechanistically. The word of caution was that bureaucracies founded on laboratory testing, whether private or a regulatory agency, will only begrudgingly accept QSAR as a strategic tool in designing chemicals and managing chemical risks. Looking back over the last three decades, the Hansch/Leo principles for QSAR development have been largely ignored, if not disputed, by the growing QSAR community, with the possible exception in Europe where QSAR acceptance criteria will require transparency and a mechanistic foundation. Only the skepticism toward QSAR itself by our testing-oriented society seems to have been steadfast over three decades. The increasing costs of testing have produced renewed interest in more strategic *in silico* methods at a time when QSAR has been freed from many early computational barriers. Now more than ever, the scientific community needs an expert summary of QSAR methods like this book.

The guiding principles for QSAR development were intended to aid in the discovery of useful and robust models. The literature is replete with more than 10,000 QSAR correlations and models, yet few of them are useful enough to sway the skeptics. Still, progress in QSAR research can be measured by its own critics and the changing nature of their skepticism. The "*yes-but*" skeptics are particularly instructive to me. In 1974, our research plans faced the criticism, "*yes, QSAR may be able to predict some chemical properties, but it will never be able to predict bioaccumulation of chemical residues.*" In 1981, we faced, "*okay, QSAR may be able to predict bioconcentration potential, but it will never be able to predict toxicity.*" When the acute toxicity models appeared, we were confronted by "*yes, QSAR may be able to predict some ecotoxicity endpoints, but it will never predict chronic toxicity in mammals.*" Today, as the first mechanistic QSAR models are emerging for chronic reproductive effects and mutagenicity, this historical perspective on the QSAR skeptics serves as benchmarks for progress, if not encouragement.

Chemical reactivity in biological systems is far more complex than 20th century computational capabilities could have allowed one to address in quantitative terms. The rapid progress in computing power over the last decade enabled a steady stream of new computational methods in QSAR to emerge. Unfortunately, these new capabilities were not matched with the generation of high-quality biological databases needed to reveal systematic variation within heterogeneous chemical inventories. While many combinatorial problems in QSAR are likely to challenge computer sciences for years, present computer capabilities are sufficient to make future QSAR progress limited mostly by the databases for relevant, well-defined endpoints.

Our QSAR program at the Duluth, MN, U.S.A., laboratory focused on well-defined ecotoxicological endpoints that could be used directly in regulatory decisions. Our proof-of-concept paper in 1979 for estimating the bioconcentration potential required only a minimal database. Since then, many researchers have contributed to the evolution of bioaccumulation models and to extend them from simple screening-level methods for new chemicals to more exact estimates of tissue residues for risk assessments. In contrast to the bioconcentration database, the creation of the Duluth ecotoxicity database involved a multimillion dollar investment and dozens of scientists over most of a decade. Finding chemicals with common toxicity pathways to build mechanistic structure-toxicity relationships required better diagnostic bioassays, including behavioral symptomology (fish acute toxicity syndromes) and joint-toxicity studies. Our first paper on acute toxicity in 1983 was delayed almost 3 years due to rejections from toxicological journals based on our use of the term "narcosis" in describing reversible, baseline lethality — a criticism that lingers today in the health research community. The dozens of more recent supporting papers on baseline toxicity and the even larger toxicity database created by Terry Schultz at the University of Tennessee (Knoxville) should be sufficient to overcome the skeptics of acute toxicity predictions so that the full attention of effects research can focus on important chronic toxicity endpoints.

The European Chemical Industry Council-led analysis of the state of QSAR in Setubal, Portugal (March 2002), concluded that QSARs for biodegradability were still the largest research gap in exposure research. Developing QSARs for important chemical properties progressed rapidly in the 1980s, but developing structure-biodegradability models has been paralyzed by a lack of systematic databases. Fortunately, in 1985 Hiroshi Tadakoro at the Hita laboratory in Japan recognized the need for a biodegradation database, and his team devoted more than a decade to systematically testing chemicals using activated sludge. Almost immediately after the Hita database was made available, the first QSAR screening models for biodegradability began to appear at scientific meetings. Again, these advances illustrate the importance of generating systematic data on crucial endpoints in the overall progress of predictive methods. Finding such endpoints and understanding how they can be reliably used in risk management is the central research challenge for QSAR. Once identified, QSAR progress seems to depend only on government funding to generate the systematic data needed to build acceptable QSARs for the respective endpoints.

The estimation of lethality and biodegradability directly from chemical structure has been one of the important first steps in applying QSAR to risk management. Shifting our focus to chronic effects and persistence of chemicals will require us to cross some exciting new frontiers, not the least of which will be the merger of metabolism and effects models as QSAR is incorporated into systems biology. To meet these challenges, scores of chronic toxicity pathways will have to be described, and "-omics" technology promises to open new doors in clustering chemicals by common toxicity pathways for QSAR modeling. With metabolic activation a critical step in many pathways, metabonomics offers unprecedented capability for identifying the key molecular initiating events for chronic effects, many being the new well-defined endpoints QSAR needs for chronic hazard identification. It is hoped that this book will play an important role in advancing QSAR in the face of healthy skepticism, and will bring greater attention to the need for high-quality data in strategic testing.

**Dr. Gilman Veith**
Duluth, MN

# Preface

The motivation for this book was stimulated by a one-day meeting, "Modelling Environmental Fate and Toxicity," organized by the BioActive Sciences Group of the Society of Chemical Industry. The meeting was chaired by Drs. Mark Cronin and Dave Livingstone and held in London on March 27, 2001. The speakers at the meeting were drawn from industry and academia and described how computational methods could be applied to predict the toxicity and fate of chemicals in the environment. The meeting itself was well attended and was particularly timely. It coincided with an upsurge of interest in this area due both to legislative changes and the commercial possibilities of predicting toxicity and fate.

We are moving into a new era that is computationally rich and data poor. Modeling of toxicity is much easier than it was a decade ago because of increased computational power and greater availability of software to calculate descriptors of molecules (some of which is freely download-able). However, we must never lose sight of the fact that good models require high quality input data, and preferably large amounts of it. Neither should we forget that predictive techniques are empirical models to be used; they should not be seen as an academic exercise. In commissioning this book we attempted to bring together a collection of chapters that would assist future modelers develop meaningful predictive techniques. This was always hoped to be a practical and didactic book, there are plenty of published reviews in all areas covered in the book. All authors were encouraged to make recommendations for the use of the methods and techniques described. The editors support the recommendations and hope they will be applied and useful to the next generation of modelers.

**Mark Cronin and Dave Livingstone**
July 2003

# Acknowledgments

The editors wish to thank the BioActive Sciences Group of the Society of Chemical Industry (London, England) for the original opportunity to put on the one-day meeting that stimulated this volume. Without the group's foresight, encouragement, and organization, none of this would have been achievable. We also wish to thank the authors who have cheerfully contributed to the book, accepted our criticism, and made helpful comments. Finally we are extremely grateful to Taylor and Francis for originally commissioning the book and CRC Press for the final opportunity to publish it.

# List of Abbreviations

| | |
|---|---|
| $\chi$ | Randić branching index, or molecular connectivity |
| $\kappa$ | Kappa shape index |
| $\sigma$ | Hammett constant |
| $\mu$ | Dipole moment |
| $\Psi$ | Wave function characterizing the state of a system state |
| **AAR** | Activity-Activity Relationship |
| **ADME** | Absorption, Distribution, Metabolism, and Excretion |
| **AFP** | $\alpha$-Feto Protein |
| **AM1** | Austin Model 1 |
| **A$_{max}$** | Maximum acceptor superdelocalizability |
| **ANN** | Artificial Neural Network |
| **AO** | Atomic Orbital |
| **AR** | Androgen Receptor |
| **ARI** | Automated Rule-Induction |
| **ATSDR** | Agency for Toxic Substances and Disease Registry |
| **BBB** | Blood-Brain Barrier |
| **BCF** | Bioconcentration factor |
| **BESS** | Biodegradability Evaluation and Simulation System |
| **BgVV** | (German) Federal Institute for Health Protection of Consumers and Veterinary Medicine |
| **BMD** | BenchMark Dose |
| **BMF** | Biomagnification Factor |
| **BRM** | Carcinogenic potency in mice |
| **BRR** | Carcinogenic potency in rats |
| **B3LYP** | Hybrid density functional theory *ab initio* calculation method |
| **c** | Cluster (molecular connectivity) |
| **C** | Concentration (of a drug or toxicant) |
| **C** | Corrosive |
| **CADD** | Computer-Aided Drug Design |
| **CAS** | Chemical Abstract Service |
| **CASE** | Computer Automated Structure Evaluation |
| **CCOHS** | Canadian Center for Occupational Health and Safety |
| **CDER** | Center for Drug Evaluation and Research |
| **Cl** | Clearance |
| **CM** | Classification Model |
| **CM1** | Charge Model 1 |
| **CM2** | Charge Model 2 |
| **CODESSA** | COmprehensive DEscriptors for Structural and Statistical Analysis |
| **CoMFA** | Comparative Molecular Field Analysis |
| **COMPACT** | Computerized Optimized Parametric Analysis of Chemical Toxicity |
| **COREPA** | Common REactivity PAttern |
| **CPSA** | Charged Partial Surface Area |
| **CRADA** | Cooperative Research and Development Agreement |
| **CT** | Classification Tree |
| **CV** | Cross Validation |
| **D** | Distribution coefficient |
| **DBP** | Disinfection By Products |
| **DEREK** | Deductive Estimation of Risk from Existing Knowledge |

| | |
|---|---|
| **DES** | Diethylstilbestrol |
| **DFT** | Density Functional Theory |
| **DSC** | Differential Scanning Calorimetry |
| **DSL** | Domestic Substance List |
| **DSS** | Decision Support System |
| **DSSTox** | Distributed Structure-Searchable Toxicity |
| **E** | Hepatic extraction ratio |
| **EA** | Electron Affinity |
| $E_{1/2}$ | Half-wave oxidation potential |
| **ΔE** | Difference in the energies of the highest occupied and lowest unoccupied molecular orbitals |
| **EC** | Extended connectivity |
| **ECB** | European Chemical Bureau |
| $EC_{50}$ | Concentration causing 50% reduction in a specified effect |
| **ECOSAR** | Syracuse Research Corporation program to predict environmental toxicities |
| **ECVAM** | European Centre for the Validation of Alternative Methods |
| **EDC** | Endocrine Disrupting Chemical |
| **EDKB** | Endocrine Disruptor Knowledge Base |
| **EDPSD** | Endocrine Disruption Priority Setting Database |
| **EDSTAC** | Endocrine Disruptors Screening and Testing Advisory Committee |
| $E_{HOMO}$ | Energy of the Highest Occupied Molecular Orbital |
| $E_{kin}$ | Kinetic energy of a system |
| $E_{LUMO}$ | Energy of the Lowest Unoccupied Molecular Orbital |
| **EN** | Electronegativity |
| **EPIWIN** | Estimations Programs Interface for Windows |
| $E_{pot}$ | potential energy of a system |
| **ER** | Estrogen Receptor |
| **e-state** | Electrotopological state index |
| $E_{tot}$ | Total energy of a system |
| **EU** | European Union |
| **FDA** | Food and Drug Administration |
| **GI** | Gastrointestinal |
| **GST** | Glutathione S-Transferase |
| **FIRM** | Formal Inference-based Recursive Modeling |
| **H** | Harary index |
| *H* | Hamilton operator |
| **HD** | Hardness |
| **HENRYWIN** | Syracuse Research Corporation program to predict Henry's law constant |
| **HESI** | Health and Environmental Sciences Institute |
| **HF** | Hartree-Fock |
| **ΔHF** | Heat of Formation |
| **HPLC** | High Performance Liquid Chromatography |
| **HPV** | High Production Volume |
| **HPVC** | High production volume chemical |
| **HQSAR** | Hologram Quantitative Structure-Activity Relationship |
| $ICG_{50}$ | Concentration causing 50% inhibition of growth |
| **ILSI** | International Life Sciences Institute |
| **IP** | Ionization potential |
| **Ipb** | Isopropylbenzene |
| *Ind* | Induction |
| **ITC** | Interagency Testing Committee |

| | |
|---|---|
| **IUCLID** | International Uniform Chemicals Information Database |
| **JME** | Java Molecular Editor |
| **JRC** | Joint Research Centre of the European Commision |
| **K** | Partition coefficient |
| **$K_a$** | Equilibrium acid ionisation |
| **KBS** | Knowledge-Based Systems |
| **$K_i$** | Inhibition constant |
| **$K_m$** | Binding constant |
| **$K_{mxa}$** | Cuticular Matrix-Air partition coefficient |
| **KNN** | *K*-Nearest Neighbors |
| **$K_{oa}$** | Octanol-Air partition coefficient |
| **$K_{oc}$** | Soil-Water partition coefficient normalised for organic carbon content |
| **$K_{om}$** | Soil-Organic matter partition coefficient |
| **$K_{ow}$** | Octanol-Water partition coefficient |
| **$K'_{ow}$** | Apparent Octanol-Water partition coefficient |
| **KOWWIN** | Syracuse Research Corporation program to predict octanol-water partition coefficient |
| **$K_p$** | Skin permeability coefficients |
| **$K_{pa}$** | Plant-Air partition coefficient |
| **$K_{vs}$** | Vegetation-Soil partition coefficient |
| **LCAO** | Linear Combinations of Atomic Orbitals |
| **$LC_{50}$** | Lethal Concentration for 50% of animals |
| **$LD_{50}$** | Lethal Dose for 50% of animals |
| **LFER** | Linear Free Energy Relationship |
| **LSER** | Linear Solvation Energy Relationship |
| **LNO** | Leave-*N*-Out |
| **Log** | Logarithm to base 10 |
| **LOO** | Leave-One Out |
| **LRA** | Linear Regression Analysis |
| **MHBP** | Molecular Hydrogen Bond Potential |
| **MLP** | Multilayer Perceptron |
| **MLPot** | Molecular Lipophilicity Potential |
| **MLR** | Multiple Linear Regression |
| **MNDO** | Modified Neglect of Diatomic Overlap |
| **MO** | Molecular Orbital |
| **MOPAC** | Molecular Orbital PACkage |
| **MP** | Melting Point |
| **MPBPVP** | Syracuse Research Corporation program to predict melting point, boiling point, and vapor pressure |
| **MR** | Molar Refractivity |
| **MS-WHIM** | Molecular-Surface Weighted Holistic Invariant Molecular |
| **MultiCASE** | Multiple Computer Automated Structure Evaluation |
| **MW** | Molecular Weight |
| **NC** | Non-corrosive |
| **NCI** | National Cancer Institute |
| **NCTR** | National Center for Toxicological Research |
| **NDDO** | Neglect of Diatomic Differential Overlap |
| **NIOSH** | National Institute for Occupational Safety and Health |
| **NN** | Neural Network |
| **NOEC** | No Observed Effect Concentration |
| **NR** | Nuclear Receptor |

| | |
|---|---|
| **NTP** | National Toxicology Program |
| **OECD** | Organization for Economic Co-operation and Development |
| **OM1** | Orthogonalization Model 1 |
| **OM2** | Orthogonalization Model 2 |
| **OPPT** | Office of Pollution Prevention and Toxics |
| **OPS** | Optimum Prediction Space |
| **ORMUCS** | Ordered MUlticategorical Classification method using the Simplex technique |
| **p** | Path (molecular connectivity) |
| **P** | Partition coefficient |
| **PAH** | PolyAromatic Hydrocarbon |
| **P$_{alkb}$** | Promoter of the *alkb* gene |
| **PBPK** | Physiologically Based PharmacoKinetic |
| **PBT** | Persistent, Bioaccumulative, and toxic |
| **pc** | Path-cluster (molecular connectivity) |
| **PC** | Principal Component |
| **PCA** | Principal Component Analysis |
| **PCB** | PolyChlorinated Biphenyl |
| **PCR** | Regression on Principal Components |
| **P-gp** | P-Glycoprotein |
| **pH** | Negative logarithm of the hydrated proton concentration |
| **PLS** | Partial Least Squares |
| **PM** | Prediction Model |
| **PM3** | Parameterized Model 3 |
| **PM5** | Parameterized Model 5 |
| **PMN** | PreManufacture Notification |
| **PNN** | Probabilistic Neural Network |
| **PNSA** | Partial Negative Surface Area |
| **PPAR** | Peroxisome Proliferator Activated Receptor |
| **PPSA** | Partial Positive Surface Area |
| **PSA** | Polar Surface Area |
| **$Q^2$** | Leave-one out or cross-validated $R^2$ |
| **$Q_A$** | Net atomic charge on atom A |
| **$Q_h$** | Hepatic blood flow |
| **QM** | Quantum Mechanical |
| **QSAR** | Quantitative Structure-Activity Relationship |
| **QSBR** | Quantitative Structure-Biodegradability Relationship |
| **QSPKR** | Quantitative Structure-Pharmacokinetic Relationship |
| **QSPR** | Quantitative Structure-Property Relationship |
| **R and $R^2$** | Multiple correlation coefficient and its square |
| **RBA** | Relative Binding Affinity |
| **REACH** | Registration, Evaluation, and Authorization of CHemicals |
| **rms** | Root-mean-square |
| **RTECS** | Registry of Toxic Effects of Chemicals |
| **SA** | (Sub-)Structural Alerts |
| **SAS** | Statistical Analysis System |
| **S$_{aq}$** | Aqueous solubility |
| **SAR** | Structure-Activity Relationship |
| **SCF** | Self-Consistent Field |
| **SDF** | Structure Data File |
| **SHBG** | Sex Hormone Binding Globulin |
| **SMILES** | Simplified Molecular Line Entry System |

| | |
|---|---|
| **SRC** | Syracuse Research Corporation |
| **t ½** | Half-life |
| **TD$_{50}$** | Dose required to halve the probability of animals remaining tumourless |
| **TGD** | Technical Guidance Document |
| **TI** | Topological Indices |
| **TOPKAT** | TOxicity Prediction by Komputer-Assisted Technology |
| **TPSA** | Topological Polar Surface Area |
| **TSCA** | Toxic Substances Control Act |
| **UNIFAC** | Uniquac Functional-group Activity Coefficient (where UNIQUAC = Universal Quasi-Chemical) |
| **U.S. EPA** | United States Environmental Protection Agency |
| **V$_d$** | Volume of distribution |
| **V$_m$** | Molecular Volume |
| ***vol*** | molar volume in $cm^3\ mol^{-1}$ |
| **W** | Wiener index |
| **WLN** | Wiswesser Line Notation |
| **WMPT** | Waste Minimization Prioritization Tool |
| **WSKOWIN** | Syracuse Research Corporation program to predict water solubility |

# SECTION 1

# Introduction

# CHAPTER 1

# Predicting Chemical Toxicity and Fate in Humans and the Environment — An Introduction

**Mark T.D. Cronin**

## CONTENTS

## I. INTRODUCTION

Chemicals are able to bring about both desirable and undesirable effects on organisms to which they are exposed, and the actions of medicines and poisons and toxic agents have been recognized for thousands of years. As a result of industrialization, modern man and the environment is now exposed to increasing numbers of chemicals. Because of their potential hazard, there is an appreciation of the requirement to assess the effects of these chemicals. Since chemical structure was elucidated (for a very brief history see Table 1.1), the relationship between chemical structure and

**Table 1.1   Summary of the Key Historical Events (Scientific and Sociological) That Have Given Rise to the Modern Science of Predictive Toxicology**

| Year/Era | Event |
| --- | --- |
| c. 5000 B.C. and earlier | Knowledge of animal venoms and poisonous plants |
| 3000 B.C. | Egyptians recognized the toxic effects of some substances; Menes, first of the Pharaohs, studied and cultivated poisonous plants |
| 1550 B.C. | Ebers Papyrus describes over 800 recipes for poisons |
| c. 300 B.C. | Theophrastus, a pupil of Aristotle, referenced poisons and was later executed by poison |
| Early 1500s | Paracelsus determined that specific chemicals were actually responsible for the toxicity of a plant or animal poison |
| Middle Ages | Poisoning an accepted fact of life; Shakespeare's Romeo begs to the apothecary, "*A dram of poison, such soon-speeding gear, As will disperse itself through all the veins, That the life-weary taker may fall dead*" |
| Early 1800s | Orfila is credited with founding toxicology (i.e., the correlation between the chemistry and biology of poisons) |
| 1860's | Transition from alchemy to chemistry: structure of benzene proposed, the periodic table determined |
| 1863 | Observation by Cros that the toxicity of alcohols decreased with their water solubility |
| 1868 | Crum-Brown, and Fraser concluded that physiological activity is a function of chemical constitution |
| 1893 | Richet observed toxicity to be inversely related to solubility |
| 1899–1901 | Meyer and Overton independently proposed that narcosis is related to partitioning between oil and water phases; assessment of partitioning using olive oil and water |
| 1939 | Ferguson proposed solubility cutoff for acute toxicity |
| 1940 | Hammett published *Physical Organic Chemistry* showing that the effects of substituents could be quantified |
| 1964 | Hansch and coworkers used regression analysis and descriptors for the hydrophobic, electronic, and steric properties of molecules to formulate a QSAR |
| 1970s | Growth in the development of QSARs |
| 1976 | The U.S. Toxic Substances Control Act acts as the spur to find methods of predicting toxicity |
| 1981 | Könemann demonstrated that the acute toxicity to fish of non-reactive narcotic compounds may be modeled by hydrophobicity |
| 1980s | Rapid expansion of computational power makes molecular graphics and modeling, as well as multivariate statistical analysis, practical; computational power allows for the commercial development of software for descriptor calculation, molecular modeling, and toxicity prediction |
| 1980s | 3D QSAR techniques allow for the modeling of receptor-mediated effects, including toxicity and metabolism |
| 1980s | Development of mechanism-of-action-based QSARs for acute toxicity, progression from narcotic mechanisms to reactive mechanisms |
| 1980s | Application of QSAR techniques to a wide variety of toxic and fate endpoints such as carcinogenicity, irritation, biodegradation, and bioaccumulation |
| 1980s | Creation of the fathead minnow toxicity database |
| Late 1980s | SMILES for molecular description developed and becomes widely used |
| 1990s | Application of QSAR techniques to a broader range of toxicological endpoints such as skin sensitisation, percutaneous absorption, skin and eye irritation, and carcinogenicity |
| 1990s | A growth in concern over animal usage, and resultant public pressure, increases the commercial and legislative requirement for alternative methods to animal testing |
| Early 1990s | The Internet becomes a reality. Storage, searching and analysis of large amounts of chemical and biological information become trivial. Desktop computing becomes the standard. |
| Mid 1990s | The National Toxicology Program's carcinogenicity prediction challenge highlights the difficulty of estimating this endpoint |
| Mid 1990s | Concern over endocrine disruption brings receptor-mediated modeling techniques into mainstream toxicity prediction |
| Late 1990s | User-friendly software to calculate large numbers of molecular descriptors from 2D structure becomes widely available both commercially and from the Internet (e.g. DRAGON) |
| 2000 | Release of the EPISUITE software freely downloadable from the Internet |
| 2000 | Solving of the human genome and application of toxicogenomics |
| 2000 | *Tetrahymena* database reaches 2000 compounds tested |
| 2000 | Pharmaceutical development regularly utilizes combinatorial chemistry, high throughput screening and virtual library design; the interest in predictive ADMET grows |
| 2001 | European Union's *White Paper on the Strategy for a Future Chemicals Policy* stimulates further interest in the validation and application of QSARs to predict toxicity and fate |
| 2001 | Bioterrorism, including the use of toxic agents, becomes a reality |

biological activity has intrigued scientists. Latterly, it has been recognized that the investigation of chemical structure — biological activity relationships (or structure-activity relationships [SARs]) is more than an academic exercise. They may provide useful tools to solve real world problems, such as the requirement for information regarding the effects of chemicals on man and the environment.

This book intends to provide a starting point for those interested in the prediction of the toxicity and fate of chemicals to humans and the environment. SARs and, more frequently, quantitative structure-activity relationships (QSARs) provide methods to predict these endpoints. A brief history of the area, the driving forces, and basis of the topic is provided in this chapter. Further chapters (2 to 7) describe the methods to develop predictive models; the application of models to human health endpoints (Chapters 8 to 11); their application to environmental toxicity and fate (Chapters 12 to 17); and the use of predictive models (Chapter 19), adoption by the regulatory authorities (Chapter 19), and validation (Chapter 20).

## A. History of Predictive Methods for Toxicology and Fate

It would be wrong to consider the history of predictive toxicology in complete isolation from other scientific and sociological events. The interest and ability to predict the toxicity and fate of chemicals has a number of drivers and has been influenced by key individuals, organizations, commercial and welfare pressures, and legislation. The prediction of effects has relied on advances in all the areas of biology, chemistry, and informatics, as well as benefiting in particular from the substantial advance in computer technology. To describe these events in detail is clearly beyond the scope of this book, and certainly exceeds the capability of this author! As a starting point for historians, various papers provide a good overview of past achievements (Kubinyi, 2002; Lipnick, 1999; Rekker, 1992; Schultz et al., 2003; van de Waterbeemd, 1992). From a personal point of view, some of the key events that influenced the science are summarized in Table 1.1 and commented upon below.

It is often overlooked that much of the basis of modern drug design can be traced back to toxicological research performed in the 1890s. Indeed a cynic might suggest that there has been little progress since the work of Richet, Meyer and Overton! Certainly the finding of Könemann (1981) reinforced and quantified these findings, but this was 80 years after the original work. From this reinvention of acute toxicological QSAR there has been progress through class-based to mechanism-of-action modeling, leading to the development of more global approaches to toxicity prediction (see Chapter 12). This progress has been underpinned by the development of reliable and diverse databases of toxicity values, those having been developed for the fathead minnow (*Pimephales promelas*) by the U.S. Environmental Protection Agency (EPA; Mid-Continent Ecology Division, Duluth) (Russom et al., 1997) and freshwater ciliate (*Tetrahymena pyriformis*) at the University of Tennessee, Knoxville (Schultz, 1997), being of considerable importance.

In other areas of predictive toxicology and fate, progress has been steady, and spurred on in recent years by many of the legislative and commercial pressures mentioned in Table 1.1. Progress and interest in the prediction of human effects and pharmacokinetics has been complemented by advances in chemo-informatics. This has resulted in a large number of commercially available expert system approaches to toxicity prediction (see Chapter 9) and algorithms for the prediction of absorption, distribution, metabolism, and excretion (ADME; see Chapters 10 and 11).

## B. Motivation for Predicting Toxicity and Fate

There is no single motivation for wishing to predict the toxicity and fate of chemicals — the desire to do so varies from user to user. The following describe some of the key reasons for which models have been developed (in no particular order). Naturally, the drivers for predicting these endpoints are related closely to the historical development of the science and most of the criteria listed below are related in some manner.

## 1.  Computer Models Provide a Prediction of Toxicity and Fate

It may seem too obvious to state, but it is fundamental that computer models allow for the effects of chemicals (i.e., physicochemical properties, toxicological activity, distribution, fate, etc.) to be predicted. These predictions may be obtained from a knowledge of chemical structure alone. For most methods, provided that the chemical structure can be described in two (or occasionally three) dimensions, the effects may be predicted. Information regarding the chemicals may be gained without chemical testing, or even the need to synthesize the chemical.

## 2.  Public Pressure to Reduce Animal Testing

For several decades there has been growing public concern regarding the use of animals in testing, especially in toxicology and medical research. The concern over animal welfare has been concentrated in Europe and in the U.K. in particular. This has resulted in the boycotting of companies, organizations, and individuals associated with animal testing. At the most radical edge, terror campaigns have been mounted to target particular workers and even financial institutions. Campaigners for animal welfare cite a number of approaches to reduce and ultimately replace animal tests. These include the use of validated alternative tests such as *in vitro* and cell culture techniques, as well as the computer-aided prediction of toxicity. The status of alternative techniques is well reviewed by Worth and Balls (2002).

There is clearly a role for predictive techniques in the replacement of animal tests, either as stand-alone methods, or more commonly as part of a tiered assessment strategy. Chapter 18 describes the integration of computational methods, in combination with the judicious use of physicochemical properties, as viable alternatives to animal testing. The strategies described in Chapter 18 are being integrated into international guidelines for the toxicological assessment of endpoints such as irritation.

## 3.  Legislation

Much of the legislation that has underpinned the use of computational methods to predict toxicity is summarized in Chapter 19. Governmental policy in both the European Union (EU) and North America has encouraged and, in some cases, mandated the use of computational techniques to predict toxicity. For instance, the EPA has utilized QSARs to assist in the pre-manufactory notification of new chemicals, especially where no toxicity data exist. This requirement for models has stimulated considerable progress in the prediction of acute toxicity for environmental endpoints. Elsewhere, EU directives decree that animal tests should not be used if a suitable, validated, alternative (which includes computer models) is available. Generally national and international regulatory agencies and other competent bodies require and utilise predictive techniques to prioritize, classify, and label compounds for testing.

## 4.  Filling Data Gaps

Approximately 100,000 separate chemicals may be released into the environment annually; it is frightening to consider that reliable toxicity data exist for only a tiny proportion of these chemicals, probably less than 5%. The percentage of chemicals with a complete set of reliable toxicity data (i.e., across a broad spectrum of environmental and human health effects) is considerably less than 5%. Computer-aided prediction of toxicity has the capability to assist in the prioritisation of chemicals for testing, and for predicting specific toxicities to allow for labeling. Chapter 19 describes these activities in more detail. As the reliability of models for toxicity prediction increases, there will undoubtedly be increased use for the filling of data gaps.

## 5. *Cost of Testing — Finance and Time*

Toxicological testing is costly financially as well as in terms of the animals used and time taken. Even a simple ecotoxicological assay may cost several thousand dollars, and a two-year carcino-genicity assay may cost several million dollars. The cost of testing impacts business in a number of ways. For existing chemicals, cost (and obtaining resources) is clearly prohibitive to the filling of data gaps for the many compounds that have not been tested. With regard to the development of new chemicals (e.g., pharmaceuticals), the cost of toxicological testing of large numbers of lead compounds is prohibitive both financially and in the time it may take to obtain a full profile of a chemical. In both these areas, there are clear advantages to the use of methods to predict toxicity and fate (i.e., costs are greatly reduced). This should allow for faster and less expensive product development, as well as assessment of environmental effects.

## 6. *Reaction to New Toxicological Problems*

As we become exposed to more xenobiotic chemicals, and we learn more about the human genome, we will almost certainly become aware of an increasing number of toxic effects. A good example is provided by the issues associated with endocrine disruption, which came to the forefront during the 1990s. The use of computer-aided modeling in this area is well described in Chapter 13. This shows that the development of computational techniques not only allows for the prediction of the potential for estrogenicity to be made, but also allows for rational direction to be given to testing programs. The modeling of estrogenicity is an excellent example of the application of tools developed primarily for drug design (i.e., 3D QSAR and Comparative Molecular Field Analysis [CoMFA]) for the purpose of toxicological prediction. As an integral part of the modeling process, hypotheses have been built about the receptor (in this case the estrogen receptor) and have been tested in a rational manner. This has greatly extended the knowledge available from testing above that which would be gained from testing of random compounds.

## 7. *Designing New Compounds*

Drug and pesticide design has sought for many years to optimize activity and efficacy. As well as designing "in" attractive features of molecules, it is now possible to design "out" toxic features, or those associated with an unwanted ADME profile. To achieve this there is increasing use of commercial expert systems (described in Chapter 9) in various industries. There are obvious benefits to this process, such as savings in time and money throughout product development, as well as making lead optimization more relevant and directed.

## 8. *Increased Understanding of the Biology and Chemistry*

An often-ignored spin-off from the development of QSARs is the increased understanding they can provide in both the biology and chemistry of active compounds. In the modeling of acute toxicological endpoints much has been gained regarding mechanisms of action. For many modeling approaches, it may be assumed that compounds fitting the same QSAR are acting by the same mechanism of action (see Chapter 12 or Schultz et al. [2003] for more details). This has allowed workers to define the chemical domain of certain mechanisms. Compounds that do not fit a particular model also become of interest. Such compounds, known as outliers, suggest that they are acting by a different mechanism of action, or may be broken down rapidly either by chemical effects (i.e., degradation) or biological effects (i.e., metabolism). There are countless examples where knowledge of biology and chemistry has been advanced by modeling in the field of toxicological and fate effects.

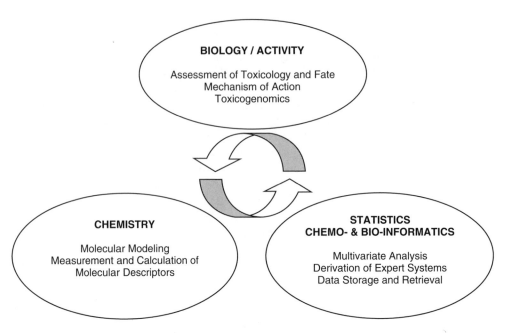

**Figure 1.1**   Interfacing sciences behind predictive toxicology.

## C.  The Cornerstones of Predictive Methods for Toxicity and Fate

One of the interesting aspects of predictive toxicology is that it brings together a large number of interfacing sciences and challenges. Figure 1.1 outlines the three key areas, and some of the important factors.

There are various definitions of what constitutes a computer-aided toxicity prediction method. It is true to say that all techniques are based upon the relationship of the biological activity of one or more molecules with some aspect of chemical structure. This broadest of definitions normally requires three components for a prediction method:

1.  Some measure of the activity (i.e., toxicity or fate) in a biological or environmental system
2.  A description of the physicochemical properties and/or structure of a molecule
3.  A form of statistical relationship to link activity and descriptors

The relationship between these three areas is given in Figure 1.2 and a brief introduction to each is provided below.

### 1.  Biological Activity

There are a remarkable number and diversity of activities that have been modeled successfully. The activity to be modeled may be a toxicity to an environmental organism or to man, the fate of a pollutant in an ecosystem, or the pharmacokinetic properties of a xenobiotic in man. To model any of these activities, relevant biological data for the endpoint are required. Chapter 2 describes how toxicological and fate information for chemicals may be obtained from external sources such as the open literature, databases, and the Internet. QSAR developers may also have their own data to model.

A key to the successful development of a predictive model for toxicity is the use of high-quality data. A definition of the quality of data is provided in Chapter 2 and discussed further by Cronin and Schultz (2003) and Schultz and Cronin (2003). However the quality of data is described, it is

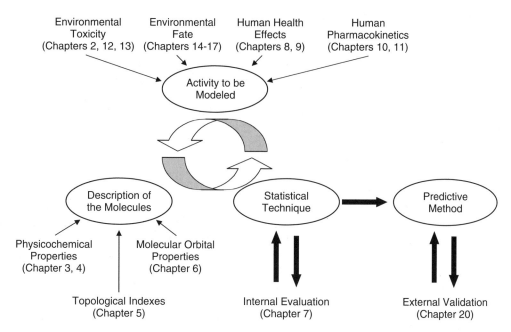

**Figure 1.2** General scheme for a computer-aided toxicological prediction method (as developed from Figure 1.1).

clear that reliable data from a single, well-defined protocol and measured to a high standard are required to build high-quality models. While the use of poor quality data is not precluded in modeling, an appreciation of the limitations of the data must be apparent to the modeler and ultimately to the user of the model.

From a broad (statistical) point of view, there are two types of activity that may be modeled. These are continuous and categorical data — as such they are statistical terms, rather than relating specifically to biological activity. Continuous data are numeral values that typically describe a concentration that elicits a particular effect. Most often this is a 50% effect concentration such as $LC_{50}$, $LD_{50}$, $EC_{50}$, $IGC_{50}$, etc. Occasionally it may be an effect brought about by a particular concentration such as the number of revertants in the Ames test. It must be stressed that when dealing with effect concentration, or equi-concentration, data, concentrations in molar units must be used. The use of molar values allows for the comparison of the effect of one molecule with another, rather than the weight of molecules. A subgroup of continuous data for modeling is not necessarily biological in nature. It includes physicochemical properties themselves (such as partition coefficient, melting point, etc.), diffusion characteristics (such as the flux through a particular membrane), fate descriptors (including persistence and bioaccumulation), pharmacokinetic properties (bioavailability and metabolism), and a multitude of other activities. As with the modeling of effects, if any of these data include a concentration, that concentration will be required in molar units.

The second type of activity data to be modeled are those data that are described as being categoric or ordinal. Typically these are yes/no data that classify a chemical as being active or inactive in a particular toxicological assay (e.g., carcinogenic or non-carcinogenic), or may be descriptive such as high or low bioavailability, or rapidly or slowly degraded. Occasionally there may be further grading (e.g., non-active, weakly active, moderately active, and highly active). For some activities and endpoints a classification may be assigned on the basis of a particular number of criteria, which may originally be developed from quantitative measures of toxicity. It is possible that a chemical may be classified differently according to its toxicity on the basis of national regulations (e.g., EU vs. U.S.). For such data it is essential that models are based on consistent and reliable data, and that the basis for the classification is consistent and understood.

## 2. Description of the Compounds

In order to make a model, methods to describe the molecules are required. There are literally thousands of descriptors of molecules (these are also termed parameters and variables in an equation). They range from simple counts of atoms, functional groups, or molecular weight, to properties based on 3D structure and requiring a super-computer for their calculation. Descriptors may require measurement of a physicochemical property (although this is not recommended for a predictive technique), or may be calculated from a knowledge of the contributions of functional groups, molecular topology or electron distribution.

Despite the descriptors' diversity, they are often considered to describe only three aspects of a molecule, namely its hydrophobic, electronic, and steric properties. It is widely considered that these three properties of a molecule are responsible for biological activity in an organism. In recent years a subgroup of descriptors has become widely available, broadly termed the topological indices. These descriptors are derived mainly from a knowledge of the connections within a molecule (and to a certain extent atom types and electronic environment) and are formulated from graph theory. They account for a number of molecular properties, mostly the steric attributes of the molecule such as molecular size; branching; and, to a certain degree, shape. More details on topological indices are available in Chapter 5 and from Todeschini and Consonni (2000).

Many modelers believe that, as far as possible, it is desirable, if not essential, that predictive techniques should based on fundamental physicochemical properties such as the partition coefficient (Cronin and Schultz, 2003; Schultz and Cronin, 2003). The use of physicochemical properties is described in more detail in Chapter 4. Physicochemical properties tend to describe fundamental molecular effects, such as the partition coefficient being a measure of hydrophobicity. There is a belief that such properties may be related to the mechanism of action of a molecule and are thus less susceptible to spurious correlation.

Molecular orbital theory has also been widely applied to the prediction of toxicity and fate (e.g., metabolism, persistence, and biochemical reactivity). One of the many advantages from the vast increase in speed of computers, and their continuing reduction in price, has been the rapid calculation of molecular orbital properties. Semiempirical methods, in particular, have become almost routine for the calculation of low energy electronic structures. Shareware such as Molecular Orbital PACkage (MOPAC) (Stewart, 1990) allows for the optimization of molecular structure followed by the calculation of numerous atomic (e.g., charges) and molecular (e.g., dipole moment, energy levels) descriptors. As described in Chapter 6 these have found widespread use in the prediction of toxicological and fate effects. The full range of descriptors available from molecular orbital theory and associated calculations is well reviewed by Karelson et al. (1996).

In addition to Chapters 3 to 6, other excellent reviews of the physicochemical and structural descriptors for use in QSAR include the works by Dearden (1990) and Livingstone (2000).

## 3. Statistical Techniques

From a statistical viewpoint, the modeling of continuous and categorical data requires different techniques. Statistical techniques that may be used to relate activity and structure are described in Chapter 7. In general terms, quantitative continuous data require quantitative statistical techniques to model them. Such techniques include regression analysis, regression on partial least squares, and neural networks. Categoric data have been assessed by pattern recognition techniques, which include cluster analysis, discriminant analysis, principal component analysis, and the use of neural networks. With a confusing range of statistical techniques to choose from, it is important that the modeler is able to decide which is the best and most appropriate to use for any given endpoint.

## D. How to Use Predictions

This book describes numerous methods for the prediction of toxicity (Chapters 8, 9, 12, and 13), environmental fate (Chapters 14 to 16), and the effects of chemicals in humans (Chapters 8, 10, 11, 13, and 17). In addition to those models reviewed in these chapters there are many more available in the open literature (see the next section). Despite these predictive models, animal tests are still being performed to assess toxicity and fate. The question then becomes, "How are we to use predictions?" The simple answer to this question is "cautiously".

As described in Chapter 9 there are an increasing number of commercial toxicological prediction systems available. Naturally these have been designed to be user friendly; most run under Microsoft Windows and use the Simplified Molecular Input Line Entry System (SMILES) as the molecular input. It is therefore possible to obtain a prediction of toxicity instantaneously, and often this may be performed for large numbers of compounds. There is a great temptation to use predicted toxicities at face value (i.e., if a compound is predicted to be non-toxic then it *must* be non-toxic). This simplistic use of predicted values should be avoided at all costs. Ideally, there are a number of criteria that should be applied when predicting toxicity. It is essential that a trained expert uses the predictive system. The user should be an expert both in the endpoint being predicted and the use of the predictive system.

It is also essential to understand how a prediction has been made. The predictive model should be transparent (i.e., the relationship between structure and activity is examinable). Another important feature to look for is that a model be based upon a recognisable mechanism of action (this is addressed in many chapters of this book). A decision must also be made as to whether the model is appropriate to make the prediction for the chemical in question. This is an extremely subtle point, and requires considerable expertise. For a prediction to be made, the compound should fall within the domain of the model (or the training set). The domain may be defined in terms of biology (e.g., the mechanism of action) and chemistry. The chemical domain of a model may again be defined in terms of physicochemical properties (*e.g.* the range of a property such a the partition coefficient or molecular weight), or in structural terms (e.g., the presence of structural groups or fragments). If a compound does not fall within the domain of the model, then the prediction should be discarded. The process of validating a predictive model should include the definition of the applicability domain of a QSAR; this is discussed further in Chapter 20.

The most successful application of structure-based predictive modeling in the future may be to specific endpoints. As described in Chapter 8 the prediction of something as broad as carcinogenicity is very difficult. If one isolates individual effects and endpoints within this category, then more success will ensue. Such an approach is likely to require the development of tiered strategies for toxicity prediction. It will also allow for the integration of test data from other assays, and where possible, human knowledge.

## E. General Reference Sources

One of the motivations for the commission of this book was the lack of a general reference source specifically for the modeling of toxicity and fate. Examples of the books that are, however, available in this area and provide valuable further information are those by Karcher and Devillers (1990) (environmental toxicology), Hansch and Leo (1995) (QSAR methodology), Livingstone (1995) (statistical analysis), Boethling and Mackay (2000) (physicochemical properties), Todeschini and Consonni (2000) (molecular descriptors), and Benigni (2003) (mutagenicity and carcinogenicity). In addition, a broad review of toxicological QSAR has been published as a special issue of *Environmental Toxicology and Chemistry* (Volume 22, Issue 8, 2003).

There are hundreds, if not thousands, of publications in the area of predictive toxicology and fate in the literature. No attempt will be made to list these here, as entry of keywords such as QSAR and toxicity into any literature retrieval service will reveal the references. Journals that have been particularly favoured for publications include *Quantitative Structure-Activity Relationships* (John Wiley), *SAR and QSAR in Environmental Research* (Taylor and Francis), and *Journal of Chemical Information and Computer Sciences* (American Chemical Society). Other important journals include *Chemosphere* (Elsevier), *Chemical Research in Toxicology* (American Chemical Society), *Environmental Toxicology and Chemistry* (Society of Environmental Toxicology and Chemistry), although this is by no means an exhaustive list.

The Internet is also a good starting point to obtain information. Using a search engine with appropriate keywords will again reveal a vast quantity of information. Currently the best starting point is the homepage of the International QSAR and Modelling Society (www.qsar.org). This is a well-developed website with considerable and constantly updated information on all aspects of the science. The site's links and resources this contains should keep the interested browser occupied for a considerable length of time.

## REFERENCES

Benigni, R., Ed., *Quantitative Structure-Activity Relationship (QSAR) Models of Mutagens and Carcinogens*, CRC Press, Boca Raton FL, 2003.

Boethling, R.S. and Mackay, D., Eds., *Handbook of Property Estimation Methods for Chemicals: Environmental and Health Sciences*, Lewis Publishers, Boca Raton FL, 2000.

Cronin, M.T.D. and Schultz, T.W., Pitfalls in QSAR, *J. Mol. Struct. (Theochem)*, 622, 39–51, 2003.

Dearden, J.C., Physico-chemical descriptors, in *Practical Applications of Quantitative Structure-Activity Relationships (QSARs) in Environmental Chemistry and Toxicology*, Karcher, W. and Devillers, J., Eds., Kluwer, Dordrecht, 1990, pp. 25–59.

Hansch, C. and Leo, A., *Exploring QSAR: Fundamentals and Applications in Chemistry and Biology*, American Chemical Society, Washington D.C., 1995.

Karelson, M., Lobanov, V.S., and Katritzky, A.R., Quantum-chemical descriptors in QSAR/QSPR studies, *Chem.Rev.*, 96, 1027–1043, 1996.

Karcher, W. and Devillers, J., Eds., *Practical Applications of Quantitative Structure-Activity Relationships (QSARs) in Environmental Chemistry and Toxicology*, Kluwer, Dordrecht, 1990.

Könemann, H., Quantitative structure-activity-relationships in fish toxicity studies. 1. Relationship for 50 industrial pollutants, *Toxicology*, 19, 209–221, 1981.

Kubinyi, H., From narcosis to hyperspace: the history of QSAR, *Quant. Struct.-Act. Relat.*, 21, 348–356, 2002.

Lipnick, R.L., Correlative and mechanistic QSAR models in toxicology, *SAR QSAR Environ. Res.*, 10, 239–248, 1999.

Livingstone, D.J., *Data Analysis for Chemists: Applications to QSAR and Chemical Product Design*, Oxford University Press, Oxford, 1995.

Livingstone, D.J., The characterization of chemical structures using molecular properties: a survey, *J. Chem. Info. Comput. Sci.*, 40, 195–209, 2000.

Rekker, R.F., The history of drug research — from Overton to Hansch, *Quant. Struct.-Act. Relat.*, 11, 195–199, 1992.

Russom, C.L., Bradbury, S.P., Broderius, S.J., Hammermeister, D.E. and Drummond, R.A., Predicting modes of toxic action from chemical structure: acute toxicity in the fathead minnow (*Pimephales promelas*), *Environ. Toxicol. Chem.*, 16, 948–967, 1997.

Schultz, T.W., Tetratox: *Tetrahymena pyriformis* population growth impairment endpoint: a surrogate for fish lethality, *Toxicol. Methods*, 7, 289–309, 1997.

Schultz, T.W. and Cronin, M.T.D., Essential and desirable characteristics of ecotoxicity QSARs, *Environ. Toxicol. Chem.*, 22, 599–607, 2003.

Schultz, T.W., Cronin, M.T.D., Walker, J.D. and Aptula, A.O., Quantitative structure-activity relationships (QSARs) in toxicology: a historical perspective, *J. Mol. Struct. (Theochem)*, 622, 1–22, 2003.

Stewart, J.J.P., MOPAC: A semiempirical molecular-orbital program, J. Comput.-Aided Mol. Design, 4, 1–45, 1990.

Todeschini, R. and Consonni, V., Eds., *Handbook of Molecular Descriptors (Methods and Principles in Medicinal Chemistry)*, Wiley-VCH, Weinheim, 2000.

van de Waterbeemd, H., The history of drug research — from Hansch to the present, *Quant. Struct.-Act. Relat.*, 11, 200–204, 1992.

Worth, A.P. and Balls, M., Alternative (non-animal) methods for chemical testing: current status and future prospects, *Alt. Lab. Animals (ATLA)*, 30 (Suppl. 1), 1–125, 2002.

# SECTION 2

# Methodology

CHAPTER **2**

# Toxicity Data Sources

**Klaus L.E. Kaiser**

## CONTENTS

0-415-27180-0/04/$0.00+$1.50
© 2004 by CRC Press LLC

## I. DATA SOURCES

Biological, or fate, data provide the basis to the development of the quantitative structure-activity relationships (QSARs) described in this book. This introduction to data sources is not meant to be comprehensive, but to provide information on some of the more easily accessible and useful data sources, particularly in the environmental field. This includes the toxic effects of chemicals on aquatic and terrestrial organisms.

### A. Books

There is a variety of books and monographs that provide listings of data of environmental relevance. For example, several handbooks should be mentioned, such as the *Handbook on Physical Properties of Organic Chemicals* (Howard and Meylan, 1997), which lists both experimental and estimated physicochemical properties for over 10,000 substances. The *CRC Handbook of Chemistry and Physics* (known universally as the *Rubber Handbook*), has a large section on organic chemicals with basic physicochemical information (Lide, 2001). Other handbooks with toxicological information include the *Handbooks of Ecotoxicological Data* (Devillers and Exbrayat, 1992; Kaiser and Devillers, 1994). In the drug field, the *Merck Index* (Budavari et al., 1996) has been a standard compendium for researchers for decades. It continues to be available in hardcover, but has recently also become available on the web to subscribers of the Dialog and other services. It can be searched by many of its entry fields describing the physical properties of the substances, but not by chemical structure element (*Dialog Merck Index,* 2003).

Japan's Ministry of International Trade has published several books with detailed information on biodegradation tests of several hundred chemicals. This information has now also been made available by the Japan Chemical Industry Ecology–Toxicology and Information Center (JETOC); see Chapter 14 for more details on the MITI biodegradation database. The JETOC also published a compendium of mutagenicity test data of several hundred chemicals. The books can be purchased from JETOC and partial data can be found at various websites, such as members.jcom.home.ne.jp/mo-ishidate/.

The Nanogen Index (2003) is a specialized database for pesticides. It has recently been updated to the *Nanogen Index 2* and only gives basic information for substances without any effect data. Other works specializing on pesticides are the *Pesticide Manual* (Worthing and Hance, 1991) and other volumes of a similar nature, including the *Handbook of Pesticide Toxicology* (Hayes and Laws, 1991) and the *Pesticide Fact Handbook* published by the U.S. Environmental Protection Agency (EPA) (1988).

In terms of property estimation methods, the *Handbook on Chemical Property Estimation Methods* (Lyman et al., 1990) has recently been succeeded by the *Handbook of Property Estimation Methods for Chemicals* (Boethling and Mackay, 2000).

### B. Internet Sources

The excellent Internet search engine Google (www.google.com) provides subdirectories with chemical database listings at directory.google.com/Top/Science/Chemistry/Chemical_Databases/?tc=1/ and toxicology databases at directory.google.com/Top/Science/ Biology/Toxicology/. Other website listings of databases include www.chemweb.com and www.chemclub.com. The QSAR and Modeling Society (QMS) also maintains a website at www.qsar.org with a listing of database and modeling software providers.

### 1. Free Sources

Formerly known as Aquatic Information and Retrieval (AQUIRE) (Hunter et al., 1990), the ECOTOX database is one of the earliest free sources of toxicological and, to an extent, physico-chemical data on the Internet. It is made available from the EPA and has undergone several modifications. Originally only accessible to U.S. government personnel and contractors, it was made freely available, without restrictions, several years ago. Its strength lies in a very detailed listing of aquatic toxicity data for approximately 6000 chemicals. It can be searched by a variety of means, including Chemical Abstract Service (CAS) number, formula, name fragment, but not by chemical structure fragment. The ECOTOX database can be accessed at www.epa.gov/ecotox/.

The ChemExper chemical directory (www.chemexper.com) provides a substructure searchable database claimed to contain 60,000 chemicals with melting or boiling points, where available. No toxicity or environmental property data are given.

MDL Information Systems, Inc., a subsidiary of Elsevier Science, Inc., also have a free Internet access to its database of commercially available substances, claimed to contain information on more than 400,000 chemicals (available from www.mdli.com). No toxicity or environmental property data are given. Similarly, the ChemBridge Corp. provides a list of over available 450,000 substances to subscribers (www.chembridge.com). Russia's ChemStar Ltd. provides a similar type of database with over 500,000 compounds available (www.chemstar.ru). The list of compounds is downloadable in structure data file (SDF) format and is being updated regularly.

Chemweb, available from www.chemweb.com, provides limited access to several databases to registrants free of charge, including basic information in chemical directories, such as the *Chapman & Hall CRC Combined Chemical Dictionary*. No toxicological information is available and access is relatively slow. Retrieval of actual data is subject to purchase.

ChemFinder is a freely available commercial database with an estimated 100,000+ chemicals. These can be searched at chemfinder.cambridgesoft.com by name or chemical structure fragment (with browser plug-ins), CAS, or molecular formula. Unfortunately, it has very little to offer in terms of physicochemical or toxicological data, although a variety of links to other databases are provided (which may or may not have any additional information). Another severe limitation of ChemFinder is that search results are limited to the first 25 substances; any results beyond that number are not given. It is interesting to note that there is a great overlap between various U.S. government databases, such as ChemIDplus, and several of these commercial products.

The European Chemical Bureau (ECB) has an equivalent collection of high- and low-production volume chemicals — the International Uniform Chemicals Information Database (IUCLID; ecb.jrc.it/existing-chemicals/). IUCLID has no longer toxicological information accessible to the user and has limited search options. There is also a CD-ROM version, available for a nominal cost.

The National Cancer Institute (NCI) provides free access to its NCI-3D database of over 250,000 substances. It can be searched by several means, including chemical structure. It does not contain any toxicological or physicochemical data and provides a great number of synonyms, particularly for drugs. It can be accessed at chem.sis.nlm.nih.gov/nci3d/and other sites, but is best accessed through the mirror server at the University of Erlangen, Germany (131.188.127.153/services/ncidb2/). The latter site allows searching by a wide selection of input variables, either alone or in combination, including substructure queries, and gives fast responses. It also provides simultaneous estimates of a variety of drug-like effects, as available from the PASS system (www.ibmh.msk.su/PASS/A.html). However, except for anti-HIV screening data and log $K_{ow}$ (octanol-water partition coefficient), it does not have any measured properties.

The National Institute of Standards and Technology's (NIST) *Chemical WebBook* (webbook.nist.gov/chemistry/) offers free access to ion energy and thermodynamic data for several thousand substances, and is also searchable by chemical substructure. No toxicity or environmental property data are given.

## 2. *Commercial Sources*

There are a number of commercial sources of chemical information that are very detailed and encompassing, but generally available only at high cost to industrial users. These include the Chemical Abstracts Service (www.cas.org), Dialog (library.dialog.com/bluesheets/html/bl0304.html), Prous Science (www.prous.com), and Derwent (www.derwent.com) databases. In recent years, some of the major scientific publishers have also begun to provide similar types of databases, some of which can at least presently be accessed on the Internet by free subscription to the ChemWeb site (www.chemweb.com). This includes Chapman & Hall's Properties of Organic Compounds database. The number of compounds available is rather limited and any factual information is only available for subscribers at a cost. ChemINDEX is a new, subscription-based service from Cambridgesoft Corp. It is similar to ChemFinder but without limitation on the number of search results.

Approximately three decades ago, the U.S. government created the Registry of Toxic Effects of Chemicals (RTECS) database (www.ccohs.ca/education/asp/search_rtecs.html). Initially available in book form only, it became later available on CD-ROM, from the National Institute of Occupational Safety and Health, USA, or affiliated vendors (e.g., the Canadian Center for Occupational Health and Safety [CCOHS]; www.ccohs.ca). This database contains information on approximately 120,000 substances, including (where available) acute and chronic toxicity data for terrestrial organisms, primarily mammalian species, such as rats, mice, rabbits, monkeys, and humans. This database will be transferred to the private sector in the near future for maintenance. RTECS cannot be searched by structure, but by name, formula, CAS, and several other means. CCOHS provides also a website which allows limited searching of the RTECS database at ccinfoweb.ccohs.ca/rtecs/search.html, but access to data is for subscribers only.

TerraBase Inc. (www.terrabase-inc.com) is a Canadian company specializing in databases for QSAR-type research. It provides the data in a normalized, logarithmic fashion for direct use in QSAR development. It has several CD-ROM products specialized to the endpoint of interest and the application of chemicals. These databases can be searched by a variety of means, including chemical structure fragments. Information includes use, physicochemical properties, and over 100 types of toxicity data to aquatic and terrestrial species. A complete list of the types of data covered is available on the company's website.

### 3.  New Data

The availability of measured data from existing sources is the subject of much interest and debate. Major chemical and pharmaceutical companies rely on their own databases, often containing measurements of many endpoints for tens or hundreds of thousands of chemicals. New compounds are constantly being synthesized and tested and their information added to such databases. Not surprisingly, this wealth of information is a fervently guarded secret and is the cornerstone of a company's success in the competitive industrial environment. Some of this information has been released in confidence to government agencies charged with the protection of human health and the environment. Generally, such data are not available to the public as their release could harm the competitive edge of the informants. At the same time, both university and government spending on measuring basic data for new compounds has severely declined, and comparatively little new information is becoming available from these traditional sources. Furthermore, increasing concern over animal testing, particularly of products developed for nonessential purposes, such as cosmetics, adds to the pressure for the development of data *in silico* rather than by testing. However, there is a question as to how far one can go before doing some tests, which could confirm or disprove predictions and theories. As observed recently by Mackay et al. (2003), there is still a real and urgent need to undertake good quality measurements of a variety of physicochemical properties and toxicological effects.

## II. CURRENT EFFORTS

There is a considerable recognition of the need for more information regarding toxicity and fate on which to build and validate models. There is also an appreciation of the need to collate all existing data to ensure that limited resources may be allocated to fill data gaps and expand our knowledge. At least two public database initiatives have been instigated in response to the need for more data to develop structure-activity relationships (Richard and Williams 2003). A consortium of industry and government sponsors has commissioned the International Life Sciences Institute (ILSI) to develop a QSAR toxicity database. ISLI is working with LHASA Ltd. to develop a database using a modified version of IUCLID. More details of the ILSI project are given in Chapter 9 and are available from www.ilsi.org.

A second initiative is being developed by Dr. Ann Richard and coworkers at the EPA. The Distributed Structure-Searchable Toxicity (DSSTox) public database network is a flexible community-supported, web-based approach for the collation of data. It is based on the SDF format for the representation of chemical structure. It is intended to enable decentralized, free public access to toxicity data files. This should allow users from different disciplines to be linked. Public, commercial, industry, and academic groups have also been asked to contribute to, and expand, the DSSTox public database network. Data from potentially any toxicological endpoint can be collated in the DSSTox public database network, including both human health, and environmental endpoints (Richard et al., 2002; Richard and Williams, 2002).

## III. DATA SEARCH PARAMETERS

It is probably correct to assume that all databases can be searched by the name of a substance, including its fragments. In addition, search capabilities by CAS numbers or molecular formulae are available in most databases. With the increase of more complex structures in such databases, and the wide variations in chemical nomenclature (both systematic and nonsystematic), names of chemicals become rapidly less useful as search parameters. In most cases, an initial search by chemical formula will help to focus the search onto a few compounds, which can then be scanned visually or by electronic means for the substances of interest. The following example from the International Nonproprietary Names (INN) List 84 (World Health Organization, 2000) demonstrates this: diflomotecanum, an antineoplastic drug, CAS 220997-97-7, with the formula $C_{21}H_{16}F_2N_2O_4$, has the systematic International Union of Pure and Applied Chemistry (IUPAC) name (5R)-5-ethyl-9,10-difluoro-1,4,5,13-tetrahydro-5-hydroxy-3H,15H-oxepino[3′,4′:6,7]indolizino[1,2-b]quinoline-3,15-dione. Name fragment searches for quinoline in ChemIDplus returns 4024 compounds. In contrast, the (exact) formula search does not result in any match, while a search for $C_{21}H_{16}$, finds 275 compounds with that number of carbon and hydrogen atoms.

The superiority of computer-based database searching becomes apparent with fragment search capability of one or more named fragments within a name (e.g., nitr), and even more so when applying chemical structure fragment search capability. With the introduction of the ISIS (www.mdli.com) and Accord (www.accelrys.com/accord/) chemical structure file systems for the spreadsheet and database formats, such as Microsoft Excel and Microsoft Access, using the SDF system, and the convertibility to and from the Simplified Molecular Line Entry System (SMILES), chemical structure information has become accessible to the common desktop computer. A number of databases are available that provide such substructure-searchable contents on CD; the Terratox products database (www.terrabase-inc.com) is one example. Several web-based databases also provide this structure-based search capability; examples include ChemFinder (chemfinder.cambridgesoft.com) and ChemIDplus (chem.sis.nlm.nih.gov/chemidplus/setupenv.html).

Books containing databases generally also contain indexes with the substance names and formula, as well as CAS registry numbers. Provided one knows one of those parameters, it is generally possible to narrow down the search to a reasonable number of entries without too much difficulty.

## IV. DATA FORMAT

### A. Typical Data Format

Toxicity data for use in the development of QSARs are normally required for a particular endpoint (i.e., a specific biological response). Toxicity data may be categoric (i.e., indicating the presence or absence of a toxicity or risk) or continuous (i.e., a 50% effect concentrations). The different methods of modeling such data are described in Chapter 7. The most common notation for toxicity data is in milligrams per liter for aquatic exposure concentrations (e.g., $EC_{50}$, $IC_{50}$, $LC_{50}$), and milligrams per kilogram (body weight) for single-dose values (e.g., $LD_{50}$), as is widely used for mammalian toxicity data. In addition, special notations may be common for certain species and endpoints, such as microgram per honeybee dose values.

Some databases use the standard prefixes of micro ($\mu$), nano (n), and pico (p) for values that would require several zeros after the decimal delimiter to indicate the correct order of magnitude. While such prefixes are correct, they can also lead to typographical mistakes (the letters m and n are beside each other on most keyboards) that may be difficult to spot. For example, an earlier version of the RTECS showed the oral rat $LD_{50}$ value for tetrachlorodibenzodioxin (TCDD) as 24,000 mg/kg, while the original source reported it as 24 ng/kg. At the same time, a number of zeros can also lead to mistakes by the addition or loss of a zero. An example of this can be found in the database by Wauchope et al. (1992), which gives a literature value for the solubility of the insecticide cyromazine as 13,600 mg/L, but then provides a recommended value of 136,000 mg/L. Only after comparison of these values will the erroneous recommended value become apparent.

### B. QSAR Data Format

One solution to detecting and avoiding erroneous values is the use of logarithmic transformed values and internal consistency. For example, in order to undertake any kind of QSAR study, all toxicity values expressed in milligrams per unit must first be converted to molar or millimolar values, which are then converted to their base-10 logarithms. For example, a substance with a molecular weight of 100 amu (or Da) and a toxicity (e.g., acute toxicity, 96-h $LC_{50}$) value of 10 mg/L, has a toxicity value of 10/100 = 0.1 mmol/L. The logarithm of that is –0.10. As most substances of interest (i.e., more toxic substances have $LC_{50}$ values of <10 mg/L), their log ($LC_{50}$) values will all be negative. This can lead to further complications and potential errors. Furthermore, when plotting the (logarithmic) toxicity (in millimoles per liter) against hydrophobicity values (most commonly, octanol/water partition coefficients), the correlation slope will also be negative, as shown in Figure 2.1. Therefore, the negative logarithm of the millimolar concentration (i.e., pT = –log[mmol/L]) has become a standard notation to use (Kaiser, 1987). This is identical to the inverse of the millimolar concentration (i.e., pT = log[l/mmol]). Using this type of notation, the slopes are positive, the number of negative values is much reduced or eliminated, and higher toxicity will be expressed with a higher value. All of these will aid in increased clarity, avoidance of typographical errors, and increased understanding. Figure 2.2 shows the resulting plot against hydrophobicity.

## V. DATA QUALITY AND COMPATIBILITY

### A. Data Quality

Data quality is an issue of great concern to many people in every area of chemistry and toxicology. It involves precision (repeatability) and accuracy (correct value) of test results. There are many national and international organizations dealing with data quality, trying to set standards, providing reference compounds, conducting round-robin studies for participating laboratories,

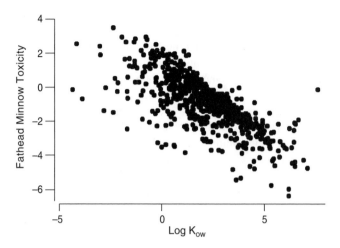

**Figure 2.1** Plot of 96-h $LC_{50}$ values for Fathead minnow (*Pimephales promelas*) in log (mmol/L) vs. the octanol/water partition coefficient (log $K_{ow}$) of 710 compounds.

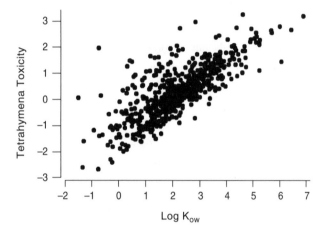

**Figure 2.2** Plot of *Tetrahymena pyriformis* $IGC_{50}$ values in log (L/mmol) vs. the octanol/water partition coefficient (log $K_{ow}$) of 576 compounds.

analysing results, and recommending test protocols. For example, the Organization for Economic Cooperation and Development (OECD), European Union, and NIST and American Society for Testing and Materials (ASTM) provide protocols and recommendations for various kinds of testing. Where possible, tests performed in accordance with such standards should provide an adequate level of quality for most research studies. However, even a claimed adherence to such standards is not necessarily a guarantee for data quality. For example, the toxicity test result obtained for the pesticide malathion in the *Daphnia magna* bioassay, claimed to be performed according to OECD standard protocols, was incorrect by many orders of magnitude, as pointed out by Kaiser (1995). Further information regarding data quality with respect to the development of QSARs is provided in Chapters 19 and 20, as well as Cronin and Schultz (2003) and Schultz and Cronin (2003).

## B.  Data Compatibility

Most researchers compiling data from the literature for one study or another are faced with the question of data compatibility. In most cases, this is of greater significance than data quality *per se*, presuming a comparable degree precision for all experiments. Whenever possible, it is desirable to compile data for a particular measurement from references originating within the same laboratory and measurement system. It is rarely possible to obtain all the desired data from one source and the question of compatibility always looms on the horizon. For example, looking at bioassays with commonly used fish species, such as rainbow trout (formerly *Salmo gairdneri*, recently renamed to *Oncorhynchus mykiss*), fathead minnow (*Pimephales promelas*), or zebrafish (*Brachydanio rerio*), there are several test conditions that influence the values obtained and may cause data incompatibility between different laboratories' results. Such variables include temperature, pH, hardness, alkalinity, and oxygen levels of the test water. Differences among laboratories, such as in the oxygen levels and water temperature, would have different consequences for the three species mentioned, as they are referred to as cold water (trout), warm water (minnow), and tropical water (zebrafish) species, respectively.

Even for studies from different sources, but where the above noted variables are identical, there may be other reasons for data incompatibility. Such reasons include the type of assay and chemical exposure control. Some tests are performed in static systems, while others are performed in flow-through systems with constant renewal of the water at a fixed rate. The latter requires a much larger setup with constant chemical addition and dilution of the water. In contrast, the former often uses no or only limited water renewal at fixed intervals and often assumes that the nominal concentrations of the test chemical added are also the actual exposure concentrations. This assumption is justified for chemicals that are well soluble in water; not highly volatile; and do not rapidly degrade, volatilize, or adsorb to the surfaces in the test system. For substances that do not fulfill these assumptions, the actual exposure levels can be substantially different from the nominal concentrations; reports of changes in the concentration (declines) by one order magnitude over a 24-h period are not uncommon.

## C. Erroneous Data

Most existing (electronic) databases use typical database formats that present all data pertaining to one compound entry (and there may be more than one entry per individual compound) in a data form. While this is a convenient way to see all the information of one particular entry, it prevents getting an overview of all entries on that compound and how the values from other entries compare with the particular one shown. Some databases allow table-type views, which can be useful to gain this overview. Alternatively, all entries may be exported and printed for a more comprehensive view with the use of another software or on paper.

### 1. Data Errors in Databases

There are many sources and types of errors that can creep into any system of organization of data. They may stem from typographical mistakes, oversight or misunderstanding of the stated concentrations (e.g., parts per billion in Europe refers to $10^{12}$, but in North America it refers to $10^9$), misunderstanding of the delimiters used (e.g., the common notation for the value five thousand three hundred in European notation may be 5.300, vs. 5300 in American English) and a variety of other causes.

### 2. Data Errors in Primary Literature

When in doubt about particular data values, it is always advisable to refer to the original literature values. Unfortunately, this does not always solve the problem since these values can have mistakes as well. One common problem can be the incorrect electronic file translation from one computer system to another. For example, certain software products (even by the same manufacturer) incorrectly convert micro from one system to milli in another. Another potential pitfall is the erroneous association of milliliter with milligram. At a density of 1.0, 1 ml of liquid weighs 1000 mg, or 1 g, not 1 mg. I suspect that a number of primary literature values suffer from this problem; one example is the $LC_{50}$ values of *N*-methylaniline and *N,N*-dimethylaniline given by Groth et al. (1993).

## D. How to Spot Errors

One of the most useful tools to spot and eliminate errors is a spreadsheet, such as Excel or QuattroPro. QSAR modelers very frequently use spreadsheets to organize data into columns and rows of standardized values of the independent and dependent parameters. Spreadsheets allow easy sorting and filtering — two important functions used to find problem data and duplicates and other errors. In addition, spreadsheets have search and replace routines, plotting, and correlation functions, which allow the data to be reviewed in various comprehensive ways. The data can also be exported to other file types, which allow analysis by other software for statistics and any types of quantitative and qualitative relationships that may exist. It cannot be emphasized enough that the typical spreadsheet functions (including graphing functions) are excellent tools to find and eliminate erroneous or questionable values, duplicates, and other problem entries.

## VI. OUTLIERS

One problem being faced by most modelers is the recognition, use, and elimination of outliers. Although the term outlier is used quite commonly, there is no one single mathematical, or, for that matter, even one single practical definition that is generally accepted. Given a normal distribution of values around the mean, outliers have for practical reasons been defined as those values that vary more than 2, 3, or 5 standard deviations from the mean, representing approximately 5, 1, or 0.1% of the sample population, respectively. However, data sets that do not follow normal distribution rules are not covered by these definitions. Since it is difficult to determine with absolute certainty the type of distribution any limited data set may have, it follows that the recognition of outliers is equally problematic. Practically speaking, many modelers describe outliers as being compounds that "do not fit the model." It should be appreciated that there is no statistical meaning or basis for such a statement. Other issues with outliers include proper ways in which to deal with them. They should not be excluded without good reason (see below), and when excluded, a statement must be presented explicitly as to which compounds are considered outliers and hence removed. So saying, it must be recognized that the identification of outliers (in terms of compounds that do not fit a specific chemical domain) has greatly assisted our appreciation of the mechanism of toxic action (Lipnick 1991).

In principle, outliers can occur for one of three reasons: (1) the values represent a true deviation from the model domain, as compound- or endpoint-specific causes for this departure are not being modeled; (2) the values are within the model domain, but are being modeled improperly because of insufficiencies of the model; and (iii) the values are incorrect because of data incompatibilities and or transcription or measurement errors. Depending on the reason for being an outlier, its recognition as such can either contribute new knowledge (if a correct data point), or impede the creation of useful structure-activity relationships (if a false data point). The recognition of such influential data and their resolution as to true or false is an important part of developing structure-activity relationships.

## VII. CHEMICAL STRUCTURE NOTATIONS

In order to store chemical information, there is a need to store chemical structures in some type of database format. For most practical purposes chemical structures are stored in 2D formats as described below.

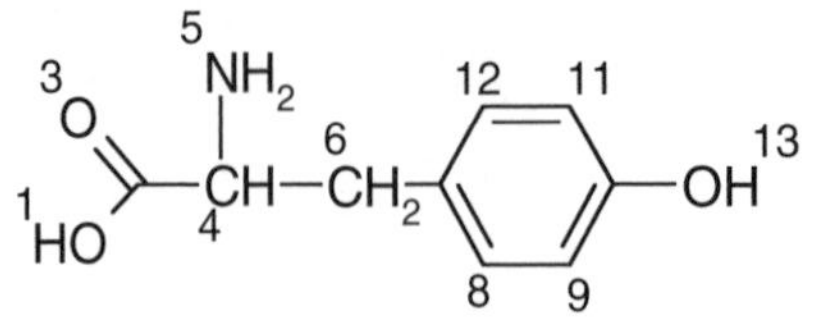

WLN: **QVYZ1R DQ**

SMILES: **OC(=O)C(N)CC1=CC=C(O)C=C1**

**Figure 2.3**   Structure, WLN, and SMILES notations for 2-amino-3-(4-hydroxyphenyl)propanoic acid.

## A.  Wiswesser Line Notation

Until the late 1970s, the Wiswesser Line Notation (WLN) was the only widely recognized format in which to code chemical structures in a computer-readable linear format. It was invented by William J. Wiswesser (1952), and his work was recognized and honored with the Skolnik Award for "outstanding contributions to and achievements in the theory and practice of chemical information science" by the American Chemical Society in 1980. The WLN was quickly adopted by major chemical companies to store and retrieve machine-readable, 3D-chemical structure information in a 2D, linear array, and is still in use today.

Figure 2.3 gives the structure, WLN, and SMILES codes for a sample chemical. It is apparent that the WLN code is more complicated and much less obvious to a chemist than the SMILES code. The numbers refer to the atom numbers used in developing the WLN code.

## B.  SMILES Notation

In the late 1970s, David Weiniger developed a new system for chemical structure encoding called SMILES. This system is much more akin to the actual structure of a substance as drawn by a chemist, with C representing a carbon atom, N a nitrogen atom, and so forth. In addition, double and triple bonds are immediately apparent by the = and # symbols. Branches are shown in parentheses and rings are opened and closed by the use of numbers. Detailed information on the SMILES code and an excellent tutorial on its use are available on the the Daylight Corp.'s website (www.daylight.com/smiles/f_smiles.html).

There are four simple rules to apply for the generation of a SMILES string for most organic compounds:

1. Atoms are represented by atomic symbols.
2. Double bonds are represented by =, and triple bonds are represented by #.
3. Branching is represented by parentheses.
4. Ring closures are indicated by pairs of matching numbers.

Note that these rules do not apply for the coding of salts and isomers, although the full SMILES language and variants can handle all types of chemical structure. These rules are illustrated in the examples below.

## 1. Rule 1

Each atom in the compound is coded by a letter or letters denoting the element: B, C, N, O, F, P, S, Cl, Br, and I. Hydrogen atoms are ignored and single bonds between atoms are implied. The following simple compounds are therefore coded as:

| Name | Formula | SMILES |
|---|---|---|
| Methane | $CH_4$ | C |
| Ethane | $CH_3CH_3$ | CC |
| Ethanol | $CH_3CH_2OH$ | CCO |
| Hexane | $CH_3CH_2CH_2CH_2CH_2CH_3$ | CCCCCC |

## 2. Rule 2

Double bonds are represented by = and triple bonds by #:

| Name | Formula | SMILES |
|---|---|---|
| Formaldehyde | CO | C=O |
| Hydrogen Cyanide | HCN | C#N |
| Carbon Dioxide | $CO_2$ | O=C=O |
| Acetylene | CHCH | C#C |

## 3. Rule 3

Branches are shown by parentheses and may be nested within one another:

| Name | Formula | SMILES |
|---|---|---|
| 2-Methylpropane | | CC(C)C |
| Acetic Acid | | CC(=O)O |
| *Tert*-butyl acetate | | CC(=O)OC(C)(C)C |

The presence of a branch in the structure raises the question of where to start coding. With the SMILES system it does not matter where one starts. A SMILES interpreter will produce the same structure from any valid SMILES coding for a compound. In some circumstances, such as the system's use in databases, it is necessary to have a unique SMILES string for a molecule. Using a set of rules it is possible to uniquify a SMILES string.

## 4. Rule 4

Ring closures are indicated by pairs of matching numbers. If one bond in the ring structure is chosen, such that it may be "broken", a number (the same one) can be appended to the atom at each end of the breakage and simply included as part of the linear string. A discussion of aromatic ring notification is provided below:

| Name | Formula | SMILES |
| --- | --- | --- |
| Cyclohexane | | C1CCCCC1 |
| 4-Chloro-benzoic Acid | | Clc1ccc(C(=O)O)cc1 |

In recent years, desktop computer-based software has become available to provide 2D and 3D structures from SMILES notations. Several companies (e.g., Accelrys, MDL, etc.) provide a variety of software utilities that allow users to visualize chemical structure in common spreadsheet (e.g., Excel) and database (e.g., Access) programs. The latter has a regular new pamphlet, which can also be accessed on the Internet (chemnews.cambridgesoft.com) with information as to databases and other tools. Typically, these programs have extensions for calculating molecular formula, molecular weight, and other basic values. More complex properties, such as various connectivity indices, octanol/water partition coefficient, and others, are also becoming standard or add-in features of these programs. Typically, these vendors also provide some databases with information for additional costs.

It should be noted here that the SMILES code is no longer an entirely universal system of notation and no longer equal between the computer program interpretation by Daylight Corp. and those by other software providers. For example, in the definitions used by the Daylight system, *any* sp2 carbon could be written either as "C=" or "c" In contrast, the Accelrys system now interprets "c" as sp2 carbons only when part of an aromatic ring (as defined by Hückel rules). While the SMILES for cyclopentadiene could previously be written as either C1C=CC=C1, c1cccc1, or c1cccC1, the recent changes will produce erroneous results in the Accelrys system when using the latter two notations; they will interpret the structures to be that of cyclopentane instead.

It is this writer's firm conviction, that the invention of the SMILES code and its convertibility with computer-readable SDF format has opened the door to great steps forward in man's ability to understand structure-activity relationships. Despite the present differences between the two major SMILES code interpreting software environments, I recommend that all students of chemistry become familiar with this code.

## REFERENCES

Boethling, R.S. and Mackay, D. Eds., *Handbook of Property Estimation Methods for Chemicals: Environmental and Health Sciences*, Lewis Publishers, Boca Raton, FL, 2000.

Budavari, S., Neil, M.J., Smith, A., Heckelman, P.E., and Kinneary, J.F., *The Merck Index*, 12th ed., Merck and Co. Inc., NJ, 1996.

Cronin, M.T.D. and Schultz, T.W., Pitfalls in QSAR, *J. Mol. Struct. (Theochem)*, 622, 39–51, 2003.

Devillers, J. and Exbrayat, J.M., *Ecotoxicity of Chemicals to Amphibians*, Gordon and Breach Science Publishers, Reading, MA, 1992.

*Dialog Merck Index*, accessed from www.dialog.com/, 2003.

Google, Google database listings, directory/google.com/top/science/chemistry/chemical_databases/.

Groth, G., Schreeb, K., Herdt, V., and Freundt, K.J., Toxicity studies in fertilized zebrafish eggs treated with N-methylamine, N,N-dimethylamine, 2-aminoethanol, isopropylamine, aniline, N-methylaniline, N,N-dimethylaniline, quinone, chloroacetaldehyde, or cyclohexanol, *Bull. Environ. Contamination Toxicol.*, 50, 878–882, 1993.

Hayes, W.J. and Laws, E.R.E., *Handbook of Pesticide Toxicology*, Academic Press, Inc., San Diego, 1991.

Howard, P.H. and Meylan, W.M., *Handbook of Physical Properties of Organic Chemicals*, Lewis Publishers, Boca Raton, FL, 1997.

Hunter, R., Niemi, G., Pilli, A., and Veith, G., Aquatic Information and Retrieval (AQUIRE) database system, in *Computer Applications for Environmental Impact Analysis*, Pillmann, W., Ed., International Society for Environmental Protection, Vienna, 1990, pp. 42–48.

Kaiser, K.L.E., QSAR of acute toxicity of 1,4-di-substituted benzene derivatives and relationships with the acute toxicity of corresponding mono-substituted benzene derivatives, in *QSAR in Environmental Toxicology — II*, Kaiser, K.L.E., Ed., D. Reidel Publishing Company, Dordrecht, 1987, pp. 169–188.

Kaiser, K.L.E., Re: QSAR models for predicting the acute toxicity of selected organic chemicals with diverse structures to aquatic non-vertebrates and humans. Calleja, M.C., Geladi, P., and Persoone, G., SAR QSAR Environ. Res. 2, 193–234, *SAR QSAR Environ. Res.*, 3, 151–159, 1995.

Kaiser, K.L.E. and Devillers, J., *Ecotoxicity of Chemicals to Photobacterium phosphoreum*, Gordon and Breach Science Publishers, Reading, MA, 1994.

Lide, D.R., *CRC Handbook of Chemistry and Physics*, CRC Press, Boca Raton, FL, 2001.

Lipnick, R.L., Outliers: their origin and use in the classification of molecular mechanisms of toxicity, *Sci. Total Environ.*, 109/110, 131–153, 1991.

Lyman, W.J., Reehl, W.F., and Rosenblatt, D.H., *Handbook of Chemical Property Estimation Methods*, American Chemical Society, Washington, D.C., 1990.

Mackay, D., Hubbarde, J., and Webster, E., The role of QSARs and fate models in chemical hazard and risk assessment, *QSAR Combinatorial Chem.*, 22, 106–112, 2003.

Richard, A.M. and Williams, C.R., Distributed Structure-Searchable Toxicity (DSSTox) database network: a proposal, *Mutation Res.*, 499, 27–52, 2002.

Richard, A.M. and Williams, C.R., Public sources of mutagenicity and carcinogenicity data: use in structure-activity relationship models in *QSARs of Mutagens and Carcinogens*, Benigni, R., Ed., CRC Press, Boca Raton, FL, 2003, pp. 151–179.

Richard, A.M., Williams, C.R., and Cariello, N.F., Improving structure-linked access to publicly available chemical toxicity information, *Curr. Opinion Drug Discovery Dev.*, 5, 136–143, 2002.

Schultz, T.W. and Cronin, M.T.D., Essential and desirable characteristics of ecotoxicity QSARs, *Environ. Toxicol. Chem.*, 22, 599–607, 2003.

The Nanogen Index 2, www.nanogens.co.uk.

U.S. Environmental Protection Agency, *Pesticide Fact Handbook*, Noyes Data Corporation, Park Ridge, NJ, 1988.

Wauchope, R.D., Buttler, T.M., Hornsby, A.G., Augustijn Beckers, P.W.M., and Burt, J.P., The SCS/ARS/CES pesticide properties database for environmental decision-making, *Rev. Environ. Contamination Toxicol.*, 123, 1–164, 1992.

World Health Organization, International non-proprietary names for pharmaceutical substances (INN): List 84, *WHO Drug Inf.*, 14, 245–280, 2000.

Worthing, C.R. and Hance, R.J., *The Pesticide Manual*, The British Crop Protection Council, Unwin Brothers Ltd, Surrey, UK, 1991.

# CHAPTER 3

# Calculation of Physicochemical Properties

**Mark T.D. Cronin and David J. Livingstone**

## CONTENTS

## I. INTRODUCTION

Physicochemical properties are required in their own right as part of the description of the characteristics of a chemical (e.g., for regulatory submission), or as one of the three components of quantitative structure-activity relationships (QSARs) (see Chapter 1, Figure 1.1 and Figure 1.2 for more details). While the properties of a pure chemical substance may be measured, the computation of physicochemical properties has many advantages. These include the speed and low cost of calculation and, more importantly, the fact that calculation may be performed for chemicals that are not available.

In principle, all physicochemical properties are calculable. This chapter aims to describe briefly how the major properties, in terms of risk assessment and chemical design, may be calculated. It is not intended to be a definitive and exhaustive review in this area. For further information the reader is referred to a number of other sources, starting with the excellent treatise from Boethling and Mackay (2000). Other excellent reviews in the calculation, use, and application of physico-chemical properties include the works of Cronin (1992), Dearden (1990), Karelson (2000), Livingstone (2000; 2003), Todeschini and Consonni (2000) and many others. A number of reliable resources are available on the internet that will also help the reader find information in this fast moving field. These include the homepage of the International QSAR and Modelling Society (www.qsar.org). The Society's site includes an excellent collection of resources with links to a large number of software packages (www.qsar.org/resource/software.htm). These links will take the reader to the homepages of the providers. A further excellent resource is provided by the Organisation for Economic Cooperation and Development's (OECD) website (www.oecd.org). The OECD database, Database on Chemical Risk Assessment Models, gives details on models to predict physicochemical properties, as well as toxicity, fugacity, and other models.

## A.  Input of Chemical Structures

In order to calculate a physicochemical property, the structure of a molecule must be entered in some manner into an algorithm. Chemical structure notations for input of molecules into calculation software are described in Chapter 2, Section VII and may be considered as either being a 2D string, a 2D representation of the structure, or (very occasionally) a 3D representation of the structure. Of this variety of methods, the simplicity and elegance of the 2D linear molecular representation known as the Simplified Molecular Line Entry System (SMILES) stands out. Many of the packages that calculate physicochemical descriptors use the SMILES chemical notation system, or some variant of it, as the means of structure input. The use of SMILES is well described in Chapter 2, Section VII.B, and by Weininger (1988). There is also an excellent tutorial on the use of SMILES at www.daylight.com/dayhtml/smiles/smiles-intro.html.

There are also a variety of methods of entering a 2D structural representation of a molecule. Many molecular modeling and graphics packages have developed capabilities to draw in a molecule, often with the software being able to fill valence, add hydrogen, and calculate a reasonable 2D shape. One very user-friendly package for the beginner is the Java Molecular Editor (JME), a web-based tool available for inspection at www.molinspiration.com/jme/. JME is a Java applet that allows the user to draw and edit molecules and reactions (including generation of substructure queries) and to depict molecules directly within a hypertext markup language (HTML) page. The editor can also generate SMILES or MDL .mol files of created structures. Some software (especially web-based packages) use JME, but it is an excellent tool for many further applications, or even checking SMILES of complex structures.

## II. OCTANOL-WATER PARTITION COEFFICIENT

The partitioning of a substance between two immiscible solvents is an important property of a molecule. If the two solvents are polar and non-polar, the ratio of the concentrations (when measured at equilibrium and below saturation in either solvent) is considered to describe the hydrophobicity of a compound. The partition coefficient ($K_{ow}$ or P) may therefore be defined as:

$$K_{ow} = \frac{\text{concentration in oil}}{\text{concentration in water}} \qquad (3.1)$$

There have been a number of solvent systems utilized in the measurement of the partition coefficient (Livingstone 2003). By far the most commonly used in the last few decades has been the octanol-water solvent pair. The octanol-water partition coefficient is the most commonly measured and applied in QSAR analysis, and it will be the subject of this discussion. By convention the logarithm to the base 10 of the partition coefficient is taken, and known as log $K_{ow}$ or log P. In a series of compounds, a higher log $K_{ow}$ represents more hydrophobic compounds (i.e., more soluble in lipid), and a lower log $K_{ow}$ represents more hydrophilic compounds (i.e., more water soluble).

Without doubt log $K_{ow}$ has been the most commonly utilized physicochemical descriptor in QSAR analysis. The reason is that it is considered to represent molecular hydrophobicity (it is incorrect, however, to consider a log $K_{ow}$ value as hydrophobicity — it merely describes it). Hydrophobicity is an extremely important property in predictive toxicology. As described in Chapter 1 hydrophobicity is assumed to account for the capability to enter, pass through, and/or accumulate in cell membranes, as well as a being one of the main forces in binding effects. Therefore hydrophobicity, as described by log $K_{ow}$, is a major driving force in predicting toxicity (Chapters 8 and 12) and distribution in humans (Chapter 11) and the environment (Chapters 14 to 16).

## A.  Measurement of the Octanol-Water Partition Coefficient

There are a large number of methods to measure log $K_{ow}$. These may be considered as either chromatographic methods (e.g., high performance liquid chromatography, [HPLC]), or more classical separation methods (e.g., slow-stir, filter probe, etc.). It is beyond the remit of this chapter to describe all methods and the reader is referred to Dearden and Bresnen (1988) and Sangster (1997) for more details. It is worth noting that the OECD has published a Guideline (117) for the measurement of partition coefficient using HPLC. A further Guideline (122) has been proposed for a pH-metric method to determine the log $K_{ow}$ of ionizable substances.

## B.  Calculation of the Logarithm of the Octanol-Water Partition Coefficient

The octanol-water partition coefficient is an additive chemical property that lends itself very well to calculation. Since the work of Rekker (1977) and Hansch and Leo (1979) a large number of methods (possibly in excess of 50) for the calculation of log $K_{ow}$ have been published. Many of the published methods have been computerized and are available as commercial software and shareware, or may be used over the Internet. A number of the better recognized software packages are summarized in Table 3.1.

With the considerable choice of methods to calculate log $K_{ow}$ it is difficult even for an experienced researcher in this area to determine the best one to use. A number of comparative studies on the performance of calculation methods have been performed (Mannhold et al., 1990; Buchwald and Bodor, 1998). It is often difficult to use the results of comparative studies, however, as it is difficult to find suitable data to establish a truly independent test set (i.e., data for compounds that have not been included in the original training set). The choice of method often becomes a subjective decision based on criteria such as ease of entry of structure and handling of the predicted values, cost, and any personal conviction or opinion on the method. From the authors' experience, methods including (but not limited to) ClogP for Windows, KOWWIN, and ACD/log P all have been shown to provide robust predictions for the most commonly encountered toxicants.

A final consideration in the calculation of log $K_{ow}$ is the role of ionization. The reality is that many molecules (especially drugs) contain one, and often more, ionizable functional groups. Methods to calculate log $K_{ow}$ assume that a molecule is uncharged. When a compound with strongly acidic or basic groups is placed in a test environment, it is unlikely to remain in the unionized form; calculations may need to take account of this. The degree of ionization is related to the pH of the test system and the intrinsic acid dissociation constant ($pK_a$) of the molecule (see Section IV).

Table 3.1  A Summary of Computational Methods Used
to Calculate log $K_{ow}$

| Program[a] | Calculation Method[b] | Supplier[c] |
|---|---|---|
| Absolv | LSER[e] | www.sirius-analytical.com |
| ACD/LogP | Fragmental-A/F | www.acdlabs.com |
| ALOGPS[d] | Topological descriptors | www.vcclabs.org |
| AutologP | Properties | j.devillers@ctis.fr |
| CERIUS[2]* | Atomic values | www.accelrys.com |
| CHEMICALC-2 | Atomic values | QCPE[f] |
| CLIP | Properties | Bernard.testa@ict.unil.ch |
| ClogP | Fragmental-HL | Www.biobyte.com |
| HINT | Properties | www.eslc.vabiotech/hint |
| IALOGP[d] | Topological descriptors | www.logp.com |
| KLOGP | Fragmental-C | www.multicase.com |
| LOGKOW | Fragmental-A/F | EPA[g] |
| MiLOGP[d] | Group contributions | www.molinspiration.com |
| PCMODELS | Fragmental-HL | www.daylight.com |
| PROLOGP | Fragmental-R | www.compudrug.com |
| SCILOGP | Properties | www.scivision.com |
| SYBYL* | Fragmental-R | www.tripos.com |
| TLOGP | Not specified | www.upstream.ch |
| TSAR* | Atomic values | www.accelrys.com |
| VLOGP* | Topological descriptors | www.accelrys.com |

[a] These are stand-alone programs except those marked with *.
[b] The fragmental methods refer to the system of Hansch and Leo (HL),
Rekker (R), computer-identified (C), and atom/fragment contributions
(A/F). Properties means that various molecular properties are used in
the calculations. Atomic values means that tables of atom-based values
are used. Topological descriptors means (usually) electrotopological
descriptors (see Chapter 5).
[c] Web addresses were correct at the time of this chapter's preparation.
[d] These programs will calculate log $K_{ow}$ on the Internet.
[e] Linear solvation energy relationship.
[f] Quantum Chemistry Program Exchange (www.osc.edu/ccl/qcpe).
[g] Environmental Protection Agency (www.epa.gov/oppt/exposure/docs/
episuitedl.htm).

From this knowledge, assuming either a test system buffered to a particular pH or modeling at a constant pH (e.g., a physiological pH), log $K_{ow}$ may be corrected to account for the degree of ionization. The corrected partition coefficient is termed the distribution coefficient (D). The relationship between $K_{ow}$ and D for basic compounds is

$$\log D = \log K_{ow} - \log_{10} (pK_a - pH) \tag{3.2}$$

Inevitably the calculation of log D relies on a knowledge of $pK_a$ values. These in turn may need to be calculated (see Section IV). Mainly because of the problems of calculating $pK_a$, there are fewer methods to calculate log D. Some of these methods are summarized in Table 3.2.

## C.  Recommendations for the Use and Calculation of Log $K_{ow}$

1. A calculation method should be chosen to meet the needs of the user in terms of ease of use, cost, and domain.
2. A consideration of the likely effects of ionization on log $K_{ow}$ should be taken into account. If necessary log $K_{ow}$ should be substituted by log D.

**Table 3.2  A Summary of Computational Methods Used to Calculate log D**

| Program | Calculation Method[a] | Supplier |
|---------|----------------------|----------|
| ACDLogD | Fragmental-A/F | www.acdlabs.com |
| PROLOGD | Fragmental-R | www.compudrug.com |
| SLIPPER | Properties | www.ipac.ac.ru/qsar |

[a] The fragmental methods refer to the system of Rekker (R) or atom/fragment contributions (A/F). Properties means that various molecular properties are used in the calculations.

3. Where possible, calculated values should be compared with measured values for some similar compounds. It is not unusual for any calculation scheme to over- or underestimate log $K_{ow}$ for a particular chemical class or combination of structural features. Such estimation errors are often quite consistent and corrections may be applied based on measured values.
4. Beware of extreme values. Remember that log $K_{ow}$ is a logarithmic scale and that both measurement and prediction for highly hydrophobic (and hydrophilic) compounds is difficult.

## III.  WATER SOLUBILITY

The water (aqueous) solubility ($S_{aq}$) of a chemical may be defined as being the maximum concentration that may be dissolved in water at equilibrium at any given temperature and pressure. It is possibly the most important fundamental physicochemical property that may be assessed. Its importance is due to a number of reasons, such as its requirement in regulatory submissions and its governing role in a number of biological processes. More specifically, if a chemical is not soluble at a biologically active concentration, it will not cause a biological effect, whether that be a toxic or pharmacological response. With regard to toxicology, compounds lacking suitable aqueous solubility fail to produce an accurate toxic endpoint (the so-called Ferguson [1939] cut-off). For acute aquatic endpoints, this is normally represented by compounds with high log $K_{ow}$ (over 5) being seen to be "not toxic at saturation." These compounds should not be included in QSAR analysis, and predictions of toxicity for compounds with very low water solubility should be treated with caution.

In terms of developing QSARs, water solubility is not a commonly used parameter in the development of quantitative models. When used it is usually in the form of the logarithm to the base 10 (log $S_{aq}$). The reasons for its low usage are probably due to difficulty in calculation and colinearity with log $K_{ow}$ (which should be used in preference), rather than its lack of meaning or relevance. Despite this, the use of log $S_{aq}$ in QSAR should not be discounted either to describe the solubility cut-off, or as a parameter in it own right.

### A.  Calculation of Water Solubility

Despite its importance, there still remain few methods to calculate log $S_{aq}$. Many methods are based on the relationship between $S_{aq}$ and hydrophobicity (log $K_{ow}$) and some measure of the enthalpy of crystallization (e.g., melting point). Other methods are based on fragment- or atom-based contributions. A full review of methods to calculate log $S_{aq}$ is provided by Livingstone (2003). Methods to calculate log $S_{aq}$ are summarized in Table 3.3.

Little is known about the accuracy of predictions of water solubility. Practically, however, the assessment of $S_{aq}$ is difficult and will be complicated by any number of considerations including ionization, formation of salts, and the inclusion of a co-solvent. All of these effects may significantly alter $S_{aq}$. As such, considerable caution should be used when utilizing calculated log $S_{aq}$ values.

**Table 3.3    A Summary of Computational Methods Used to Calculate Water Solubility**

| Program | Calculation Method | Supplier |
|---|---|---|
| Absolv | LSER[b] | www.sirius-analytical.com |
| ACD/Solubility | Database and properties | www.acdlabs.com |
| ADME Boxes | Properties and similarity | www.ap-algorithms.com |
| ALOGPS[a] | Topological descriptors | www.vcclab.org/lab/alogps/ |
| C[2] ADME module | Not specified | www.accelrys.com |
| EPIWIN | From log $K_{ow}$ | EPA[c] |
| LogW[a] | Topological descriptors | www.logp.com |
| ToxAlert | Group contribution | Multicase, Inc.[d] |

[a] These programs are available on the Internet.
[b] Linear solvation energy relationship.
[c] Environmental Protection Agency (www.epa.gov/oppt/exposure/docs/episuit-edl.htm).
[d] Multicase Inc., 23715 Mercantile Rd., Ste B104, Beachwood, OH, 4412.

Within a homologous series it may be better to rank relative water solubility (rather than relying on specific calculated values). Their use with heterogeneous data sets is even more fraught, and calculated values should be used with considerable circumspection.

## B.  Recommendations for the Calculation of Water Solubility

1. Calculations of log $S_{aq}$ must be used cautiously and conservatively.
2. As for log $K_{ow}$, comparison of calculated values with measured values is often instructive and it is wise to be aware of extremes.

## IV. IONIZATION OR DISSOCIATION CONSTANT ($K_a$, $pK_a$)

The Brønsted and Lowry theory states that an acid is a proton donor and a base is a proton acceptor. Since equilibrium exists between what are considered the unionized (neutral) and ionized forms of a compound, a constant can be determined. This is termed the equilibrium acid ionization ($K_a$) and expresses the ratio of concentrations for the reaction:

$$HA + H_2O \rightarrow H_3O^+ + A^-  \tag{3.3}$$

By convention it is assumed that the concentration of water is constant and it is absorbed into the definition. $K_a$ may therefore be defined as:

$$K_a = \frac{[H_3O^+][A^-]}{[HA]}  \tag{3.4}$$

Again, by definition, the negative logarithm to the base 10 of $K_a$, termed $pK_a$, is usually reported. $pK_a$ is a fundamental chemical property and the subject of countless physical chemistry textbooks; its theory will not be defined further here. In defining $pK_a$ we should also define $pK_b$ (the base dissociation constant):

$$pK_b = -\log K_b = -\log \frac{[BH^+][OH^-]}{[B]}  \tag{3.5}$$

**Table 3.4  A Summary of Computational Methods Used to Calculate pK$_a$**

| Program | Calculation Method | Supplier |
|---------|-------------------|----------|
| ACD/pK$_a$ | Database and properties | www.acdlabs.com |
| ADME Boxes | Properties and similarity | www.ap-algorithms.com |
| PKalc | Properties | www.ipac.ac.ru/qsar/ |
| PROLOGD | Fragmental-R | www.compudrug.com |
| SPARC | Not specified | ibmlc2.chem.uga.edu/sparc/ |
| Jaguar | *Ab Initio* quantum chemistry | Schrodinger[a] |

[a] Schrodinger, Inc., 1500 SW First Avenue, Suite 1180, Portland, OR, 97201.

pK$_a$ is one of the most fundamental properties available to describe a molecule. In terms of QSAR, it has a number of applications. As a raw value it can be used as a parameter in its own right to describe acid or base strength. This is important in terms of effects such as skin and eye corrosivity, where strong acids and bases may be assumed to be corrosive without the need for further testing (see Chapter 18). More frequently it is used to describe the degree of ionization of a compound at a particular pH. As noted in Section III.B, ionization plays a vital role in the distribution and transport of molecules, and ultimately their toxic potency.

## A. Calculation of pK$_a$

The prediction of ionization constants as defined by Hammett (1940) is still regarded as a milestone in the development of modern predictive toxicology (see Table 1.1). Hammett defined a constant ($\sigma$) that related the electron withdrawing and releasing characteristics of substituents on benzoic acid. The constant was derived directly from pK$_a$ and enabled prediction for further molecules. Methods for the calculation of pK$_a$ based on the Hammett constant were further described by Perrin et al. (1981). Thus, within well-defined series of aromatic compounds, it is possible to calculate a pK$_a$ value, or at least to put a series of compounds in rank order. The prediction of pK$_a$ has not progressed considerably since the original work of Hammett. Predicting pK$_a$ is still very difficult for compounds with multiple ionizable groups, which may have a number of pK$_a$ and pK$_b$ values.

Following on from the substituent constant methods, a number of other approaches have been applied to the prediction of pK$_a$. The main prediction methods for pK$_a$ are summarized in Table 3.4. Of the methods to calculate pK$_a$ some are derived from atom and fragment values, others are derived from molecule orbital properties. Because of the problems of modeling ionization constants for molecules with multiple ionizable functional groups, the accuracy and predictivity of these methods remains questionable.

## B. Recommendations for the Calculation of pK$_a$

1. pK$_a$ may be calculated with reasonable accuracy within a congeneric series of aromatic molecules by methods such as the Hammett equation.
2. Calculations of pK$_a$ must be used cautiously and conservatively, especially when there are multiple ionizable groups.
3. Compare calculations with measured values, as well as for all experimental properties.

## V. OTHER PHYSICOCHEMICAL PROPERTIES

There are a number of other physicochemical properties that may be usefully calculated. Properties relating to phase transitions such as melting and boiling points are commonly predicted, and methods to do so will be described briefly here. Other important properties not covered in this

**Table 3.5   A Summary of Computational Methods Used to Calculate Melting Point**

| Program | Calculation Method | Supplier |
|---|---|---|
| ChemOffice | Fragmental | www.cambridgesoft.com |
| MPBPWIN | Fragmental | EPA[a] |
| ProPred | Fragmental | www.capec.kt.dtu.dk/main/software/propred/propred.html |

[a] Environmental Protection Agency (www.epa.gov/oppt/exposure/docs/episuitedl.htm).

**Table 3.6   A Summary of Computational Methods Used to Calculate Boiling Point**

| Program | Calculation Method | Supplier |
|---|---|---|
| ACD/boiling point | Database and properties | www.acdlabs.com |
| ADME Boxes | Properties and similarity | www.ap-algorithms.com |
| MPBPWIN | Fragmental | EPA[a] |
| SPARC | Not specified | ibmlc2.chem.uga.edu/sparc/ |

[a] Environmental Protection Agency (www.epa.gov/oppt/exposure/docs/episuitedl.htm).

section, but for which models are available, include vapor pressure (Dearden, 2003) and Henry's law constant (Dearden and Schüürmann, 2003).

Normal melting and boiling points are the temperatures at which melting and vaporization occur at 1 atm, respectively. In theory, these are thermodynamically related properties and should be related directly to the enthalpies of fusion and vaporization, respectively. These predicted properties are seldom used as descriptors in QSAR analysis and are usually predicted for use in their own right. A knowledge of melting and boiling points, for example, will allow a modeler to determine the physical state of a substance at room, or any other, temperature.

## A.  Calculation of Melting Point

There are a number of approaches to the calculation of melting point, and the main methods are summarized in Table 3.5. Recent methods in this area are well reviewed by Dearden (1999; 2003). Generally speaking, it is possible to calculate melting points with reasonable accuracy, although most predictions must be considered with reasonable error limits (e.g., ±20 K).

## B.  Calculation of Boiling Point

A small number of approaches to calculate boiling point are available and are summarized in Table 3.6. Recent advances in this area are well reviewed by Dearden (2003). Boiling point is more difficult to model than melting point because a variety of factors can affect passage from the liquid to gaseous phases. Predictions of boiling point (and also related properties such as vapor pressure) should be treated with a degree of caution and not be expected to reach the accuracy of, for instance, melting point.

## C.  Recommendations for the Calculation of Melting and Boiling Points

1. Melting points may be calculated for simple molecules with a reasonable accuracy.
2. The boiling point of a substance is often less accurately predicted, and predictions of this property must be treated with caution.

**Table 3.7   A Summary of Packages Used to Calculate Physicochemical Properties and Other Descriptors for QSAR analysis**

| Product Name | Provider | Website | Properties and Descriptors Calculated |
|---|---|---|---|
| CoMFA, Molconn-Z, Volsurf (and other related products) | Tripos Inc. | www.tripos.com | A variety of descriptors, including comparative molecular field analysis; physicochemical properties; and molecular, structural, and topological descriptors |
| EPISUITE | Syracuse Research Corporation | esc.syrres.com | Log $K_{ow}$, solubility, melting and boiling points, vapor pressure, assorted other properties for fate assessment |
| MDL QSAR | MDL Inc. | www.mdli.com | Physicochemical properties and topological descriptors |
| PhysChem Batch | Advanced Chemistry Development Inc. | www.acdlabs.com | $pK_a$, log $K_{ow}$, log D, $K_{oc}$, bioconcentration factor, solubility at a certain pH, boiling point, vapor pressure, enthalpy of vaporization, flash point, macroscopic properties |
| QSAR Builder | Pharma Algorithms | www.ap-algorithms.com | Log $K_{ow}$, hydrogen bonding parameters, molecular properties |
| TSARBatch for Windows | Accelrys Inc. | www.accelrys.com | Log $K_{ow}$, various topological, structural, 3D and molecular orbital descriptors |

## VI. SOFTWARE FOR THE CALCULATION OF PHYSICOCHEMICAL PROPERTIES AND OTHER DESCRIPTORS

There is a considerable variety of products available to calculate properties and structural descriptors. An exhaustive review is beyond the scope of this chapter, but useful links are given in the resources section of the International QSAR and Modeling Society homepage. Table 3.7 lists a selection of available software; the packages listed represent only a selection, and other products are available. The systems listed in Table 3.7 have been selected because of their relevance, and they offer the possibility of calculating properties in some form of batch mode. This is important for speeding up QSAR development and screening large databases.

With the exception of the EPISUITE package, which is freely downloadable, all the products listed in Table 3.7 are commercial and charges are associated with their use. Most enable large numbers of compounds to be input at one time, using most common file formats such as SMILES, .mol, and .pdb. Descriptors may be extracted easily from most of these packages and transferred into spreadsheets for statistical analysis. Some of the products include some form of statistical analysis, although the use of dedicated external statistical packages is recommended in most cases.

The packages listed in Table 3.7 (and others not listed) offer the user the capability to calculate a large variety of descriptors rapidly and efficiently. This is an excellent facility in terms of QSAR development, but the user must always remember that these are calculated values. More specifically, while a complete data sheet may be produced, calculated properties may not be valid if the compound is outside the domain of the original model. The user must resist the temptation to take any calculated value as a correct value.

## VII. GENERAL RECOMMENDATIONS FOR THE CALCULATION OF PHYSICOCHEMICAL PROPERTIES

1. When measured physicochemical data are available for registering chemicals or developing QSARs, these data should be as high quality as possible, preferably performed to Good Laboratory Practice (GLP) standards and using the appropriate OECD Guideline.
2. As with all predictive methods, the calculation of physicochemical properties should only be performed within the domain of the training set for that model.
3. For the large scale calculation of physicochemical values (e.g., screening databases or developing QSARs for large data sets), a computational method that allows for the simple and easy entry of chemical structure by, for instance, SMILES notation is recommended.

## REFERENCES

Boethling, R.S. and Mackay, D., *Handbook of Property Estimation Methods for Chemicals: Environmental and Health Sciences*, CRC Press, Boca Raton, FL, 2000.

Buchwald, P. and Bodor, N., Octanol-water partition: searching for predictive model, *Curr. Med. Chem.*, 5, 353–380, 1998.

Cronin, M.T.D., Molecular descriptors of QSAR, in *Quantitative Structure-Activity Relationships (QSAR) in Toxicology,* Coccini, T., Giannoni, L., Karcher, W., Manzo, L., and Roi, R., Eds., Commission of the European Communities, Brussels, 1992, pp. 43–54.

Dearden, J.C., Physico-chemical descriptors, in *Practical Applications of Quantitative Structure-Activity Relationships (QSAR) in Environmental Chemistry and Toxicology*, Karcher, W. and Devillers, J., Eds., Commission of the European Communities, Brussels, 1990, pp. 25–59.

Dearden, J.C., The prediction of melting point, in *Advances in Quantitative Structure Property Relationships*, Charton, M. and Charton, I., Eds., JAI Press, Stamford, CT 1999, pp. 127–175.

Dearden, J.C., Quantitative structure-property relationships for prediction of boiling point, vapor pressure and melting point, *Environ. Toxicol. Chem.,* 22, 1696–1709, 2003.

Dearden, J.C. and Bresnen, G.M., The measurement of partition coefficients, *Quant. Struct.-Act. Relat.*, 7, 133–144, 1988.

Dearden, J.C. and Schüürmann, G., Quantitative structure-property relationships for predicting Henry's law constant from molecular structure, *Environ. Toxicol. Chem.*, 22, 1755–1770, 2003.

Ferguson, J., The use of chemical potentials as indices of toxicity, *Proc. R. Soc. London, Ser. B: Biol. Sci.*, 127, 387–404, 1939.

Hammett, L.P., *Physical Organic Chemistry*, 1st ed., McGraw-Hill, New York, 1940.

Hansch, C. and Leo, A.J., *Substituent Constants for Correlation Analysis in Chemistry and Biology*, John Wiley and Sons, New York, 1979.

Karelson, M., *Molecular Descriptors in QSAR/QSAR*, John Wiley and Sons, London, 2000.

Livingstone, D.J., The characterisation of chemical structures using molecular properties: a survey, *J. Chem. Inf. Comput. Sci.*, 40, 195–209, 2000.

Livingstone, D.J., Theoretical property predictions, *Curr. Top. Med. Chem.* 3, 1171–1192, 2003.

Mannhold, R., Dross, K.P., and Rekker, R.F., Drug lipophilicity in QSAR practice. I. A comparison of experimental with calculative approaches, *Quant. Struct.-Act. Relat.*, 9, 21–28, 1990.

Perrin, D.D., Dempsey, B., and Serjeant, E.P., *pKa Prediction for Organic Acids and Bases*, Chapman and Hall, London, 1981.

Rekker, R.F., *The Hydrophobic Fragmental Constant*, Elsevier, Amsterdam, 1977.

Sangster, J., *Octanol-Water Partition Coefficients: Fundamentals and Physical Chemistry*, John Wiley and Sons, Chichester, UK, 1997.

Todeschini, R. and Consonni, V., Eds., *Handbook of Molecular Descriptors (Methods and Principles in Medicinal Chemistry)*, Wiley-VCH, Weinheim, 2000.

Weininger, D., SMILES. 1. Introduction and Encoding Rules, *J. Chem. Inf. Comput. Sci.*, 28, 31–36, 1988.

CHAPTER **4**

# Good Practice in Physicochemical Property Prediction

Peter R. Fisk, Louise McLaughlin, Rosalind J. Wildey

## CONTENTS

0-415-27180-0/04/$0.00+$1.50

# I. INTRODUCTION

This chapter is a review of practical and easily accessible techniques in physicochemical property prediction. It is not intended to be comprehensive, or be a guide to every kind of approach that can be adopted. The scientific basis of the various methods has been reviewed frequently and recently (Boethling and Mackay, 2000; Fisk 1995). The intention is to set out approaches that can be readily applied, and to give guidance on good practice. The very fact of the ready availability of computerized methods of well-established reliability can give a false sense of security. The examples show that the off-the-peg methods are remarkably effective, which, somewhat ironically, increases the risk of bad practice creeping in.

The review is largely of rather traditional techniques — fragment methods and correlation between properties. More modern techniques based on wholly *a priori* computational approaches have not yet yielded methods that are robust; in fact, much published material in this area is singularly unconvincing. Why should that be? It is because physiochemical properties involve such matters as solvation and intermolecular forces that computational methods frequently fail; the energy differences that need to be understood are small and not easily predicted computationally.

Another reason is that this topic is one that is not receiving much attention at the cutting edge of computational chemistry. The importance of physiochemical properties in the understanding of the behavior of xenobiotics will just not go away. Predictive methods for drug design are improved when these properties are included. The topic is vital in the modeling of the environmental behavior of chemicals. Uptake, distribution, and bioavailability studies in pharmacokinetics and pesticide behavior rely on the understanding of physicochemical properties. Measurements are not expensive, but the demand for them outstrips supply, so prediction is widely used. Prediction will be necessary as the high throughput methods, such as combinatorial chemistry and virtual screening, common in the pharmaceutical industry, become more widely adopted.

This chapter first outlines some general principles of good practice, and then summarizes methods. The Syracuse Research Corporation (SRC) package of programs is in such wide use that some information about them is included as an appendix (see Appendix 1, Section 4.VII). They are freely downloadable from the U.S. Environmental Protection Agency (EPA).

# II. GENERAL PRINCIPLES OF ASSESSMENT, EVALUATION, AND VALIDATION OF METHODS FOR PHYSICOCHEMICAL PROPERTY ESTIMATION

The user should see estimation as an experiment. Just as a good experiment is planned, performed, and accurately repeated, so should estimations be. There is a danger in computer-based methods that non-experts or even non-chemists can use. Because the programs can give decent answers without any fine-tuning or validation, it does not mean that best practice can be abandoned. The best experiments are reported with a confidence interval for a dependent variable, with hopefully much smaller uncertainties concerning the independent variable. This is harder to achieve when we are performing an estimation. Consider an example that will be returned to later: the estimation of octanol-water partition coefficients ($K_{ow}$). Take a method such as the SRC software KOWWIN to

**Table 4.1   Techniques for Measurement of Physicochemical Properties**

| Property | Method | Comment |
|---|---|---|
| Melting point | Capillary | Standard |
| | DSC | Standard |
| Boiling point | | Many methods, essential to make sure the pressure was specified |
| Vapor pressure | Many methods | Beware results operated by gross value methods on impure samples |
| | GC estimation | Not usable |
| Water solubility | Shake flask | Not valued for solubility below 10 mg/L without high levels of care |
| | Column elution | For solids of solubility <10 mg/L |
| Partition coefficient | Shake flask | Acceptable although extreme values present experiment difficulties |
| | Slow stirring | Acceptable |
| | HPLC estimation | Not usable |
| $pK_a$ | Many methods | All acceptable provided nonaqueous solvent content not high |
| Henry's law constant[a] | $VP/S_{aq}$ estimation | Acceptable provided VP $<10^3$ Pa and $S_{aq} < 10$ g/L[a] |

[a] Note that obtaining HLC from $VP/S_{aq}$ is acceptable for most purposes. In the absence of these measurements, estimation of HLC by a program such as HENRYWIN is usually more reliable than its estimation form (estimated VP/estimated $S_{aq}$). Again, the validation story will guide. DSC = Differential Scanning Calorimetry; GC = Gas Chromatography; HPLC = High Performance Liquid Chromatography.

predict $K_{ow}$. A general error of 0.32 is stated for this method. This is potentially misleading in that it ignores the error inherent in the measured values (and there are occasions when an experienced chemist will trust a prediction more than the measured value). The authors are trying, in general terms, to quantify the success of the fragment values assigned. For the user, the fragment values are given and fixed, and very few users will have the knowledge or time to start changing those values.

As a general approach, we propose that users should, as far as possible, examine graphically and statistically the measured value as a function of the predicted one. The predicted value is at least a number with a defined origin, whereas the measured ones are of uncertain heritage. Let us imagine that we need to estimate the $K_{ow}$ of an alkyl ether, where the alkyls are linear. By some form of literature-searching measured data and associated predictions are obtained for this group, and their relationship is examined graphically and statistically, with the prediction as the independent variable ($x$) and the measured as dependent ($y$). Examples of this will follow.

In common with good experimental design, users of prediction techniques should beware of extrapolation (i.e., performing a prediction that is outside the validated range of the method). This can occur when:

1. Values are predicted that are numerically larger or smaller than the training set that the method was based on.
2. The test structure contains a combination of structural features that the method may not recognize. This is often exhibited for cases of internal hydrogen bonding or delocalization leading to non-standard behavior of the functional groups involved.

## A.   Source Data for Assessment, Evaluation, and Validation

Various techniques exist to measure certain physicochemical properties. Table 4.1 sets out some views on their admissibility for reference data. This section discusses the problems of using data for evaluating predictive methods.

### 1.   When No Validation Data Are Available

In novel areas of chemistry it is to be expected that no immediately obvious validation data are available. What can be done? The aim here must be to reduce the amount of extrapolation inherent in the prediction. Consider a substance containing three different substituted heterocyclic rings linked together. No validation data exist, so all that can be done is to see how well a method works for each ring either singly or perhaps in pairs. Starting from very simple substructures, and then building up toward the target, the performance of the method should be checked.

## 2. Propagated Error

Some methods of estimation use the relationships between properties; for example, water solubility may be modeled as a function of log $K_{ow}$ and melting point. Should it be the case that log $K_{ow}$ itself has been estimated, then it is frequently pointed out that there are two estimations involved in obtaining the water solubility, increasing uncertainty. This may not necessarily be the case if sufficient validation data are available. If a water solubility is required, and several close analogs are known with measured water solubility, it might well be possible to model water solubility as a function of $K_{ow}$ from KOWWIN and the measured melting point. It might not matter at all whether KOWWIN is good at predicting log $K_{ow}$ for these examples, because it is merely acting as a molecular descriptor whose value is precisely known.

## 3. Understanding of the Basis of a Method

Validation apart, there are other reasons why chemical knowledge needs to be applied to obtain the best results. Some of the reasons come out in the examples given in the sections on each property. One common issue is the ability to interpret information. The majority of computer-based methods provide a report on how the calculation was performed. That needs to be examined for its relevance and appropriateness, and whether some parts of the model (e.g., fragment values, are less well founded than others). The authors of programs used high levels of knowledge to put them together, but they cannot have envisaged every use to which their method might be used. A common example concerns prediction of the properties of acids and bases. Any method used should ensure that all the calculations concern structures in the same ionization state. For simple acids and bases, it is usually possible to perform estimations for the non-ionized form and then use standard equations to make corrections to give the value at the pH of interest. For example, where $K'_{ow}$ is the apparent partition coefficient at the pH of the aqueous phase, for acids:

$$K'_{ow} = \frac{K_{ow}}{1 + 10^{(pH-pKa)}} \qquad (4.1)$$

Where no such reference data exist (e.g., for zwitterions or very strong acids and bases), it is still essential to ensure that the methods used are consistent.

Another general concern is the order of magnitude of the result. Values can be generated that are not only outside the validated range of the method, outside the range of what would ever be achieved, and outside the range of any conceivable application, but also outside the range of what is scientifically possible! It therefore makes some sense to set limits on what is a practical value. Such ignorance can lead to serious propagated errors. A predicted log $K_{ow}$ of 15 (a scientific nonsense) being used in predictions of $K_{oc}$, water solubility, or bioconcentration factor (BCF) is one example of such an error.

## 4. Stability

Chemical stability is discussed in Chapter 10 (with regard to effects *in vivo*) and Chapter 14 (for environmental effects). It does have an impact on physiochemical property prediction. With all the various types of degradation, only two are important in the present context:

1.  Rapid hydrolysis — This makes measurement of solubility in water and octanol-water partition coefficient impossible. Consider an example: isocyanates have half-lives in water at normal temperatures and pH of a few seconds. There can be no valid reference data for them, nor any practical application of a partition coefficient. That is a clear case; others will depend upon the half-life, the intended use of the result, and any validation required.
2.  Thermal stability — Certain structural classes have such instability that prediction of other physiochemical properties would be meaningless.

## III. OVERVIEW OF PHYSIOCHEMICAL PROPERTY PREDICTION METHODS

It is useful to understand what type of method is being adopted; several classes of method may be identified.

### A. Correlation of Property A with Property B

Physicochemical property prediction is only one example of this most familiar of approaches to structure-property relationships.

### B. Methods Based on Fundamental Equations or Physical Models

There are very few useful examples in this category since most substances, particularly complex structures such as active components of pesticides or pharmaceuticals, frequently fall outside the scope of ideal equations. One example is the Antoine modification of the Clausius-Clapeyron equation to predict vapor pressure (VP) and its temperature dependence.

### C. Methods That Use Molecular Fragment Constants

This is the main easily accessible method for direct prediction of a property from chemical structure alone. Fragment methods view a molecule as composed of specified parts, which contribute individually to the compound property.

### D. Statistical Methods

This approach employs statistical methods that use no obvious theory-derived basis, but which derive usable relationships from realistic inputs. It is beyond the scope of this review to describe the methods and their validation in detail. Useful reviews are available (Livingstone, 2000; 2003) and more details are provided in Chapter 3. The methods may be divided into two classes, often referred to as those derived from supervised and unsupervised learning. In the latter, the techniques used are more free to explore relationships between variables, and are therefore less likely to produce chance effects.

### 1. Neural Networks

An unpublished example of the power of this method is provided by the recent work of SciMetrics (www.scimetrics.com). VP values of 653 compounds were obtained from the SRC database of physicochemical properties, PHYSPROP, covering 16 orders of magnitude. The Simplified Molecular Line Entry System (SMILES) strings were converted to connectivity matrices, providing topological and atom/fragment count indices. Sixty percent of the substances were used for the training set, and 40% were used for the test set. The training set $r^2$ was 0.961, with a standard error of 0.035 and zero intercept and unit gradient. The test set performance was good, with $r^2$ of

0.906 and a standard error of 0.035. This is an interesting development in that it appears not to require any input or prediction of boiling point.

### E. Molecular Modeling Methods

A review of this topic is also beyond the scope of this article, but some background is given because some modeling is being used in descriptions of molecular shape, volume, and area. Molecular volume and area calculated by modeling methods are being used particularly in prediction of solubility and partition coefficient. This is because, in any fundamental understanding of these properties, it is necessary to consider the cavity that has to form in water to accommodate the solute.

## IV. IS PROPERTY PREDICTION APPLICABLE TO REAL SUBSTANCES OR JUST TO IDEAL COMPOUNDS?

All chemical substances are impure — there is no such thing as 100% purity. However, many substances are pure enough for use. Why is the word substance used here? Apart from its regulatory definition, it is to distinguish practical from theoretical. The substance is the sample available to the experimenter. A compound is seen here as a theoretical concept, a material having molecules of only the intended substance within it. Various classes of substance can be distinguished, which would be treated differently both in respect of validation and estimation itself:

1. Pure — Very high purity, >99.9% of the stated substance, with known impurities
2. Pure (effectively) — >95% of the stated substance, but with the impurities not affecting the measurement
3. Impure — No composition range, but with one component dominant, impurities not necessarily known
4. Complex — A fractionation product, or a substance derived from multicomponent starting materials, with perhaps in excess of 50 components.

It is immediately obvious that there is a gradation here, and it may be hard to fit a particular substance into a definite group. The principle here is if measurement would not be valid, then estimation will be difficult (although possibly very useful). For melting, boiling, and vapor pressure, impurities have large effects. The measurements may be useful, but apply in only a limited sense. For water solubility, minor components can affect the main component. What is experimentally achievable depends upon the analytical technique available. In the extreme of a liquid complex mixture, a solubility study becomes a multiple partition coefficient study, but with each individual component partitioning out the whole substance into water. In that sense water solubility is thermodynamically meaningless, but may have some practical value. Estimated values can come to the rescue, as they would provide an upper limit of the solubility of each component. With the octanol-water partition it is perfectly possible to study a complex mixture. Here, as for water solubility, a single value exists for the melting point. Estimations can therefore only be the same — a value for each known component.

## V. STANDARD TECHNIQUES FOR PROPERTY PREDICTION

This section summarizes some of the main findings from recent reviews (see also Chapter 3).

### A. Melting Point

To predict the melting point of organic compounds, Tesconi and Yalkowsky (2000) recommend a method that uses the group contribution method based on the works of Simamora and Yalkowsky

(1994) and Krzyzaniak et al. (1995) to estimate the enthalpy of melting ($\Delta H_m$). They also recommend the method be used with the method of Dannenfelser and Yalkowsky (1996) to estimate the entropy of melting ($\Delta S_m$).

An alternative method is an adaptation of Joback and Reid (1987), used in the melting point, boiling point, and vapor pressure (MPBPVP) program available from SRC; this method is more generally applicable though it is not always as accurate. MPBPVP offers another alternative: the Gold and Ogle method, recommended by Lyman (1985), which derives melting temperature from boiling temperature using a simple equation.

## B. Boiling Point

The method used by the MPBPVP program available from SRC has been adapted from the Stein and Brown method (1994). This is a group contribution method, which has an average error of 4.3%.

Lyman (2000) recommends several methods. The most accurate is said to be the non-linear group contribution method of Lai et al. (1987), which is generally applicable to most organics, including multifunctional compounds. The average error is 1.29%.

## C. Vapor Pressure

Both SRC and Sage and Sage (2000) recommend the use of two different methods depending on the physical state of the substance. The method developed by Antoine is suitable for liquids and gases, while the Grain (1982) method is suitable for solids, liquids, and gases. The SRC program MPBPVP also takes the average of the two methods for liquids and gases, which has proven to be an appropriately accurate approach.

Both methods derive the VP based on a known (or estimated) boiling point. For solids, the melting point must also be known.

## D. Acid Dissociation Constant

Various methods are available for predicting the acid dissociation constant ($pK_a$) within homologous series. Most are based on the Hammett equation (benzene derivatives) and the Taft correlation (aliphatics and alicyclics). A comprehensive review of methods was published by Perrin et al. (1981).

## E. Octanol-Water Partition Coefficient

The SRC program KOWWIN uses an atom/fragment contribution method to predict log $K_{ow}$. This is a reductionist method (the fragment coefficients were derived by multiple regression from a development set of reliably measured log $K_{ow}$ values). The other main software tool, ClogP for Windows (Leo, 1993) is a constructionist method (the fragment coefficients are evaluated from the simplest examples in which they occur). Both methods have a high level of accuracy and are widely accepted as the best tools available.

## F. Solubility in Water

Mackay (2000) recommends two methods, one of which uses the value of log $K_{ow}$ to derive solubility. The SRC program WSKOWWIN also predicts solubility from the value of log $K_{ow}$ (Meylan et al. 1996). Molecular weight and melting point (if known, for solids) are also inputs to WSKOWWIN.

The other approach recommended by Mackay (2000) is a group contribution method to derive the molar activity coefficient, calculating the solubility (the AQUAFAC method, Myrdal et al. [1992; 1993; 1995]).

Table 4.2  Water Solubility of Some Alkenes

| Substance Name | CAS Number | Water Solubility (mg/L) | Log $K_{ow}$ | KOWWIN | WSKOWWIN (mg/L) | Adjusted Prediction (mg/L) |
|---|---|---|---|---|---|---|
| Propene | 115-07-1 | 200.0 | 1.77 | 1.68 | 1162 | 962.6 |
| 1-Butene | 106-98-9 | 221.0 | 2.4 | 2.17 | 354.8 | 267.7 |
| 2-Methylpropene | 115-11-7 | 263.0 | 2.34 | 2.23 | 399.2 | 304.0 |
| cis-2-Butene | 590-18-1 | 659.0 | 2.33 | 2.09 | 423.5 | 324.0 |
| 2-Butene trans | 624-64-6 | 511.0 | 2.31 | 2.09 | 423.5 | 324.0 |
| 1-Pentene | 109-67-1 | 148.0 | | 2.66 | 210.0 | 152.0 |
| cis-2-Pentene | 627-20-3 | 203.0 | | 2.58 | 245.1 | 179.6 |
| trans-2-Pentene | 646-04-8 | 203.0 | | 2.58 | 245.1 | 179.6 |
| 3-Methyl-2-butene | 513-35-9 | 193.0 | 2.67 | 2.64 | 206.1 | 149.0 |
| 3-Methyl-1-butene | 563-45-1 | 130.0 | | 2.59 | 242.7 | 177.7 |
| 2-Methyl-1-butene | 563-46-2 | 130.0 | | 2.72 | 188.1 | 135.0 |
| 1-Hexene | 592-41-6 | 50.0 | 3.39 | 3.15 | 47.46 | 30.6 |
| 4-Methyl-1-pentene | 691-37-2 | 48.0 | | 3.08 | 87.64 | 59.2 |
| 2-Methyl-1-pentene | 763-29-1 | 78.0 | | 3.21 | 67.91 | 45.0 |
| 1-Heptene | 592-76-7 | 18.1 | 3.99 | 3.64 | 13.45 | 7.84 |
| 2-Heptene trans | 14686-13-6 | 14.5 | | 3.56 | 31.08 | 19.4 |
| 1-Octene | 111-66-0 | 4.1 | 4.57 | 4.13 | 3.885 | 2.05 |
| 1-Decene | 872-05-9 | 0.115 | 5.7 | 5.12 | 1.037 | 0.49 |

## G.  Henry's Law Constant

Mackay et al. (2000) recommend the bond contribution method of Meylan and Howard (1991), one of two methods of predicting the Henry's law constant used by the SRC program HENRYWIN. These are developed from the work of Hine and Mookerjee (1975). HENRYWIN also predicts the Henry's law constant based on a group contribution method.

Mackay et al. (2000) also recommend the method of Nirmalakhandan and Speece (1988), which uses the molecular connectivity index and polarizability. If the water solubility and VP are known, the Henry's law constant can be approximated by the ratio of the two.

## VI.  EXAMPLES OF GOOD PRACTICE IN ESTIMATING PHYSICOCHEMICAL PROPERTIES

This chapter concludes by giving some very simple examples of the principles.

## A.  Example 1: Water Solubility of Some Alkenes

In this set (see Table 4.2), the water solubility is calculated by WSKOWWIN from calculated KOWWIN and molecular weight alone. The data are taken from the SRC PHYSPROP database and no measured values for higher molecular weights were found. Immediately, a limitation on valid prediction is established; however, it would be perfectly reasonable to suggest that, for example, the estimated water solubility of 1-dodecene is less than that of 1-decene.

Figure 4.1 shows there is a deviation from unit gradient and zero intercept. Inspection shows no real difference for linear or branched alkenes.

## B.  Example 2: Boiling Point of Some Aniline Derivatives

A set of 55 aniline derivatives were found within PHYSPROP (see Table 4.3); there were many others that could have been taken. The SRC MPBPVP program was used to obtain the predicted boiling point.

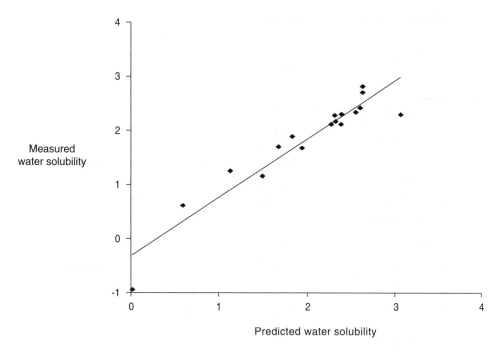

**Figure 4.1**   Measured and estimated (from WSKOWWIN) water solubility (mg/L) of some alkenes. The line of best fit is $y = 1.08x - 0.32$ ($r^2 = 0.90$).

The relationship between estimated and measured boiling points is shown in Figure 4.2. Two outliers have electron-withdrawing groups in the four-position relative to the amino group. There is therefore delocalization of the amino lone pair, and the dipolarity of the molecule is increased. The group contribution method used by the program is not parameterized for this interaction and under-predicts the boiling point. Also, N-alkylsubstituted anilines tend to be underpredicted, and ring-substituted ones are over-predicted. This is an example where inspection of the data set can lead to much better predictions; the $r^2$ for the N-alkyl examples is 0.98 (s.e. = 5.9), compared to $r^2 = 0.83$ and s.e. = 17 for the whole set.

## C.  Example 3: Vapor Pressure of Ethers

The VPs at 25°C for ethers noted in Table 4.4 are taken from PHYSPROP. The estimated VPs from MPBPVP do not use measured boiling points as inputs, but the estimated boiling point.

The relationship between estimated and measured VPs is shown in Figure 4.3. The predictions are very impressive at the high VP end, reflecting the fact that these substances have boiling points not very much higher than 25°C; the extrapolation (based on fundamental chemical principles) from boiling point to VP is therefore relatively small. Similarly, the estimation error can clearly be seen to increase at lower VP. A single value for the standard error of estimation is not applicable to the whole set, but it is possible to set errors for ranges of values of the calculated VP.

## D.  Example 4: Henry's Law Constant for Some Organophosphorus Insecticides

Henry's law constant is an important property in understanding environmental fate of substances. Henry's law constant for a set of organophosphorus insecticides has been collected from *The Pesticide Manual* (Tomlin, 2000) and is reported in Table 4.5.

The relationship between measured Henry's law constants and those estimated from HENRYWIN is shown in Figure 4.4. There appear to be some obvious candidates as outliers. The data set is

**Table 4.3   Boiling Point of Some Aniline Derivatives**

| Name | CAS Number | Predicted BP (°C) | Measured BP (°C) |
|------|-----------|-------------------|------------------|
| Aniline | 62-53-3 | 184.0 | 184.1 |
| 2,3-Dimethylaniline | 87-59-2 | 223.4 | 221.5 |
| 2,4-Dimethylaniline | 95-68-1 | 223.4 | 214 |
| 2,5-Dimethylaniline | 95-78-3 | 223.4 | 214 |
| 2,6-Dimethylaniline | 87-62-7 | 223.4 | 215 |
| 3,5-Dimethylaniline | 108-69-0 | 223.4 | 220.5 |
| 2,4,5-Trimethylaniline | 137-17-7 | 234.5 | 234.5 |
| 2-Ethylaniline | 578-54-1 | 223.4 | 209.5 |
| 3-Ethylaniline | 587-02-0 | 223.4 | 214 |
| 4-Ethylaniline | 589-16-2 | 223.4 | 217.5 |
| 2-Propylaniline | 1821-39-2 | 241.8 | 226 |
| 4-Isopropylaniline | 99-88-7 | 230.8 | 225 |
| 4-Octylaniline | 16245-79-7 | 320.1 | 310 |
| *N*-Methylaniline | 100-61-8 | 167.2 | 196.2 |
| *N,N*-Dimethylaniline | 121-69-7 | 169.4 | 194.1 |
| *N,N*,4-Trimethylaniline | 99-97-8 | 190.2 | 211 |
| *N*-Ethylaniline | 103-69-5 | 188.1 | 203 |
| *N,N*-Diethylaniline | 91-66-7 | 210.1 | 216.3 |
| *N*-Propylaniline | 622-80-0 | 208.1 | 222 |
| *N*-Butylaniline | 1126-78-9 | 227.2 | 243.5 |
| *N*-Benzylaniline | 103-32-2 | 298.3 | 306.5 |
| 4-Fluoroaniline | 371-40-4 | 179.2 | 182 |
| 3-Fluoroaniline | 372-19-0 | 179.2 | 188 |
| 2,4-Difluoroaniline | 367-25-9 | 174.4 | 170 |
| 2,3,4,5,6-Pentafluoroaniline | 771-60-8 | 159.6 | 153 |
| 2-Chloroaniline | 95-51-2 | 216.1 | 208.8 |
| 3-Chloroaniline | 108-42-9 | 216.1 | 230.5 |
| 4-Chloroaniline | 106-47-8 | 216.1 | 232 |
| 2,3-Dichloroaniline | 608-27-5 | 245.8 | 252 |
| 2,4-Dichloroaniline | 554-00-7 | 245.8 | 245 |
| 2,5-Dichloroaniline | 95-82-9 | 245.8 | 251 |
| 3,4-Dichloroaniline | 95-76-1 | 245.8 | 272 |
| 2,4,6-Trichloroaniline | 634-93-5 | 273.1 | 262 |
| 2,4,5-Trichloroaniline | 636-30-6 | 273.1 | 270 |
| 2,3,4-Trichloroaniline | 634-67-3 | 273.1 | 292 |
| 2-Bromoaniline | 615-36-1 | 235.4 | 229 |
| 3-Bromoaniline | 591-19-5 | 235.4 | 251 |
| 2,6-Dibromoaniline | 608-30-0 | 280.5 | 263 |
| 2-Methoxyaniline | 90-04-0 | 224.2 | 224 |
| 4-Methoxyaniline | 104-94-9 | 224.2 | 243 |
| 2,5-Dimethoxyaniline | 102-56-7 | 260.6 | 270 |
| 2-Ethoxyaniline | 94-70-4 | 242.5 | 232 |
| 4-Ethoxyaniline | 156-43-4 | 242.5 | 254 |
| 3-Phenoxyaniline | 2688-84-8 | 323.9 | 308 |
| 2-Nitroaniline | 88-74-4 | 272.6 | 284 |
| 4-Nitroaniline | 100-01-6 | 272.6 | 332 |
| 2,4-Dinitroaniline | 97-02-9 | 340.3 | 333.6 |
| 3-Trifluoromethylaniline | 98-16-8 | 190.0 | 187 |
| 2-Chloro-*N,N*-dimethylaniline | 698-01-1 | 202.5 | 205 |
| 3-Methyl-4-chloroaniline | 7149-75-9 | 234.8 | 241 |
| 2-Methyl-3-chloroaniline | 87-60-5 | 234.8 | 245 |
| 3-Methyl-4-bromoaniline | 6933-10-4 | 253.2 | 240 |
| 2-Bromo-4,6-dichloroaniline | 697-86-9 | 289.5 | 273 |
| 2,6-Dichloro-4-ethoxyaniline | 51225-20-8 | 295.4 | 275 |
| 4-Cyano-*N,N*-dimethylaniline | 1197-19-9 | 247.8 | 318 |

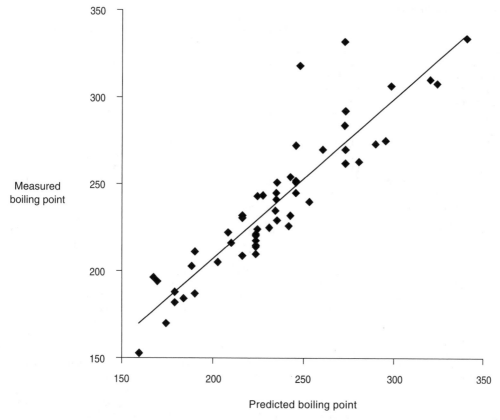

**Figure 4.2** Measured and estimated (from MPBPVP) boiling point (°C) of some aniline derivatives. The line of best fit is $y = 0.92x + 22.6$ ($r^2 = 0.83$).

based on measured VP and water solubility from many laboratories. The s value of 1.05 and the 95% confidence interval of the intercept (±0.5) set against the 8 orders of magnitude in the results indicate that the predictions are fit for the purpose of use in understanding environmental fate.

## E. Example 5: Octanol-Water Partition Coefficient for Some Pyrethroids

This final example illustrates a limitation of fragment-based methods such as KOWWIN. Apart from the experimental difficulty of producing log $K_{ow}$ values (assumed, but not conclusively known, to be all shake-flask results), fragment methods cannot usually account for the shape that a molecule adopts in solution. Molecules of higher molecular weight in aqueous solution can adopt shapes in which lipophilic centers cluster together, effectively reducing the overall hydrophobicity. In this case, this is illustrated by the KOWWIN value being the upper limit of the experimental ones, (i.e., the predicted values are generally too high). This data set illustrates an important point in that the pyrethroids cover a narrow property range (see Table 4.6). How could the values for novel pyrethroids be validated? One approach (not really a validation) is provided in the program, called the experimental value adjusted. A known $K_{ow}$ is taken as the base and the test substance value calculated by difference. The relationship between measured and estimated log $K_{ow}$ values for the pyrethroids is shown in Figure 4.5.

**Table 4.4   Vapor Pressure of Ethers**

| Name | CAS Number | Measured VP (Pa) | Measured log VP (Pa) | Estimated VP (Pa) | Estimated log VP (Pa) |
|---|---|---|---|---|---|
| Diethyl ether | 60-29-7 | 71,554 | 4.85 | 72,115.3 | 4.86 |
| Diphenyl ether | 101-84-8 | 2.9925 | 0.48 | 2.2661 | 0.36 |
| Dibenzyl ether | 103-50-4 | 0.13699 | −0.86 | 0.28926 | −0.54 |
| Di-isopropyl ether | 108-20-3 | 19,817 | 4.30 | 20,128.3 | 4.31 |
| Ethyl vinyl ether | 109-92-2 | 67,963 | 4.83 | 69,582.6 | 4.84 |
| Divinyl ether | 109-93-3 | 89,110 | 4.95 | 90,110.8 | 4.95 |
| Butyl vinyl ether | 111-34-2 | 6530.3 | 3.81 | 7344.83 | 3.87 |
| Di(n-propyl) ether | 111-43-3 | 8312.5 | 3.92 | 8637.84 | 3.94 |
| bis(2-Chloroethyl) ether | 111-44-4 | 206.15 | 2.31 | 138.632 | 2.14 |
| Diethylene glycol monomethyl ether | 111-77-3 | 33.25 | 1.52 | 14.9296 | 1.17 |
| Diethylene glycol monoethyl ether | 111-90-0 | 16.758 | 1.22 | 12.5035 | 1.10 |
| 2-methoxyethylether | 111-96-6 | 393.68 | 2.60 | 401.233 | 2.60 |
| Diethylene glycol mono-n-butyl ether | 112-34-5 | 2.9127 | 0.46 | 1.45297 | 0.16 |
| bis (2-Ethoxy ethyl) ether | 112-36-7 | 69.2265 | 1.84 | 121.436 | 2.08 |
| Di-n-hexyl ether | 112-58-3 | 6.1313 | 0.79 | 22.661 | 1.36 |
| Diethylene glycol dibutyl ether | 112-73-2 | 3.46598 | 0.54 | 4.33225 | 0.64 |
| Dimethyl ether | 115-10-6 | 591,850 | 5.77 | 513,205 | 5.71 |
| Isopropenyl methyl ether | 116-11-0 | 68,761 | 4.84 | 70,915.6 | 4.85 |
| Phenyl glydidyl ether | 122-60-1 | 1.33 | 0.12 | 5.26535 | 0.72 |
| Di-n-butyl ether | 142-96-1 | 799.33 | 2.90 | 1053.07 | 3.02 |
| Triethylene glycol butyl ether | 143-22-6 | 0.3325 | −0.48 | 0.06905 | −1.16 |
| Benzyl ethyl ether | 539-30-0 | 123.025 | 2.09 | 96.1093 | 1.98 |
| Ethyl methyl ether | 540-67-0 | 198,170 | 5.30 | 185,287 | 5.27 |
| bis(Chloromethyl) ether | 542-88-1 | 3910.2 | 3.59 | 3852.37 | 3.59 |
| Diisopentyl ether | 544-01-4 | 186.2 | 2.27 | 249.271 | 2.40 |
| Methyl propyl ether | 557-17-5 | 61,898.2 | 4.79 | 61,051.4 | 4.79 |
| Methyl isopropyl ether | 598-53-8 | 80,451.7 | 4.91 | 82,779.3 | 4.92 |
| Ethylene glycol monobenzyl ether | 622-08-2 | 2.66 | 0.42 | 0.29459 | −0.53 |
| Methyl isobutyl ether | 625-44-5 | 28,063 | 4.45 | 29,459.3 | 4.47 |
| Methyl n-butyl ether | 628-28-4 | 18,460.4 | 4.27 | 18,928.6 | 4.28 |
| Ethyl propyl ether | 628-32-0 | 24,019.8 | 4.38 | 24,793.8 | 4.39 |
| Di-isobutyl ether | 628-55-7 | 2044.21 | 3.31 | 2199.45 | 3.34 |
| Butyl ethyl ether | 628-81-9 | 6862.8 | 3.84 | 7864.7 | 3.90 |
| Ethyl tert-butyl ether | 637-92-3 | 16,465.4 | 4.22 | 16,929.1 | 4.23 |
| Di-n-pentylether | 693-65-2 | 113.981 | 2.06 | 110.639 | 2.04 |
| Ethyl(tert-amyl) ether | 919-94-8 | 6650 | 3.82 | 5305.34 | 3.72 |
| Methyl-tert-amyl ether | 994-05-8 | 10,002.93 | 4.00 | 10,024.2 | 4.00 |
| Allyl propyl ether | 1471-03-0 | 8081.08 | 3.91 | 8291.26 | 3.92 |
| Methyl tert-butyl ether | 1634-04-4 | 33,250 | 4.52 | 33,591.6 | 4.53 |
| Glycidyl n-butyl ether | 2426-08-6 | 425.6 | 2.63 | 291.927 | 2.47 |
| Dinonyl ether | 2456-27-1 | 0.006992 | −2.16 | 0.09398 | −1.03 |
| Ethylene glycol monopropyl ether | 2807-30-9 | 414.561 | 2.62 | 182.621 | 2.26 |
| Di (tert-butyl) ether | 6163-66-2 | 4334.47 | 3.64 | 4278.93 | 3.63 |
| sec-Butyl methyl ether | 6795-87-5 | 27,677.3 | 4.44 | 23,060.9 | 4.36 |
| Di (sec-butyl) ether | 6863-58-7 | 2161.25 | 3.33 | 2972.59 | 3.47 |
| 4-chlorophenyl phenyl ether | 7005-72-3 | 0.3591 | −0.45 | 0.2546 | −0.59 |
| Tripropylene glycol methyl ether | 20324-33-8 | 2.66 | 0.42 | 0.43323 | −0.36 |
| Dipropylene glycol methyl ether | 34590-94-8 | 73.15 | 1.86 | 19.7284 | 1.30 |

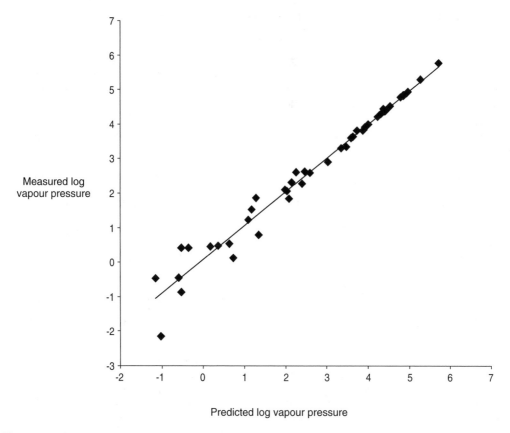

**Figure 4.3**   Measured and estimated (from MPBPVP) vapor pressure (Pa) of some ethers. The line of best fit is $y = 0.98x + 0.095$ ($r^2 = 0.97$).

## VII.  APPENDIX 1. COMMONLY AVAILABLE METHODS FOR THE PREDICTION OF PHYSICOCHEMICAL PROPERTIES: SRC EPIWIN SOFTWARE

Methods for the prediction of the physicochemical properties of molecules are summarized in Chapter 3. The SRC methods form an integrated package run through an integrated estimation programs interface (EPIWIN). Because they are very widely used, some aspects of their function are discussed further below.

### A.  Prediction of Vapor Pressure: MPBPWIN

VP estimation is performed by the MPBPWIN program using three separate methods: (1) the Antoine method, (2) the modified Grain method, and (3) the Mackay method. All three use the normal boiling point to estimate VP. Unless the user enters a boiling point on the data-entry screen, MPBPWIN uses the estimated boiling point from the adapted Stein and Brown method. When a boiling point is entered on the data entry screen, MPBPWIN uses it. Each VP method is discussed below.

1. Antoine Method — See Dearden (2003) for a basic description of the Antoine method used by MPBPWIN. It was developed for gases and liquids. The PC-CHEM (formerly CHEMEST) program of EPA's Graphical Exposure Modeling System uses the Antoine method to estimate VP for gases

**Table 4.5   Henry's Law Constant and Other Physicochemical Data for Some Organophosphorus Insecticides**

| Substance Name | CAS Number | Vapour Pressure at 20°C (mPa) Measured | Water Solubility (mg/L) Measured | Henry's Constant (Pa m³/mol) Calculated as VP/$S_{aq}$ | Predicted from HENRYWIN |
|---|---|---|---|---|---|
| Fenitrothion | 122-14-5 | 15 | 21 | $1.98 \times 10^{-1}$ | $1.88 \times 10^{-2}$ |
| Fenthion | 55-38-9 | 0.74 | 4.2 | $4.90 \times 10^{-2}$ | $1.39 \times 10^{-1}$ |
| Dichlorvos | 62-73-7 | $2.10 \times 10^{+3}$ | $1.80 \times 10^{+4}$ | $2.58 \times 10^{-2}$ | $8.69 \times 10^{-2}$ |
| Isofenphos | 25311-71-1 | 0.22 | 18 | $4.22 \times 10^{-3}$ | $1.60 \times 10^{-3}$ |
| Methidathion | 950-37-8 | $2.50 \times 10^{-1}$ | 200 | $3.78 \times 10^{-4}$ | $7.19 \times 10^{-4}$ |
| Parathion | 56-38-2 | 0.89 | 11 | $2.36 \times 10^{-2}$ | $3.00 \times 10^{-2}$ |
| Parathion-methyl | 298-00-0 | 0.2 | 55 | $9.57 \times 10^{-4}$ | $1.70 \times 10^{-2}$ |
| Phoxim | 14816-18-3 | 2.1 | 1.5 | $4.18 \times 10^{-1}$ | $7.49 \times 10^{00}$ |
| Pirimiphos-ethyl | 23505-41-1 | 0.68 | 2.3 | $9.86 \times 10^{-2}$ | $4.52 \times 10^{-1}$ |
| Pyridaphenthion | 119-12-0 | $1.47 \times 10^{-3}$ | 100 | $5.00 \times 10^{-6}$ | $1.19 \times 10^{-5}$ |
| Quinalphos | 13593-03-8 | 0.346 | 17.8 | $5.80 \times 10^{-3}$ | $4.02 \times 10^{-4}$ |
| Sulfotep | 3689-24-5 | 14 | 10 | $4.51 \times 10^{-1}$ | $1.25 \times 10^{-1}$ |
| Sulprofos | 35400-43-2 | 0.084 | 0.31 | $8.74 \times 10^{-2}$ | $1.62 \times 10^{-1}$ |
| Tebupirimfos | 96182-53-5 | 5 | 5.5 | $2.89 \times 10^{-1}$ | $1.41 \times 10^{-2}$ |
| Chlorpyrifos | 2921-88-2 | 2.7 | 14 | $6.76 \times 10^{-2}$ | $2.55 \times 10^{-1}$ |
| Chlorpyrifos-methyl | 5598-13-0 | 3 | 2.6 | $3.72 \times 10^{-1}$ | $1.45 \times 10^{-1}$ |
| Isazofos | 42509-80-8 | 7.45 | 168 | $1.39 \times 10^{-2}$ | $6.01 \times 10^{-2}$ |
| Tolclofos-methyl | 57018-04-9 | 57 | 1.1 | $1.56 \times 10^{+1}$ | $2.61 \times 10^{00}$ |
| Dimethoate | 60-51-5 | 0.25 | $2.38 \times 10^{+4}$ | $2.41 \times 10^{-6}$ | $2.14 \times 10^{-6}$ |
| Disulfoton | 298-04-4 | 7.2 | 25 | $7.90 \times 10^{-2}$ | $2.13 \times 10^{-1}$ |
| Ethion | 563-12-2 | 0.2 | 2 | $3.85 \times 10^{-2}$ | $1.71 \times 10^{-2}$ |
| Formothion | 2540-82-1 | 0.113 | $2.60 \times 10^{+3}$ | $1.12 \times 10^{-5}$ | $1.50 \times 10^{-5}$ |
| Malathion | 121-75-5 | 5.3 | 145 | $1.21 \times 10^{-2}$ | $8.50 \times 10^{-5}$ |
| Phorate | 298-02-2 | 85 | 50 | $4.43 \times 10^{-1}$ | $1.60 \times 10^{-1}$ |
| Terbufos | 13071-79-9 | 34.6 | 4.5 | $2.22 \times 10^{00}$ | $2.82 \times 10^{-1}$ |
| Chlorfenvinphos | 470-90-6 | 0.53 | 145 | $1.31 \times 10^{-3}$ | $5.24 \times 10^{-3}$ |
| Dimethylvinphos | 2274-67-1 | 1.3 | 130 | $3.32 \times 10^{-3}$ | $2.97 \times 10^{-3}$ |
| Dicrotophos | 141-66-2 | 9.3 | $1.00 \times 10^{+6}$ | $2.21 \times 10^{-6}$ | $1.22 \times 10^{-7}$ |
| Methacrifos | 62610-77-9 | 160 | 400 | $9.61 \times 10^{-2}$ | $1.82 \times 10^{-1}$ |
| Phosphamidon | 13171-21-6 | 2.2 | $1.00 \times 10^{+6}$ | $6.59 \times 10^{-7}$ | $1.54 \times 10^{-7}$ |
| Prophetamphos | 31218-83-4 | 1.9 | 110 | $4.86 \times 10^{-3}$ | $5.95 \times 10^{-3}$ |
| Tetrachlorvinphos | 22248-79-9 | 0.0056 | 11 | $1.86 \times 10^{-4}$ | $2.20 \times 10^{-3}$ |
| Acephate | 30560-19-1 | 0.226 | $7.90 \times 10^{+5}$ | $5.24 \times 10^{-8}$ | $2.85 \times 10^{-7}$ |
| Iprobenfos | 26087-47-8 | 0.247 | 430 | $1.66 \times 10^{-4}$ | $3.48 \times 10^{-3}$ |
| Profenfos | 41198-08-7 | $1.24 \times 10^{-1}$ | 28 | $1.65 \times 10^{-3}$ | $3.54 \times 10^{-3}$ |
| Fenamiphos | 38260-54-7 | 0.12 | 400 | $9.10 \times 10^{-5}$ | $2.45 \times 10^{-3}$ |
| Fosamine | 59682-52-9 | 0.53 | 283.3 | $3.18 \times 10^{-4}$ | $1.76 \times 10^{-9}$ |
| Heptenophos | 23560-59-0 | 170 | $2.20 \times 10^{+3}$ | $1.94 \times 10^{-2}$ | $1.19 \times 10^{-1}$ |
| Azinphos-ethyl | 2642-71-9 | 0.32 | 4 | $2.76 \times 10^{-2}$ | $5.11 \times 10^{-5}$ |
| Azinphos-methyl | 86-50-0 | $5.00 \times 10^{-4}$ | 28 | $5.67 \times 10^{-6}$ | $2.90 \times 10^{-5}$ |
| Diazinon | 333-41-5 | 12 | 60 | $6.09 \times 10^{-2}$ | $8.85 \times 10^{-3}$ |
| Etrimfos | 38260-54-7 | 6.5 | 40 | $4.75 \times 10^{-2}$ | $2.45 \times 10^{-3}$ |
| Triazophos | 24017-47-8 | 0.21 | $3.90 \times 10^{+1}$ | $1.69 \times 10^{-3}$ | $7.84 \times 10^{-6}$ |
| Cadusafos | 95465-99-9 | $1.20 \times 10^{+2}$ | 248 | $1.31 \times 10^{-1}$ | $5.53 \times 10^{-2}$ |
| Ethoprophos | 13194-48-4 | 46.5 | 700 | $1.61 \times 10^{-2}$ | $3.14 \times 10^{-2}$ |
| Trichlorfon | 52-68-6 | 0.21 | $1.20 \times 10^{+5}$ | $4.50 \times 10^{-7}$ | $2.69 \times 10^{-7}$ |
| Edifenphos | 17109-49-8 | 0.032 | 56 | $1.77 \times 10^{-4}$ | $4.03 \times 10^{-4}$ |
| Fonofos | 944-22-9 | 28 | 13 | $5.31 \times 10^{-1}$ | $1.13 \times 10^{+1}$ |
| Cyanophos | 2636-26-2 | 105 | 46 | $5.55 \times 10^{-1}$ | $4.17 \times 10^{-2}$ |
| Phoxim | 14816-18-3 | 2.1 | 1.5 | $4.18 \times 10^{-1}$ | $7.49 \times 10^{00}$ |
| Coumaphos | 56-72-4 | 0.013 | 1.5 | $3.14 \times 10^{-3}$ | $1.10 \times 10^{-2}$ |
| Demeton-*S*-methyl | 919-86-8 | 40 | $2.20 \times 10^{+4}$ | $4.19 \times 10^{-4}$ | $2.60 \times 10^{-04}$ |

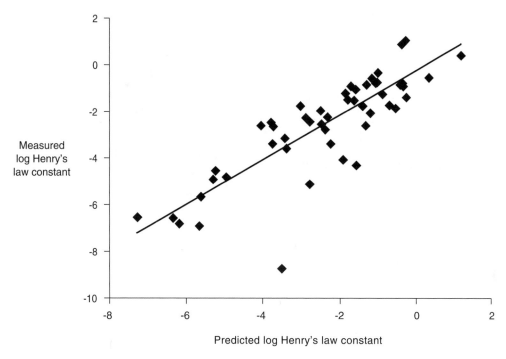

**Figure 4.4** Measured and estimated (from HENRYWIN) Henry's law constant of some organophosphrous insecticides. The line of best fit is $y = 0.96x - 0.25$ ($r^2 = 0.70$).

**Table 4.6 Physicochemical Properties of Selected Pyrethroids**

| Substance Name | CAS Number | Molecular Weight | log $K_{ow}$ Predicted | Measured |
|---|---|---|---|---|
| Acrinathrin | 101007-06-1 | 541.4 | 6.73 | 5.00 |
| Allethrin | 584-79-2 | 302.4 | 5.52 | 4.96 |
| Cyhalothrin | 68085-85-8 | 449.9 | 6.85 | 6.80 |
| α-Cypermethrin | 67375-30-8 | 416.3 | 6.38 | 6.94 |
| Deltamethrin | 52918-63-5 | 505.2 | 6.18 | 4.60 |
| Flucythinrate | 70124-77-5 | 451.4 | 6.59 | 6.20 |
| Imiprothrin | 72963-72-5 | 318.4 | 2.98 | 2.90 |
| Permethrin | 52645-53-1 | 391.3 | 7.43 | 6.10 |
| Prallethrin | 23031-36-9 | 300.4 | 4.88 | 4.49 |
| Resmethrin | 10453-86-8 | 338.4 | 7.11 | 5.43 |
| Tefluthrin | 79538-32-2 | 418.7 | 7.19 | 6.50 |
| Tetramethrin | 7696-12-0 | 331.4 | 5.54 | 4.60 |
| Tralomethrin | 66841-25-6 | 665.0 | 7.56 | 5.00 |
| Transfluthrin | 118712-89-3 | 371.2 | 6.17 | 5.46 |

and liquids. MPBPWIN has extended the Antoine method to make it applicable to solids by using the same methodology as the modified Grain method to convert a super-cooled liquid VP to a solid-phase VP.

2. Modified Grain Method — See Dearden (2003) for a basic description of the modified Grain method used by MPBPWIN. This method is a modification and significant improvement of the modified Watson method (which is currently used by PC-PCHEM to estimate VP for solids). It is applicable to solids, liquids, and gases. It is probably the best all-round VP estimation method currently available.

3. Mackay Method — See Dearden (2003) for a basic description of the Mackay method used by MPBPWIN. Mackay derived the following equation to estimate VP:

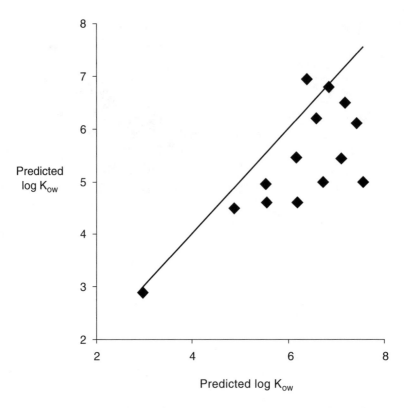

**Figure 4.5**   Measured and estimated (from KOWWIN) octanol-water partition coefficient for some pyrethroids. The line represents the ideal fit (unit gradient).

$$\ln VP = -(4.4 + \ln T_b)[1.803(T_b/T - 1) - 0.803 \ln(T_b/T)] - 6.8(T_m/T - 1) \qquad (4.2)$$

where $T_b$ is the normal boiling point (K), T is the VP temperature (K), and $T_m$ is the melting point (K). The melting point term is ignored for liquids. It was derived from two chemical classes: hydrocarbons (aliphatic and aromatic) and halogenated compounds (again aliphatic and aromatic).

MPBPWIN reports the VP estimate from all three methods. It then reports a suggested VP. For solids, the modified Grain estimate is the suggested VP. For liquids and gases, the suggested VP is the average of the Antoine and the modified Grain estimates. The Mackay method is not used in the suggested VP because its application is currently limited to its derivation classes.

## B.  Prediction of Water Solubility: WSKOWWIN

The program WSKOWWIN estimates the water solubility ($S_{aq}$) of an organic compound using the compound's log $K_{ow}$. WSKOWWIN requires only a chemical structure to estimate water solubility. The estimation methodology is:

$$\log S_{aq} \text{ (mol/L)} = 0.796 - 0.854 \log K_{ow} - 0.00728 \text{ MW} + \text{Corrections} \qquad (4.3)$$

$$\log S_{aq} \text{ (mol/L)} = 0.693 - 0.96 \log K_{ow} - 0.0092(T_m - 25) - 0.00314 \text{ MW} + \text{Corrections} \quad (4.4)$$

where MW is the molecular weight and $T_m$ is melting point (MP) in °C (used only for solids). Equation (4.4) is used when a measured MP is available.

Corrections are applied to 15 structural types (e.g., alcohols/acids; selected phenols, nitros, amines, alkyl pyridines, amino acids, PAHs, multi-nitrogen types, etc.). Application and magnitude depend on available MP.

WSKOWWIN version 1.4 includes an experimental water solubility database of 6230 compounds. When experimental data are available for the SMILES being estimated, the data are retrieved and shown in the program's results window of the program.

## C. Prediction of the Octanol-Water Partition Coefficient: KOWWIN

KOWWIN estimates the logarithmic octanol-water partition coefficient of organic compounds. The octanol-water partition coefficient is a physical property used extensively to describe a chemical's lipophilic or hydrophobic properties. It is the ratio of a chemical's concentration in the octanol-phase to its concentration in the aqueous phase of a two-phase system at equilibrium. Since measured values range from $<10^{-4}$ to $>10^8$ (at least 12 orders of magnitude), the logarithm (log $K_{ow}$) is commonly used to characterize its value. KOWWIN uses a fragment constant methodology to predict log $K_{ow}$. In a fragment constant method, a structure is divided into fragments (atoms or larger functional groups) and coefficient values of each fragment or group are summed together to yield the log $K_{ow}$ estimate.

To estimate log $K_{ow}$, KOWWIN initially separates a molecule into distinct atoms/fragments. In general, each non-hydrogen atom (e.g., carbon, nitrogen, oxygen, sulfur, etc.) in a structure is a 'core' for a fragment; the exact fragment is determined by what is connected to the atom. Several functional groups are treated as core atoms; these include carbonyl (C=O), thiocarbonyl (C=S), nitro (–NO$_2$), nitrate (ONO$_2$), cyano (–C≡N), and isothiocyanate (–N=C=S). Connections to each core atom are either general or specific; specific connections take precedence over general connections. For example, aromatic carbon, aromatic oxygen, and aromatic sulfur atoms have nothing but general connections (i.e., the fragment is the same no matter what is connected to the atom). In contrast, there are five aromatic nitrogen fragments:

1. If there is a five-member ring
2. If there is a six-member ring
3. If the nitrogen is an oxide-type (i.e., pyridine oxide)
4. If the nitrogen has a fused-ring location (i.e., indolizine)
5. If the nitrogen has a +5 valence (i.e., N-methyl pyridinium iodide)

Since the oxide-type is most specific, it takes precedence over the other four. The aliphatic carbon atom is another example; it does not matter what is connected to — –CH$_3$, –CH$_2$–, or –CH< — because the fragment is the same. An aliphatic carbon with no hydrogens has two possible fragments: (1) if there are four single bonds with three or more carbon connections and (2) any other not meeting the first criteria.

Results of two successive multiple regressions (first for atoms/fragments and second for correction factors) yield the following general equation for estimating log $K_{ow}$ of any organic compound:

$$\log K_{ow} = \sum(f_i n_i) + \sum(c_j n_j) + 0.229 \tag{4.5}$$

$$n = 2413, \ r^2 = 0.981, \ sd = 0.219, \ \text{mean error} = 0.161$$

where: $\sum(f_i n_i)$ is the summation of $f_i$ (the coefficient for each atom/fragment) times $n_i$ (the number of times the atom/fragment occurs in the structure), and $\sum(c_j n_j)$ is the summation of $c_j$ (the coefficient for each correction factor) times $n_j$ (the number of times the correction factor occurs [or is applied] in the molecule).

# REFERENCES

Boethling, R.S. and Mackay, D., Eds., *Handbook of Property Estimation Methods for Chemicals*, Lewis Publishers, Boca Raton, FL, 2000.

Dannenfelser, R.M. and Yalkowsky, S., Estimation of entropy of melting from molecular structure: a non-group contribution method, *Ind. Eng. Chem. Res.*, 35, 1483–1487, 1996.

Dearden, J.C., Quantitative structure-property relationships for prediction of boiling point, vapor pressure, and melting point, *Environ. Toxicol. Chem.*, 22, 1696–1709, 2003.

Fisk, P.R., Methods of estimation of physicochemical properties, in *Environmental Chemistry of Pesticides*, Roberts, T.R. and Kearney, P.C., Eds., John Wiley, Chichester, England 1995.

Grain, C.F., Vapour pressure, in *Handbook of Chemical Property Estimation Methods*, Lyman, W.J., Reehl, W.F., and D.H Rosenblatt Eds., McGraw Hill, New York, 1982, pp. 14.1–14.20.

Hine, J. and Mookerjee, P.K., The intrinsic hydrophilic character of organic compounds. Correlations in terms of structural contributions, *J. Org. Chem.*, 40, 292–298, 1975.

Joback, K.G. and Reid, R.C., Estimation of pure-component properties from group-contributions, *Chem. Eng. Comm.*, 57, 233–243, 1987.

Krzyzaniak, J., Myrdal, P., Simamora, P., and Yalkowsky, S., Boiling point and melting point prediction for aliphatic, non-hydrogen-bonding compounds, *Ind. Eng. Chem. Res.*, 34, 2530–2535, 1995.

Lai, W.Y., Chen, D.H., and Maddox, R.N., Application of a nonlinear group-contribution model to the prediction of physical constants. 1. Predicting normal boiling points with molecular structure, *Ind. Eng. Chem. Res.*, 26, 1072–1079, 1987.

Leo, A.J., Calculating $logP_{oct}$ from structures, *Chem. Rev.*, 93, 1281–1306, 1993.

Livingstone, D.J., The characterisation of chemical structures using molecular properties: a survey, *J. Chem. Inf. Comput. Sci.*, 40: 195-209.

Livingstone, D.J., Theoretical property predictions, *Curr. Top. Med. Chem.*, 3, 1171–1192, 2003.

Lyman, W.J., Boiling point, in *Handbook of Property Estimation Methods for Chemicals*, Boethling, R.S. and Mackay, D., Eds., Lewis Publishers, Boca Raton, FL, 2000, pp. 29–51.

Lyman, W.J., Estimation of physical properties, in *Environmental Exposure From Chemicals*, Neely, W.B. and Blau, G.E., Eds., CRC Press, Boca Raton, FL, 1985, pp. 13–47.

Mackay, D., Solubility in water, in *Handbook of Property Estimation Methods for Chemicals*, Boethling, R.S. and D. Mackay, D., Eds., Lewis Publishers, Boca Raton, FL, 2000, pp. 125–139.

Mackay, D., Shiu, W.Y., and Ma, K.C., Henry's law constant, in *Handbook of Property Estimation Methods for Chemicals*, Boethling, R.S. and D. Mackay, D., Eds., Lewis Publishers, Boca Raton, FL, 2000, pp. 69–87.

Meylan, W.M. and Howard, P.H., Bond contribution method for estimating Henry's Law Constants, *Environ. Toxicol. Chem.*, 10, 1283–1293, 1991.

Meylan, W.M., Howard, P.H., and Boethling, R.S., Improved method for estimating water solubility from octanol/water partition coefficient, *Environ. Toxicol. Chem.*, 15, 100–106, 1996.

Myrdal, P.B., Manka, A.M., and Yalkowsky, S.H., Aquafac 3: Aqueous functional group activity coefficients, application to estimation of aqueous solubility, *Chemosphere*, 30, 1619–1637, 1995.

Myrdal, P.B., Ward, G.H., Dannenfelser, R.M., Mishra, D., and Yalkowsky, S.H., AQUAFAC: aqueous functional group activity coefficients; application to hydrocarbons, *Chemosphere*, 24:1047–1061, 1992.

Myrdal, P.B., Ward, G.H., Simamora, P., and Yalkowsky, S., AQUAFAC: aqueous functional group activity coefficients, *SAR QSAR Environ. Res.*, 1, 53–61, 1993.

Nirmalakhandan, N.N. and Speece, R.E., QSAR model for predicting Henry's law constant, *Environ. Sci. Technol.*, 22, 1349–1357, 1988.

Perrin, D.D., Dempsey, B., and Serjeant, E.P., *pKa Prediction for Organic Acids and Bases*, Chapman and Hall, London, 1981.

Sage, M.L. and Sage, G.W., Vapor pressure, in *Handbook of Property Estimation Methods for Chemicals*, Boethling, R.S. and D. Mackay, D., Eds., Lewis Publishers, Boca Raton, FL, 2000, pp. 53–65.

Simamora, P. and Yalkowsky, S., Group contribution methods for predicting the melting points and boiling points of aromatic compounds, *Ind. Eng. Chem. Res.*, 33, 1405–1409, 1994.

Stein, S.E. and Brown R.L., Estimation of normal boiling points from group contributions, *J. Chem. Inf. Comput. Sci.*, 34, 581–587, 1994.

Tesconi, M. and Yalkowsky S.H., Melting point, in *Handbook of Property Estimation Methods for Chemicals*, Boethling, R.S. and Mackay, D., Eds., Lewis Publishers, Boca Raton, FL, 2000, pp. 3–27.

Tomlin, C., Ed., *The Pesticide Manual*, 12th ed., British Crop Protection Council Publications, Bracknell, England, 2000.

CHAPTER **5**

# Whole Molecule and Atom-Based Topological Descriptors

Tatiana I. Netzeva

## CONTENTS

## I. INTRODUCTION

The manner in which chemical structure is described depends upon the concept applied. A chemical can variously be considered as a microscopic ensemble of nuclei and electrons and so may be described with energy functions. Alternatively, it can be regarded as a macroscopic collection of molecules and characterized with physicochemical properties. The term chemical structure is also related to the molecular formula (i.e., the atoms of which the molecule is composed, and the way in which those atoms are connected). According to Kier and Hall (2001), the structure is "the count of each atom, identified as its element, along with a description, enumeration, or characterization of the set of connections between the atoms." This formalism in the definition of chemical structure has a fundamental role in an original theory, namely, Graph Theory, adopted for the presentation and description of chemical structure.

The structure of a chemical is responsible for the presence and magnitude of its properties. The properties can be energy levels and their derivatives, as well as physicochemical, or biological, properties. To avoid discrimination between the different properties (which are the subject of detailed consideration in Chapters 3, 4, and 6) in the context of quantitative relationships between structural descriptors (i.e., topological indices [TIs]) and various properties, we will use the broader abbreviation QSPR (Quantitative Structure-Property Relationship).

TIs are sometimes referred to as examples of 2D descriptors because they are derived from 2D chemical graphs (Zheng and Tropsha, 2000). TIs have been shown to have many advantages in QSPR studies. For example, they do not require conformational analysis and 3D optimization of the structure, alignment or any 3D pharmacophore hypothesis (Golbraikh et al., 2001). They are also easily computed (Randić and Zupan, 2001) and adaptable to extremely rapid algorithms even on slow computers (Boyd, 2001). TIs also allow for the quantification of global, as well as local or specific, structural properties (Sabljić, 2001). In addition, there are many different methods (i.e., calculation algorithms) to generate each descriptor (Estrada, 2001). The improvement of existing and the development of new indices has yielded a large number of descriptors available for use in QSPR studies (Randić and Zupan, 2001). An extensive review of topological indices and other descriptors can be found in the excellent books by Devillers and Balaban (1999), and Todeschini and Consonni (2000).

The aims of this chapter are to illustrate to the reader (1) the calculation of some commonly used TIs and structural descriptors, (2) the current status of the interpretation of topological indices, and (3) to make some recommendations regarding the application of the structural descriptors in QSPR studies.

## II. CALCULATION OF TOPOLOGICAL INDICES

The calculation of topological indices follows the depiction of the molecule as a structural graph (i.e., as a collection of dots [vertices] that are connected to each other by lines [edges]). The dots in any structural graph may have different locations in terms of position and metric distance between them. The lines can be short or long, straight or curved. The first publication on Graph Theory by Leonard Euler in 1736 provides an excellent example to illustrate that the structural graph is not a geometric object. Euler discussed the possibility of walking around the town Königsberg in East Prussia, a plan of which is shown in Figure 5.1, crossing each of its bridges across the River Pregel only once. On the schematic description of the town (Figure 5.1a) the land

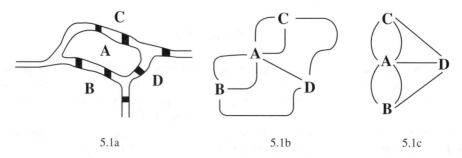

5.1a                                    5.1b                                    5.1c

**Figure 5.1**   Schematic presentation of town Königsberg with its seven bridges and connections. The plan of the town (a) is converted into graphs (b and c), which are completely equivalent representation of the object in (5.1a) and between themselves in Graph Theory.

**Table 5.1 Glossary of Terms Used in the Graph Theory**

| Term | Explanation |
|---|---|
| Graph[a] | A representation consisting of points (vertices) connected by lines (edges). A graph is a topological, but not geometric, object. |
| Topological index[a] | A numerical value associated with chemical constitution that can be used to correlate chemical structure with various physical properties, chemical reactivity, or biological reactivity. The numerical basis for topological indices is provided (depending on how a molecular graph is converted into a numerical value) by either the adjacency matrix or the topological distance matrix. In the latter, the topological distance between two vertices is the number of edges in the shortest path between these. |
| Distance matrix[a] | In the topological distance matrix, the topological distance between two vertices is the number of edges in the shortest path between these. |
| Adjacency matrix[a] | A matrix which consists of entries $a_{ij} = 1$ for adjacent vertices and $a_{ij} = a_{ii} = 0$ otherwise. The matrix is isomorphic to the bonds drawn in simple molecular representation. |
| Path | A path is a sequence of consecutive edges in a graph. The length of the path (k) is the number of edges traversed. The repetition of the vertices is not allowed. |
| Walk | A path in which both vertices and edges may be repeated. |

[a] The definitions are taken from Minkin (1999).

(dots) is presented as capital letters and these are linked together by bridges (lines). Euler began to search for an answer to this task using one initial assumption that the shapes and sizes of the land and bridges are not important. Thus, he constructed the first structural graph. The objects, shown on Figure 5.1b and Figure 5.1c, actually repeat the object from Figure 5.1a, from a graph theoretical point of view.

In the present day, Graph Theory has moved on substantially from the task of crossing bridges, coloring the map, and the problems of the traveling salesman. The schematic representation of objects as dots with connections between them is used in many areas of science, including computer technology, social science, ecology, biology, geography, and even psychology. The implementation of Graph Theory over 50 years ago in the field of chemistry provided and contributed to a number of indices to describe chemical structure. Many of these, as well as their siblings, are finding extensive use in QSPR studies today.

To assist the reader in the comprehension of TIs, some simple terms related to their use and calculation are listed in Table 5.1. For example, the molecule 2,3-dimethylhexane contains 8 vertices (atoms) and 7 edges (bonds). The adjacency matrix and the vertex degree matrix are shown in Figure 5.2a and Figure 5.2b. The degree of a vertex (also called the vertex valence) is the number of edges with which it is connected. It can also be obtained as a row sum from the adjacency matrix as shown in Figure 5.2b. The distance matrix for 2,3-dimethylhexane is shown on Figure 5.2c. By definition, the adjacency and distance matrices are square matrices, symmetric with respect to the main diagonal. The number of row and column elements is equal to the number of the vertices in the graph. The diagonal contains only zero values, which for clarity are replaced with the numbered carbon atoms on Figure 5.2a and Figure 5.2c. For the calculation of some TIs (e.g., Wiener index) only a half of a matrix (in the case of Wiener index, the distance matrix) is considered.

Path and walk are other basic terms applied in Graph theory. A path of length 1 represents 2 vertices, i and j, connected with a bond. The number of paths of length 1 ($p_1$) can be obtained as a total count of all bonds in the molecule. A path of length 2 indicates that there are 2 bonds between i and j in the shortest pathway between them. The sum of all possible patterns, containing 2 joint bonds, gives the count of paths of length 2 ($p_2$), and so on. Thus, for 2,3-dimethylhexane, $p_1 = 7$, $p_2 = 8$, $p_3 = 7$, $p_4 = 4$, and $p_5 = 2$. Paths of no longer than five atoms can be constructed for this molecule. The possible paths of length 2 and length 4 for 2,3-dimethylhexane are shown in Figure 5.3a and Figure 5.3b, respectively. In the construction of the paths, the repetition of vertices is forbidden. Conversely, in the construction of walks, it is possible to go back and forward and visit vertices repeatedly. It is easier for the human eye to recognize the paths rather than the

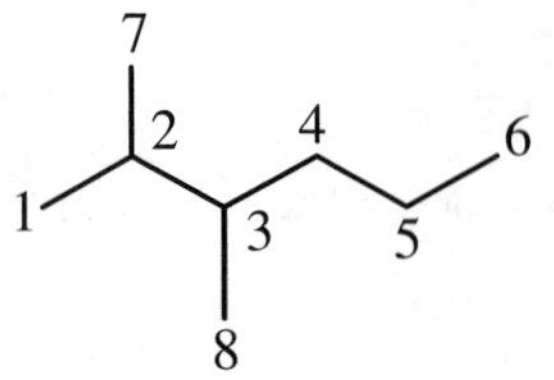

| $C_1$ | 1 | 0 | 0 | 0 | 0 | 0 | 0 | | 1 |
|---|---|---|---|---|---|---|---|---|---|
| 1 | $C_2$ | 1 | 0 | 0 | 0 | 1 | 0 | | 3 |
| 0 | 1 | $C_3$ | 1 | 0 | 0 | 0 | 1 | | 3 |
| 0 | 0 | 1 | $C_4$ | 1 | 0 | 0 | 0 | | 2 |
| 0 | 0 | 0 | 1 | $C_5$ | 1 | 0 | 0 | | 2 |
| 0 | 0 | 0 | 0 | 1 | $C_6$ | 0 | 0 | | 1 |
| 0 | 1 | 0 | 0 | 0 | 0 | $C_7$ | 0 | | 1 |
| 0 | 0 | 1 | 0 | 0 | 0 | 0 | $C_8$ | | 1 |

5.2a 　　　　　　　　　　　　　　　　　5.2b

| $C_1$ | 1 | 2 | 3 | 4 | 5 | 2 | 3 |
|---|---|---|---|---|---|---|---|
| 1 | $C_2$ | 1 | 2 | 3 | 4 | 1 | 2 |
| 2 | 1 | $C_3$ | 1 | 2 | 3 | 2 | 1 |
| 3 | 2 | 1 | $C_4$ | 1 | 2 | 2 | 3 |
| 4 | 3 | 2 | 1 | $C_5$ | 1 | 4 | 3 |
| 5 | 4 | 3 | 2 | 1 | $C_6$ | 5 | 4 |
| 2 | 1 | 2 | 2 | 4 | 5 | $C_7$ | 3 |
| 3 | 2 | 1 | 3 | 3 | 4 | 3 | $C_8$ |

5.2c

**Figure 5.2** (a) Adjacency matrix, (b) vertex degree matrix, and (c) distance matrix for 2,3-dimethylhexane.

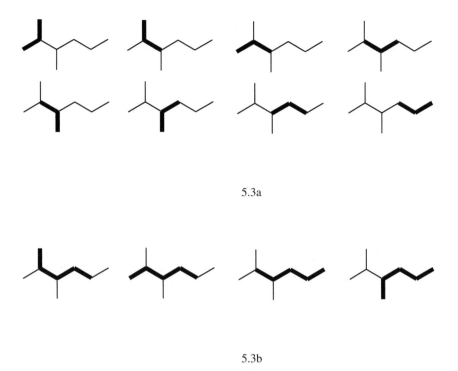

5.3a

5.3b

**Figure 5.3** Paths of (a) length 2 and (b) length 4 in 2,3-dimethylhexane.

walks. However, there are many ways to calculate the number of atomic and molecular walks; probably the easiest way is to use the concept for the extended connectivities (ECs). By definition ECs (or atomic walk counts) can be obtained from the degree of all vertices by iterative summation over all neighbors (Morgan's summation procedure) (Rücker and Rücker, 1993):

$$EC_k = awc_k(i) = \sum_{j(i)} awc_{k-1}(j) \qquad (5.1)$$

where awc is the abbreviation for atomic walk count and k is the length of the walk. Subsequently, the molecular walk count (mwc, w) can be calculated as a sum of the atomic walk counts. Practically, the number of atomic walks of length 1 is equal to the vertex degree (Figure 5.4a). For calculation of the number of walks of length 2 (k = 2, Figure 5.4b), summation of all vertex degrees, assigned in the walk of length 1 (k = 1, Figure 5.4a), is required. For example, $C_1$ in 2,3-dimethylhexane (from Figure 5.2) is connected only to $C_2$, for which $awc_1 = 3$ (Figure 5.4a). Hence, for $C_1$, $awc_2 = 3$ (Figure 5.4b). $C_2$ is connected to $C_1$ (with $awc_1 = 1$), $C_3$ (with $awc_1 = 3$) and $C_7$ (with $awc_1 = 1$). Therefore, for $C_2$, $awc_2 = 1 + 3 + 1 = 5$.

To calculate walk counts of length 3 (k = 3, Figure 5.4c), the extended connectivities, obtained for walks of length 2 (k = 2, Figure 5.4b), are needed, and so on. The summation of $awc_k$, where $k = 1 \div 5$, for 2,3-dimethylhexane gives the following walk counts: $w_1 = 14$ (Figure 5.4a), $w_2 = 30$ (Figure 5.4b), $w_3 = 60$ (Figure 5.4c), $w_4 = 126$ (Figure 5.4d), and $w_5 = 258$ (Figure 5.4e). The path counts give only the number of neighbors at increasing distances. The walk counts show not only the number of more remote neighbors but also their relative distribution (Rücker and Rücker, 1993). The sum of ratios $p_k/w_k$ over all possible path lengths (Randić, 1998) and both path lengths and

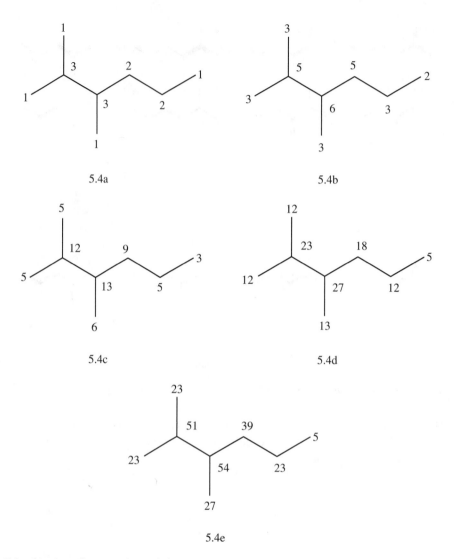

**Figure 5.4**   Atomic walk counts (extended connectivities) for walks of (a) length 1 (b) length 2, (c) length 3, (d) length 4, and (e) length 5 in 2,3-dimethylhexane.

vertices (Randić and Zupan, 2001), is related to the molecular shape and could be used as a separate descriptor in QSPRs.

The next sections give an overview of the major TIs for application in chemistry listed roughly in chronological order.

## A.  Wiener Index (W)

The first non-trivial structural invariant (i.e., object that is independent of the particular drawing or numbering of the vertices) was proposed over 50 years ago by Harold Wiener (Wiener 1947). By definition the Wiener number (W) is obtained as follows: "for each bond multiply the number of carbon atoms on each side of the bond, and sum all bond contributions," (i.e., it consists of summed bond contributions). Using 2,3-dimethylhexane as an example gives the following:

$$W = 1 \times 7 + 3 \times 5 + 3 \times 5 + 2 \times 6 + 1 \times 7 + 1 \times 7 + 1 \times 7 = 70 \qquad (5.2)$$

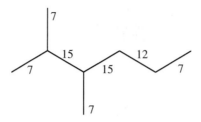

**Figure 5.5**  Bond distribution of Wiener index (W) for 2,3-dimethylhexane.

Alternatively, W can be obtained as a half sum of the off-diagonal elements of the distance matrix (see Figure 5.2c for example), or as the product of the path vector P and the path length vector L. For 2,3-dimethylhexane, as discussed above, P = (7, 8, 7, 4, 2) and L = (1, 2, 3, 4, 5). According to that definition:

$$W = 7 \times 1 + 8 \times 2 + 7 \times 3 + 4 \times 4 + 2 \times 5 = 70 \tag{5.3}$$

One contra-intuitive feature of the Wiener index is that more distant atom pairs make a larger contribution to W than adjacent atom pairs (Randić and Zupan, 2001). A typical characteristic of W is that the central C-C bonds make a greater contribution than the peripheral bonds (Figure 5.5). For many bond additive physicochemical properties, such as boiling point, the opposite may be true, (i.e., the terminal bonds are considered more important to determine the magnitude of the property).

## B.  Harary Index (H)

To overcome the disadvantage of the Wiener index, Plavšić et al. (1993) and Ivanciuc et al. (1993) independently proposed the use of a reciprocal distance matrix. The index derived from this approach was called the Harary index (H), in honor of Frank Harary, who introduced the distance matrix to Graph Theory. It is obtained by considering the sum of contributions of each bond in paths of different length (for the bond contributions see Table 5.2), multiplied by factors of 1, $(1/2)^2$, ..., $(1/k_i)^2$, ..., $(1/m)^2$. Here, k is the number of atoms in the path of length i and m is the number of atoms in the longest path that can be identified in a molecule. The ratio $(1/k)$ is derived from the assumed reciprocal dependence on distance. The first factor $(1/k_1)^2$ is always equal to 1 because it is related to an individual bond (path of length 1) and $k_1 = 1$. Using 2,3-dimethylhexane as an example, the longest path in the molecule consists of 5 atoms, and the Harary index is calculated as follows:

$$H = 1 \times 7 + (1/2)^2 \times 16 + (1/3)^2 \times 21 + (1/4)^2 \times 16 + (1/5)^2 \times 10 = 14.7333 \tag{5.4}$$

The formal distribution of H between the bonds (Figure 5.6) again shows the smaller contribution to the index of terminal bonds, than for internal bonds. The desirable ratio between the weights of internal to terminal bonds can be found in the Hosoya Z index, which, by definition, has no apparent bond contribution.

## C.  Hosoya Z Index

The Hosoya Z index was introduced in 1971 as the count of all possible patterns of k disjoint bonds (i.e., bonds without a common atom) in a molecule (Hosoya, 1971). Using as an example 2,3-dimethylhexane, there are 7 patterns involving 1 disjoint bond as illustrated in Figure 5.7a, 13 patterns involving 2 disjoint bonds (Figure 5.7b), and 6 patterns involving 3 disjoint bonds (Figure 5.7c), or:

**Table 5.2** **Bond Contributions of the Individual Bonds in 2,3-Dimethylhexane for Calculation of Harary Index**

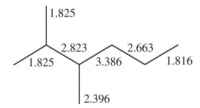

| Bond/Path Length (l) | k = 1 | k = 2 | k = 3 | k = 4 | k = 5 |
|---|---|---|---|---|---|
| a | 1 | 2 | 2 | 1 | 1 |
| b | 1 | 4 | 5 | 3 | 2 |
| c | 1 | 3 | 5 | 4 | 2 |
| d | 1 | 2 | 3 | 4 | 2 |
| e | 1 | 1 | 1 | 2 | 2 |
| f | 1 | 2 | 2 | 1 | 1 |
| g | 1 | 2 | 3 | 1 | 0 |
| Sum | 7 | 16 | 21 | 16 | 10 |

*Note:* The contributions are obtained by counting and summation of how many times the bond participates in a path of particular length.

**Figure 5.6** Bond distribution of Harary index (H) for 2,3-dimethylhexane.

$$Z = 1 + 7 + 13 + 6 = 27 \tag{5.5}$$

The +1 in Equation 5.5 can be considered an empirical factor, which always has the same value. A formal bond weight distribution, ignoring the leading +1, as calculated by Randić and Zupan (2001) is shown in Figure 5.8.

## D. Zagreb Indices (M)

The Zagreb group was the first to propose indices ($M_1$ and $M_2$) that were based directly on the graph adjacency matrix (Hall and Kier, 2001). $M_1$ and $M_2$ are defined as the sum of the squared vertex degrees (i.e., the number of edges with which it is connected, $a_i$), and the sum of vertex degrees products ($a_i a_j$) over all pairs of adjacent vertices, respectively (Gutman et al., 1975):

$$M_1 = \sum_i a_i^2 \tag{5.6}$$

$$M_2 = \sum_{ij} a_i a_j \tag{5.7}$$

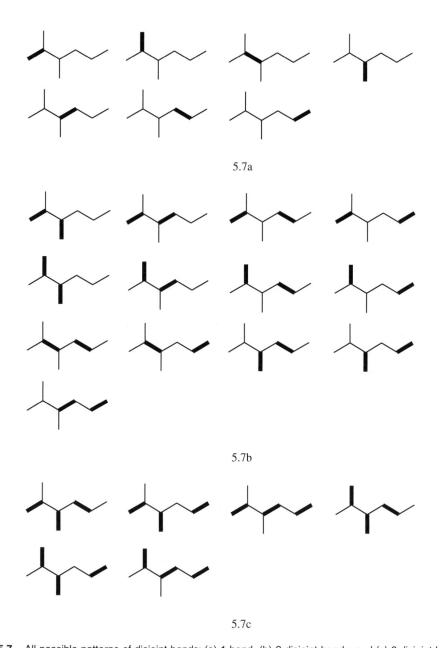

5.7a

5.7b

5.7c

**Figure 5.7** All possible patterns of disjoint bonds: (a) 1 bond, (b) 2 disjoint bonds, and (c) 3 disjoint bonds in 2,3-dimethylhexane.

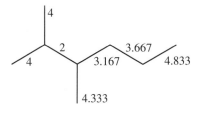

**Figure 5.8** Bond distribution of Hosoya Z index for 2,3-dimethylhexane.

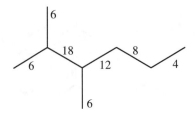

**Figure 5.9**   Bond distribution of Zagreb $M_2$ index for 2,3-dimethylhexane.

The calculation of $M_1$ and $M_2$ is again illustrated with 2,3-dimethylhexane:

$$M_1 = 1^2 + 3^2 + 3^2 + 2^2 + 2^2 + 1^2 + 1^2 + 1^2 = 30 \tag{5.8}$$

$$M_2 = 2 \times (1 \times 3 + 3 \times 3 + 3 \times 2 + 2 \times 2 + 2 \times 1 + 1 \times 3 + 1 \times 3) = 60 \tag{5.9}$$

In Equation 5.9, the sum in the brackets equals the vertex degree products for half of the adjacency matrix. It is multiplied by two in order to obtain summation over all pairs of adjacent vertices. Note that by definition $M_2$ is not equal to $2 \times M_1$. Although it is difficult to derive bond contributions for the index based only on the vertex degrees ($M_1$), the formal bond distribution of the Zagreb index $M_2$ (Figure 5.9) shows that the terminal bonds are again underestimated, although by a different amount.

## E.  Randić Index ($\chi$)

Milan Randić (1975) proposed a branching index ($\chi$), commonly referred to now as the connectivity index:

$$\chi(G) = \sum_{\text{allbonds}} \left(a_i a_j\right)^{-1/2} \tag{5.10}$$

Here, G is the hydrogen-suppressed (i.e., the hydrogen atoms are ignored) graph of the molecule.

The direct summation of vertex degree products in $M_2$ has been changed in $\chi$ (G) to a summation of inverse-square-root terms. This specific function selection has been made to provide better correlations of $\chi$ with the properties of isomeric alkanes. This shows the high sensitivity of the new molecular descriptor to variations in molecular structure. More recently, some restrictions in the applicability of the inverse-square-root function to compounds with large numbers of atoms (Gutman and Lepović, 2001) and new values for the exponent were investigated (Gutman and Lepović, 2001; Estrada, 2002).

An illustration how $\chi$ (G) is calculated for 2,3-dimethylhexane is given below:

$$\chi(G) = \left(1 \times 3\right)^{-1/2} + \left(3 \times 3\right)^{-1/2} + \left(3 \times 2\right)^{-1/2} + \left(2 \times 2\right)^{-1/2} + \left(2 \times 1\right)^{-1/2} + \left(1 \times 3\right)^{-1/2} + \left(1 \times 3\right)^{-1/2} \tag{5.11}$$

$$= 3.679$$

The participation of individual bonds in the branching index is shown in Figure 5.10.

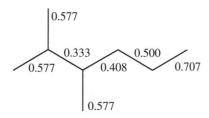

**Figure 5.10** Bond distribution of the branching index ($^1\chi$) in 2,3-dimethylhexane.

## F. Kier and Hall Indices

### 1. Connectivity Indices

The skeletal branching index, published by Randić, motivated significant research to enlarge its applicability to chemical systems other than alkanes and to predict properties other than boiling point. Kier and coworkers developed the molecular connectivity idea into a full paradigm for the representation of molecular structure (Hall and Kier, 2001).

Kier et al. (1975) developed the concept that greater sensitivity to structural variation, other than that obtained by merely counting atoms, can be achieved by the adoption of an algorithm, similar to the branching index algorithm, to subgraphs of different size. They included subgraphs in the structure-encoded algorithm. These contained single atoms, single edges, and sets of connected edges (paths), in which no vertex is repeated (Hall and Kier, 2001). An example of possible subgraphs is shown in Figure 5.11. The extension of the chemical information included in the subgraphs with increasing complexity led to the following definition of the $\chi$ indices (Bonchev, 2001):

$$^k\chi_c(G) = \sum_{\text{all } k-\text{edge subgraphs}} \left(a_i a_j \ldots a_{k+1}\right)^{-1/2} \tag{5.12}$$

The subscript "c" partitions the set of subgraphs with a fixed number of edges into a subgraph (e.g., path [p], cluster [c], path-cluster [pc], etc.) as shown in Figure 5.11.

### 2. Valence Connectivity Indices

The definition of the valence connectivity indices is related closely to the definition of simple ($\delta$) and valence ($\delta^v$) value. In the molecular connectivity formalism:

$$\delta = \sigma - h \tag{5.13}$$

$$\delta^v = \sigma + \pi + n - h \tag{5.14}$$

where $\sigma$ is the number of electrons in $\sigma$-orbitals, $\pi$ is the number of electrons in $\pi$-orbitals, n is the number of electrons in lone pair orbitals, and h is the number of hydrogen atoms bonded to one heavy atom.

Note that the given formula for $\delta^v$ is valid only for second row atoms. For third row atoms a modified calculation method to account for the effect of the core electrons should be used (Hall and Kier, 2001). The introduction of $\delta$ and $\delta^v$ values led to the following generalization for simple and valence $\chi$ indices:

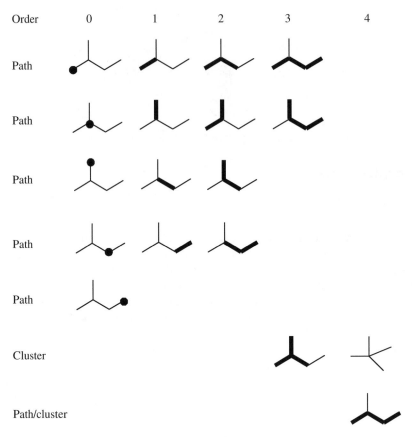

**Figure 5.11**   Subgraphs of different order and type shown for 2-methylbutane. The cluster index of fourth order is illustrated with 2,2-dimethylpropane.

$$ {}^{m}\chi_{t} = \sum_{1} \prod_{i} \left(\delta_{i}\right)_{1}^{-1/2} \tag{5.15} $$

$$ {}^{m}\chi_{t}^{v} = \sum_{1} \prod_{i} \left(\delta_{i}^{v}\right)_{1}^{-1/2} \tag{5.16} $$

The large number and the variety of connectivity indices have imposed the need for common nomenclature. It is accepted that the Greek letter $\chi$ is used to represent the index itself. The left-side superscript (0 or real positive integer) is used to designate the order of the index, and the right-side superscript (the Greek letter v) differentiates between simple and valence type indices. The right-side subscript (a, b, c, pc, etc.) specifies the subclass of the index (atom, bond, cluster, path/cluster etc.). If no right-side subscript is specified, then the path type index is assumed.

The calculation of the first-order valence molecular connectivity index ($^{1}\chi^{v}$) is demonstrated for 2,3-dimethylpentanol, which differs from 2,3-dimethylhexane only by 1 oxygen atom (replacing the carbon atom at the first position):

$$ {}^{1}\chi^{v} = \left(3 \times 5\right)^{-1/2} + \left(3 \times 3\right)^{-1/2} + \left(3 \times 2\right)^{-1/2} + \left(2 \times 2\right)^{-1/2} + \left(2 \times 1\right)^{-1/2} + \left(1 \times 3\right)^{-1/2} + \left(1 \times 3\right)^{-1/2} \tag{5.17} $$

$$ = 3.362 $$

The two molecules have the same topology, however, the alcohol, which contains an oxygen atom, has a lower $^1\chi^v$ value.

## 3. Kappa Shape Indices

Kappa indices are designed to account for molecular shape where similarity or cavity filling ability estimation is required. To enable the comparison between molecules of different size, the Kappa shape indices are related to reference structures within the pool of possible structural isomers of a molecule, containing a certain number of atoms. The first reference structure is the non-branched isomer (linear graph) with a shape that can be described as a cylinder or an ellipse. It always is a real structure. The second reference structure, which may be either real or hypothetical, is the so-called star graph, a graph in which all vertices are connected to a central vertex (Randić, 1998). The shape of the star graphs can be roughly described as a sphere. All non-cyclic molecules in an isomeric series are considered to have a shape within the limits set by these two extreme graphs.

For the calculation of the Kappa shape indices, the count of paths of different length in the hydrogen-suppressed graph of the molecule is used. If we assume that the linear graph has a minimum number of paths of certain length ($^mp_{min}$) and the star graph has the maximum possible number of paths of the same length ($^mp_{max}$), for any molecule i within the isomeric series, $^mp_i$ should have a value between $^mp_{min}$ and $^mp_{max}$ (i.e., $^mp_{min} \leq {}^mp_i \leq {}^mp_{max}$). Based on this assumption, a general formula for calculation of Kappa shape indices is derived (Kier, 1985):

$$^m\kappa = \frac{2\ ^mp_{max}\ ^mp_{min}}{\left(^mp_i\right)^2} \tag{5.18}$$

The $^mp_{min}$ and $^mp_{max}$ values are predictable and depend on the number of atoms in the molecule, a. Their substitution for paths of different length has yielded the following equations for the calculation of $^m\kappa$:

$$^1\kappa = a(a-1)^2 \big/ \left(^1p_i\right)^2 \tag{5.19}$$

$$^2\kappa = (a-1)(a-2)^2 \big/ \left(^2p_i\right)^2 \tag{5.20}$$

$$^3\kappa = (a-3)(a-2)^2 \big/ \left(^3p_i\right)^2, \text{ when a is even} \tag{5.21}$$

$$^3\kappa = (a-1)(a-3)^2 \big/ \left(^3p_i\right)^2, \text{ when a is odd} \tag{5.22}$$

The calculation of $^2\kappa$, for example, is illustrated for 2,3-dimethylhexane, which has 8 paths of length 2 in the molecule (see Figure 5.3b). For any molecule containing 8 carbon atoms, $^2p_{min} = a - 2 = 6$ (6 paths of length 2 in the linear graph, Figure 5.12a) and $^mp_{max} = (a - 1)(a - 2)/2 = 21$ (21 paths of length 2 in the star graph with 8 vertices, Figure 5.12b). Then, for 2,3-dimethylhexane:

$$^2\kappa = \left(2 \times 21 \times 6\right) \big/ 64 = 7 \times 36 / 64 = 3.938 \tag{5.23}$$

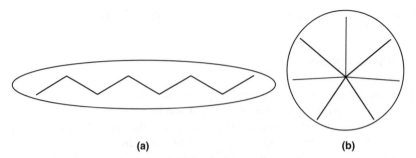

**(a)**                                                                 **(b)**

**Figure 5.12**  Extreme cases for a graph, containing eight vertices: (a) path (linear) graph, (b) star graph.

The influence of atoms other than carbons in a sp³ hybrid state on the molecular shape is accounted for by the Kappa alpha shape indices. They can be obtained by modifying each a and $^m p_i$ in Equation 5.19 to Equation 5.22, with a value:

$$\alpha = r(x) \big/ r\!\left(C_{(sp3)}\right) - 1 \tag{5.24}$$

where $r(x)$ is the covalent radius of atom x and $(C_{(sp3)})$ is the covalent radius of a carbon atom in the sp³ hybrid state. Subsequently, a is replaced by a + $\alpha$, and $^m p_i$ is replaced by $^m p_i$ + $\alpha$. The resulting indices are designated as $^m\kappa^\alpha$.

### 4.  Electrotopological State Indices

   The electrotopological state (E-state) indices are a group of atom level descriptors that are calculated for each atom (such as >C<, >N–, =O) or hydride group (such as –CH₃, >NH, –OH) in the molecule. For simplicity, these groups are all termed atoms (Rose et al., 2002). The E-state index for an atom i in a molecule, S(i), is composed of an intrinsic state, $I_i$, plus the sum of perturbations, $\Delta I_{ij}$, from all other atoms in the molecule:

$$S(i) = I_i + \sum_j \Delta I_{ij} \tag{5.25}$$

The intrinsic state and the perturbation term are calculated as follows:

$$I_i = \left(\left(2/N_i\right)2\delta_i^v + 1\right)\big/\delta_i, \tag{5.26}$$

   Here, N is the principal quantum number of valence electrons, and

$$\Delta I_{ij} = \left(I_i - I_j\right)\big/r_{ij}^2 \tag{5.27}$$

where $r_{ij}$ is the topological distance between atoms, given as the number of atoms in the shortest path between atoms i and j. An example of how this index is calculated is shown in Figure 5.13.
   The E-state value of each atom contains both electronic and topological structural information. An additional portion of information (i.e., the presence or absence and count of a particular atom type) is added to the classical E-state indices by summation of individual atom level E-State indices for that particular atom type t (such as >C<, >C=, >CH–, =CH₂, =CH–, >N–, >NH, –NH₂, =O, –OH, –CH₃, etc.):

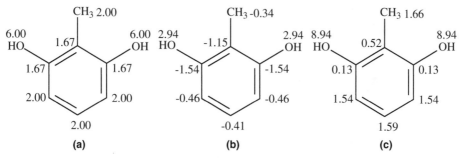

**Figure 5.13** (a) Intrinsinc states, (b) sum of perturbations, and (c) the product E-state indices (5.13c) in 2,5-dimethylphenol.

$$s^T(t) = \sum s(t) \tag{5.28}$$

These values are also known as atom-type E-state indices (Hall and Kier, 1995). In the topological state formalism, hydrogen atoms, which were neglected for many years in the Graph Theory, can also be represented as hydrogen atom-type E-state indices (Maw and Hall, 2000; 2001).

## III. INTERPRETATION OF TOPOLOGICAL INDICES

There is a common opinion that the interpretation of the topological indices, as constructed using refined mathematical operations with graphs, is complex and difficult (Estrada, 2001). The lack of understanding of these indices can lead to problems in the choice of TIs and their justification in QSPRs (Stankevich et al., 1995).

The first attempt to interpret TIs may be made implicitly from the word branching. This arises from branching index — the name originally given by Randić to his topological index in 1975 (Randić, 1975). Later, with the development of the extended connectivity indices, this interpretation is associated more with the higher order path, as well as cluster and path/cluster indices. They are considered to reflect, to a greater extent, local properties, while it is suggested that the lower-order indices describe global molecular properties such as bulk and molecular size (Sabljić, 2001). The boundary between global and local indices is not defined strictly and it depends, to a certain degree, on the chemical class under consideration. However, it is anticipated that the second- and third-order indices are somewhere in the gray zone between the global and local properties.

To illustrate this point, and particularly the reported correlation of lower order connectivity indices with molecular bulk (Dearden et al., 1988), a simple experiment was conducted. For 39 molecules (ethane, propane, butane, pentane, hexane, heptane, octane, and all their acyclic isomers) the zero- to fourth-order simple path connectivity indices were calculated. The connectivity indices were compared with the number of atoms (N) as a structural descriptor for molecular size, and between themselves (Figure 5.14). As can be seen from Figure 5.14, $^0\chi$ has the highest correlation coefficient with N (n = 39, $r^2$ = 0.983). Conversely, $^4\chi$ correlates only marginally with the number of atoms in the molecule (n = 29, $r^2$ = 0.355. The compounds, having a value of zero for $^4\chi$ were excluded from the analysis.) It is worth noting that the correlation between $^0\chi$ and N is higher for the smaller molecules and lower for the larger ones. This indicates that the interpretation of $^0\chi$ as a descriptor of molecular size may have some limitations. From Figure 5.14 it can also be seen that $^1\chi$ correlates relatively strongly with both N and $^0\chi$, and hence can be related to the molecular size. However, the correlation of the increasingly higher-order connectivity indices with N is poorer, there being a lower correlation with higher path lengths.

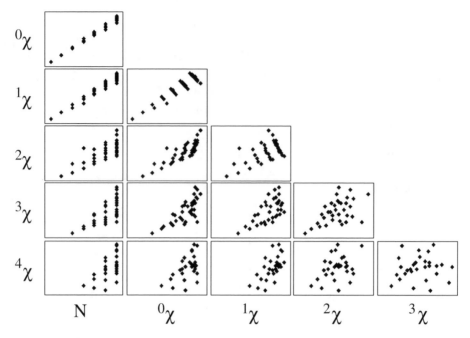

**Figure 5.14**  Matrix plot of the relationships between the numbers of atoms in acyclic alkanes ($C_nH_{2n+2}$, where n = 2 – 8) and their connectivity indices up to fourth order. The zero value cells are omitted from the plot.

Considerable efforts have been made by many workers to interpret TIs. One logical direction was to investigate the relationship with physicochemical properties, such as boiling point, melting point, molar refraction, surface tension, water solubility (log $S_{aq}$), and the octanol-water partition coefficient (log $K_{ow}$), with topological indices. A separation of physicochemical properties depending on the relative weights of their dynamic and static parameters was recently proposed. This has facilitated the interpretation of the relationship of different thermodynamic properties with different topological indices (Hosoya et al., 1999). One cannot expect to describe successfully properties as diverse as boiling point and the density of liquid using the same topological descriptors. This is because of the different factors determining the magnitude of the two properties (the dynamic behavior of individual molecules in the case of boiling point, and the denseness of packing for the density of a liquid). However, as Randić (2001) emphasizes in a recent monograph, to request that topological indices must have physicochemical interpretation is unreasonable because the models and concepts of the Graph Theory are distinct from those in physical chemistry.

Continuing to search for interpretation within the theoretical model considered (i.e., Graph Theory), Randić and coworkers (Randić and Zupan, 2001; Randić et al., 2001) related the success of various indices in structure-property studies to the degree with which they differentiate the contribution of terminal and interior C-C bonds. The rationale for this conclusion is the fact that the terminal bonds are more exposed to interactions with other small molecules of the same type, or with biological macromolecules, and hence should be weighted more heavily in the construction of topological indices. Thus, it was found that the descriptors giving greater weights to terminal rather than to interior C-C bonds, such as Hosoya Z index and molecular connectivity index, are more successful to predict boiling point. The understanding of the role of the different weights of the terminal and interior bonds was used as a basis for the modification of some well-known indices (e.g., Wiener, Harary, Hosoya Z indices). It subsequently allowed for the development of better relationships of the boiling points of octane isomers (Randić and Zupan, 2001). Later, recommendations for the improvement of the performance of novel indices were made (Randić et al., 2001). However, a warning was issued regarding the applicability of the different weights in the modeling of different properties (i.e., the trends); although they may be valid for the boiling point, they may not be valid for other properties.

Based on the premise that a useful description of molecular structure represents the molecule in a surrounding of other structures, Kier and Hall developed the encounter concept to provide a theoretical basis for the interpretation of the molecular connectivity indices (Kier and Hall, 2000; 2001). According to this concept, the measurement of a physical (or biological) property reflects the collective influence of the encounters between each molecule and other molecules in its direct environment. In the light of this theoretical concept, the interpretation of the $\delta$ value is that it encodes the relative accessibility of a bond to encounter another bond in another molecule, the resultant encounter which may lead to an intermolecular interaction. The term accessibility is defined as the topological and electronic availability of one bond to engage in some interaction with another bond.

The relation between this definition and the mathematical expression of $\delta$ and $\delta^v$ values (Equation 5.13 and Equation 5.14) can be easily seen. The simple $\delta$ represents the vertex valence (a number of skeletal neighbors for each vertex). It can be presented as both $v = \delta = k - h$, and $\delta = \sigma - h$, after the substitution of the number of valence electrons k with the number of electrons assigned to sigma orbitals $\sigma$. It is evident from Equation 5.15 that the greater the number of skeletal neighbors, the larger the $\delta$ value and the lower the connectivity index. Recently, new arguments were evaluated in support of the thesis that the molecular connectivity indices represent molecular accessibility areas and volumes (Estrada, 2002).

The substitution of k with $\sigma$ in the definition of $\delta$, introduces electronic information in the connectivity indices. The significance of this switch between mathematics and chemistry can be seen clearly when considering the $\delta^v$ value. According to Equations 5.13 and 5.14, $\delta^v = \delta + \pi + n$. This expression reflects the fact that the $\pi$- and lone-pair electrons are more exposed and more interactive than the $\sigma$-electrons, and hence more important to noncovalent intermolecular interactions. The greater reactivity of the $\pi$- and lone-pair electrons can be described by the difference between the simple and the corresponding valence $\chi$ indices. The resulting delta index ($\Delta^m\chi_t = {}^m\chi_t - {}^m\chi_t^v$), and particularly the linear combination of $\Delta^0\chi$, $\Delta^1\chi$ and $\Delta^2\chi$ indices, was shown to correlate strongly with the ionization potential for a series of alkyl amines, alcohols, and ethers (Hall and Kier, 2001). In the relationship, described by the authors, $\Delta^0\chi$ is zero for all atoms but the heteroatoms, which have a larger contribution to the ionization potential than the carbon atoms. The term $\Delta^1\chi$ represents the differentiation among primary, secondary, and tertiary amines as well as between alcohols and ethers. The $\Delta^2\chi$ adds information representing the refinement of structure in carbon atoms in $\alpha$- or $\beta$-position to the heteroatoms, encoding branching in the alkyl parts of the molecules.

Strong support for the relationship between connectivity indices and quantum chemical formalism in the presentation of molecular structure is given by the comparison of Randić index with an energy function (Hamiltonian function) describing the $\pi$-electron properties of conjugated hydrocarbons (Stankevich et al., 1995). The authors showed that the minimum of this function is equal to the total $\pi$-electron energy of the system and is defined by an expression depending on the $\chi$ index and the number of vertices in the molecular graph ($E_\pi = E_\pi [\chi, n]$). Despite the fact that a linear relationship between $E_\pi$ and $\chi$ was suggested *a priori*, a non-linear dependence for large classes of compounds was anticipated.

## IV. APPLICATION OF TIs TO DEVELOP QSPRs

Structural descriptors are implemented extensively in the development of prediction models for a large number of endpoints related to the fate and toxicity of organic chemicals in the environment and to human health. The TIs were reported for the modeling of properties of the pure substances such as boiling point (Basak et al., 1996), vapor pressure (Liang and Gallagher, 1998) and water solubility (Thomsen et al., 1999). Such properties are very important in determining the risk of environmental contamination with organic substances. Various partitioning properties, such as the octanol-water partition coefficient (Thomsen et al., 1999; Sabljić, 2001), Henry's law constant (Sabljić, 1990), soil sorption (Sabljić, 1990; Thomsen et al., 1999; Tao and Lu, 1999; Sabljić, 2001), bioaccumulation (Sabljić, 1990) and bioconcentration (Lu et al., 1999; 2000; Sabljić, 2001) were modeled with TIs. The partitioning properties are of great importance to predict the distribution of the pollutants in the atmosphere, hydrosphere, pedosphere, and biosphere. Of all endpoints, the TIs are probably used most extensively in the development of models for acute toxicity (Gute and Basak, 1997; Basak et al., 2000; Grodnitzky and Coats, 2002; Huuskonen, 2003). However, models for some other important for hazard assessment endpoints such as genotoxicity (Cash, 2001), carcinogenicity (Gini et al., 1999), mutagenicity (Llorens et al., 2001), skin permeability (Dearden et al., 2000), blood-brain barrier penetration (Rose et al., 2002), and similar effects, have also been evaluated with topological indices alone, or in combination with other descriptors.

## V.  LIMITATIONS OF TIs

The successful application of the topological indices (some software for their calculation is listed in Table 5.3) to different problems in QSPR analyses depends significantly on the critical selection of descriptors and proper evaluation of the quantitative models. To this end, some knowledge regarding the limitations of topological indices is beneficial.

A recognized problem in the application of TIs in QSPR analyses is that different molecules often display the same magnitude for a given descriptor. This paradox may cause difficulties in the interpretation of QSPRs. This is especially a concern if a solution of the inverse problem — determining a structure when several of its descriptors are given, is required (Randić, 2001). For example, the 2-, 3-, and 4-isomers of dihydroxybenzene, as expected, cannot be distinguished by $^0\chi_a$ (because it is an atom-based index), or by 3 other connectivity indices $^1\chi_b$, $^3\chi_c$ and $^6\chi$. For more complex isomeric structures such as cata-condensed benzoid hydrocarbons consisting of five-membered rings, polycyclic hydrocarbons consisting of six-membered rings (polyphenylenes), and two simple rings connected by a chain with arbitrary length, it is noted that the Randić index has the same value (Stankevich et al., 1995). Because such drawbacks of the topological indices are more or less predictable from the algorithms used for their calculation, they can be easily avoided by the selection of the appropriate descriptor for the given series of compounds. Thus, to model 2-, 3-, and 4-substitution in the benzene ring, $^4\chi_{pc}$ has been shown to be a useful descriptor. Its value increases sharply with the degree of substitution, while in the isomeric classes of substituted benzenes it increases with the proximity of substituents (Sabljić, 2001). Even if a descriptor provides the desired discrimination between chemical structures, other problems in its application may arise. As Randić (2001) correctly observed, it is not meaningful to use the connectivity index, which is a bond additive quantity, to describe ortho-meta-para effects, which are known not to be bond additive. This example supports the understanding that different properties of the same compounds may require different descriptors, and that there are no inherently unique descriptors that will satisfy different models for different properties equally well.

A further shortcoming of the most commonly used topological indices is the inability to take into account stereo-specific properties of molecules, such as atomic chiralities, enantiomers (R- and S-isomers), and $\sigma$- and $\pi$-diastereomers (Z- and E-isomers). Topological descriptors proposed to fill the gap to account for chirality and ZE-isomerism can be found in Schultz et al. (1995), de Julián-Ortiz et al. (1998), and Golbraikh et al. (2001) as well as in Lekishvili (1997; 2001), and Golbraikh et al., (2002), respectively. It is probably a question of availability and time until the power of these descriptors is demonstrated and they begin to be used routinely in QSPR analysis.

A typical characteristic of the molecular connectivity indices (with exception of the zero-order term $^0\chi$), which may or may not be a disadvantage, is that they increase with molecular size and

**Table 5.3 Software for Calculation of Topological Indices (in Alphabetical Order)**

| Software | Supplier | Link | Descriptors |
|---|---|---|---|
| C$^2$-Descriptor+ | Accelrys Inc. | www.accelrys.com/cerius2/descriptor.html | Topological indices, information content indices, fingerprint descriptors, physicochemical properties, charged partial surface area descriptors, shadow indices |
| CODESSA | Semichem Inc. | www.semichem.com/codessa.html | Topological, geometric, constitutional, thermodynamic, electrostatic, quantum mechanical descriptors |
| DRAGON | Milano Chemometrics and QSAR Research Group | www.disat.unimib.it/chm/Dragon.htm (free to download) | Constitutional, topological, geometrical, charge descriptors, molecular walk counts, BCUT descriptors, Galvez topological charge indices, 2D autocorrelations, aromaticity indices, Randic molecular profiles, RDF, 3D-MoRSE, WHIM, GETAWAY descriptors, functional groups, atom-centered fragments, empirical descriptors, properties |
| MDL®QSAR | MDL Information Systems, Inc. | www.mdl.com/products/predictive/qsar/index.jsp | Molecular property descriptors, total topological descriptors, E-state indices |
| MOE/QuaSAR | Chemical Computing Group Inc. | www.chemcomp.com/fdept/prodinfo.htm | 2D (physical properties, subdivided surface areas, atom and bond counts, connectivity indices, adjacency and distance matrix descriptors, pharmacophore feature, partial charge descriptors), and 3D descriptors |
| MOLCONN-Z | EduSoft, LC | www.eslc.vabiotech.com/molconn/ | Molecular connectivity, molecular connectivity difference, and kappa shape indices, E-state indices, atom-type and group-type E-state indices, topological equivalence classification of atoms, other topological indices, counts of subgraphs: paths, rings, clusters, etc.; vertex eccentricities |
| PreADME | Computer Aided Molecular Design Research Centre | camd.ssu.ac.kr/adme (free, on-line calculations) | Constitutional, topological, physicochemical, and geometrical descriptors. |
| TOPIX | Software Development Lohninger | www.lohninger.com/topix.html | Simple constitutional descriptors, topological indices, descriptors of substructures (augmented atoms) |

decrease with molecular complexity. Consequently, they are well suited to describe isomeric molecules, but not very effective in series of molecules that vary in size. The answer to this criticism is the development of the overall connectivity indices (Bonchev, 2001). Overall connectivity combines the basic ideas of the classical concept of molecular connectivity with that of molecular complexity. The resulting overall connectivity descriptors are designed in such a way that they increase with all elements of molecular complexity, first with increasing molecular size and subsequently by enhancing typical complexity patterns such as branching, cyclicity, and centrality.

A limitation of the TIs, especially in accounting for molecular shape, arises from the underlying assumptions in Graph Theory. The independence of 2D descriptors from molecular conformation, which is undoubtedly an advantage for the whole class of descriptors, appears to be a disadvantage when estimating a typical 3D property such as shape (Kier, 1985). In the case of modeling an activity that involves interaction of the lock and key type, the 3D complementarity is of great importance. It is worth noting that this complementarity is most often not only steric, but also involves hydrophobic and electrostatic interactions between the ligand and the target structure. It would be naive to believe that such complexity can be captured by a single value (Rücker and Rücker, 1993). The 2D shape indices can be useful shape estimators considering a limited structural domain (i.e., in the case of strongly congeneric series [Kier, 1985]). If a more detailed picture of the interaction is required, or a broader chemical domain is investigated, the modeler should look for other approaches for QSPR analysis.

## VI. GOOD PRACTICE IN THE USE OF TIs

Just as there are no unique descriptors for all QSPR problems, so there is no unique approach for the application of TIs in the development of quantitative models to predict chemical and biological properties. Some good practices for the use of TIs can be recommended, but the choice of the best structural descriptor or modeling strategy is always dependent on the chemical series and task at hand. A good model to predict toxicity can be obtained combining physicochemical, molecular-orbital, and topological descriptors when there is not a significant correlation among them (Cronin et al., 2002; Schultz et al., 2002). Such a combination is believed to reflect the partitioning as well as the electronic and steric properties of the molecules.

Another good rule of thumb, especially in the modeling of partitioning properties, is to select one index for each global molecular property, and two to three indices for local and specific molecular features (Sabljić, 2001). However, when many descriptors are available, many statistically equivalent models could be derived. Such a situation leads to the question of what is the best or most preferred model. To avoid this problem and the issue of the intercorrelation between the topological indices, the construction of orthogonal descriptors as performed by principal component analysis (PCA) or partial least squares (PLS), is recommended (Randić, 2001; Sabljić, 2001). A simpler solution to the problem of inter-correlation between variables is offered by the analysis of the correlation matrix between descriptors, before the statistical analysis. A good practice is also to avoid mixing indices that describe the same, or similar, molecular property. Graphical analysis of the relationship between descriptors and the property being modeled can also help to find non-linearities (most often quadratic relationships) or other fluctuations in the data.

The combination of structural descriptors and non-linear approaches such as neural networks (NN) is a valuable tool for the derivation of quantitative models when large and structurally diverse data sets are considered. The robustness of the prediction using NN is highly dependent on the training process (Niculescu et al., 2000). A number of potential problems in multivariate modeling, especially with NN, were reported in Cronin and Schultz (2003). There are many reasons to suggest that initially one should attempt more transparent approaches such as multivariate linear regression and PLS analysis. The non-linear statistical techniques should be preferred only if there are clear indications that the linear methods give unsatisfactory results.

Using topological indices, and especially connectivity and E-state indices for modeling, consideration should also be given to the proportion of the zero to non-zero values of a descriptor. It is recognized that a descriptor containing a large proportion of zero values may not have sufficient variability and statistical significance for modeling purposes. The cut-off for whether to use a descriptor usually varies about at 90 to 95% of zero values per descriptor but those with a conservative nature may wish to use a considerably lower value (e.g., 25 to 50%). It should be mentioned that if the modeler intends to perform a kind of Free-Wilson analysis with electrotopological atom-type descriptors, the deletion of any non-zero indices may result in a loss of important information.

Finally, some authors (Lu et al., 2000; Sabljić, 2001) suggest that due to the significant correlation between the topological indices and physicochemical properties, such as log $K_{ow}$, the former alone or in principal components can replace the latter in the QSPR models. The strongest support of this point of view is probably the fact that the TIs do not bring experimental or calculation error into the models. For the successful replacement of physicochemical properties such as log $K_{ow}$, more than one structural descriptor is often required, which reflects the stability and interpretability of the resulting models.

## VII. CONCLUSION

This chapter describes some frequently used topological indices, important aspects of their interpretation, as well as some recommendations for their rational application. It should provide the understanding that Graph Theory offers a different approach to structural description from physical chemistry and quantum mechanics. It is beautiful, elegant, and attractive because of the original presentation of molecular structure and the relative simplicity in the calculation algorithms. The topological indices can be used alone, or in combination with other descriptors, for the modeling of properties related to the fate and toxicity in environmental chemistry. However, critical and purpose-oriented exploitation is highly recommended.

## REFERENCES

Basak, S.C., Gute, B.D. and Grunwald, G.D., A comparative study of topological and geometrical parameters in estimating boiling point and octanol/water partition coefficient, *J. Chem. Inf. Comput. Sci.*, 36, 1054–1060, 1996.

Basak, S.C., Gute, B.D., Lučić, B., Nikolić, S. and Trinajstić, N., A comparative QSAR study of benzamidines complement-inhibitory activity and benzene derivatives acute toxicity, *Comput. Chem.*, 24, 181–191, 2000.

Bonchev, D., Overall connectivity: a next generation molecular connectivity, *J. Mol. Graphics Modelling*, 20, 65–75, 2001.

Boyd, D.B., Introduction and foreword to the Special issue commemorating the 25th anniversary of molecular connectivity as a structure description system: editorial, *J. Mol. Graphics Modelling*, 20, 1–3, 2001.

Cash, G.G., Prediction of the genotoxicity of aromatic and heteroaromatic amines using electrotopological state indices, *Mutation Res.*, 491, 31–37, 2001.

Cronin, M.T.D. and Schultz, T.W. Pitfalls in QSAR, *J. Mol. Struct. (Theochem)*, 622, 39–52, 2003.

Cronin, M.T.D., Dearden, J.C., Duffy, J.C., Edwards, R., Manga, N., Worth, A.P., and Worgan, A.D.P., The importance of hydrophobicity and electrophilicity descriptors in mechanistically based QSARs for toxicological endpoints, *SAR QSAR Environ. Res.*, 13, 167–176, 2002.

de Julián-Ortiz, J.V., de Gregorio Alapont, C., Ríos Santamarina, I., García-Doménech, R. and Gálvez, J. Prediction of properties of chiral compounds, *J. Mol. Graphics Modelling*, 16, 14–18, 1998.

Dearden, J.C., Bradburne, S.J.A., Cronin, M.T.D., and Solanki, P., The physical significance of molecular connectivity, in *Proc. Third Int. Workshop Quant. Struct.-Act. Relat. Environ. Chem.*, Turner, J.E., England, M.W., Schultz, T.W., and Kwaak, N.J., Eds., U.S. Department of Energy, Oak Ridge, TN, 1988, pp.43–50.

Dearden J.C., Cronin, M.T.D., Patel, H., and Raevsky, O.A., QSAR prediction of human skin permeability coefficients, *J. Pharm. Pharmacol.*, 52 (Suppl.), 221, 2000.

Devillers J. and Balaban, A.T., Eds., *Topological Indices and Related Descriptors in QSAR and QSPR*. Gordon and Breach Science Publishers, Amsterdam, 1999.

Estrada, E., Generalisation of topological indices, *Chem. Phys. Lett.*, 336, 248–252, 2001.

Estrada, E., Physicochemical interpretation of molecular connectivity indices, *J. Phys. Chem.*, 106, 9085–9091, 2002.

Gini, G., Lorenzini, M., Benfenati, E., Grasso, P., and Bruschi, M., Predictive carcinogenicity: a model for aromatic compounds, with nitrogen-containing substituents, based on molecular descriptors using artificial neural network, *J. Chem. Inf. Comput. Sci.*, 39, 1076–1080, 1999.

Golbraikh, A., Bonchev, D., and Tropsha, A., Novel chirality descriptors derived from molecular topology, *J. Chem. Inf. Comput. Sci.*, 41, 147–158, 2001.

Golbraikh, A., Bonchev, D., and Tropsha, A., Novel ZE-isomerism descriptors derived from molecular topology and their application to QSAR analysis, *J. Chem. Inf. Comput. Sci.*, 42, 769–787, 2002.

Grodnitzky, J.A. and Coats, J.R., QSAR evaluation of monoterpenoids' insecticidal activity, *J. Agric. Food Chem.*, 50, 4576–4580, 2002.

Gute, B.D. and Basak, S.C., Predicting acute toxicity (LC50) of benzene derivatives using theoretical molecular descriptors: a hierarchical QSAR approach, *SAR QSAR Environ. Res.*, 7, 117–131, 1997.

Gutman, I. and Lepović, M., Choosing the exponent in the definition of the connectivity index, *J. Serbian Chem. Soc.*, 66, 605–611, 2001.

Gutman, I., Ruščić, B., Trinajstić, N. and Wilcox, C.W., Jr., Graph theory and molecular orbitals. Part 12. Acyclic polyenes, *J. Chemical Physics* 62: 3399-3405, 1975.

Hall, L.H. and Kier, L.B., Electrotopological state indices for atom types: a novel combination of electronic, topological and valence state information, *J. Chem. Inf. Comput. Sci.*, 35, 1039–1045, 1995.

Hall, L.H., and Kier, L.B., Issues in representation of molecular structure. The development of molecular connectivity, *J. Mol. Graphics Modelling*, 20, 4–18, 2001.

Hosoya, H., Topological index. A newly proposed quantity characterizing the topological nature of structural isomers of saturated hydrocarbons, *Bul. Chem. Soc. Jpn.*, 44, 2332–2339, 1971.

Hosoya, H., Gotoh, M., and Ikeda, S., Topological index and thermodynamic properties. 5. How can we explain the topological dependency of thermodynamic properties of alkanes with the topology of graphs?, *J. Chem. Inf. Comput. Sci.*, 39, 192–196, 1999.

Huuskonen, J., QSAR modeling with the electrotopological state indices: predicting the toxicity of organic chemicals, *Chemosphere*, 50, 949–953, 2003.

Ivanciuc, O., Balaban, T.S., and Balaban, A.T., Design of topological indices. Part 4. Reciprocal distance matrix, related local vertex invariants and topological indices, *J. Math. Chem.*, 12: 309-318, 1993.

Kier, L.B., A shape index from molecular graphs, *Quant. Struct.-Act. Relat.*, 4, 109–116, 1985.

Kier, L.B. and Hall, L.H., Intermolecular accessibility: the meaning of molecular connectivity, *J. Chem. Inf. Comput. Sci.*, 40, 792–795, 2000.

Kier, L.B. and Hall, L.H., Molecular connectivity: intermolecular accessibility and encounter simulation, *J. Mol. Graphics Modelling*, 20: 76-83, 2001.

Kier, L.B., Hall, L.H., Murray, W.J., and Randić, M., Molecular connectivity. Part I. Relationship to nonspecific local anesthesia, *J. Pharm. Sci.*, 64, 1971–1974, 1975.

Lekishvili, G., On the characterization of molecular stereostructure. 1. Cis-Trans isomerism, *J. Chem. Inf. Comput. Sci.*, 37, 924–928, 1997.

Lekishvili, G., On the characterization of molecular stereostructure. 2. The invariants of the two-dimensional graphs, *Match*, 43, 135–152, 2001.

Liang, C. and Gallagher, D.A., QSPR prediction of vapour pressure from solely theoretically-derived descriptors, *J. Chem. Inf. Comput. Sci.*, 38, 321–324, 1998.

Llorens, O., Perez, J.J., and Villar, H.O., Toward the design of chemical libraries for mass screening biased against mutagenic compounds, *J. Med. Chem.*, 44, 2793–2804, 2001.

Lu, X., Tao, S., Cao, J., and Dawson, R.W., Prediction of fish bioconcentration factors of nonpolar organic pollutants based on molecular connectivity indices, *Chemosphere*, 39, 987–999, 1999.

Lu, X., Tao, S., and Dawson, R.W., Estimation of bioconcentration factors of non-ionic compounds in fish by molecular connectivity indices and polarity correction factors, *Chemosphere*, 41, 1675–1688, 2000.

Maw, H.H. and Hall, L.H., E-state modelling of dopamine transporter binding: validation of model for small data set, *J. Chem. Inf. Comput. Sci.*, 40, 1270–1275, 2000.

Maw, H.H. and Hall, L.H., E-state modelling of corticosteroid binding affinity: validation of model for small data set, *J. Chem. Inf. Comput. Sci.*, 41, 1248–1254, 2001.

Minkin, V.I., Glosary of terms used in theoretical organic chemistry (IUPAC recommendations, 1999), *Pure Appl. Chem.*, 71: 1919-1981, 1999.

Niculescu, S.P., Kaiser, K.L.E., and Schultz, T.W., Modeling the toxicity of chemicals to *Tetrahymena pyriformis* using molecular fragment descriptors and probabilistic networks, *Arch. Environ. Contamination Toxicol.*, 39, 289–298, 2000.

Plavšić, D., Nicolić, S., Trinajstić, N., and Mihalić, Z., On the Harary index for the characterization of chemical graphs, *J. Math. Chem.*, 12, 235–250, 1993.

Randić, M., On characterization of molecular branching, *J. Am. Chem. Soc.*, 97, 6609–6615, 1975.

Randić, M., On characterisation of molecular attributes, *Acta Chimica Slovenia*, 45, 239–252, 1998.

Randić, M., The connectivity index 25 years after, *J. Mol. Graphics Modelling*, 20, 19–35, 2001.

Randić, M. and Zupan, J., On interpretation of well-known topological indices, *J. Chem. Inf. Comput. Sci.*, 41, 550-560, 2001.

Randić, M., Balaban, A.T., and Basak, S.C., On structural interpretation of several distance related topological indices, *J. Chem. Inf. Comput. Sci.*, 41, 593–601, 2001.

Rose, K., Hall, L.H., and Kier, L.B., Modeling blood-brain barrier partitioning using the electrotopological state, *J. Chem. Inf. Comput. Sci.*, 42: 651–666, 2002.

Rücker, G. and Rücker, C., Counts of all walks as atomic and molecular descriptors, *J. Chem. Inf. Comput. Sci.*, 33, 683–695, 1993.

Sabljić, A., Topological indices and environmental chemistry, in *Practical Applications of Quantitative Structure-Activity Relationships (QSAR) in Environmental Chemistry and Toxicology*, Karcher, W., and Devillers, J., Eds., Kluwer Academic Publishing, Dordrecht, 1990, pp. 61–82.

Sabljić, A., QSAR models for estimating properties of persistent organic pollutants required in evaluation of their fate and risk, *Chemosphere*, 43, 363–375, 2001.

Schultz, H.P., Schultz, E.B., and Schultz, T.P., Topological organic chemistry. 9. Graph theory and molecular topological indices of stereoisomeric organic compounds, *J. Chem. Inf. Comput. Sci.*, 35, 864–870, 1995.

Schultz, T.W., Cronin, M.T.D., Netzeva, T.I., and Aptula, A.O., Structure-toxicity relationships for aliphatic chemicals evaluated with *Tetrahymena pyriformis*, *Chem. Res. Toxicol.*, 15, 1602–1609, 2002.

Stankevich, I.V., Skvortsova, M.I., and Zefirov, N.S., On a quantum chemical interpretation of molecular connectivity indices for conjugated hydrocarbons, *J. Mol. Struct. (Theochem)*, 342, 173–179, 1995.

Tao, S. and Lu, X., Estimation of organic carbon normalised sorption coefficient (Koc) for soils by topological indices and polarity factors, *Chemosphere*, 39, 2019–2034, 1999.

Thomsen, M., Rasmussen, A.G. and Carlsen, L., SAR/QSAR approaches to solubility, partitioning and sorption of phthalates, *Chemosphere*, 38, 2613–2624, 1999.

Todeschini, R. and Consonni, V., *Handbook of Molecular Descriptors*, Wiley-VCH, Weinheim, Germany, 2000.

Wiener, H., Structural determination of paraffin boiling points, *J. Am. Chem. Soc.*, 69, 17–20, 1947.

Zheng, W. and Tropsha, A., Novel variable selection quantitative structure-property relationship approach based on the k-nearest-neighbor principle, *J. Chem. Inf. Comput. Sci.*, 40, 185–194, 2000.

# CHAPTER 6

# Quantum Chemical Descriptors in Structure-Activity Relationships — Calculation, Interpretation, and Comparison of Methods

Gerrit Schüürmann

## CONTENTS

## I. INTRODUCTION

The electronic structure of a molecule holds a vast amount of information. Harnessing this information in a usable form is of immense value for predictive toxicology. It is of particular use for the prediction of electrostatic and covalent interactions such as those resulting in mutagenicity and carcinogenicity (Chapter 8), metabolic capability (Chapter 10), and specific acute toxicity (Chapter 12).

Methods to calculate molecular geometries and associated electronic structures have undergone substantial development in the last 30 years. Widely applicable all valence electron semiempirical methods were introduced in the 1970s. These were fitted to reproduce, as well as possible, measured geometric, energetic and electronic structure properties (Bingham et al., 1975; Dewar and Thiel, 1977). A fruitful interaction of theory and experiment became possible for a broad range of problems. In particular, such methods allowed, at that time, calculations for molecules consisting of more than 80 atoms, being faster than *ab initio* methods by many orders of magnitude. Today, semiempirical quantum chemistry enables large-scale applications to analyze compound sets with more than 50,000 chemical structures (Beck et al., 1998; Brüstle et al., 2003).

The approximations associated with the semiempirical level of calculation (as discussed in Section III) put significant constraints on the quality of the results. This is still true for the more recent members of the prominent family of methods based on the neglect of differential diatomic overlap (NDDO) approach (Dewar et al., 1985; Stewart, 1989a; Stewart, 1989b; Stewart, 2002). In this context, important developments concern the implementation of *d* orbitals for the main group elements in the third and higher rows of the periodic table (Thiel and Voityuk, 1996), and the introduction of orthogonalization corrections that improve the quantification of conformational energies (Thiel, 1998; 2000; Weber and Thiel, 2000). The latter, however, is currently parameterized only for the four basic elements C, H, N, and O, and the former is an extension of the modified neglect of diatomic overlap (MNDO) scheme (Dewar and Thiel, 1977) that does not address hydrogen bonding (see Section III).

With respect to toxic modes of action of chemical substances, quantum chemical descriptors have great potential to unravel underlying mechanisms (cf. Schüürmann, 1998a; Soffers et al., 2001). Besides quantitative structure-activity relationships (QSARs) that aim to predict toxic potencies, such parameters can also be used to model the categorical allocation of compounds to given property profiles such as antibacterial activity (Aptula et al., 2003) or general drug-likeness (Brüstle et al., 2002). Moreover, inclusion of solute-solvent interactions through the use of dielectric continuum-solvation methods (Cramer and Truhlar, 1999) offers the prediction of molecular dissociation constants ($pK_a$) (Mujika et al., 2003; Schüürmann, 1996; 1998b; Schüürmann et al., 1998) as an important parameter for the bioavailability and toxicity profile of ionogenic compounds; local reactivity parameters defined on the molecular surface provide an alternative quantum chemical route to predict $pK_a$ (Gross et al., 2001). Interestingly, continuum-solvation models can also be used to calculate environmental fate parameters such as Henry's law constant (Schüürmann, 1995; Dearden and Schüürmann, 2003); solution-phase parameters are not considered further in this chapter.

While there are reviews of the application of various quantum chemical parameters in QSARs (Karelson et al., 1996; Famini and Wilson, 2002), little attention has been paid so far to the dependence of descriptor values on the level of theory. This holds true in particular with respect to potential discrepancies between semiempirical and *ab initio* methods when calculating parameters such as frontier orbital energies and descriptors that characterize the molecular charge distribution.

This chapter discusses the background, mechanistic meaning, and method dependence of quantum chemical descriptors. Starting with the fundamentals of quantum mechanics (Section 6.II), the approximations associated with semiempirical parameterizations are introduced (Section 6.III). Subsequently, common reactivity parameters are presented, focussing on their interpretation in the context of QSAR applications (Section 6.IV). Descriptors based on frontier orbital energies, molecular orbital wavefunctions, surface areas, and on the electronic charge distribution are calculated and comparatively analyzed using 3 semiempirical and 2 *ab initio* quantum chemical methods for 607 organic compounds for which acute toxicities to the fish fathead minnow are available (Russom et al., 1997). The results reveal systematic differences between the different levels of theory for certain parameters and types of chemical functionalities, which hold true in particular for the characterization of the electronic charge distribution in terms of net atomic charges (cf. population analysis schemes as discussed in Section IV.E, below) and associated electrostatic molecular surface interaction terms (Section IV.E; Grigoras, 1990; Stanton and Jurs, 1990). The comparative analysis leads to criteria for a judicious selection of molecular descriptors and associated calculation methods in the context of computational toxicology.

## II. QUANTUM MECHANICAL BACKGROUND

In classic mechanics, the total energy of a system ($E_{tot}$) is the sum of its kinetic ($E_{kin}$) and potential energy ($E_{pot}$):

$$E_{tot} = E_{kin} + E_{pot} \tag{6.1}$$

For a single particle with mass m and (average) velocity v, the kinetic energy is given by:

$$E_{kin} = \frac{m}{2} v^2 = \frac{p^2}{2m} \tag{6.2}$$

where $p = m\,v$ denotes the momentum. When considering a molecule that consists of N nuclei and n electrons, the total kinetic energy would be the sum of all kinetic energies of all $N + n$ particles:

$$E_{kin} = \sum_{A=1}^{N} \frac{p_A^2}{2m_A} + \sum_{i=1}^{n} \frac{p_i^2}{2m_e} \tag{6.3}$$

and the electrostatic potential energy is given by the sum of all attractive interactions between electrons and nuclei as well as by all nucleus-nucleus and electron-electron repulsions:

$$E_{pot} = \sum_{A<B}^{N} \frac{1}{4\pi\varepsilon_0} \frac{Z_A Z_B e^2}{\left|R_A - R_B\right|} - \sum_{A=1}^{N}\sum_{i=1}^{n} \frac{1}{4\pi\varepsilon_0} \frac{Z_A e^2}{\left|R_A - r_i\right|} + \sum_{i<j}^{n} \frac{1}{4\pi\varepsilon_0} \frac{e^2}{\left|r_i - r_j\right|} \tag{6.4}$$

In Equation 6.3 and Equation 6.4, subscript A refers to the N nuclei with masses $m_A$, momenta $p_A$, core charge numbers $Z_A$, and spatial coordinates $R_A = (x_A, y_A, z_A)$, and the n electrons are characterized by mass $m_e$, momenta $p_i$, charge e, and positions $r_i = (x_i, y_i, z_i)$. Moreover, $\varepsilon_0$ is the vacuum permittivity; according to classical electrostatics, an electric charge q gives rise to a Coulomb potential $q/(4\pi\varepsilon_0 r)$ at distance r in the vacuum (cf. Atkins, 1983).

For the respective quantum mechanical description of a molecule in a stationary state, a few additional aspects need to be addressed. First, the system state is characterized by a wavefunction $\Psi$, and system properties, such as the total energy or dipole moment, are calculated through integration of $\Psi$ with the relevant operator in a distinct way. Note that an operator is simply an instruction to do some mathematical operation such as multiplication or differentiation, and generally (but not always) the order in which such calculations are performed affects the final result. Second, the wavefunctions $\Psi$ obey the Schrödinger equation:

$$H\,\Psi = E\,\Psi \tag{6.5}$$

where H denotes the Hamiltonian operator that yields the total energy of the system through calculation of the expression:

$$E = \frac{\iint \Psi^* H \, \Psi \, dR \, dr}{\iint \Psi^* \Psi \, dR \, dr} \equiv \frac{\langle \Psi | H | \Psi \rangle}{\langle \Psi | \Psi \rangle} \tag{6.6}$$

(for further reading see textbooks such as Atkins [1983] and Springborg [2000]). On the right-hand side of Equation 6.6, the so-called bracket notation, $\langle | \rangle$, is used as a convenient shortcut notation for integral expressions. Note further that $\Psi^*$ denotes the complex-conjugate form of $\Psi$, and that the denominator is 1 (and then could be skipped) in case of respectively normalized wavefunctions. Equation 6.6 shows that the energy is calculated as the expectation value of the Hamiltonian operator. Similarly, other observable properties are calculated as expectation values of their respective operators. This shows some parallel to expectation values in statistics.

For a molecular system, the wavefunction would generally depend on both the nuclei and electrons, $\Psi = \Psi(R, r)$. However, the mass of a nucleus proton or neutron is ca. 2000 times greater than the mass of an electron, results in much lower velocities for the nuclei. As a consequence, the electrons can follow the movement of the nuclei easily, and the nucleus and electron movements can thus, to a good degree, be treated separately (when considering stationary states). In this Born-Oppenheimer approximation, the total wavefunction is written as product of nuclear and electronic wavefunctions:

$$\Psi(R, r) = \Psi_{nuc}(R) \, \Psi_{el}(r; R) \tag{6.7}$$

and the electronic energy, $E_{el}$, as eigenvalue of the electronic part of the Hamiltonian operator, $H_{el}$, depends also on the nuclear coordinates:

$$H_{el}(R) \, \Psi_{el}(r; R) = E_{el}(R) \, \Psi_{el}(r; R) \tag{6.8}$$

In Equation 6.8, which is the electronic Schrödinger equation, the nuclei positions are included as parameters, not as variables, and this equation forms the theoretical basis for the geometry optimization of molecular systems. A detailed discussion of problems associated with the Born-Oppenheimer approximation when considering a more elaborated level of description is given in the literature (Sutcliffe, 1997).

For the specification of $H_{el}$, atomic units are now applied, where, for example, $4\pi\varepsilon_0 = 1$, $m_e = 1$ and $e = 1$ hold true. Moreover, the momentum operator of a given electron is defined (in atomic units) as a partial derivative with respect to position (multiplied by $1/i$ with $i = \sqrt{-1}$), which for one spatial coordinate x reads:

$$p_x = \frac{1}{i} \frac{\partial}{\partial x} \tag{6.9}$$

and yields the following expression for the kinetic energy:

$$\frac{1}{2}\left(p_x^2 + p_y^2 + p_z^2\right) = -\frac{1}{2}\left(\frac{\partial^2}{\partial x^2} + \frac{\partial^2}{\partial y^2} + \frac{\partial^2}{\partial z^2}\right) \equiv -\frac{1}{2}\nabla^2 \tag{6.10}$$

(cf. Atkins, 1983; Springborg, 2000). Considering Equations 6.3, 6.4, and 6.10, the electronic Hamiltonian operator is now defined as:

$$H_{el} = H_{el,kin} + H_{el,pot} \tag{6.11}$$

and for a molecule with N nuclei and n electrons it can be written as:

$$H_{el} = -\frac{1}{2}\sum_{i=1}^{n}\nabla_i^2 - \sum_{i=1}^{n}\sum_{A=1}^{N}\frac{Z_A}{r_{iA}} + \sum_{i<j}^{n}\frac{1}{r_{ij}} \tag{6.12}$$

where $r_{iA} = |r_i - R_A|$ denotes the distance between electron i and nucleus A, and $r_{ij} = |r_i - r_j|$ the distance between electrons i and j, respectively.

In Equation 6.12, the first term represents the kinetic energy of the electrons, the second term represents the Coulomb electron-nucleus attraction, and the third term represents the electron-electron repulsion. Note further that the present form of $H_{el}$ does not address relativistic effects (which are neglected throughout this chapter), and that it refers to time-independent states of the molecule (as is also assumed throughout this chapter, and which is indicated implicitly by omitting the time variable t in all expressions and equations).

For a given set of nuclear coordinates R and a stationary electronic wavefunction ($\Psi_{el}$) normalised to 1 (i.e., $\langle\Psi_{el}|\Psi_{el}\rangle = 1$ holds true), the electronic energy can be calculated in the Born-Oppenheimer approximation as the expectation value of the electronic Hamiltonian operator as follows:

$$E_{el} = \langle\Psi_{el}|H_{el}|\Psi_{el}\rangle \tag{6.13}$$

(cf. Equation 6.6). Addition of the nuclear repulsion energy:

$$E_{nuc} = E_{nuc,pot} = \sum_{A<B}^{N}\frac{Z_A Z_B}{R_{AB}} \tag{6.14}$$

yields the total energy of the molecular system:

$$E_{tot} = E_{el} + E_{nuc} = \langle\Psi_{el}|H_{el}|\Psi_{el}\rangle + \sum_{A<B}^{N}\frac{Z_A Z_B}{R_{AB}} \tag{6.15}$$

So far the form of $\Psi_{el}$ has not been specified. While the Schrödinger equation defines clear conditions for valid state functions of the system, it does not offer a practical method to actually obtain the exact solution $\Psi$ or $\Psi_{el}$. As a consequence, iterative procedures have been developed that start with some trial wavefunction, and subsequently improve their quality until some predefined convergence criterion is fulfilled.

The rationale behind this approach is the variational principle. This principle states that for an arbitrary, well-behaved function of the coordinates of the system (e.g., the coordinates of all electrons in case of the electronic Schrödinger equation) that is in accord with its boundary conditions (e.g., molecular dimension, time-independent state, etc.), the expectation value of its energy is an upper bound to the respective energy of the true (but possibly unkown) wavefunction. As such, the variational principle provides a simple and powerful criterion for evaluating the quality of trial wavefunctions: the lower the energetic expectation value, the better the associated wavefunction.

A general way to construct $\Psi_{el}$ is through a linear combination of trial wavefunctions, $\Phi_K$,

$$\Psi_{el} = \sum_{K}C_K\Phi_K \tag{6.16}$$

In Equation 6.16, $\Phi_K$ represent certain electronic configurations, and usually have the form of so-called Slater determinants that are in accord with the Pauli principle and the antisymmetry condition (the sign of $\Phi_K$ changes if the coordinates of two electrons are interchanged; see Atkins [1983], Jensen [1998], Klessinger [1982], Levine [1991], Springborg [2000]). These $n$-electron Slater determinants are built from orbitals as one-electron functions $\phi_i$, the latter of which are multi-center molecular orbitals in the case of molecular orbital schemes (as opposed to, for example, valence-bond schemes where the one-electron functions are much more localized).

In a symbolic notation, the Slater determinant for the electronic ground-state configuration, $\Phi_0$, as function of $n$ electrons can be written in the closed-shell case as:

$$\Phi_0 (1, 2, ..., n) = \left| \phi_1(1) \, \overline{\phi_1}(2) \, \phi_2(3) \, \overline{\phi_2}(4) ... \phi_{n/2}(n-1) \, \overline{\phi_{n/2}}(n) \right| \qquad (6.17)$$

In Equation 6.17, each orbital $\phi_i$ listed in the Slater determinant contains two electrons of opposite spin (in accordance with the Pauli principle), and the occupancies with $\beta$ spin electrons are indicated by overlinings. $\Phi_0$ thus contains $n/2$ doubly occupied spatial orbitals.

In the often used linear combination of atomic orbitals–molecular orbital (LCAO-MO) approach, the molecular orbitals $\phi_i$ are built from atomic orbitals $\chi_\mu$ according to:

$$\phi_i = \sum_{\mu=1}^{m} c_{\mu i} \chi_\mu \qquad (6.18)$$

where $c_{\mu i}$ quantifies the contribution of atomic orbital $\chi_\mu$ to the molecular orbital $\phi_i$. While an iterative solution of the electronic Schrödinger equation (Equation 6.8) could, in principle, include an optimization of the expansion coefficients $C_K$ (Equation 6.16) and $c_{\mu i}$ (Equation 6.18) as well as of the functional form of the atomic orbitals, $\chi_\mu$, typical calculations are confined to modifying only one type of parameter.

For common closed-shell systems, such as most organic molecules, the electronic ground state can be sufficiently well described by reducing the sum in Equation 6.16 to only one term — the ground-state configuration function as actually specified in Equation 6.17. Moreover, a given calculation often employs a fixed set of atomic orbitals as basis functions, leaving the expansion coefficients $c_{\mu i}$ as the only parameters to be actually optimized during the iterative (approximate) solution of the Schrödinger equation. Note, however, that not every molecule with an even number of electrons is a closed-shell system; e.g., the oxygen molecule $O_2$ with 16 electrons has a triplet ground state with 2 unpaired electrons, which requires to go beyond the one-determinant closed-shell form of Equation 6.17.

Even when confining the variation of the trial wavefunction to the LCAO-MO coefficients $c_{\mu i}$, the respective approximate solution of the Schrödinger equation is still quite complex and may be computationally very demanding. The major reason is that the third term of the electronic Hamiltonian, $H_{el}$ (Equation 6.12), the electron-electron repulsion, depends on the coordinates of two electrons at a time, and thus cannot be broken down into a sum of one-electron functions. This contrasts with both the kinetic energy and the electron-nucleus attraction, each of which are functions of the coordinates of single electrons (and thus are written as sums of $n$ one-electron terms). At the same time, orbitals are one-electron functions, and molecular orbitals can be more easily generated as eigenfunctions of an operator that can also be separated into one-electron terms.

Accordingly, a common further approximation for calculating the molecular orbitals $\phi_i$ is to replace the $H_{el}$ term containing the pair-wise electron-electron interactions by an effective potential that treats the interaction of a given electron with all other electrons in an average way. The resultant operator includes the one-electron terms of $H_{el}$:

$$h_{el} = -\frac{1}{2}\sum_{i=1}^{n}\nabla_i^2 - \sum_{i=1}^{n}\sum_{A=1}^{N}\frac{Z_A}{r_{iA}} \tag{6.19}$$

as well as an effective one-electron potential, $V_{eff}$, that represents a mean-field approximation of the electron-electron repulsion:

$$V_{eff} = \sum_{i=1}^{n}\left(2J_i - K_i\right) \tag{6.20}$$

and is called the Fock (F) operator:

$$F = h_{el} + V_{eff} = \sum_{i=1}^{n}\left(-\frac{1}{2}\nabla_i^2 - \sum_{A=1}^{N}\frac{Z_A}{r_{iA}} + \left(2J_i - K_i\right)\right) \tag{6.21}$$

(e.g., Levine 1991). In Equations 6.20 and Equation 6.21, the terms $J_i$ and $K_i$ represent the effective (mean-field approximated) Coulomb and (non-classical) exchange interaction of a given electron with all other electrons. More precisely, the averaged repulsive interaction between two electrons with different spin is given by $J_i$, while the averaged repulsion between two electrons with equal spin is lowered by $K_i$ and thus is represented by $(J_i - K_i)$. The latter can be understood as reflecting the fact that in accordance with the Pauli principle, electrons with equal spin avoid each other more effectively (and thus have a greater average distance from each other than electrons with opposite spin).

Application of the Fock operator yields so-called Hartree-Fock (HF) orbitals $\phi_i$:

$$F\phi_i = \varepsilon_i\phi_i \tag{6.22}$$

The eigenvalues $\varepsilon_i$ are the well-known orbital energies, and represent (in absolute value) the amount of energy necessary to remove the electron from the respective orbital. According to Koopmans' theorem, $\varepsilon_i$ corresponds to the negative of the respective ionization potential (see Section IV).

Introduction of the LCAO-MO expansion (Equation 6.18) into Equation 6.22 leads to secular equations (for every molecular orbital $\phi_i$) in the atomic orbital basis, which are called the Roothaan-Hall equations:

$$\sum_{v}\left(F_{\mu v} - \varepsilon_i S_{\mu v}\right)c_{vi} = 0 \tag{6.23}$$

(e.g., Levine, 1991). Here, $S_{\mu v}$ denotes the overlap integral between two atomic orbitals:

$$S_{\mu v} = \int \chi_\mu^* \chi_v \, dr \equiv \langle \chi_\mu | \chi_v \rangle \tag{6.24}$$

and the matrix element of the Fock operator in the AO basis, $F_{\mu v}$, can be written as:

$$F_{\mu v} = h_{\mu v} + \sum_{\rho}\sum_{\sigma}P_{\rho\sigma}\left[2\left(\mu v|\rho\sigma\right) - \left(\mu\sigma|\rho v\right)\right] \tag{6.25}$$

In Equation 6.25), $h_{\mu\nu}$ is the one-electron integral with $h_{el}$ (Equation 6.19):

$$h_{\mu\nu} = \int \chi_\mu^* \, h_{el} \, \chi_\nu \, dr \equiv \left\langle \chi_\mu \middle| h_{el} \middle| \chi_\nu \right\rangle \tag{6.26}$$

which quantifies, for $\mu \neq \nu$, the kinetic energy of a single electron in the overlap region of atomic orbitals $\chi_\mu$ and $\chi_\nu$ together with its potential energy in the field of the core charges of all nuclei. $P_{\rho\sigma}$ is the density matrix:

$$P_{\rho\sigma} = 2 \sum_{j}^{occ} c_{\rho j} c_{\sigma j} \tag{6.27}$$

(whose diagonal and off-diagonal elements can be interpreted as electron charge densities and measures of bond orders, respectively), and $(\mu\nu|\rho\sigma)$ is a shortcut notation for the following two-electron integral (where the electrons are numbered by 1 and 2, respectively):

$$\left(\mu\nu|\rho\sigma\right) \iint \chi_\mu^*(1) \, \chi_\rho^*(2) \, \frac{1}{r_{12}} \, \chi_\nu(1) \, \chi_\sigma(2) \, dr_1 \, dr_2 \equiv \left\langle \chi_\mu \chi_\rho \middle| \frac{1}{r_{12}} \middle| \chi_\nu \chi_\sigma \right\rangle \tag{6.28}$$

For a given set of atomic orbitals $\chi_\mu$, the one- and two-electron integrals (Equation 6.26 and Equation 6.28) can be calculated, and all terms of the Fock matrix in the AO basis, $F_{\mu\nu}$ (Equation 6.25), would be known except for the density matrix, $P_{\rho\sigma}$, which depends on the LCAO-MO coefficients as solutions of the secular equation (Equation 6.23) and thus on F. Accordingly, the density matrix depends on the evaluation of the Fock matrix, which depends on the elements of the density matrix.

To solve this problem of mutual dependency, an initial trial density matrix is formed (often employing a separate formalism according to the one-electron Hückel theory), and the associated (still crude) LCAO-MO coefficients $c_{\mu i}$ (Equation 6.18) are used to calculate a first version of $F_{\mu\nu}$ (Equation 6.25). Diagonalization of this Fock matrix yields a first solution of the Roothaan-Hall secular equation, and generates a revised set of LCAO-MO coefficients and thus a modified density matrix, which can be used to calculate a revised Fock matrix, and to go through the iteration cycle again and again. Note that in every iteration, a set of eigenvalues, $\varepsilon_i$ (the orbital energies), and associated eigenvectors, $\phi_i$, defined through the LCAO-MO coefficients $c_{\mu i}$, is generated, and the cycle is stopped when the differences between the $\varepsilon_i$ and $c_{\mu i}$ of two consecutive cycles are below preset thresholds. The associated solution is then called self-consistent, and the overall procedure is called LCAO-MO SCF approach (with SCF denoting self-consistent field).

A major result of a SCF calculation is the total electronic energy, which can be obtained according to:

$$E_{el} = \sum_{j}^{occ} \left(I_j + \varepsilon_j\right) = \frac{1}{2} \sum_{\mu} \sum_{\nu} P_{\mu\nu} \left(h_{\mu\nu} + F_{\mu\nu}\right) \tag{6.29}$$

from the orbital energies $\varepsilon_j$ and the one-electron integrals in the MO basis:

$$I_j = \int \phi_j^* \, h_{el} \, \phi_j \, dr \equiv \left\langle \phi_j \middle| h_{el} \middle| \phi_j \right\rangle \tag{6.30}$$

or alternatively, as shown in the second part of Equation 6.29, in the AO basis from the density matrix, the respective one-electron terms (Equation 6.26) and the Fock matrix (e.g., Klessinger,

1982). Note that $E_{el}$ is not simply the sum of the orbital energies, which results from the fact that a given $\varepsilon_i$ includes the averaged repulsive interaction of the electron in orbital $\phi_i$ with all other electrons, such that for every pair of electrons, this repulsive interaction would be counted twice.

According to Equation 6.15, the total energy of the molecular system in the Born-Oppenheimer approximation, $E_{tot}$, is then obtained by adding the nucleus-nucleus repulsion terms, $E_{nuc}$ (Equation 6.14), and in general both $E_{nuc}$ and $E_{el}$ depend on the actual molecular geometry. A geometry optimization thus involves the calculation of $E_{el}$ and $E_{nuc}$ for different trial coordinates and (today) makes use of analytical gradient techniques to find geometric coordinates for which $E_{tot}$ (within preset thresholds) achieves a minimum value (cf. Schlegel, 2003).

Due to the mean-field approximation of the Fock operator (Equation 6.21), the instantaneous electron-electron repulsion is not properly accounted for. On the average, the electrons in a molecular system are further apart from each other and avoid each other more efficiently than according to the Hartree-Fock model. This neglect of the correlation of the electron movement — the lack of electron correlation — constitutes an important deficiency of the HF SCF level of calculation, and improved descriptions of the local electron-electron interaction require an exsension of the wavefunction in regions that are not properly sampled by the SCF eigenfunction, the latter of which spans only the space associated with the occupied molecular orbitals. For different ways to go beyond the HF model, such as configuration interaction and Møller-Plesset perturbation theory as well as the density functional theory (DFT; see also Section V) as an alternative approach to the calculation of the molecular electronic structure, the reader is referred to the literature (Baerends, 2000; Barends and Gritsenko, 1997; Jensen, 1999; Koch and Holthausen, 2001; Kohn, 1999; Levine, 1991; Raghavachari and Anderson, 1996), including the Nobel Prize lecture of Pople (1999).

## III. SEMIEMPIRICAL QUANTUM CHEMICAL METHODS

In *ab initio* schemes, all relevant one- and two-electron integrals are calculated using a basis set of atomic orbitals (e.g., Hehre et al., 1986), while semiempirical methods are defined through the neglect of certain types of integrals, and through the use of parameterized functions that allow a substantially simplified (and in particular much faster) calculation of the remaining integrals. The error introduced by the various approximations is then partly corrected through fitting the results to experimental values.

The popular semiempirical methods, MNDO (Dewar and Thiel, 1977), Austin Model 1 AM1; Dewar et al., 1985), Parameterized Model 3 (PM3; Stewart 1989a; 1989b), and Parameterized Model 5 (PM5; Stewart, 2002), are all confined to treating only valence electrons explicitly, and employ a minimum basis set (one $s$ orbital for hydrogen, and one $s$ and three $p$ orbitals for all heavy atoms). Most importantly, they are based on the NDDO approximation (Stewart, 1990a, 1990b; Thiel, 1988, 1996; Zerner, 1991):

$$\chi_\mu^A \chi_\nu^B = \delta_{AB} \chi_\mu^A \chi_\nu^A \tag{6.31}$$

In Equation 6.31, A and B indicate atomic centers, and $\delta_{AB}$ is the Kronecker function, yielding 1 for A = B, and 0 for A ≠ B. The NDDO approximation implies that (with orthonormal AOs) the overlap matrix $S_{\mu\nu}$ reduces to the unit matrix, and that all two-electron integrals with charge clouds arising from the overlap of AOs from two different atomic centers are ignored:

$$\left(\mu^A \nu^B \middle| \rho^C \sigma^D\right) = \delta_{AB}\, \delta_{CD} \left(\mu^A \nu^A \middle| \rho^C \sigma^C\right) \tag{6.32}$$

(cf. Equation 6.28). This results in the neglect of all three- and four-center integrals (see Stewart [1990a; 1990b] for details). With regard to the NDDO formalism as actually employed in MNDO,

AM1, PM3, and PM5, there is one important exception to Equation 6.31 when evaluating the one-electron, two-center integrals (the so-called resonance integrals) that describe the energy gain through a bonding overlap between two AOs from different atomic centers:

$$h_{\mu\nu}^{AB} = \left\langle \chi_\mu^A \middle| h_{el} \middle| \chi_\nu^B \right\rangle = S_{\mu\nu}^{AB} \frac{1}{2} \left( \beta_\mu^A + \beta_\nu^B \right) \tag{6.33}$$

As can be seen from Equation 6.33, the one-electron part of the Fock operator (cf. Equation 6.25 and Equation 6.26) is set to be proportional to the overlap between atoms A and B with respect to the two specified AOs. Note that this approach violates the NDDO formalism; consequently, the approach was called MNDO, and the successor models AM1, PM3, and PM5 are in fact MNDO-type methods. The one-electron matrix element of Equation 6.33 represents the kinetic and potential energy of the electron in the field of the two nuclei (represented by their core charges), and the $\beta$ terms are the respective single-atom resonance integrals.

The two-electron integrals (Equation 6.32) are determined from atomic experimental data in the one-center case, and are evaluated from a semiempirical multipole model in the two-center case that ensures correct classical behavior at large distances and convergence to the correct one-center limit. Interestingly, this parameterization results in damped effective electron-electron interactions at small and intermediate distances, which reflects a (however less regular) implicit partial inclusion of electron correlation (Thiel, 1998). In this respect, semiempirical methods go beyond the HF level, and may accordingly be superior to HF *ab initio* treatments for certain properties that have a direct or indirect connection to the parameterization procedure.

Known deficiencies of MNDO as the oldest of the general-purpose NDDO schemes include the inability to model hydrogen bonding, the overestimation of steric crowding (resulting in corresponding problems with small-ring structures), and an inadequate treatment of hypervalent compounds (cf. Stewart, 1990a). AM1 provides a significant improvement in the overall accuracy when compared with MNDO, and in particular reasonable hydrogen bond energetics (with some problems regarding the associated geometries), and has much less problems with steric crowding.

PM3 yields a still improved overall accuracy with particular enhancements for hypervalent structures (within the constraints imposed by using the minimum basis set), and generally better hydrogen bond geometries. However, hydrogen bond energetics may also be inferior to AM1 results (Dannenberg 1997), and there have been observed problems associated with the PM3 parameterization of the core repulsion function, leading to an unphysical H···H attraction between 1.8 and 2.0 Å (Csonka, 1993) as well as to spurious oscillations in the O···H, N···H, and N···O interactions (Csonka and Ángyán, 1997). Moreover, it was reported that the charge distribution associated with electrostatic potentials derived from PM3 calculations is inferior to MNDO and AM1 results (Alemán et al., 1993), and nitrogen atoms appear to be less well treated with PM3 (too positive net atomic charge, preference of pyramidal structures).

For PM5 (Stewart, 2002), no publication presenting the method is yet available. Interestingly, a recent comparative analysis of different NDDO methods reported that the PM5 mean absolute error in the heat of formation for a set of 622 compounds was slightly inferior to PM3, but superior to AM1 (Repasky et al., 2002b).

In the last decade, significant progress has been made to address some of the shortcomings inherent in the NDDO approach (Thiel, 1996; 2000). On the one hand, MNDO was extended to $d$ orbitals as a better basis for heavier elements and in particular for hypervalent structures (MNDO/d), with parameters available for the elements of the second large row except Ar (Na to Cl), as well as for Br, I, and the Zn group (Thiel and Voityuk, 1992, 1996). The inclusion of $d$ atomic orbitals leads to a substantial increase in the total number of distinct two-center two-electron integrals, which rises from 22 in the $sp$ basis set (Stewart 1990a) to 204 in the $spd$ basis set (Thiel, 1996). Recently, a similar extension of AM1 has been reported for phosphorus (AM1/$d$; Lopez and York, 2003), incorporating the advantage of AM1 over MNDO with regard to hydrogen bonding.

Another quite significant development concerns the introduction of orthogonalization corrections (Kolb and Thiel, 1993; Weber and Thiel, 2000). The starting point is the reinterpretation of the NDDO formalism as referring to an orthogonalised basis (following a respective analysis of the still cruder zero-differential-overlap [ZDO] approximation that neglects the integral expressions containing products $\chi_\mu \chi_\nu$ also for different AOs $\chi_\mu$ and $\chi_\nu$ located at the same atomic center; Löwdin 1950, 1970; Fischer-Hjalmars, 1965), and a Taylor expansion of the relevant transformation matrix (which is actually the inverse of the square root of the overlap matrix, $S^{-\frac{1}{2}}$) leads to additional terms that have been evaluated for the one-electron part of the Fock operator. A valence-shell orthogonalization correction in the one-center part of $h_{el}$ leads to an explicit treatment of the Pauli exchange repuslion in a method (Kolb and Thiel, 1993) later called OM1 (Orthogonalization Model 1); Thiel 2000). The more recent Orthogonalization Model 2 (OM2) method (Weber and Thiel, 2000) results in three-center contributions to the resonance integrals (note that conventional NDDO is usually confined to one- and two-center interaction terms).

So far, OM1 is parameterized for C, H, N, O and F, and leads to a substantial improvement in the calculation of excitation energies; OM2 is parameterized for C, H, N, and O, and provides a distinct improvement with regard to the stereochemical environment of electron pair bonds, which is important for modeling conformational properties (Thiel, 2000). Recent comparative analyses of the performance of various NDDO-based schemes include also results for OM1 and OM2 (Thiel, 1998; 2000), and there are also other routes of improving the NDDO scheme (Repasky et al., 2002a, 2002b).

Despite the significant progress made in computer power, semiempirical schemes are still widely used. This holds true particularly in applied areas such as the development of qualitative or quantitative structure-activity relationships. One reason is the fact that *ab initio* calculations still require a prohibitive amount of computer time when screening quantum chemical parameters for large compound sets, particularly when dealing with drug-like compounds or other medium-sized or even large-sized molecules. Apart from that, it was noted recently that in the immediate future, only semiempirical methods will allow more comprehensive analyses of molecular structures with regard to conformational degrees of freedom and to the problem of multiple minima with more complex compounds (Clark, 2000). Interestingly, the conformational flexibility of molecular structures was introduced in the QSAR paradigm already some years ago (Ivanov et al., 1994; Lipkowitz et al., 1989; Mekenyan et al., 1994), and has been developed significantly over the last years (Mekenyan et al., 1999; 2003).

A final note concerns the heat of formation at 298 K as the primary result of the popular MNDO-type methods. In the valence electron approximation, the total energy, $E_{tot}$ (Equation 6.15), with $E_{el}$ according to Equation 6.29 (which is the electronic energy calculated with a wavefunction built from Hartree-Fock molecular orbitals) quantifies the energy gain in building the molecule from single atom cores (atom ions where all valence electrons have been removed) and single electrons, all of which have been at infinite separation from each other. Note that the reference state is the (theoretical) state where all considered particles (electrons and atom cores, the latter of which include the inner-shell electrons) do not interact with each other.

Clearly, a more practical reference state would be the one in which the elements occur under standard conditions (e.g., at 298 K and 1 atm). When starting from the latter, the formation of a molecule would require the following three steps:

1. Generation of free (non-interacting) atoms from the elements in their standard state (e.g., when aiming at the formation of $CH_4$, first split four moles of $H_2$ molecules into four moles of H atoms, and remove one mole of C atoms from a graphite block)
2. Removal of all valence electrons from the free atoms to generate the atom cores
3. Formation of the molecule from the free (noninteracting) atom cores and the free (noninteracting) electrons.

The first step requires the sum of the heats of formation of all atoms involved in the target molecule ($\Sigma_A \Delta H_f^A$, where A denotes the individual atoms). For the second step, the relevant ionization energies are required for all atoms ($\Sigma_A E_{core}{}^A$) that are the negatives of the respective electronic energies of the valence electrons in the isolated atomic states ($-\Sigma_A E_{el}{}^A$). The final step is to sum the kinetic energies of the electrons and all core-core, core-electron, and electron-electron interactions in the molecular system as actually calculated through $E_{tot}$ (cf. Equation 6.12 to Equation 6.15). Accordingly, the heat of formation can be written as:

$$\Delta H_f = E_{tot} + \sum_{A=1}^{N} \left( \Delta H_f{}^A + E_{core}{}^A \right) = E_{tot} + \sum_{A=1}^{N} \left( \Delta H_f{}^A - E_{el}{}^A \right) \tag{6.34}$$

For the atomic heats of formation, experimental data are taken, while the energy to remove all valence electrons also depends on parameters of the semiempirical model. In Equation 6.34, the electronic energies of the atoms, $E_{el}{}^A$, have to be calculated in their corresponding ground states.

## IV. QUANTUM CHEMICAL MOLECULAR DESCRIPTORS

For aromatic compounds, the classic description of the variation in reactivity with varying substituents is given by the Hammett equation:

$$\log \frac{k}{k_0} = \rho \bullet \sigma \tag{6.35}$$

where: $k$ and $k_0$ denote the rate constants of a substituted and parent compound, respectively; $\rho$ is a constant specific for the reaction and molecular system under investigation; $\sigma$ is a parameter quantifying the electronic substituent effect.

In Equation 6.35, $\sigma$ is called the Hammett constant and is defined as the logarithmic ratio of the equilibrium constants of the parent compound and derivative, $K_0$ and $K$, according to:

$$\sigma = \log \frac{K}{K_0} \tag{6.36}$$

Electron acceptor substituents such as $-NO_2$, $-CN$, $-COOR$, and $-SOR$ have positive $\sigma$ values, enhancing reactions that have an excess electronic charge at the reaction site in the activated complex, which in turn are characterized by reaction constants $\rho > 0$ (resulting in $k > k_0$). The increase in reactivity can be traced back to the substituent-mediated delocalization of the excess charge, which lowers the energy demand of the reaction. By contrast, electron donor substituents such as $-NH_2$, $-OH$, and alkyl groups have negative Hammett constants and increase the electron density at the reaction site.

Although the Hammett equation was, and still is, very successful in quantifying electronic interactions between different molecular moieties and has led to a certainly large compendium of Hammett constants generated so far (Hansch et al., 1991), this approach is principally limited to certain reaction types, and among these to those compounds where substituent constants are available. More precisely, the application of the Hammett equation requires that the molecular system of interest can be broken down into some parent compound and substituents, such that for both the relevant reaction (referring to some specific reaction site) and the substituents parameters must be available.

By contrast, quantum chemical methods allow, in principle, the analysis of the electronic structures of molecules with regard to chemical attacks at every site. Moreover, global reactivity parameters such as molecular electronegativity (see below) can be quantified. In addition, besides the calculation of various reactivity indices quantum chemistry can, for example, also be used to monitor reaction pathways and evaluate their energetic patterns.

## A.  Frontier Orbital Energies $E_{HOMO}$ and $E_{LUMO}$

The chemical reaction between two molecular species is always accompanied by a rearrangement of electron density. For a given pair of two reacting agents, the more electronegative compound will gain some electronic charge upon forming a covalent bond with the reaction partner, which, in turn, loses the respective amount of electron density. It follows that the general tendency of a molecule to gain or lose electronic charge may serve as a global reactivity parameter in the context of QSAR investigations. When considering the formally simple (but energetically not trivial) reaction of removing a single electron from a molecular species X to infinity:

$$X \rightarrow X^+ + e^- \qquad (6.37)$$

the respective change in the heat of formation is called the adiabatic ionization potential ($IP_{adiabatic}$):

$$IP_{adiabatic} = \Delta H_f\left(X^+\right) - \Delta H_f(X) + \Delta H_f\left(e^-\right) \qquad (6.38)$$

In Equation 6.38, the heat of formation terms, $\Delta H_f$ (X) and $\Delta H_f$ (X$^+$), refer to the neutral species, X, and the associated radical cation, X$^+$, in their respective ground states; they would both usually require separate quantum chemical calculations, including geometry optimizations.

An approximate quantification of the energy change associated with the ionization reaction of Equation 6.37) is offered by Koopmans' theorem, which states that the vertical ionization potential (IP) of an electron in the one-electron MO wavefunction $\phi_i$ as HF solution for the neutral molecular species (cf. Equation 6.18 and Equation 6.22) is equal to the negative of its orbital energy $\varepsilon_i$:

$$IP_i = E\left(X_i^+\right) - E(X) = -\varepsilon_i \qquad (6.39)$$

(the energy of a free, non-interacting electron is 0, and for convenience the vertical ionization potential is often simply called ionization potential).This approach benefits from the approximate cancellation of two errors:

1. The rearrangement of the electronic structure upon ionization is neglected (because $\varepsilon_i$ refers to a wavefunction optimized for the neutral species X). This tends to yield too large an energy value for electron abstraction. Note that in accordance with the variational principle as discussed in Section II, relaxation of the wavefunction describing the cation would lower its energy when compared with using the wavefunction of the neutral species, lowering the energy difference between X$^+$ and X.
2. The error associated with the HF level of theory with regard to the neglect of electron correlation (see Section 6.II) tends to lower the energy difference between X$^+$ and X. This is because the correlation effects are larger for the neutral species that has one more electron than the cation (a more comprehensive quantum chemical treatment, including electron correlation would lower the energy of X more than the one of X$^+$).

Formally, Koopmans' theorem also applies to the opposite process of attaching an additional electron (taken from infinity) to the molecule:

$$X + e^- \rightarrow X^- \qquad (6.40)$$

In this case, the energy associated with forming the radical anion is approximated by the negative of the energy of a previously unoccupied molecular orbital, $\phi_k$, that serves as initial one-electron wavefunction of the incoming electron:

$$EA_k = E(X) - E(X_k^-) = -\varepsilon_k \tag{6.41}$$

which is called the (vertical) electron affinity (EA) of the molecule. Unfortunately, however, orbital relaxation and electron correlation both tend to lower the anion energy as compared to that of the neutral species, such that the neglect of both effects results in a systematic underestimation of the electron affinity. Note that a positive electron affinity implies that the anion $X^-$ is energetically preferred over the neutral molecule X (indicating that the additional net electron-electron repulsion is smaller than the additional electron-nucleus attraction).

To characterize the global readiness of molecules to donate or accept electron charge, the lowest ionization potential and greatest electron affinity (that are often simply called ionization potential and electron affinity) would be the best parameters when referring to Equation 6.37 and Equation 6.40 as (highly simplified) model reactions for nucleophilic and electrophilic interactions of a compound with endogenous reaction partners, and the associated MO energy values:

$$IP = -\varepsilon_{HOMO} \equiv -E_{HOMO} \tag{6.42}$$

$$EA = -\varepsilon_{LUMO} \equiv -E_{LUMO} \tag{6.43}$$

are the often used frontier orbital energies of the highest occupied molecular orbital ($E_{HOMO}$), and lowest unoccupied molecular orbital ($E_{LUMO}$). As can be seen from Equation 6.42 and Equation 6.43, the use of IP and EA is (within the approximations imposed by Koopmans' theorem) equivalent to employing $E_{HOMO}$ and $E_{LUMO}$ as global reactivity parameters for QSAR investigations.

Early QSAR applications of $E_{HOMO}$ and $E_{LUMO}$ were confined to Hückel $\pi$-electron calculations (Snyder and Merril, 1965; Yoneda and Nitta, 1964), and the probably first applications of frontier orbital energies as reactivity descriptors in environmental toxicological QSARs were in the late 1980s (cf. Gough and Kaiser, 1988; Purdy, 1988; Schüürmann, 1990). $E_{LUMO}$ is now a well-established parameter to model global electrophilicity (Cronin et al., 1998; 2001; Schmitt et al., 2000; Schüürmann, 1998a; Veith and Mekenyan, 1993). The Hammett parameters are also still used (cf. Mekapati and Hansch, 2002) as well as molecular delocalizability as outlined below. Both $E_{HOMO}$ and $E_{LUMO}$ also serve as (again global) parameters to encode the covalent portion of the hydrogen bond donor and acceptor capacity of biological agents in aqueous solution when considering both the compound of interest and the water molecule (Famini et al., 1992; Cramer et al. 1993). Interestingly, $E_{LUMO}$ was recently shown to allow a discrimination between different modes of toxic action of phenols (Schüürmann et al., 2003), and $E_{HOMO}$, representing the global electron donor capacity of a molecule, turned out as a suitable parameter to model dye-fiber affinities of azo dyes (Schüürmann and Funar-Timofei, 2003).

## B.  Molecular Electronegativity and Hardness

Another way of characterizing the readiness of molecules to gain or lose electrons upon interaction is based on the concepts of molecular electronegativity and hardness (Berkowitz and Parr, 1988; Parr et al., 1978; Pearson, 1986; 1991). The starting point is the consideration that both the extent and ease of electronic deformation will affect the reactivity of a chemical compound (cf. Schüürmann, 1998a). The electronegativity (EN) characterizes the tendency of atoms and molecules to attract electrons, and upon interaction with a reaction partner electron charge will be transferred from lower EN (the donor molecule) to higher EN (the acceptor molecule). It follows that EN defines the direction of net electron transfer between interacting molecules, while hardness (HD) characterizes the resistance of the electronic structure to undergo any change. The lower the HD, the more easily a molecule will accept or donate electron charge. As such, molecular HD is inversely related to the polarizability of molecules.

The rationalization of EN and HD in terms of easily accessible energetic quantities begins with the finding that, analogous to the classical definition of the chemical potential as a partial derivative of the Gibbs free energy with respect to the number of particles, an electronic chemical potential $\mu$ can be correspondingly defined in terms of the electronic energy, $E_{el}$, and the number of electrons, n (Parr et al., 1978):

$$\mu = \frac{\partial E_{el}}{\partial n} \qquad (6.44)$$

According to Equation 6.44), $\mu$ represents the change in electronic energy of a molecular system upon changing its total number of electrons. When considering a neutral organic molecule, its energy would usually decrease when adding a fractional amount of electron charge. For $E_{el}$ as a function of n, it follows that around the n referring to the neutral system, $\mu$ as slope of the respective curve is negative. The negative of $\mu$ was then identified as absolute electronegativity (Parr et al., 1978), and approximating the slope through a finite difference evaluated at n − 1 (with energy difference IP) and n + 1 (with energy difference EA) led to:

$$EN = -\mu = -\frac{\partial E_{el}}{\partial n} \approx \frac{1}{2}(IP + EA) = -\frac{1}{2}\left(E_{HOMO} + E_{LUMO}\right) \qquad (6.45)$$

where the right-hand part of the equation follows from applying Koopmans' theorem in the case of closed-shell systems. Note that a respective definition of the atomic electronegativity as negative derivative of $E_{el}(n)$ and its relationship to Mulliken's definition of electronegativity (see below) has been developed earlier (Iczkowski and Margrave, 1961).

Equation 6.44 and Equation 6.45 also provide a formal basis for the principle of electronegativity equalization that was introduced 50 years ago on different grounds (Sanderson, 1951; 1988), and has formed a basis for empirical schemes to calculate net atomic charges in molecules (Bultinck et al., 2002a; 2002b; 2003; Gasteiger and Marsili, 1980; Mortier et al., 1985; Rappé and Goddard, 1991). This principle relates the electronegativity of an isolated atom to the partial charge of the atom in a molecule. It states that in a molecule, atoms that are initially more electronegative have attracted additional negative charge and correspondingly lowered their actual electronegativity, and that initially less electronegative atoms have acquired a partial positive charge and correspondingly increased their actual electronegativity. These individual adjustments stop when the final electronegativity has equalized throughout the molecule.

Considering the analogy of $\mu$ (Equation 6.44) with the thermodynamic chemical potential, electrons will move, within a molecule, from regions of high potential to regions of low potential, until all regions have achieved the same chemical potential. A corresponding equalization also holds true for EN as the negative of the electronic chemical potential. Interestingly, the original Mulliken definition of electronegativity as the average of the valence-state ionization potential and electron affinity (Mulliken 1934) is (almost) the same as that stated in Equation 6.45, while according to Pauling's definition, the electronegativity is a property of atoms, not of molecules (see also Perks and Liebman [2000] for a recent discussion of Pauling's relationship between atomic electronegativities and bond energies).

Since the function $E_{el}(n)$ is non-linear, $\mu$ and consequently also EN vary, for a given molecular system, with the number of electrons. Evaluation of this second derivative of $E_{el}(n)$ as its curvature then gave:

$$HD = \frac{1}{2}\frac{\partial^2 E_{el}}{\partial n^2} = \frac{1}{2}\frac{\partial \mu}{\partial n} \approx \frac{1}{2}(IP - EA) = -\frac{1}{2}\left(E_{HOMO} - E_{LUMO}\right) \qquad (6.46)$$

where again Koopmans' theorem (assuming the case of closed-shell molecules) was applied for the right-most part of the equation. Equation 6.45 and Equation 6.46 thus show that in molecular orbital theory, EN and HD can be calculated approximately from the frontier orbital energies. EN increases with decreasing $E_{HOMO}$ and $E_{LUMO}$, and HD increases with an increasing $E_{HOMO}$-$E_{LUMO}$ gap. The electronic structure of so-called soft molecules (that have a low HD value) can be changed more easily, and soft molecules should be generally more reactive than hard molecules. The latter shows that the $E_{HOMO}$-$E_{LUMO}$ gap can be used as a rough measure of chemical reactivity, besides its (again rough) proportionality to electronic excitation energies. For more thorough discussions of these and further relationships as well as for an extension of EN and HD to local properties, the reader is referred to the literature (Ayers and Parr, 2000; Bergmann and Hinze, 1996; Chandrakumar and Pal, 2002a; 2002b; 2002c; Chattarai et al., 2003; Chermette, 1999; Cohen, 1996; Geerlings et al., 2003; Pal and Chandrakumar, 2000; Thanikaivelan et al., 2002).

## C. Molecular Dipole Moment

A somewhat different global molecular property is the dipole moment ($\mu$), which reflects the molecular charge distribution in the ground state of the compound. The dipole moment characterizes the average charge separation in a molecular system, and thus generally increases with both increasing heterogeneity of the charge distribution and increasing molecular size. As a physically observable quantity, the dipole moment is calculated as expectation value of its quantum mechanical operator (employing atomic units):

$$\mu^{op} = -\sum_i r_i + \sum_A Z_A R_A \qquad (6.47)$$

neglecting the vector notation that would usually be required. In the electronic ground state of the molecule as described by the wavefunction $\Psi_{el} = \Phi_0$ (cf. Equation 6.16 and Equation 6.17), the dipole moment is calculated according to:

$$\mu = \left\langle \Phi_0 \left| -\sum_i r_i + \sum_A Z_A R_A \right| \Phi_0 \right\rangle = -2\sum_i^{occ} \left\langle \phi_i |r| \phi_i \right\rangle + \sum_A Z_A R_A \qquad (6.48)$$

where $\phi_i$ denotes the doubly occupied MOs, r is the coordinate of the associated (delocalized) electrons, and $R_A$ is the coordinate of nucleus A with core charge $Z_A$.

Because of convention, the symbols for the chemical potential, used in Equation 6.44 and Equation 6.45, and the dipole moment are the same. Further evaluation of Equation 6.48 proceeds through introduction of the LCAO-MO expansion (Equation 6.18) and, dependent on the level of theory, consideration of relevant approximations such as the NDDO formalism (Equation 6.31) in the case of semiempirical MNDO-type methods. Because the calculation of the dipole moment is usually considered a somewhat demanding test of the quality of the wavefunctions employed in the quantum chemical model, this property is included in the comparative statistical analysis of various methods to calculate molecular descriptors as presented in Section V.

According to classic electrostatics, a point charge (pc) representation of the dipole moment is given by the vector sum of net atomic charges, $Q_A$, located at the nuclei positions $R_A$:

$$\mu^{pc} = \sum_A Q_A R_A \tag{6.49}$$

where QA can be obtained from some population analysis scheme (see below), and is defined as the difference between the core charge at atom A, and the associated and generally fractional number of electrons, QA = ZA – nA (see Equation 6.57).

Dependent on the type and quality of QA, the point charge model can be used as a more or less crude approximation of the full quantum chemical calculation of $\mu$ according to Equation 6.48. As outlined below, semiempirical mapping procedures have been developed to derive QA such that Equation 6.49 yields a best fit (within the model parameters employed) to exact (experimental) dipole moments.

## D. Local Intermolecular Interaction

When two molecules approach each other, the initial interaction is dominated by (long-range) Coulomb forces. With decreasing intermolecular distance, the mutual polarization of the electronic structures of the compounds will increase, and at a reactive distance between the atomic centers A and B of the two molecules a partial delocalization of electronic charge between the two reaction sites takes place. Following a perturbational MO treatment as introduced by Klopman (1968), the energy change upon interaction between the atomic site A of a donor molecule (the nucleophile) and the atomic site B of an acceptor molecule (the electrophile), separated by the distance $R_{AB}$, can be written as:

$$\Delta E_{AB} = -\frac{Q_A Q_B}{R_{AB}} + 2 \sum_i^{occ} \sum_k^{vac} \sum_{\mu(A)} \sum_{v(B)} \frac{c_{\mu i}^2 c_{vk}^2}{\varepsilon_i - \varepsilon_k} \beta_{\mu v} \tag{6.50}$$

when neglecting solvation effects as well as the interaction-driven perturbation of the MO energies. In Equation 6.50 the first term represents the Coulomb interaction between the reaction sites through their net atomic charges $Q_A$ and $Q_B$, and the second term quantifies the orbital interaction resulting from a partial overlap of the wavefunctions at A and B. Here, $c_{\mu i}$ and $c_{vk}$ denote the LCAO-MO coefficients of the occupied MOs $\phi_i$ and vacant (in the ground-state configuration unoccupied) MOs $\phi_k$ (cf. Equation 6.18), $\varepsilon_i$ and $\varepsilon_k$ the respective MO energies, and $\beta_{\mu v}$ denotes the resonance integrals between the AOs $\chi_\mu$ of site A and $\chi_v$ of site B (cf. Equation 6.33).

According to Equation 6.50, chemical reactions may proceed in a charge-controlled manner if the Coulomb term is significantly greater than the orbital term. This is favored by both a large energy gap between the occupied donor MOs and the vacant acceptor MOs and by large opposed net atomic charges, and leads to the preference of hard-hard interactions. Alternatively, large amplitudes of the MOs at the reaction sites A and B and a small gap between the relevant MO energies, as well as small net atomic charges, favor orbital control of the reaction; this is the case for soft-soft interactions between nucleophiles and electrophiles.

More recently, the interaction energy between different molecular systems has been expressed in terms of the chemical potential and molecular hardness (cf. Section IV.B) based on the local hard-soft acid-base principle (Chandrakumar and Pal, 2003), and respective applications include the calculation of acidities and bond dissociation energies of 4-substituted phenols (Romero and Méndez, 2003a, 2003b).

## E. Net Atomic Charge $Q_A$

When considering the intrinsic disposition of molecules for charge-controlled (hard-hard) or orbital-controlled (soft-soft) interactions, the Coulomb term of Equation 6.50 reduces to net atomic charges $Q_A$, and the orbital term is replaced by a single-molecule term reflecting its donor or acceptor strength as outlined below. It follows that net atomic charges offer a simple way to model site-specific Coulomb interactions and the charge-controlled reactivity of molecules in particular. Moreover, atomic charges have also been used to capture the electrostatic component of the hydrogen bonding capability of molecules. Here, the maximum negative atomic charge found in the molecule was taken as a parameter for the electrostatic contribution to hydrogen bond basicity, and the maximum positive charge of hydrogen atoms as a parameter for hydrogen bond acidity (Famini et al., 1992; Cramer et al., 1993).

In the widespread Molecular Orbital PACkage (MOPAC) implementation (Stewart, 1990a; 2002) of the NDDO-type models MNDO, AM1, PM3, and PM5, the standard result for the net atomic charge of an atom in a molecule is based on the Coulson density matrix (Coulson and Longuett-Higgins [1947], Chirgwin and Coulson [1950]; note, however, that the net atomic charge is not physically observable, but a concept according to chemical intuition, and that there is no theoretically rigorous partitioning of the molecular electron density into atomic contributions). For a molecular orbital $\phi_i$, the normalization condition applied to the LCAO-MO expansion (Equation 6.18) reads:

$$\langle \phi_i | \phi_i \rangle = \sum_{\mu} c_{\mu i}^2 + \sum_{\mu \neq v} c_{\mu i} c_{vi} S_{\mu v} = 1 \qquad (6.51)$$

where $S_{\mu v}$ denotes the overlap integral between atomic orbitals $\chi_\mu$ and $\chi_v$ (cf. Equation 6.24).

In the NDDO approximation, the second sum in Equation 6.51 vanishes (zero diatomic overlap between AOs from different atomic centers and orthogonality of different AOs at the same site), and thus the total number of electrons in a closed-shell system with n/2 doubly occupied MOs is given by:

$$n = 2 \sum_{i}^{occ} \sum_{\mu} c_{\mu i}^2 = \sum_{\mu} P_{\mu\mu} = \sum_{A} \sum_{\mu(A)} P_{\mu\mu} \qquad (6.52)$$

where $P_{\mu\mu}$ is a diagonal element of the density matrix (cf. Equation 6.27). In the last part of Equation 6.52, the summation over AOs is grouped according to the respective atom centers A.

For an atom A with a core charge $Z_A$ (which is the charge obtained at atom A when all its valence electrons are removed; e.g., $Z_A = 1$ for hydrogen, $Z_A = 4$ for carbon, and $Z_A = 5$ for nitrogen), the net atomic charge can thus be calculated as the difference between $Z_A$ and the actual amount of electronic charge at A as provided by the associated AOs according to the LCAO-MO expansion:

$$Q_A = Z_A - \sum_{\mu(A)} P_{\mu\mu} = Z_A - 2 \sum_{i}^{occ} \sum_{\mu(A)} c_{\mu i}^2 \qquad (6.53)$$

Because Equation 6.53, which is sometimes called the Coulson scheme for net atomic charges, is based on the NDDO approximation, it cannot be applied properly for *ab initio* methods. Here, a standard scheme to quantify net atomic charges is Mulliken population analysis (Mulliken, 1955; 1962), despite some well-known deficiences such as its strong dependence on the basis set and its apparent lack of convergence with increasing basis set size. Nonetheless, Mulliken charges may also be calculated with the MOPAC implementation of the MNDO-type methods, in which case the AOs are deorthogonalized using the Löwdin transformation (Fischer-Hjalmars, 1965; Löwdin, 1950, 1970; see also Zerner [1991], Thiel [2000], and Weber and Thiel [2000] for a discussion of orthogonalization corrections in the NDDO formalism as mentioned in Section 6.III).

When the NDDO approximation is not applied, Equation 6.51 remains as it stands. The total number of electrons, n, is then obtained as:

$$n = 2 \sum_{i}^{occ} \langle \phi_i | \phi_i \rangle = 2 \sum_{i}^{occ} \left( \sum_{\mu} c_{\mu i}^2 + \sum_{\mu \neq v} c_{\mu i} c_{v i} S_{\mu v} \right) \tag{6.54}$$

In this case, the electron density covered by the off-diagonal terms of the second sum in the parentheses cannot be uniquely attributed to individual atoms, but refer to overlap regions between AOs $\chi_\mu$ and $\chi_v$ of different atomic centers. When assuming that the electronic charge of each overlap region is partitioned into the respective atoms according to the LCAO-MO contributions, the number of electrons of a particular atom A, $n_A$, can be defined as:

$$n_A = 2 \sum_{i}^{occ} \left( \sum_{\mu(A)} c_{\mu i}^2 + \sum_{B \neq A} \sum_{\mu(A), v(B)} c_{\mu i} c_{v i} S_{\mu v}^{AB} \right) \tag{6.55}$$

Equation 6.55 represents the Mulliken population analysis, where the first sum in the parentheses containing only quadratic terms is called net atomic population, the second sum is the overlap population, and $n_A$ is the total sum of all electrons associated with atom A, the gross atomic population (Mulliken, 1955, 1962). Note the difference between Equation 6.54 and Equation 6.55: in the former the summation includes all AOs of the molecule, while in the latter the sum is confined to those AOs that belong to atom A. Obviously:

$$n = \sum_A n_A \tag{6.56}$$

and the Mulliken net atomic charge of atom A can be written in compact form as:

$$Q_A = Z_A - n_A \tag{6.57}$$

(actually, the form of Equation 6.57 holds also true for the Coulson net atomic charge, except that $n_A$ is then defined differently [Equation 6.53]).

Besides these and other LCAO-MO-based population analysis schemes, as well as empirically defined net atomic charges such as those based on the electronegativity equalization principle as mentioned above, there are also methods that allow the decomposition of the molecular electron density into atomic contributions through calculations with a physically observable property. A prominent example is the approach to fit atomic charges to the molecular electrostatic potential (Alemán et al., 1993; Bakowies and Thiel, 1996; Besler et al., 1990; Breneman and Wiberg, 1990; Francl and Chirlian, 2000; Williams, 1991). Because such charges are designed to reproduce the electrostatic potential at intermediate distances, they appear attractive in modeling Coulomb interactions between molecules. It was noted recently that this type of atomic charges is numerically unstable, particularly for atoms further inside the molecular framework (Francl and Chirlian, 2000).

A comparative analysis between different types of charge partition schemes revealed that potential-derived charges had the largest differences from all other sets considered (Wiberg and

Rablen, 1993). This suggests that their information content is indeed somewhat different from conventional atomic charges. With regard to differences in performance between MNDO-type models, comparisons with *ab initio* results, as well as with calculated and experimental dipole moments (which means comparing the point-charge model results according to Equation 6.49 with those of Equation 6.48 or with experimental data), suggest that potential-derived atomic charges calculated with MNDO wavefunctions are significantly superior to the ones of AM1, which, in turn, appears to perform slightly better than PM3 (Alemán et al., 1993; Besler et al., 1990).

More recently, the so-called class IV charges were introduced (Storer et al., 1995; Li et al., 1998; Winget et al., 2002; Thompson et al., 2003). Here, Mulliken or Löwdin charges from an initial quantum chemical calculation are subjected to a mapping in order to reproduce experimental dipole moments through a point-charge representation (equation 6.49) as well as possible. With the respective Charge Model 1 (CM1; Storer et al., 1995) based on Mulliken charges, root-mean-square (rms) deviations of only 0.30 (CM1-AM1) and 0.36 D (CM1-PM3) were achieved for a test set of 195 organic compounds containing C, H, N, O, halogen, S, and Si.

The successor model CM2 (Li et al., 1998) employed a slightly different mapping procedure and was also parameterized for phosphorus. For a training set of 198 compounds, CM2-AM1 and CM2-PM3 yielded rms deviations of 0.25 and 0.23 D, respectively. Inclusion of a secondary set of 13 more complex compounds resulted in rms deviations of 0.30 (CM2-AM1) and 0.23 D (CM2-PM3). CM2 was also parameterized for *ab initio* and density functional theory (DFT) schemes.

More recently, a CM3 model was presented (Winget et al., 2002) that was initially confined to some DFT methods, yielding rms deviations of 0.26 to 0.40 Debye for a training set of 398 compounds containing C, H, N, O, F, Cl, Br, Si, S, and P. In the meantime, a refined CM3 model became available that is now also parameterized for AM1 and PM3 as well as for the popular B3LYP hybrid DFT method (Thompson et al., 2003). With additional parameters for compounds containing N and O as well as an additional PM3-specific parameter for N bonded to C, the overall performance has improved considerably as compared to CM2, with rms errors of 0.39 (AM1), 0.32 (PM3) and 0.26 D (B3LYP/6-31G*) for a set of 382 compounds built from the above-mentioned 10 elements. With regard to PM3 and the DFT methods, CM3 is also parameterized for Li, but so far the halogen I is not included.

## F.  Acceptor and Donor Delocalizabilities, $D^N$ and $D^E$

As mentioned above, practical applications of the Klopman relationship (Equation 6.50) often require further simplifications, particularly in cases where the relevant MO properties are known only for one of the two reaction partners. The latter is the typical situation in QSAR investigations that aim to elucidate the impact of molecular reactivity characteristics on the toxic potential of chemical agents. While the Coulomb interaction term can be reduced to calculate net atomic charges (see above), a possible candidate to replace the orbital interaction term of Equation 6.50 by a one-molecule property is Fukui's delocalizability, as was pointed out over 35 years ago (Cammarata, 1968; Cammarata and Rogers, 1971).

The delocalizability parameter is based on a perturbational MO treatment of a partial electron delocalization between two molecules when they approach each other (Fukui, 1970). More precisely, it characterizes the site-specific energy stabilization due to a fractional electron transfer to or from a reagent. Dependent on the donor or acceptor capability of the molecule of interest, the occupied or unoccupied MO energies and wavefunction contributions at the reaction site are considered.

Initially this concept was developed to explain site-specific reactivities of conjugated compounds, and thus was confined to the π-electron part of their electronic structure. The respective parameter was called superdelocalizability (Fukui et al., 1954, 1957), and subsequently the approach was extended to saturated compounds, leading to the delocalizability as a generalized reactivity index (Fukui et al., 1961; Fukui, 1970). Note that in the environmental toxicology QSAR literature, the term superdelocalizability has often been used for all-valence-electron applications that clearly go beyond the $\pi$ electron approximation, and thus actually refer to the delocalizability parameter.

For a closed-shell molecule, its susceptibility at the atomic site A for an attack by a nucleophile can be characterized by the nucleophilic or acceptor delocalizability:

$$D_A^N = 2 \sum_k^{vac} \sum_{\mu(A)} \frac{c_{\mu k}^2}{\alpha - \varepsilon_k} \qquad (6.58)$$

where the summation includes the local contributions of all vacant MOs according to the LCAO-MO expansion (cf. Equation 6.18), weighted by the respective orbital energies (cf. Equation 6.22). The term $\alpha$ represents the relevant frontier orbital energy of the attacking (unspecified) nucleophile, which could be approximately quantified as $-IP$ (following Koopmans' theorem, cf. Equation 6.39 and Equation 6.42). In practice, $\alpha$ is either skipped or assigned some reasonable but arbitrary value (e.g., 10 eV), avoiding accidentally small denominators through an upward shift of the vacant MO energies. Note that the historical terminology may lead to some confusion, because the term nucleophilic actually refers to the attacking agent, while Equation 6.58 characterizes the site-specific capability of the molecule under investigation to act as an electrophile, or its readiness to react with a nucleophile. Accordingly, the term acceptor delocalizability appears to be preferred.

When an (again unspecified) electrophile attacks the molecule of interest at site A, the energetic response can be expressed through the respective electrophilic or donor delocalizability:

$$D_A^E = 2 \sum_i^{occ} \sum_{\mu(A)} \frac{c_{\mu i}^2}{\varepsilon_i - \alpha} \qquad (6.59)$$

Here, the summation involves the site-specific contributions of all (doubly) occupied MOs weighted by their energies, and $\alpha$ could be taken as $-EA$ of the attacking agent (cf. Koopmans' theorem, Equation 6.43), but is usually skipped when focusing on the intrinsic reactivity of the molecule of interest.

Early QSAR applications of the delocalizability parameters go back to the mid-1960s (Cammarata, 1968; Cammarata and Rogers, 1971; Snyder and Merril, 1965; Yoneda and Nitta, 1964), and there are various examples of their use in toxicological and ecotoxicological studies (Cronin et al., 1988, 2001; Lewis, 1988, 1989; Mekenyan et al., 1993, 1997; Purdy, 1991; Schüürmann, 1990; Veith and Mekenyan, 1993). In particular, the maximum acceptor delocalizability has evolved as a second major parameter, besides $E_{LUMO}$, to model global electrophilicity. Recently, the maximum donor delocalizability turned out as a discriminator (besides other parameters) between antibacterials and non-antibacterials (Aptula et al., 2003). The corresponding parameter confined to aromatic carbon atoms was found to be a useful descriptor in a stepwise classification scheme for toxic modes of action of phenols (Schüürmann et al., 2003a).

Returning to the superdelocalizability as a $\pi$-electron reactivity index, a recent study has shown that this parameter diverges, under specific structural conditions, with molecular size (Hosoya and Iwata, 1999). Note that this finding refers to the single-site values of the parameter, not to the sum over all atom-specific values that is still larger (and is used as global delocalizability parameter, see Section V and Table 6.1), and that a full understanding in terms of more general structural implications is still to be developed. For the time being, the apparent instability of the delocalizability parameters suggests that these descriptors may be used preferably for comparisons of different reaction sites within a given molecule, or as QSAR parameters for series of compounds that are reasonably related to each other in terms of structural features and molecular size.

**Table 6.1   Quantum Chemical Descriptors Used for the Comparative Analysis.**

| Symbol | Description |
|---|---|
| | **Descriptors Based on Molecular Orbital Energies (Section IV.A and B)** |
| $E_{HOMO}$ | Energy of the highest occupied molecular orbital (Equation 6.42) |
| $E_{LUMO}$ | Energy of the lowest unoccupied molecular orbital (Equation 6.43) |
| EN | Molecular electronegativity (Equation 6.45) |
| HD | Molecular hardness (Equation 6.46) |
| | **Descriptors Based on the Charge Distribution (Sections IV.C and E)** |
| $\mu$ | Molecular dipole moment (Equation 6.48) |
| $Q_{Ymax}^{+}$ | Maximum positive net atomic charge of atoms of type Y in the molecule (Y: Z = heavy atom, H, C, N, X = halogen; Equations 6.53, 6.55 and 6.57) |
| $Q_{Yavp}^{+}$ | Average positive net atomic charge of all positively charged atoms of type Y in the molecule (cf. Explanation of $Q_{Ymax}^{+}$) |
| $Q_{Ymax}^{-}$ | Maximum negative net atomic charge of atoms of type Y in the molecule (Y: Z = heavy atom, C, N, O, X = halogen) |
| $Q_{Yavn}^{-}$ | Average negative net atomic charge of all negatively charged atoms of type Y in the molecule (cf. explanation of $Q_{Ymax}^{-}$) |
| | **CPSA Descriptors Based on Molecular Surface Areas and Net Atomic Charges (Section IV.G)** |
| PPSA-1 | Partial positive surface area (Equation 6.66) |
| PPSA-2 | Total charge weighed partial positive surface area (Equations 6.68 and 6.70) |
| PPSA-3 | Atomic charge weighted partial positive surface area (Equation 6.72) |
| PNSA-1 | Partial negative surface area (Equation 6.67) |
| PNSA-2 | Total charge weighted partial negative surface area (Equations 6.69 and 6.71) |
| PNSA-3 | Atomic charge weighted partial negative surface area (Equation 6.73) |
| PPSA-1Y | Partial positive surface area confined to atoms of type Y (Y: Z = heavy atoms, H) |
| PPSA-2Y | Total charge weighted partial positive surface area confined to atoms of type Y |
| PPSA-3Y | Atomic charge weighted partial positive surface area confined to atoms of type Y |
| PNSA-1Y | Partial negative surface area confined to atoms of type Y |
| PNSA-2Y | Total charge weighted partial negative surface area confined to atoms of type Y |
| PNSA-3Y | Atomic charge weighted partial negative surface area confined to atoms of type Y |
| | **Descriptors Based on Molecular Orbital Wave Functions and Energies (Section IV.F)** |
| $D_{max}^{N}$ | Maximum acceptor delocalizability (toward attack from nucleophile) of all atomic sites in a molecule (Equation 6.58) |
| $D_{av}^{N}$ | Average acceptor delocalizability (toward attack from nucleophile) of all atomic sites in a molecule |
| $D_{max}^{E}$ | Maximum donor delocalizability (toward attack from electrophile) of all atomic sites in a molecule (Equation 6.59) |
| $D_{av}^{E}$ | Average donor delocalizability (toward attack from electrophile) of all atomic sites in a molecule |
| $D_{Y-max}^{N}$ | Maximum acceptor delocalizability (toward attack from nucleophile) of all atomic sites of type Y in a molecule (Y: C, H, N) |
| $D_{Y-av}^{N}$ | Average acceptor delocalizability (toward attack from nucleophile) of all atomic sites of type Y in a molecule (Y: C, H, N) |
| $D_{Y-max}^{E}$ | Maximum donor delocalizability (toward attack from electrophile) of all atomic sites of type Y in a molecule (Y: C, N, O) |
| $D_{Y-av}^{E}$ | Average donor delocalizability (toward attack from electrophile) of all atomic sites of type Y in a molecule (Y: C, N, O) |

## G.  Charged Partial Surface Area Descriptors

The Coulomb term of the Klopman relationship (Equation 6.50) is based on a point charge model. It does not consider potential differences in the geometric availability of the spatially extended atomic sites for polar intermolecular interactions (which holds true in particular when employing the site-specific net atomic charge, $Q_A$ [Equation 6.57], as a simple descriptor for the propensity of a molecule to undergo a charge-controlled reaction at site A; see Section IV.E). More precisely, the quantification of the electrostatic potential energy (in atomic units) between two atomic sites A and B through the point charges $Q_A$ and $Q_B$ located at the geometric centers of A and B according to:

$$E_{pot}(A, B) = \frac{Q_A Q_B}{R_{AB}}$$ (6.60)

neglects that the atomic surfaces (taken at some suitable radius, e.g., the van der Waals radius) may be partially screened because of the presence of other atoms in their spatial vicinity, and that their local charge densities may also vary because of the overall electronic structure of the molecule.

For a given (spherical) surface of atom A with surface area $SA_A$, a (still constant but atom-specific) surface charge density, $\sigma_A$, can be defined such that:

$$Q_A = \sigma_A \cdot SA_A$$ (6.61)

When considering a spatially varying charge distribution, the surface charge density depends on the spatial position, and for a particular (infinitesimally small) surface area element at location $r_A$, $dSA_A$, the respective $\sigma_A(r_A)$ yields:

$$dQ_A(r_A) = \sigma_A(r_A) \, dSA_A$$ (6.62)

as local contribution to QA, which, in turn, is then given through integration of $dQ_A(r_A)$ over the whole associated atomic surface:

$$Q_A = \int_A \sigma_A(r_A) \, dSA_A$$ (6.63)

Note that in the expressions considering spatially varying charge distributions, the coordinate representing the geometric center of atom A, $R_A$ (cf. Equation 6.60), is replaced by $r_A$ that covers the relevant atomic surface. Introduction of Equation 6.63 into Equation 6.60 yields the electrostatic potential energy between 2 atoms A and B with continuous surface charge densities:

$$E_{pot}(A, B) = \int_A \int_B \frac{\sigma_A(r_A) \sigma_B(r_B)}{r_{AB}} \, dSA_A \, dSA_B$$ (6.64)

Equation 6.64 refers to the electrostatic interaction of one pair of atoms, A and B, and the overall Coulomb interaction energy between two molecules, X and Y, would include all possible pairs of atoms from the different molecules:

$$E_{pot} = \sum_{A \in X} \sum_{B \in Y} E_{pot}(A, B)$$ (6.65)

By construction, Equation 6.65 is a global property built from local (pair-wise) contributions according to Equation 6.64.

A direct evaluation of Equation 6.64 and Equation 6.65 would require a more elaborate calculation. Moreover, typical QSAR investigations focus on single molecules without specifying the electronic structure of reaction partners explicitly. Accordingly, all terms referring to atom B are omitted, including the (unkown) distance $r_{AB}$, leaving the spatially varying surface charge density of a given atom A and its associated surface area for further consideration.

From this viewpoint, the Charge Partial Surface Area (CPSA) descriptors (Stanton and Jurs, 1990) offer an interesting way to model — in a still simplified manner — the molecular disposition of polar interactions while taking into account the actual geometric availability of the relevant atomic surfaces. As outlined below, the CPSA descriptors, as introduced originally, represent global molecular properties constructed through summations of local properties. However, a local character can be achieved through restrictions of the atoms actually involved in the calculation, for example, when considering only a specific atom type (which holds true correspondingly with pure net atomic charge descriptors). The original idea to introduce electrostatic molecular surface interaction terms was developed to model bulk properties of pure substances, employing net atomic charges from extended Hückel calculations (Grigoras, 1990).

In the CPSA approach, the fundamental molecular properties are the surface area contributions of the positively and negatively charged atoms r and s, $SA_r^+$ and $SA_s^-$, and the associated net atomic charges, $Q_r^+$ and $Q_s^-$. The simple partial positive surface area (PPSA) is defined as sum of all positively charged surface areas and is called PPSA-1, and the respective partial negative surface area (PNSA) as sum of all negatively charged surface areas is called PNSA-1:

$$PPSA-1 = \sum_r SA_r^+ \tag{6.66}$$

$$PNSA-1 = \sum_s SA_s^- \tag{6.67}$$

As such, PPSA-1 and PNSA-1 quantify the actually available surface area portion provided by all positively or all negatively charged atoms. This may be useful to characterize the general molecular disposition for respective polar interactions without addressing the extent and variation of the local charges explicitly. With respect to Equation 6.64 and Equation 6.65, PPSA-1 and PNSA-1 neglect the surface charge density, except that its atom-specific sign (positive or negative, assumed to be constant for the whole atomic surface according to the point charge calculated for the atomic center) is taken into account for partitioning the total surface area into the positively and negatively charged portions as outlined above.

Consideration of the degree of overall charge separation in the molecule in terms of the sums of all positive and negative net atomic charges:

$$Q^+ = \sum_r Q_r^+ \tag{6.68}$$

$$Q^- = \sum_s Q_s^- \tag{6.69}$$

leads to the total charge weighted PPSA and PNSA:

$$PPSA-2 = Q^+ \cdot \sum_r SA_r^+ \tag{6.70}$$

$$PNSA - 2 = Q^- \cdot \sum_s SA_s^- \tag{6.71}$$

The respective atomic charge weighted PPSA and PNSA are defined as:

$$PPSA - 3 = \sum_r Q_r^+ \cdot SA_r^+ \tag{6.72}$$

$$PNSA - 3 = \sum_s Q_s^- \cdot SA_s^- \tag{6.73}$$

In Equation 6.70 and Equation 6.71, the multiplication of the relevant surface area portion with the respective total atomic charge (Equation 6.68 and Equation 6.69) emphasizes the global character of PPSA-2 and PNSA-2. It also makes these parameters less sensitive to changes of molecular structure; an increase in the molecular extension may be numerically compensated by an overall decrease in the net atomic charges, and vice versa. Although a corresponding compensation may, in principle, also occur for PPSA-3 and PNSA-3, these latter descriptors are likely to respond more distinctly to structural changes, because here each surface area contribution is weighted by its locally associated net atomic charge.

Among further CPSA descriptors are the differences between PPSA and PNSA, and fractional values as ratios of PPSA or PNSA and the total molecular surface area. For the original list of 25 CPSA descriptors as well as for recent extensions, the reader is referred to the literature (Stanton and Jurs, 1990; Aptula et al., 2003; Mattioni et al., 2003; Mosier et al., 2003). It should be further noted that, in contrast to the initially introduced electrostatic molecular surface interaction terms (Grigoras, 1990), the CPSA descriptors were actually defined using solvent accessible surface areas instead of simple van der Waals surface areas (Stanton and Jurs, 1990), a fact that is ignored in the present discussion for the sake of simplicity.

When compared with the terms involved in the actual Coulomb interaction according to Equation 6.64 and Equation 6.65, PPSA-3 and PNSA-3 take into account both the local net atomic charges and the geometric availability of the respective atomic centers for Coulomb interactions. However, they ignore (like all descriptors based on net atomic charges) the local spatial variation of the charge distribution on the geometric surfaces associated with specific atoms. Moreover, both the total charge weighted and the atomic charge weighted PPSA and PNSA depend on the type of calculation of the net atomic charges (cf. Section IV.E), and also within semiempirical NDDO schemes different population analyzes (e.g., Coulson analysis [Equation 6.53] vs. Mulliken analysis [Equation 6.55 and Equation 6.57]) are likely to yield different numerical values for the charge-based CPSA descriptors. The simple, total charge weighted and atomic charge weighted PPSA and PNSA descriptors encode different aspects of the molecular disposition for electrostatic interactions with electron-rich (nucleophilic) or electron-poor (electrophilic) sites, and the PNSA parameters may also reflect the (polar component of the) molecular hydrogen bond acceptor ability. Correspondingly, PPSA descriptors confined to hydrogen atoms that are attached to suitable heavy atoms may serve as parameters for the hydrogen bond donor capability of the compounds.

Recent applications of CPSA descriptors include QSAR investigations of the genotoxicity of thiophene derivatives (Mosier et al., 2003) as well as of secondary and aromatic amines (Mattioni et al., 2003) and a study to classify phenols with respect to toxic modes of action (Aptula et al., 2003). A somewhat different route has been explored with the concept of dynamic molecular surface areas (Lipkowitz et al., 1989) that represent Boltzmann-weighted means of surface areas of different conformations within a preset energy window (e.g., within an excess of 10.5 kJ/mol above the lowest energy found for the particular molecule). Following this strategy, so-called dynamic polar and non-polar surface areas have been defined and used to model absorption properties of drugs (Palm et al., 1996, Stenberg et al., 1999, Bergström et al., 2003). Here, the polar surface area is defined as the area occupied by nitrogen and oxygen plus any attached hydrogens, and as such would be related to PPSA-1 and PNSA-1 if the latter are evaluated separately for the respective atom types.

## V. COMPARATIVE ANALYSIS OF DESCRIPTOR VALUES FROM QUANTUM CHEMICAL METHODS

Quantum chemical descriptors, such as those outlined in the previous section, provide a valuable means to characterize the intrinsic potential of molecules for metabolic transformations as well as for electrostatic and covalent interactions with endogenous target sites. The numerical descriptor values depend on the level of theory used for their calculation (e.g., without or with explicit consideration of electron correlation), and also within the NDDO formalism of semiempirical methods on the specific parameterization employed (e.g., MNDO vs. AM1). Note further that the MNDO-type methods have been parameterized to reproduce, as well as possible, four gas-phase properties of organic molecules: heat of formation (cf. Section III, Equation 6.34), dipole moment (cf. Section IV.C, Equation 6.48), ionization potential (cf. Section IV.A, Equation 6.42), and molecular geometry (bond lengths, bond angles, and dihedral angles). As a consequence, the performance of these methods for other molecular properties as well as for descriptors related to the charge distribution and site-specific molecular reactivity is not clear *a priori*.

To elucidate the dependence of popular quantum chemical descriptors on the level of computation, descriptor values were calculated for 607 compounds for which fathead minnow toxicity data were available (Russom et al., 1997). The methods employed included three semiempirical methods AM1 (Dewar et al. 1985), PM3 (Stewart, 1989a; 1989b), and PM5 (Stewart 2002), as implemented in MOPAC (MOPAC93, 1994; Stewart, 2002). In addition, two *ab initio* methods, HF/6-31G** and B3LYP/6-31G**, as provided by the Gaussian98 package (1999). All calculations included geometry optimization, employing CORINA (CORINA 2001; Gasteiger et al., 1990; Sadowski and Gasteiger, 1993) to generate initial 3D structures from Simplified Molecular Line Entry System (SMILES) codes, followed by SYBYL force field geometry optimizations (SYBYL, 2001) before starting the quantum chemical calculations. Because no conformational search routines were employed it may well be that, in individual cases, the finally achieved geometry does not represent the energy minimum of the conformational space according to the Hamiltonian employed. The geometries can be considered as optimized on a screening level that is considered sufficient for the purpose of the present comparative analysis.

In the above-given specification of the *ab initio* methods, HF denotes the Hartree-Fock SCF scheme (cf. Section II, Equation 6.22 to Equation 6.28) without electron correlation, and 6-31G** is a shortcut notation of a polarized double-zeta split valence basis set (Hariharan and Pople, 1973; Hehre et al., 1986). B3LYP (Becke, 1993; Lee et al., 1988) is a hybrid functional of the density functional theory (DFT) approach, which has become a standard method to include the dynamic electron correlation. B3LYP has proven superior to several other DFT functionals, and is often considered competitive to high-level *ab initio* schemes with regard to the computation of molecular geometries, energies (Bauschlicher, 1995), and ionization potentials (Curtiss et al., 1998). This hybrid functional was also shown to yield good electron affinities (Curtiss et al., 1998; de Proft and Geerlings, 1997) and dipole moments (de Proft et al., 2000), and it provides net atomic charges in agreement with high-level *ab initio* results (de Proft et al., 1996). However, it should be considered that the available tests of the prediction performance of B3LYP and other DFT models refer to smaller sets of atoms and rather simple organic compounds. Moreover, the choice of the basis set may be critical for some properties and systems (de Proft et al., 2000; Scheiner et al., 1997; Wang

**Table 6.2    Structural Characteristics of the Set of Compounds[a]**

| Structural Feature | Number of Compounds |
|---|---|
| C, H (hydrocarbons) | 21 |
| + Halogen | 39 |
| + N | 107 |
| + O, but without N | 265 |
| + N, O together | 133 |
| + S, P | 42 |
| Non-aromatic | 276 |
| Aromatic | 331 |
| Total | 607 |

[a] Two tin compounds of the fathead minnow data set (Russom et al., 1997) were omitted because of missing AM1 parameters.

and Wilson, 2003), which appear to be particularly important for compounds with higher-row elements.

According to recent investigations, the so-called Kohn-Sham (KS) orbitals employed in typical DFT calculations can be interpreted in a similar way like conventional HF orbitals, and are even considered to be superior to HF orbitals because of their connection with the exact molecular density (Baerends, 2000; Baerends and Gritsenko, 1997; Bickelhaupt and Baerends, 2000; Stowasser and Hoffmann, 1999). Because of the dependence of common DFT functionals on (very) few parameters fit to reproduce experimental data (e.g., thermochemical data of the G1 data set [Curtiss et al., 1990; Pople et al., 1989]), DFT is, strictly speaking, not a pure *ab initio* scheme; nonetheless, the integrals of all electron-electron and electron-nucleus interactions are calculated on a true *ab initio* level. For convenience, this formal difference between DFT and pure *ab initio* methods is ignored in the context of the present discussion. Instead, B3LYP is considered essentially competitive to the Møller-Plesset perturbation theory level 2 (MP2) (Raghavachari and Anderson, 1996; Pople 1999) level of *ab initio* theory particularly with regard to its inclusion of electron correlation; as such it is clearly superior to the HF model that neglects the instantaneous correlation of the motion of electrons with antiparallel spin. As such, B3LYP/6-31G** represents the reference level of calculation for the present comparative analysis. For more information regarding DFT as an alternative method to compute the geometric and electronic structure of molecules, the reader is referred to a recent textbook (Koch and Holthausen, 2001) and reviews (Baerends, 2000; Baerends and Gritsenko, 1997; Bickelhaupt and Baerends, 2000; Geerlings et al., 2003; Kohn et al., 1996; Parr and Yang, 1995) as well as to the Nobel Prize lecture of Kohn (1999).

The descriptors under comparative analysis are listed in Table 6.1, and the data set of 607 organic compounds is characterized in terms of major structural features in Table 6.2. The compound set includes the atom types C, H, N, O, P, S, F, Cl, Br, and I, and the molecular weight varies from 32 (methanol) to 488.6 Da (pentabromophenol). The smallest number of atoms per molecule is five (chloroform, dichloromethane, iodoform and carbon tetrachloride), and *tris*-(2-butoxyethyl)-phosphate is the compound with the largest number of atoms (65). The respective structures are shown in Figure 6.1, which also contains the chemical structures of all compounds mentioned explicitly in the subsequent discussion.

## A.  Molecular Geometries

A statistical analysis of the bond lengths between heavy atoms as well as between heavy atoms and hydrogen atoms is summarized in Table 6.3. With AM1, the largest bond length of 2.1520 Å is found for the single bond RS–P(=O)(OR)$_2$ in the malathion molecule; here, the other methods

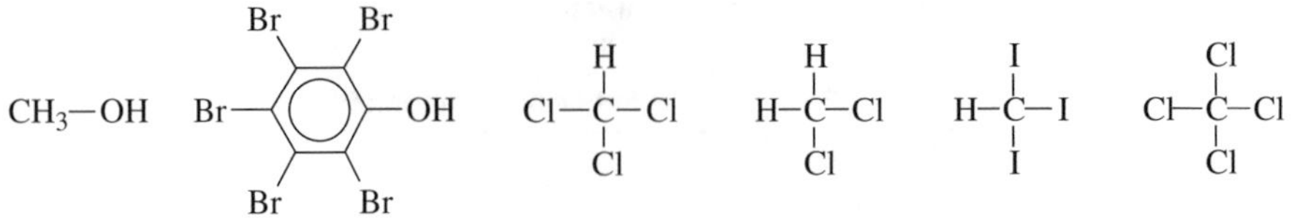

Methanol    pentabromophenol    chloroform  dichloromethane  iodoform    carbon
                                                                        tetrachloride

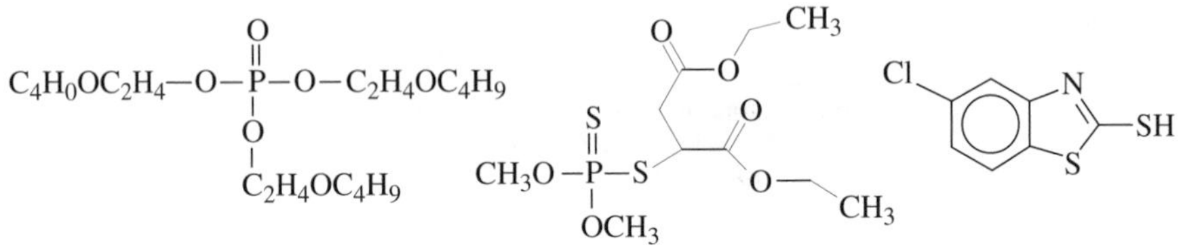

tris-(2-butoxyethyl)-phosphate              malathion              5-chloro-2-
                                                            mercaptobenzothiazole

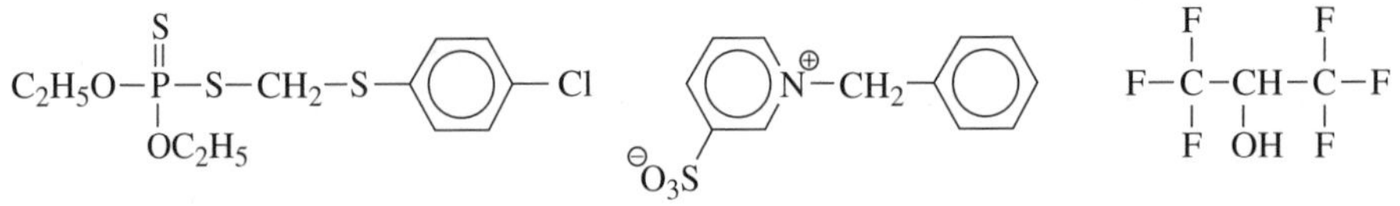

Carbophenothion                    1-benzylpyridinium-        1,1,1,3,3,3-hexafluoro-
                                      3-sulphonate                  2-propanol

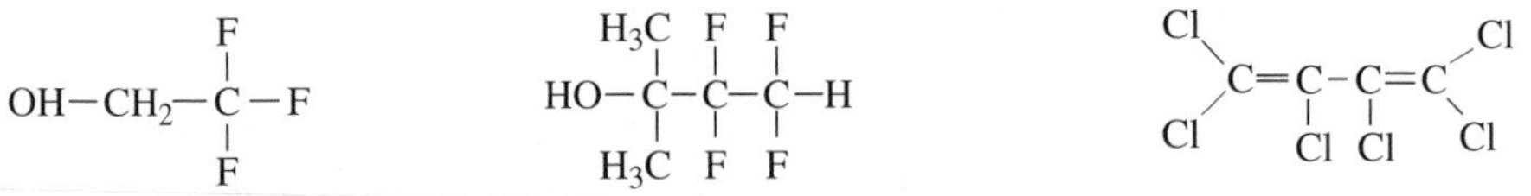

2,2,2-trifluoroethanol      2-methyl-3,3,4,4-tetrafluoro-    hexachloro-1,3-butadiene
                                    2-butanol

**Figure 6.1**  Chemical structures of compounds discussed in Section V (continued on next page).

provide values of 2.1566 (PM3), 2.1277 (PM5), 2.0927 (HF/6-31G**), and 2.1275 Å (B3LYP/6-31G**). According to B3LYP, the overall largest bond length of 2.1739 Å is found for the single bond I–C(H)I$_2$ in iodoform, where the other methods yield values of 2.0386 (AM1), 1.9612 (PM3), 2.1041 (PM5), and 2.1563 Å (HF). With regard to the overall trend of the bond distances between heavy atoms, the upper quartiles in comparison with the means and medians indicate that B3LYP has more cases with larger bond lengths than the other methods. The largest bond distance of a hydrogen atom is found with all five methods for the RS–H bond in 5-chloro-2-mercaptobenzothiazole (see Figure 6.1), which is the only thiol compound of the data set. Overall, the present analysis demonstrates quite substantial variations in individual bond lengths across the five methods, which need not affect correspondingly global molecular properties such as frontier orbital energies and related reactivity indices.

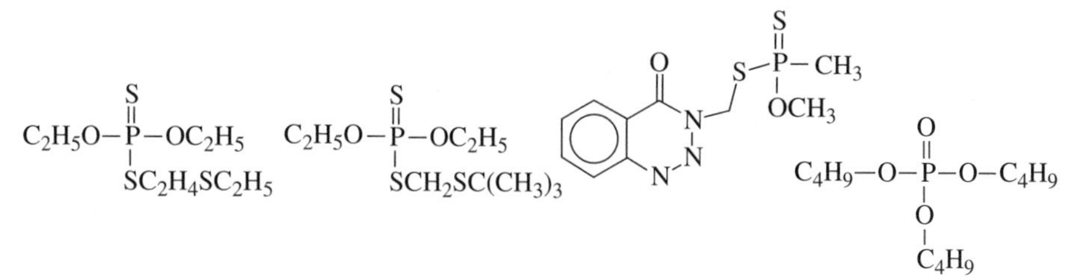

disulfoton            terbufos            azinphos-methyl        tributyl phosphate

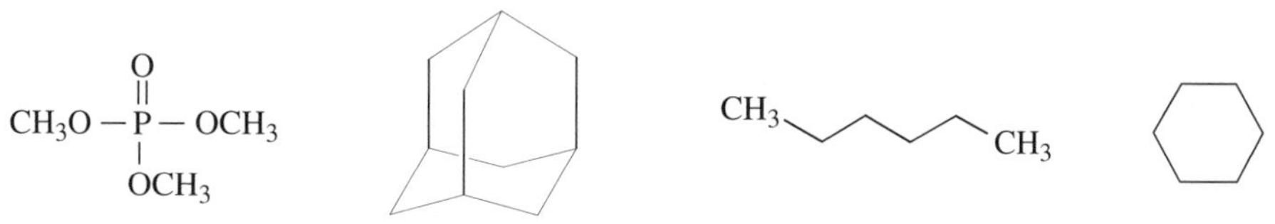

trimethyl phosphate      adamantane                hexane              cyclohexane

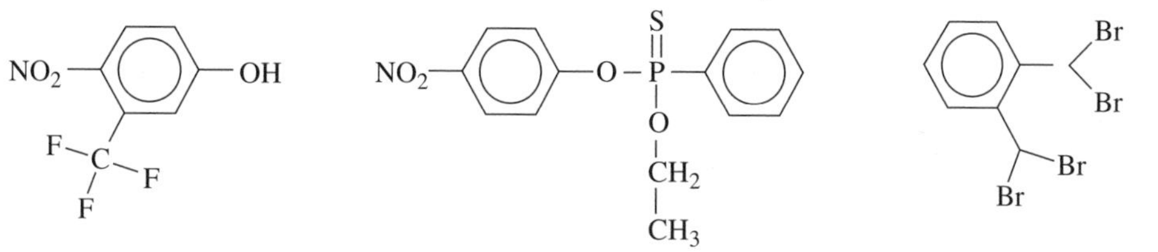

4-nitro-3(trifluoromethyl)-        O-ethyl-O-(p-nitrophenyl)        α,α,α′,α′-tetrabromo-
phenol                  phenylphosphonothioate           o-xylene

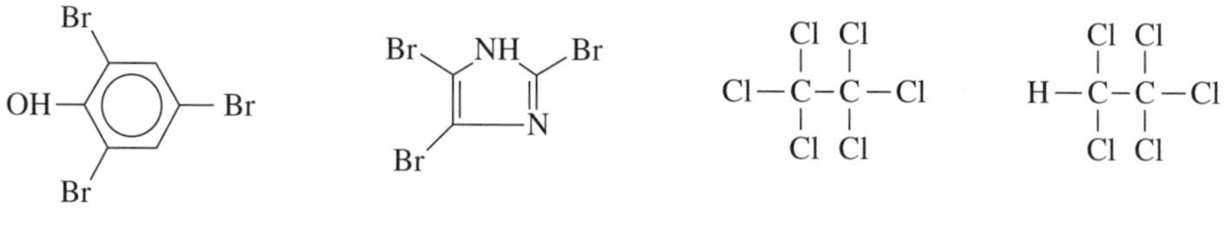

2,3,6-tribromophenol    2,4,5-tribromoimidazole    hexachloroethane    pentachloroethane

**Figure 6.1**   (continued).

## B.  Descriptors Based on Frontier Orbital Energies and the Molecular Charge Distribution

In Table 6.4, methods are compared in terms of $R^2$ values for the frontier orbital energies; molecular EN; HD and dipole moment; and for the maximum, as well as average, net atomic charges confined to heavy atoms (Z, which means all atoms except hydrogen), to carbon, hydrogen, nitrogen, oxygen, and halogen. Taking $E_{HOMO}$ as an example, AM1 correlates significantly more with PM5 ($R^2 = 0.940$) than with PM3 (0.896), and, as compared to PM3 and PM5, yields the

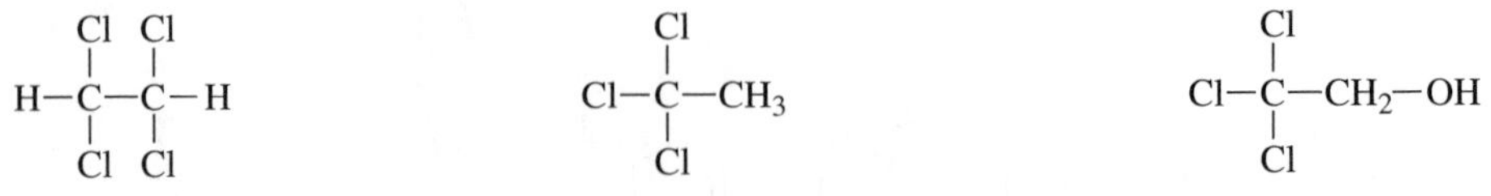

1,1,2,2-tetrachloroethane          1,1,1-trichloroethane                2,2,2-trichloroethanol

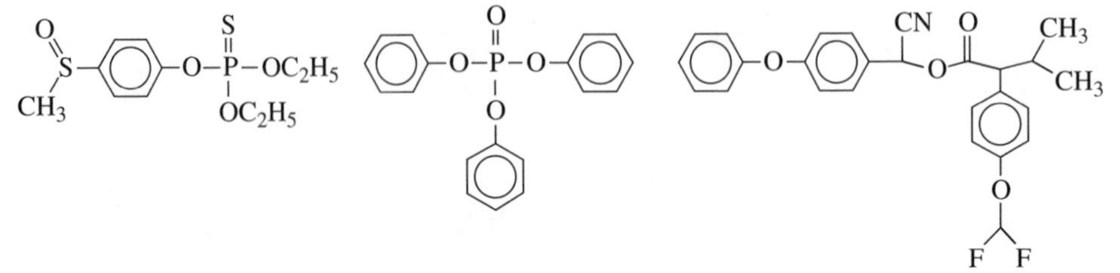

fensulfothion                triphenyl phosphate                  flucythrinate

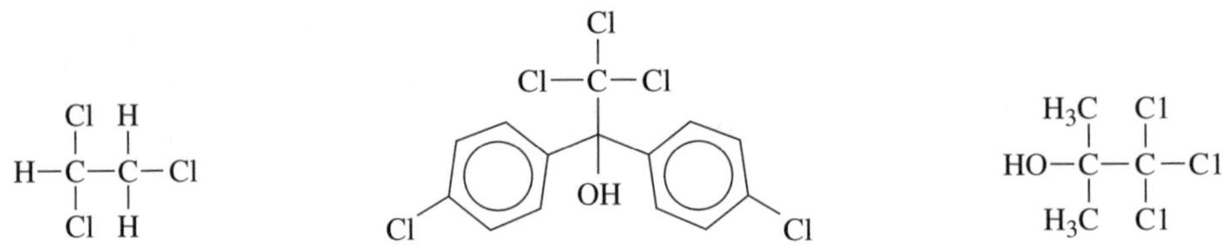

1,1,2-trichloroethane                  dicofol                  1,1,1-trichloro-2-methyl-
                                                                  2-propanol

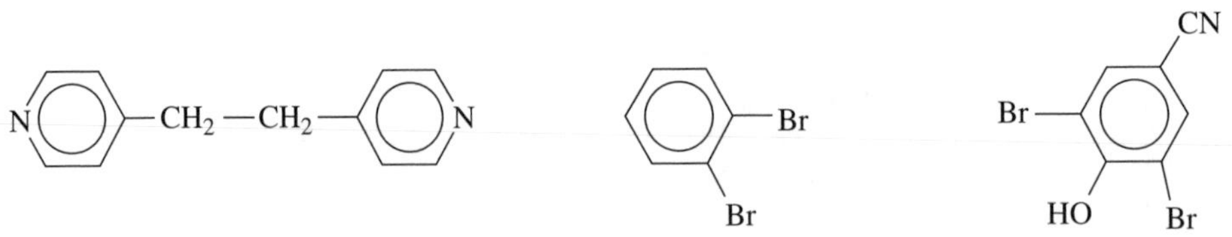

1,2-bis(4-pyridyl)ethane        1,2-dibromobenzene              3,5-dibromo-4-
                                                                hydroxybenzonitrile

**Figure 6.1** (continued).

largest correlations with the *ab initio* methods HF/6-31G** (0.882) and B3LYP/6-31G** (0.830). The latter holds also true for $E_{LUMO}$, EN, HD and the dipole moment, and suggests that AM1 appears to be slightly preferred over the other MNDO-type methods with regard to these molecular descriptors.

With respect to molecular dipole moment, a previous comparative analysis of the experimental data of 125 (mostly quite simple) compounds yielded average absolute errors of 0.38 (PM3), 0.35 (AM1), and 0.45 D (MNDO) (Stewart, 1989b). For a compound set of 256 (again generally more simple) compounds with experimental first ionization potentials (as one of the target properties of

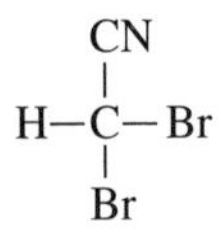

| dibromoacetonitrile | 2,4,6-trichlorophenol | 3,6-dithiaoctane |

| 2,4,6-triiodophenol | saccharin | tripropargylamine |

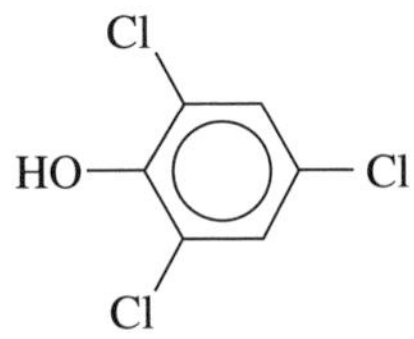

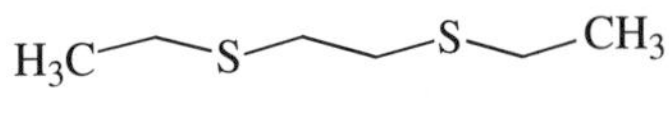

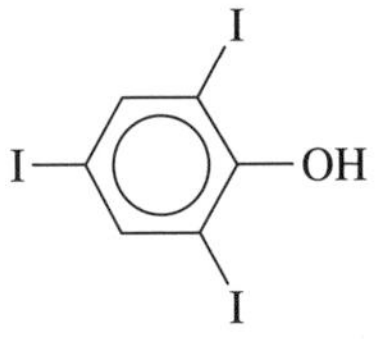

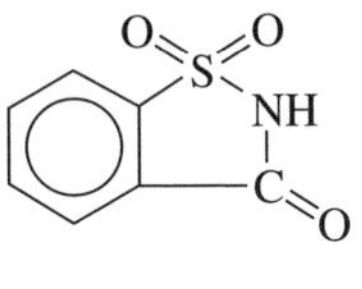

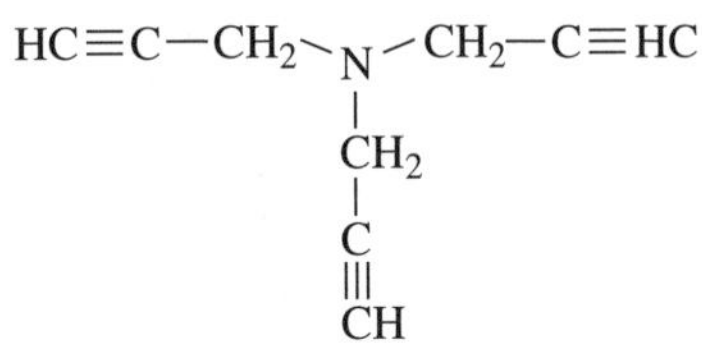

**Figure 6.1**   (continued).

the semiempirical parameterization procedure, see above), the average errors of PM3, AM1, and MNDO were 0.57, 0.61, and 0.78 eV, respectively, suggesting a slightly improved performance of PM3 as compared to AM1 with respect to this molecular property (Stewart, 1989b). Accordingly, a more definite evaluation would require a detailed analysis, considering experimental data and possibly also explicit high-level calculations of energy differences between cations and respective neutral molecules, which goes beyond the scope of the present study.

With respect to absolute values of the frontier orbital energies, AM1, PM3 and PM5 yield similar values for the means and ranges of $E_{HOMO}$ (–9.820, –9.815, and –9.686 eV; 4.607, 4.796, and 4.604 eV) and $E_{LUMO}$ (0.4640, 0.3148, and –0.0418 eV; 6.413, 6.121, and 6.388 eV), except that PM5 provides systematically lower $E_{LUMO}$ values. The HF/6-31G** $E_{HOMO}$ values have a similar mean, but a significantly greater range (–9.635 eV, 6.705 eV), and the corresponding B3LYP/6-31G** results are apparently scaled somewhat differently (–6.547 eV, 4.852 eV), as was noted earlier (Stowasser and Hoffmann, 1999). For the purpose of QSAR investigations, the parallel use of frontier orbital energies and related descriptors from different quantum chemical schemes is not warranted because of the demonstrated differences in scaling as well as in relative trends.

For $E_{HOMO}$ and $E_{LUMO}$, the data distributions of all pair-wise comparisons with B3LYP are shown in Figure 6.2 and Figure 6.3. With AM1 (top left plot of Figure 6.2), the greatest over-estimation of $E_{HOMO}$ as compared to the B3LYP trend is found for carbophenothion (cf. Figure 6.1), an organophosphorus compound with the central phosphorus bound to sulfur, $(RO)_2P(=S)–S–R$ (a phosphorodithioate; cf. Schüürmann, 1992), and the greatest respective underestimation occurs for 1-benzylpyridinium-3-sulphonate (Figure 6.1), a structure with a formal charge separation ($N^+–R–SO_3^-$). Interestingly, the calculated net atomic charges do not correspond at all to the formal charges of the pyridinium nitrogen, and differ substantially between AM1 ($Q_N = -0.060$ au), PM3 (0.439 au), PM5 (–0.004 au), HF (–0.619 au) and B3LYP (–0.397 au), the latter of which holds also true for the sulfonate sulfur (AM1: $Q_S = 2.884$ au; PM3: 2.402 au, PM5: 2.679 au, HF: 1.612 au, B3LYP: 1.167 au) and oxygen (AM1: $Q_O = -1.056$ au, PM3: –0.928 au, PM5: –0.989 au, HF: –0.733 au, B3LYP: –0.583 au, taking the maximum value of the three oxygen atoms). These findings are in accord with the general experience that MNDO-type methods employing a minimum $sp$ basis set are less reliable for hypervalent compounds (cf. Section 6.3; Thiel 1996).

Table 6.3   Bond Distances (Å) in the Set of 607 Compounds as Calculated with AM1, PM3, PM5, HF/6-31G** and B3LYP/6-31G**.

| Method | No. Bond Distances | Mean | Median | Minimum | Maxiumum | Range | Lower Quartile | Upper Quartile | Interquartile Range | Standard Deviation |
|---|---|---|---|---|---|---|---|---|---|---|
| Bond Distances between Heavy Atoms | | | | | | | | | | |
| AM1 | 6337 | 1.446 | 1.415 | 1.161 | 2.152 | 0.991 | 1.394 | 1.511 | 0.117 | 0.113 |
| PM3 | 6337 | 1.447 | 1.409 | 1.158 | 2.157 | 0.999 | 1.391 | 1.517 | 0.126 | 0.119 |
| PM5 | 6337 | 1.445 | 1.406 | 1.155 | 2.128 | 0.972 | 1.386 | 1.521 | 0.135 | 0.121 |
| HF | 6337 | 1.440 | 1.397 | 1.133 | 2.156 | 1.023 | 1.382 | 1.524 | 0.142 | 0.130 |
| B3LYP | 6337 | 1.452 | 1.410 | 1.159 | 2.174 | 1.015 | 1.393 | 1.529 | 0.136 | 0.127 |
| Bond Distances between Heavy Atoms and Hydrogen Atoms | | | | | | | | | | |
| AM1 | 6677 | 1.106 | 1.117 | 0.964 | 1.327 | 0.364 | 1.102 | 1.122 | 0.020 | 0.034 |
| PM3 | 6677 | 1.094 | 1.098 | 0.947 | 1.315 | 0.368 | 1.096 | 1.108 | 0.012 | 0.033 |
| PM5 | 6677 | 1.098 | 1.104 | 0.951 | 1.299 | 0.347 | 1.097 | 1.110 | 0.013 | 0.032 |
| HF | 6677 | 1.076 | 1.084 | 0.941 | 1.325 | 0.384 | 1.076 | 1.087 | 0.011 | 0.029 |
| B3LYP | 6677 | 1.087 | 1.095 | 0.962 | 1.348 | 0.386 | 1.086 | 1.097 | 0.011 | 0.027 |

**Table 6.4  Correlation (in terms of $R^2$) between AM1, PM3, PM5, HF/6-31G**, and B3LYP/6-31G** for Molecular Descriptors Based on Orbital Energies and Net Atomic Charges[a]**

| Descriptor | No. Compounds[b] | AM1 vs. | | | | PM3 vs. | | | PM5 vs. | | HF vs. |
|---|---|---|---|---|---|---|---|---|---|---|---|
| | | PM3 | PM5 | HF | B3LYP | PM5 | HF | B3LYP | HF | B3LYP | B3LYP |
| $E_{HOMO}$ | 607 | 0.896 | 0.940 | 0.882 | 0.830 | 0.898 | 0.803 | 0.726 | 0.806 | 0.829 | 0.689 |
| $E_{LUMO}$ | 607 | 0.978 | 0.859 | 0.832 | 0.860 | 0.851 | 0.803 | 0.822 | 0.691 | 0.765 | 0.939 |
| EN | 607 | 0.941 | 0.905 | 0.790 | 0.891 | 0.868 | 0.708 | 0.830 | 0.810 | 0.774 | 0.856 |
| HD | 607 | 0.962 | 0.899 | 0.878 | 0.844 | 0.913 | 0.839 | 0.809 | 0.747 | 0.779 | 0.887 |
| $\mu$ (dip.m.) | 607 | 0.921 | 0.891 | 0.871 | 0.847 | 0.878 | 0.824 | 0.795 | 0.822 | 0.792 | 0.978 |
| $Q_{Zmax}^{+}$ | 435 | 0.733 | 0.932 | 0.528 | 0.493 | 0.826 | 0.349 | 0.306 | 0.598 | 0.558 | 0.961 |
| $Q_{Zavp}^{+}$ | 435 | 0.580 | 0.787 | 0.367 | 0.299 | 0.578 | 0.175 | 0.142 | 0.389 | 0.322 | 0.646 |
| $Q_{Zmax}^{-}$ | 607 | 0.523 | 0.720 | 0.214 | 0.157 | 0.541 | 0.116 | 0.064 | 0.384 | 0.303 | 0.904 |
| $Q_{Zavn}^{-}$ | 607 | 0.336 | 0.781 | 0.301 | 0.255 | 0.367 | 0.215 | 0.179 | 0.384 | 0.350 | 0.871 |
| $Q_{Cmax}^{+}$ | 377 | 0.769 | 0.955 | 0.771 | 0.702 | 0.775 | 0.429 | 0.364 | 0.710 | 0.648 | 0.941 |
| $Q_{Cavp}^{+}$ | 377 | 0.594 | 0.782 | 0.440 | 0.204 | 0.471 | 0.242 | 0.085 | 0.363 | 0.244 | 0.492 |
| $Q_{Cmax}^{-}$ | 593 | 0.164 | 0.751 | 0.205 | 0.091 | 0.188 | (−)0.006 | (−)0.031 | 0.319 | 0.184 | 0.858 |
| $Q_{Cavn}^{-}$ | 593 | 0.127 | 0.728 | 0.364 | 0.246 | 0.207 | 0.007 | 0.002 | 0.287 | 0.225 | 0.860 |
| $Q_{Hmax}^{+}$ | 602 | 0.647 | 0.879 | 0.825 | 0.709 | 0.839 | 0.632 | 0.566 | 0.834 | 0.757 | 0.958 |
| $Q_{Havp}^{+}$ | 602 | 0.779 | 0.966 | 0.899 | 0.560 | 0.836 | 0.666 | 0.338 | 0.908 | 0.611 | 0.825 |
| $Q_{Nmax}^{+}$ | 39 | 0.865 | 0.627 | 0.274 | 0.075 | 0.466 | 0.392 | 0.109 | 0.003 | (−)0.007 | 0.734 |
| $Q_{Navp}^{+}$ | 39 | (−)0.083 | 0.595 | 0.121 | 0.011 | (−)0.0005 | (−)0.068 | (−)0.033 | (−)0.043 | (−)0.098 | 0.763 |
| $Q_{Nmax}^{-}$ | 152 | 0.056 | 0.941 | 0.696 | 0.430 | 0.112 | 0.011 | 0.001 | 0.638 | 0.404 | 0.718 |
| $Q_{Navn}^{-}$ | 152 | 0.025 | 0.944 | 0.691 | 0.394 | 0.072 | 0.002 | (−)0.0004 | 0.659 | 0.405 | 0.695 |
| $Q_{Omax}^{-}$ | 424 | 0.661 | 0.880 | 0.109 | 0.050 | 0.548 | 0.005 | (−)0.0003 | 0.143 | 0.093 | 0.847 |
| $Q_{Oavn}^{-}$ | 424 | 0.685 | 0.911 | 0.066 | 0.026 | 0.560 | (−)0.033 | (−)0.057 | 0.119 | 0.080 | 0.901 |
| $Q_{Xmax}^{+}$ | 33 | (−)0.033 | 0.839 | 0.362 | 0.205 | 0.031 | 0.307 | 0.331 | 0.651 | 0.468 | 0.878 |
| $Q_{Xavp}^{+}$ | 33 | (−)0.044 | 0.861 | 0.380 | 0.202 | 0.011 | 0.301 | 0.329 | 0.600 | 0.405 | 0.837 |
| $Q_{Xmax}^{-}$ | 50 | 0.094 | 0.780 | 0.192 | 0.172 | 0.358 | 0.467 | 0.497 | 0.477 | 0.470 | 0.995 |
| $Q_{Xavn}^{-}$ | 50 | 0.065 | 0.760 | 0.152 | 0.129 | 0.346 | 0.446 | 0.471 | 0.459 | 0.448 | 0.994 |

[a] The Coulson net atomic charges (Equation 6.53) are used for AM1, PM3, and PM5, while the Mulliken population analysis (Equation 6.55 and Equation 6.57) is employed for HF/6-31G** and B3LYP/6-31G**. A minus sign in parentheses such as for AM1 vs. PM3 with respect to $Q_{Navp}^{+}$ indicates a negative correlation.

[b] For evaluating the $Q_{Zmax}^{+}$ statistics, all compounds without a positively charged heavy atom according to any of the five quantum chemical methods were excluded. All other statistics with atom type-specific net atomic charges are defined correspondingly.

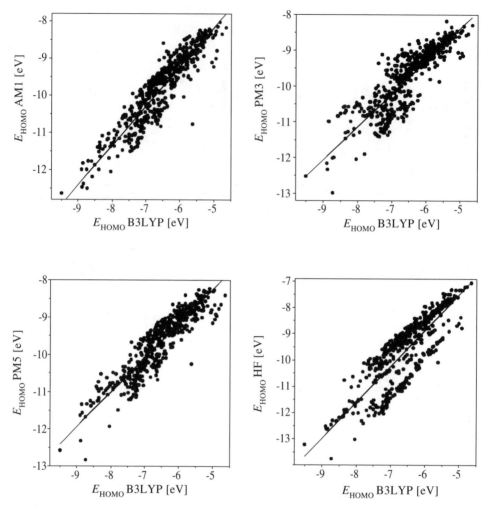

**Figure 6.2** $E_{HOMO}$ comparison between AM1 (top left), PM3 (top right), PM5 (bottom left), and HF/6-31G** (bottom right) against B3LYP/6-31G** for the set of 607 compounds.

Returning to Figure 6.2, the largest PM3 underestimation of $E_{HOMO}$ when compared with B3LYP was for 1,1,1,3,3,3-hexafluoro-2-propanol, and among the further compounds with large under-estimations are 2,2,2-trifluoroethanol and 2-methyl-3,3,4,4-tetrafluoro-2-butanol, all of which are simple hydrocarbon derivatives with a larger number of fluorine atoms. Hexachloro-1,3-butadiene yields the largest corresponding over-estimation, and 1-benzylpyridinium-3-sulphonate still belongs to the group with substantially increased $E_{HOMO}$ values when compared with the relative B3LYP trend. Interestingly, hexafluoropropanol, trifluoroethanol, and benzylpyridinium-sulfonate are also corresponding outliers with PM5 (bottom left in Figure 6.2) and HF (bottom right).

The $E_{LUMO}$ data distributions in Figure 6.3 show a greater number of outliers for all three semiempirical methods, while the HF-B3LYP correlation is greatly improved as compared to $E_{HOMO}$ (cf. Table 6.3). Very large AM1 underestimations (in relation to the B3LYP trend) are obtained for the phosphorodithioates $((RO)_2P(=S)-SR)$ disulfoton, terbufos, malathion, and carbophenothion (that is, at the same time, the greatest positive $E_{HOMO}$ outlier, see above), and other still significant outliers include 1,1,1,3,3,3-hexafluoro-2-propanol (see PM3 $E_{HOMO}$ performance above) and the phosphorodithioate azinophos-methyl. Here, iodoform belongs to the greatest positive outliers.

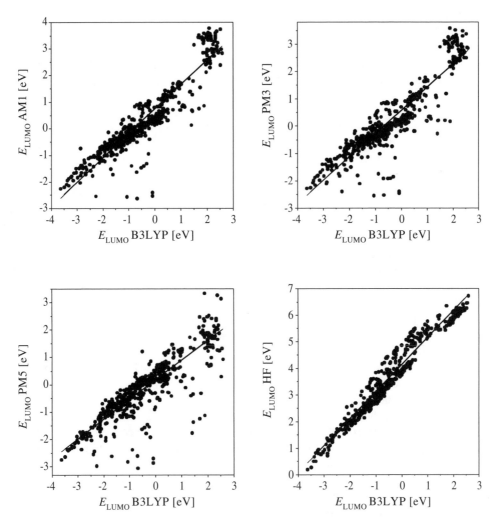

**Figure 6.3** $E_{LUMO}$ comparison between AM1 (top left), PM3 (top right), PM5 (bottom left), and HF/6-31G** (bottom right) against B3LYP/6-31G** for the set of 607 compounds.

The very large negative PM3 and PM5 outliers include the AM1 outliers mentioned above and additional compounds — the organophosphates ($(RO)_3P=O$) tris-(2-butoxyethyl)phosphate, tributyl phosphate and trimethyl phosphate (cf. Figure 6.1) for PM5. Among the positive outliers are adamantane (a cyclic $C_{10}$ hydrocarbon), hexane, and cyclohexane (cf. Figure 6.1). When taking B3LYP as reference, the comparative analysis of the frontier orbital energies shows that the MNDO-type schemes yield particularly large deviations for hypervalent structures including organophosphorus compounds. Surprisingly, PM3 and PM5 appear to be not superior to AM1 in this respect, despite an improved parameterization of PM3 for hypervalent molecules (Stewart, 1989a; 1989b) and a probably still improved overall parameterization of PM5 (Stewart 2002). For a set of 106 hypervalent (but typically small) compounds, the average PM3 error for the heat of formation was a factor of 6 below the ones of MNDO and AM1 (Stewart, 1989b); respective statistics are not yet available for PM5 as mentioned above.

The net atomic charge statistics in Table 6.4 reveal surprisingly large differences between the individual methods, and in particular between the semiempirical results based on the Coulson scheme (Section IV.E, Equation 6.53) and the *ab initio* results derived through the Mulliken population analysis (Equation 6.55 and Equation 6.57). At the same time, the charge distributions of AM1 and PM5 are more similar to each other than that of PM3 (compare columns 3 and 4 of Table 6.4). Taking the maximum positive charge of any heavy atom ($Q_{Zmax}^+$) as an example, the correlation of AM1 with PM3 is significantly lower than with PM5 ($R^2$: 0.733 vs. 0.932), but all three semiempirical methods show only moderate to low correlations with HF (AM1: 0.528, PM3: 0.349, PM5: 0.598) and B3LYP (AM1: 0.493, PM3: 0.306, PM5: 0.558). The respective average positive charge, taking into account only positively charged heavy atoms within each of the molecules with at least one such atom ($Q_{Zavp}^+$), yields still inferior statistics, and in this case there are also significant differences between HF and B3LYP (last column in Table 6.4: $R^2 = 0.646$).

For the 377 compounds with at least one positively charged carbon atom according to all 5 methods, there is a strikingly large difference between the maximum values per molecule ($Q_{Cmax}^+$) and the average confined to positively charged carbons ($Q_{Cavp}^+$) when comparing AM1 with B3LYP (0.702 vs. 0.204), with smaller $R^2$ values for PM5 (0.648 vs. 0.244) and still decreased $R^2$ values for PM3 (0.364 vs. 0.085). The corresponding negative charge descriptors, $Q_{Cmax}^-$ and $Q_{Cavn}^-$, yield still lower intercorrelations (e.g., AM1-B3LYP: 0.091 and 0.246). With respect to maximum or average (positive or negative) nitrogen charges, the PM3 trend across the molecules is essentially orthogonal to the ones of all other methods except for $Q_{Nmax}^+$ as compared to AM1 ($R^2 = 0.865$), PM5 (0.466), and HF (0.392). At the same time, there is also no correlation for $Q_{Nmax}^+$ and $Q_{Navp}^+$ between B3LYP and AM1 (0.075, 0.011) as well as PM5 ([–]0.007, [–]0.098).

The results for the 424 compounds with at least one (negatively charged) oxygen are particularly striking. All three semiempirical methods provide values with trends completely different from the ones according to HF and B3LYP, and the only high intercorrelations for $Q_{Omax}^-$ and $Q_{Oavn}^-$ are found for AM1 vs. PM5 (0.880 and 0.911) and HF vs. B3LYP (0.847 vs. 0.901). With regard to halogens, the trends of the semiempirical net atomic charges are again significantly different from the ones provided by the *ab initio* methods.

In Table 6.5, the method comparisons with respect to net atomic charges are summarized for certain atom types, considering the calculated charges of all respective atoms of all compounds. As can be seen from the first row, AM1 correlates for all 4880 carbon atoms more with PM5 ($R^2 = 0.951$) than with PM3 (0.781), and more with HF (0.811) than with B3LYP (0.683). Interestingly, aliphatic carbon yields significantly higher intercorrelations than aromatic carbon (e.g., AM1-B3LYP: 0.819 vs. 0.454), and the hydrogen net atomic charges show somewhat lower $R^2$ values (AM1-B3LYP: 0.483) with a surprisingly large difference between non-polar and polar hydrogen (AM1-B3LYP: 0.104 vs. 0.619).

As a further unexpected finding, the 6251 non-polar hydrogens yield also quite substantial differences in the relative trends of HF and B3LYP ($R^2 = 0.398$). Moreover, there is a systematic difference in the HF and B3LYP trends between hydrogens attached to aliphatic and aromatic carbon (0.801 vs. 0.951), while aliphatic carbons show a higher intercorrelation than aromatic carbons (0.967 vs. 0.892). Iodine provides the second smallest HF-B3LYP correlation ($R^2 = 0.781$), but this result is based on only eight atomic sites. Overall, the comparison between the HF and B3LYP results (both of which refer to the same basis set and the Mulliken population analysis) suggest that when describing the molecular charge distribution, electron correlation is more important for non-polar hydrogens, aromatic structures, and possibly iodine. Interestingly, electron correlation is also more important for oxygen, where the $R^2$ value referring to 823 atomic sites (0.860) is clearly below the intercorrelations for all other heteroatoms except iodine.

Coming back to the semiempirical schemes, AM1 shows a generally higher agreement with PM5 than PM3, as was also the case with the maximum and average charge descriptors presented in Table 6.4. Moreover, AM1 and PM5 correlate more with B3LYP than PM3 with only few notable exceptions such as phosphorus (AM1: 0.180, PM3: 0.874, PM5: 0.785) and the combined halogens (AM1: 0.445, PM3: 0.736, PM5: 0.588). Interestingly, the separate statistics for fluorine, chlorine, bromine, and iodine indicate greater (but overall only moderate to almost negligible) correlations of B3LYP with AM1 and PM5 than with PM3, which indicates that the individual trends vary more

**Table 6.5 Correlation in Terms of $R^2$ between AM1, PM3, PM5, HF/6-31G\*\*, and B3LYP/6-31G\*\* for Net Atomic Charges[a]**

| Atom Type[b] | No. Atoms | AM1 vs. PM3 | PM5 | HF | B3LYP | PM3 vs. PM5 | HF | B3LYP | PM5 vs. HF | B3LYP | HF vs. B3LYP |
|---|---|---|---|---|---|---|---|---|---|---|---|
| C | 4880 | 0.781 | 0.951 | 0.811 | 0.683 | 0.771 | 0.576 | 0.397 | 0.812 | 0.704 | 0.932 |
| $C_{aliphatic}$ | 2578 | 0.884 | 0.967 | 0.890 | 0.819 | 0.850 | 0.782 | 0.676 | 0.888 | 0.843 | 0.967 |
| $C_{aromatic}$ | 2302 | 0.675 | 0.916 | 0.636 | 0.454 | 0.754 | 0.335 | 0.156 | 0.631 | 0.436 | 0.892 |
| H | 6677 | 0.701 | 0.948 | 0.837 | 0.483 | 0.780 | 0.505 | 0.205 | 0.829 | 0.506 | 0.827 |
| $H_{non-polar}$ | 6251 | 0.905 | 0.953 | 0.824 | 0.104 | 0.872 | 0.670 | 0.020 | 0.796 | 0.112 | 0.398 |
| $H_{polar}$ | 426 | 0.346 | 0.708 | 0.743 | 0.619 | 0.826 | 0.753 | 0.845 | 0.897 | 0.887 | 0.967 |
| $H(C_{aliphatic})$ | 4810 | 0.801 | 0.886 | 0.749 | 0.409 | 0.693 | 0.537 | 0.193 | 0.721 | 0.445 | 0.801 |
| $H(C_{aromatic})$ | 1434 | 0.824 | 0.882 | 0.758 | 0.782 | 0.864 | 0.574 | 0.560 | 0.587 | 0.617 | 0.951 |
| N | 336 | 0.836 | 0.982 | 0.954 | 0.925 | 0.909 | 0.839 | 0.842 | 0.949 | 0.929 | 0.977 |
| O | 823 | 0.616 | 0.910 | 0.023 | 0.008 | 0.507 | 0.142 | 0.150 | 0.048 | 0.036 | 0.860 |
| P | 15 | 0.469 | 0.576 | 0.289 | 0.180 | 0.971 | 0.946 | 0.874 | 0.880 | 0.785 | 0.982 |
| S | 55 | 0.972 | 0.939 | 0.956 | 0.948 | 0.985 | 0.955 | 0.947 | 0.937 | 0.934 | 0.997 |
| Halogen | 330 | 0.635 | 0.946 | 0.509 | 0.445 | 0.799 | 0.768 | 0.736 | 0.647 | 0.588 | 0.985 |
| F | 53 | 0.983 | 0.935 | 0.030 | (–)0.005 | 0.953 | 0.026 | (–)0.009 | 0.103 | 0.005 | 0.923 |
| Cl | 223 | 0.982 | 0.987 | 0.810 | 0.650 | 0.966 | 0.761 | 0.588 | 0.845 | 0.697 | 0.949 |
| Br | 46 | 0.992 | 0.991 | 0.762 | 0.524 | 0.985 | 0.740 | 0.491 | 0.755 | 0.514 | 0.914 |
| I | 8 | 0.427 | 0.615 | 0.592 | 0.151 | 0.928 | 0.004 | 0.127 | 0.071 | 0.027 | 0.781 |

[a] All atoms of all 607 compounds were considered for selecting subsets according to atom types as specified. Net atomic charges refer to the Coulson scheme (AM1, PM3, PM5; Equation 6.53) and Mulliken population analysis (HF/6-31G\*\*, B3LYP/6-31G\*\*; Equation 6.55 and Equation 6.57), respectively. A minus sign in parentheses such as for AM1 vs. B3LYP with respect to F indicates a negative correlation.

[b] $C_{aliphatic}$ covers all non-aromatic carbon atoms. $H_{non-polar}$ represents all hydrogen atoms attached to carbon, and $H_{polar}$ includes all hydrogen atoms attached to oxygen, nitrogen, sulphur, or phosphorus. $H(C_{aliphatic})$ and $H(C_{aromatic})$ denote the two subsets of hydrogen atoms attached to non-aromatic and aromatic carbon atoms, respectively.

in their slopes with the former two methods than with PM3. It is particularly striking that for both fluorine (53 atomic sites) and iodine (8 sites, see above), the semiempirical results appear to have little to do with the *ab initio* counterparts. A further strong discrepancy concerns oxygen, where all three methods differ almost completely from both HF and B3LYP (AM1: 0.023, 0.008; PM3: 0.142, 0.150; PM5: 0.048, 0.036).

The method-specific value distributions of the net atomic charges are characterized in Table 6.6. Interestingly, HF yields an outlying maximum positive charge for carbon ($Q_C$ = 1.249 au), which refers to the $-CF_3$ carbon of 4-nitro-3(trifluoromethyl)-phenol (cf. Figure 6.1; note that for $-CF_3$ attached to aromatic carbon, HF yields typically net atomic charges around 1.2 au). At the same time, the HF and B3LYP interquartile ranges are similar to each other (0.202 au vs. 0.222 au), and significantly greater than the ones of AM1, PM3, and PM5 (0.108 au vs. 0.077 au vs. 0.162). However, when differentiating between aliphatic and aromatic carbon atoms, B3LYP yields a greater interquartile range for the latter group than HF (and all semiempirical schemes), indicating a greater susceptibility of the local charge distribution of the aromatic carbon skeleton for changes in the molecular structure.

A particularly striking feature is seen for the nitrogen charge, where PM3 yields a positive average nitrogen charge (0.189 au) in contrast to all other methods (e.g., B3LYP: –0.383); correspondingly, the maximum nitrogen charge is much larger for PM3 (1.341) than for the other schemes. PM3 yields also the greatest respective range (1.853), but at the same time the smallest interquartile range (0.140). Keeping in mind the nonetheless signficant correlation between PM3 and B3LYP when considering all 336 nitrogen sites among the atoms of the 607 compounds (Table 6.5: $R^2$ = 0.842), it follows that the PM3 nitrogen charge is systematically shifted upward as compared to other quantum chemical methods, and that the PM3 modus of the nitrogen charge may be misleading when analyzing the susceptibility of nitrogen sites for electrophilic or nucleophilic attacks.

**Table 6.6 Statistical Characteristics of the Net Atomic Charges (au) According to AM1, PM3, PM5, HF/6-31G\*\*, and B3LYP/6-31G\*\*a**

| Atom Type[b] | Method | No. Atoms | Mean | Median | Minimum | Maximum | Range | Lower Quartile | Upper Quartile | Interquartile Range | Standard Deviation |
|---|---|---|---|---|---|---|---|---|---|---|---|
| C | AM1 | 4880 | −0.093 | −0.126 | −0.923 | 0.474 | 1.397 | −0.166 | −0.058 | 0.108 | 0.136 |
| | PM3 | 4880 | −0.062 | −0.098 | −0.606 | 0.459 | 1.065 | −0.121 | −0.044 | 0.077 | 0.134 |
| | PM5 | 4880 | −0.112 | −0.154 | −0.950 | 0.679 | 1.629 | −0.212 | −0.050 | 0.162 | 0.177 |
| | HF | 4880 | −0.070 | −0.149 | −0.577 | 1.249 | 1.826 | −0.220 | −0.018 | 0.202 | 0.275 |
| | B3LYP | 4880 | −0.053 | −0.099 | −0.508 | 0.852 | 1.360 | −0.175 | 0.047 | 0.222 | 0.227 |
| $C_{aliphatic}$ | AM1 | 2578 | −0.096 | −0.157 | −0.816 | 0.474 | 1.290 | −0.197 | −0.063 | 0.134 | 0.157 |
| | PM3 | 2578 | −0.039 | −0.100 | −0.462 | 0.459 | 0.921 | −0.116 | −0.026 | 0.090 | 0.148 |
| | PM5 | 2578 | −0.121 | −0.185 | −0.718 | 0.679 | 1.397 | −0.252 | −0.064 | 0.188 | 0.206 |
| | HF | 2578 | −0.091 | −0.211 | −0.577 | 1.249 | 1.826 | −0.309 | −0.005 | 0.304 | 0.321 |
| | B3LYP | 2578 | −0.097 | −0.174 | −0.508 | 0.852 | 1.360 | −0.301 | −0.010 | 0.291 | 0.262 |
| $C_{aromatic}$ | AM1 | 2302 | −0.091 | −0.108 | −0.923 | 0.405 | 1.328 | −0.143 | −0.055 | 0.088 | 0.106 |
| | PM3 | 2302 | −0.088 | −0.094 | −0.606 | 0.444 | 1.050 | −0.132 | −0.049 | 0.083 | 0.111 |
| | PM5 | 2302 | −0.102 | −0.126 | −0.950 | 0.507 | 1.457 | −0.180 | −0.042 | 0.138 | 0.136 |
| | HF | 2302 | −0.046 | −0.132 | −0.360 | 1.096 | 1.456 | −0.162 | −0.036 | 0.126 | 0.209 |
| | B3LYP | 2302 | −0.003 | −0.086 | −0.307 | 0.787 | 1.094 | −0.110 | 0.082 | 0.192 | 0.165 |
| H | AM1 | 6677 | 0.109 | 0.094 | 0.035 | 0.282 | 0.247 | 0.079 | 0.134 | 0.055 | 0.041 |
| | PM3 | 6677 | 0.072 | 0.056 | −0.003 | 0.276 | 0.279 | 0.047 | 0.102 | 0.055 | 0.039 |
| | PM5 | 6677 | 0.136 | 0.120 | 0.059 | 0.356 | 0.297 | 0.100 | 0.163 | 0.063 | 0.048 |
| | HF | 6677 | 0.147 | 0.130 | 0.069 | 0.428 | 0.359 | 0.112 | 0.157 | 0.045 | 0.056 |
| | B3LYP | 6677 | 0.118 | 0.105 | 0.045 | 0.376 | 0.331 | 0.091 | 0.121 | 0.030 | 0.050 |
| $H_{non-polar}$ | AM1 | 6251 | 0.102 | 0.091 | 0.035 | 0.274 | 0.239 | 0.079 | 0.129 | 0.050 | 0.031 |
| | PM3 | 6251 | 0.069 | 0.055 | −0.003 | 0.196 | 0.199 | 0.047 | 0.099 | 0.052 | 0.031 |
| | PM5 | 6251 | 0.128 | 0.117 | 0.059 | 0.316 | 0.257 | 0.099 | 0.157 | 0.058 | 0.035 |
| | HF | 6251 | 0.135 | 0.128 | 0.069 | 0.364 | 0.295 | 0.112 | 0.152 | 0.040 | 0.031 |
| | B3LYP | 6251 | 0.106 | 0.103 | 0.045 | 0.309 | 0.264 | 0.091 | 0.118 | 0.027 | 0.022 |
| $H_{polar}$ | AM1 | 426 | 0.208 | 0.217 | 0.049 | 0.282 | 0.233 | 0.190 | 0.234 | 0.044 | 0.038 |
| | PM3 | 426 | 0.125 | 0.129 | 0.018 | 0.276 | 0.258 | 0.035 | 0.197 | 0.162 | 0.083 |
| | PM5 | 426 | 0.250 | 0.267 | 0.100 | 0.356 | 0.256 | 0.192 | 0.294 | 0.102 | 0.057 |
| | HF | 426 | 0.325 | 0.334 | 0.096 | 0.428 | 0.332 | 0.301 | 0.352 | 0.051 | 0.041 |
| | B3LYP | 426 | 0.288 | 0.296 | 0.104 | 0.376 | 0.272 | 0.261 | 0.317 | 0.056 | 0.038 |
| $H(C_{aliphatic})$ | AM1 | 4810 | 0.089 | 0.085 | 0.035 | 0.231 | 0.196 | 0.078 | 0.096 | 0.018 | 0.020 |
| | PM3 | 4810 | 0.054 | 0.051 | −0.003 | 0.196 | 0.199 | 0.044 | 0.062 | 0.018 | 0.018 |
| | PM5 | 4810 | 0.114 | 0.108 | 0.059 | 0.267 | 0.208 | 0.096 | 0.123 | 0.027 | 0.024 |
| | HF | 4810 | 0.124 | 0.120 | 0.069 | 0.263 | 0.194 | 0.109 | 0.134 | 0.025 | 0.023 |
| | B3LYP | 4810 | 0.108 | 0.105 | 0.045 | 0.241 | 0.196 | 0.093 | 0.119 | 0.026 | 0.021 |

| Group | Method | N | | | | | | | | | |
|---|---|---|---|---|---|---|---|---|---|---|---|
| H(C_aromatic) | AM1 | 1434 | 0.147 | 0.143 | 0.125 | 0.224 | 0.099 | 0.134 | 0.156 | 0.022 | 0.016 |
| | PM3 | 1434 | 0.116 | 0.114 | 0.096 | 0.178 | 0.082 | 0.106 | 0.122 | 0.016 | 0.012 |
| | PM5 | 1434 | 0.178 | 0.175 | 0.151 | 0.263 | 0.112 | 0.163 | 0.189 | 0.026 | 0.017 |
| | HF | 1434 | 0.169 | 0.162 | 0.126 | 0.281 | 0.155 | 0.151 | 0.182 | 0.031 | 0.024 |
| | B3LYP | 1434 | 0.100 | 0.095 | 0.059 | 0.190 | 0.131 | 0.085 | 0.113 | 0.028 | 0.021 |
| N | AM1 | 336 | -0.126 | -0.257 | -0.859 | 0.588 | 1.447 | -0.337 | -0.056 | 0.281 | 0.323 |
| | PM3 | 336 | 0.189 | -0.026 | -0.512 | 1.341 | 1.853 | -0.063 | 0.077 | 0.140 | 0.491 |
| | PM5 | 336 | -0.141 | -0.305 | -0.953 | 0.743 | 1.696 | -0.364 | -0.112 | 0.253 | 0.387 |
| | HF | 336 | -0.467 | -0.641 | -0.919 | 0.556 | 1.475 | -0.743 | -0.456 | 0.288 | 0.446 |
| | B3LYP | 336 | -0.383 | -0.490 | -0.720 | 0.422 | 1.142 | -0.603 | -0.412 | 0.192 | 0.344 |
| O | AM1 | 823 | -0.325 | -0.305 | -1.200 | -0.107 | 1.093 | -0.346 | -0.267 | 0.079 | 0.135 |
| | PM3 | 823 | -0.347 | -0.310 | -0.928 | -0.059 | 0.869 | -0.382 | -0.256 | 0.126 | 0.147 |
| | PM5 | 823 | -0.403 | -0.389 | -0.990 | -0.215 | 0.775 | -0.426 | -0.359 | 0.067 | 0.092 |
| | HF | 823 | -0.601 | -0.627 | -0.802 | -0.405 | 0.397 | -0.654 | -0.546 | 0.108 | 0.078 |
| | B3LYP | 823 | -0.490 | -0.491 | -0.644 | -0.337 | 0.307 | -0.537 | -0.458 | 0.079 | 0.058 |
| P | AM1 | 15 | 2.510 | 2.517 | 2.236 | 3.038 | 0.802 | 2.265 | 2.622 | 0.357 | 0.230 |
| | PM3 | 15 | 1.859 | 1.844 | 1.558 | 2.182 | 0.624 | 1.573 | 2.156 | 0.583 | 0.248 |
| | PM5 | 15 | 1.809 | 1.797 | 1.459 | 2.074 | 0.615 | 1.572 | 2.055 | 0.483 | 0.223 |
| | HF | 15 | 1.381 | 1.399 | 1.093 | 1.726 | 0.633 | 1.115 | 1.676 | 0.561 | 0.242 |
| | B3LYP | 15 | 0.970 | 0.989 | 0.740 | 1.285 | 0.545 | 0.757 | 1.232 | 0.475 | 0.211 |
| S | AM1 | 55 | 0.087 | 0.050 | -1.277 | 2.884 | 4.161 | -0.585 | 0.219 | 0.804 | 0.926 |
| | PM3 | 55 | 0.063 | -0.048 | -0.712 | 2.402 | 3.114 | -0.282 | 0.091 | 0.373 | 0.679 |
| | PM5 | 55 | 0.098 | -0.099 | -0.698 | 2.865 | 3.563 | -0.275 | 0.146 | 0.421 | 0.795 |
| | HF | 55 | 0.134 | 0.116 | -0.525 | 1.690 | 2.215 | -0.046 | 0.159 | 0.205 | 0.513 |
| | B3LYP | 55 | 0.097 | 0.074 | -0.392 | 1.254 | 1.646 | -0.023 | 0.117 | 0.140 | 0.383 |
| Halogen | AM1 | 330 | -0.015 | -0.001 | -0.178 | 0.191 | 0.369 | -0.065 | 0.030 | 0.095 | 0.079 |
| | PM3 | 330 | 0.030 | 0.063 | -0.149 | 0.205 | 0.354 | -0.052 | 0.099 | 0.151 | 0.090 |
| | PM5 | 330 | -0.052 | -0.026 | -0.248 | 0.172 | 0.420 | -0.138 | 0.014 | 0.152 | 0.098 |
| | HF | 330 | -0.050 | 0.010 | -0.407 | 0.175 | 0.582 | -0.100 | 0.058 | 0.158 | 0.155 |
| | B3LYP | 330 | -0.047 | -0.007 | -0.309 | 0.124 | 0.433 | -0.094 | 0.036 | 0.130 | 0.116 |
| F | AM1 | 53 | -0.130 | -0.145 | -0.178 | -0.046 | 0.132 | -0.162 | -0.098 | 0.064 | 0.039 |
| | PM3 | 53 | -0.110 | -0.127 | -0.149 | -0.050 | 0.099 | -0.135 | -0.087 | 0.048 | 0.032 |
| | PM5 | 53 | -0.208 | -0.221 | -0.248 | -0.138 | 0.110 | -0.231 | -0.195 | 0.036 | 0.032 |
| | HF | 53 | -0.367 | -0.363 | -0.407 | -0.331 | 0.076 | -0.378 | -0.356 | 0.022 | 0.017 |
| | B3LYP | 53 | -0.272 | -0.266 | -0.309 | -0.241 | 0.068 | -0.285 | -0.260 | 0.025 | 0.017 |
| Cl | AM1 | 223 | -0.008 | 0.001 | -0.125 | 0.091 | 0.216 | -0.029 | 0.025 | 0.054 | 0.051 |
| | PM3 | 223 | 0.068 | 0.084 | -0.080 | 0.205 | 0.285 | 0.039 | 0.108 | 0.069 | 0.063 |
| | PM5 | 223 | -0.032 | -0.018 | -0.201 | 0.085 | 0.286 | -0.058 | 0.008 | 0.066 | 0.068 |
| | HF | 223 | 0.030 | 0.036 | -0.119 | 0.175 | 0.294 | 0.004 | 0.070 | 0.066 | 0.059 |
| | B3LYP | 223 | 0.014 | 0.018 | -0.102 | 0.124 | 0.226 | -0.014 | 0.048 | 0.062 | 0.050 |

Table 6.7 (continued) Statistical Characteristics of the Net Atomic Charges (au) According to AM1, PM3, PM5, HF/6-31G**, and B3LYP/6-31G***[a]

| Atom Type[b] | Method | No. Atoms | Mean | Median | Minimum | Maximum | Range | Lower Quartile | Upper Quartile | Interquartile Range | Standard Deviation |
|---|---|---|---|---|---|---|---|---|---|---|---|
| Br | AM1 | 46 | 0.052 | 0.065 | −0.052 | 0.166 | 0.218 | 0.001 | 0.093 | 0.092 | 0.062 |
| | PM3 | 46 | −0.001 | 0.018 | −0.140 | 0.140 | 0.280 | −0.068 | 0.053 | 0.121 | 0.081 |
| | PM5 | 46 | 0.006 | 0.031 | −0.160 | 0.172 | 0.332 | −0.058 | 0.077 | 0.135 | 0.095 |
| | HF | 46 | −0.097 | −0.089 | −0.197 | −0.006 | 0.191 | −0.127 | −0.055 | 0.072 | 0.056 |
| | B3LYP | 46 | −0.100 | −0.104 | −0.166 | −0.010 | 0.156 | −0.131 | −0.068 | 0.063 | 0.042 |
| I | AM1 | 8 | 0.173 | 0.171 | 0.163 | 0.191 | 0.028 | 0.167 | 0.178 | 0.011 | 0.010 |
| | PM3 | 8 | 0.055 | 0.052 | 0.024 | 0.101 | 0.077 | 0.024 | 0.081 | 0.057 | 0.031 |
| | PM5 | 8 | 0.102 | 0.094 | 0.084 | 0.139 | 0.055 | 0.085 | 0.117 | 0.033 | 0.020 |
| | HF | 8 | 0.095 | 0.105 | 0.067 | 0.118 | 0.051 | 0.075 | 0.106 | 0.031 | 0.020 |
| | B3LYP | 8 | 0.054 | 0.057 | 0.014 | 0.079 | 0.065 | 0.034 | 0.079 | 0.046 | 0.026 |

[a] Net atomic charges refer to the Coulson scheme (AM1, PM3, PM5; Equation 6.53) and Mulliken population analysis (HF/6-31G**, B3LYP/6-31G**; Equation 6.55 and Equation 6.57), respectively.

[b] $C_{aliphatic}$ includes all nonaromatic carbon atoms. $H_{non-polar}$ represents all hydrogen atoms attached to carbon, and $H_{polar}$ includes all hydrogen atoms attached to oxygen, nitrogen, sulfur, or phosphorus. $H(C_{aliphatic})$ and $H(C_{aromatic})$ denote the two subsets of hydrogen atoms attached to non-aromatic and aromatic carbon atoms, respectively.

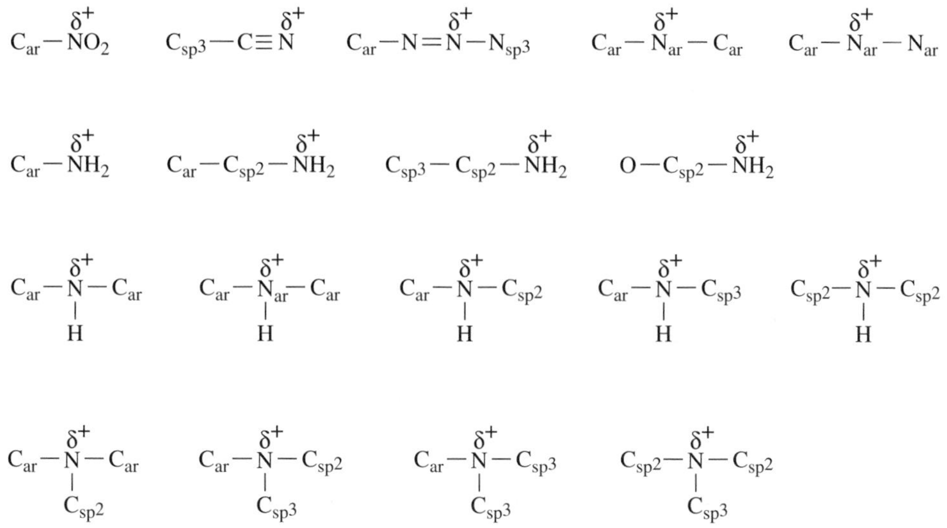

$C_{ar}—\overset{\delta+}{N}O_2 \qquad C_{sp3}—C\equiv\overset{\delta+}{N} \qquad C_{ar}—N=\overset{\delta+}{N}—N_{sp3} \qquad C_{ar}—\overset{\delta+}{N}_{ar}—C_{ar} \qquad C_{ar}—\overset{\delta+}{N}_{ar}—N_{ar}$

$C_{ar}—\overset{\delta+}{N}H_2 \qquad C_{ar}—C_{sp2}—\overset{\delta+}{N}H_2 \qquad C_{sp3}—C_{sp2}—\overset{\delta+}{N}H_2 \qquad O—C_{sp2}—\overset{\delta+}{N}H_2$

$C_{ar}—\underset{H}{\overset{\delta+}{N}}—C_{ar} \qquad C_{ar}—\underset{H}{\overset{\delta+}{N}_{ar}}—C_{ar} \qquad C_{ar}—\underset{H}{\overset{\delta+}{N}}—C_{sp2} \qquad C_{ar}—\underset{H}{\overset{\delta+}{N}}—C_{sp3} \qquad C_{sp2}—\underset{H}{\overset{\delta+}{N}}—C_{sp2}$

$C_{ar}—\underset{C_{sp2}}{\overset{\delta+}{N}}—C_{ar} \qquad C_{ar}—\underset{C_{sp3}}{\overset{\delta+}{N}}—C_{sp2} \qquad C_{ar}—\underset{C_{sp3}}{\overset{\delta+}{N}}—C_{sp3} \qquad C_{sp2}—\underset{C_{sp3}}{\overset{\delta+}{N}}—C_{sp2}$

**Figure 6.4** Structural features with positive PM3 net atomic charges at nitrogen.

Figure 6.4 contains all structural features of the present data set where PM3 yields a positively charged nitrogen, and Figure 6.5 shows the respective substructures with positive AM1 nitrogen charges as well as some respective examples. The first 3 of the substructures listed in Figure 6.5 apply also for PM5, while according to HF and B3LYP, only the nitro nitrogen (attached to aromatic carbon) carries a positive net atomic charge. Returning to Table 6.6, the quite substantial differences in (negative) mean and median nitrogen charges between AM1, PM5, HF, and B3LYP are also remarkable.

Another surprising feature is the fact that PM3 yields a negative (though small) value for the minimum hydrogen charge, again in contrast to all other methods. The respective compound, trimethylphosphate, is shown in Figure 6.6 together with the relevant hydrogen sites. Note further that when comparing hydrogen attached to aliphatic carbon (4810 sites) and to aromatic carbon (1434 sites), all methods except B3LYP yield significantly greater mean and median values for the latter.

With oxygen, the *ab initio* methods provide systematically larger negative charges than the semiempirical schemes, except for individual structural features with excessively large negative charges according to AM1 and (in a somewhat reduced manner) the other semiempirical methods. This is seen by the minimum values (which are the maximum negative values, ranging from −1.200 au for AM1 to −0.644 au for B3LYP) and ranges (from 1.093 au for AM1 to 0.397 au for B3LYP) in connection with inconspicuous interquartile ranges for the semiempirical schemes (from 0.067 to 0.126 au).

With regard to sulfur, AM1, PM3, and PM5 yield much larger maximum positive values (2.884, 2.402, and 2.865 au) than HF and B3LYP (1.690 and 1.254 au), and interestingly the largest negative sulfur charge is also provided by AM1 (−1.277 au, applying for the P=S sulfur atom of O-ethyl-O-(4-nitrophenyl)phenylphosphorothionate), which in this case results in values for the range and interquartile range that are greatest for AM1 (4.161 au, 0.804 au) and in fact much larger than for B3LYP (1.646 au, 0.140 au). Note that for the sulfur atom of phosphorothionate groups $(RO)_2P(=S)OR$, the typical AM1 net atomic charge is around −1.2 au when compared with HF and B3LYP values of ca. −0.5 au and −0.38 au, respectively. The results indicate that the individual methods provide systematically different scales for the sulfur net atomic charge, while the relative trends as summarized in Table 6.5 show a quite high degree of agreement ($R^2$ values from 0.934 for PM5 vs. B3LYP to 0.997 for HF vs. B3LYP).

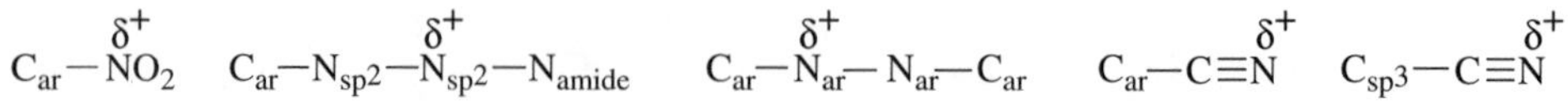

$$\overset{\delta+}{C_{ar}}-NO_2 \qquad \overset{\delta+}{C_{ar}}-N_{sp2}-N_{sp2}-N_{amide} \qquad \overset{\delta+}{C_{ar}}-N_{ar}-N_{ar}-C_{ar} \qquad \overset{\delta+}{C_{ar}}-C\equiv N \qquad \overset{\delta+}{C_{sp3}}-C\equiv N$$

Some examples:

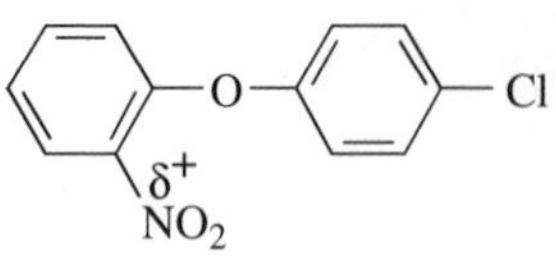

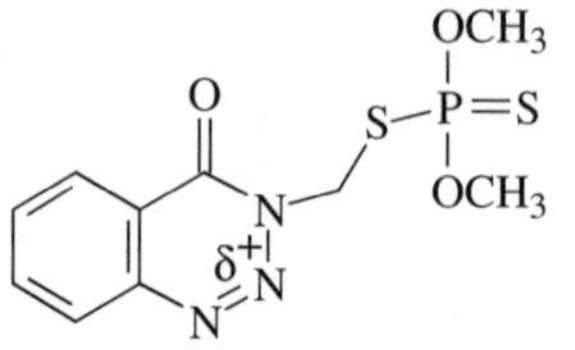

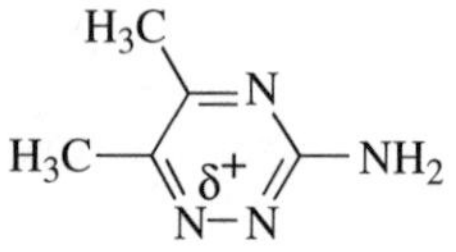

p-chlorophenyl-o-nitrophenyl ether　　　azinphosmethyl　　　3-amino-5,6-dimethyl-1,2,4-
1,2,4-triazine

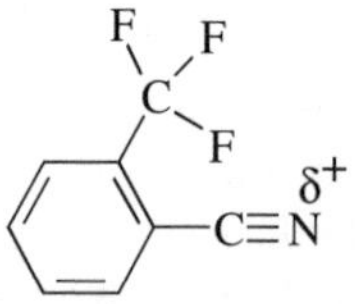

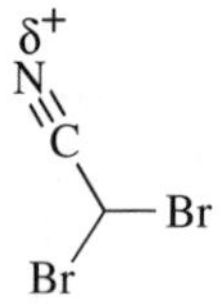

α,α,α-trifluoro-o-toluonitrile　　　　dibromoacetonitrile

**Figure 6.5**　Structural features with positive AM1 net atomic charges at nitrogen (top) with respective examples (bottom).

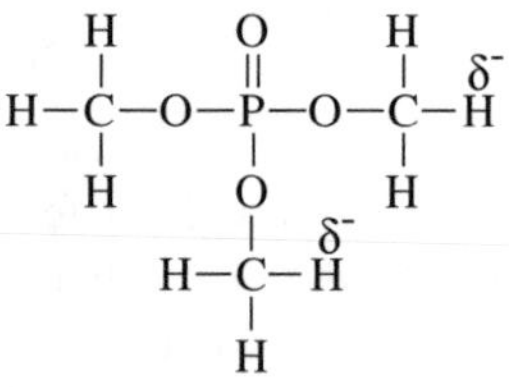

**Figure 6.6**　Trimethylphosphate as only compound with a negative PM3 net atomic charge at hydrogen.

Among the halogenated compounds, systematic differences between net atomic charges are found in particular for chlorine (223 atomic sites) and bromine (46 sites). For the former, AM1 and PM5 yield negative mean values (–0.008 and –0.032 au) as opposed to the other three methods, and with bromine the situation is reversed. It suggests that according to PM3 and the *ab initio* methods, chlorine atoms in molecules, when averaging over all functionalities, are less electronegative than with AM1 and PM5. By contrast, the electronegativity of bromine is more pronounced with PM3 and the *ab initio* methods than with AM1 and PM5, and here both HF and B3LYP do not yield any positive net atomic charges. Accordingly, the ranges are significantly smaller for the *ab initio* methods than for the MNDO-type schemes (except that in this respect, AM1 is relatively close to HF). As discussed in the following section, the systematic differences found for chlorinated and brominated compounds relate to the net atomic charges of chlorine attached to aliphatic carbon in particular, and to bromine attached to aromatic carbon.

## C. CPSA Descriptors

The comparison between the method-specific trends for the partial positive and negative surface areas (PPSA-1, PNSA-1) as well as their total charge weighted and atomic charge weighted counterparts (PPSA-2, PNSA-2; PPSA-3, PNSA-3) is summarized in terms of $R^2$ values in Table 6.7 together with respective results for the three PPSA descriptors confined to heavy atoms (PPSA-1Z, PPSA-2Z, PPSA-3Z) and to hydrogen (PPSA-1H, PPSA-2H, PPSA-3H).

As before, the AM1 trends are more similar to the ones of PM5 than of PM3, but this time the situation is more complex with regard to the correlations between semiempirical and *ab initio* methods. Considering PPSA-1, which quantifies the positively charged part of the molecular surface area, the greatest degree of agreement with B3LYP is achieved by PM3 ($R^2 = 0.920$); for both PPSA-2 (which is PPSA-1 multiplied by the sum of all positive net atomic charges) and PPSA-3 (where each atom-centered surface area portion is weighted by the respective net atomic charge) AM1 and PM5 are closer to B3LYP than PM3. Restriction of the surface area to the contributions from heavy atoms lowers the $R^2$ values significantly, and now PM3 appears to be slightly superior to both AM1 and PM5 for PPSA-1Z ($R^2 = 0.717$) and PPSA-2Z (0.597), while for PPSA-3Z the order of relative agreement with B3LYP is PM5 (0.616) followed by PM3 (0.567) and AM1 (0.524). In contrast, the respective surface areas confined to hydrogen (PPSA-1H, PPSA-2H, PPSA-3H) yield highly significant correlations among the semiempirical methods as well as with HF and B3LYP, except for PM3 with regard to PPSA-3H.

For the negatively charged surface (PNSA-1), its total charge weighted value (PNSA-2) and the sum of surface area contributions weighted locally by the respective net atomic charges (PNSA-3), the intercorrelations are, in most cases, inferior to the ones with the PPSA counterparts. Taking PNSA-3 as an example, the PM5 trend is in good (but not excellent) agreement with HF ($R^2 = 0.848$) and B3LYP (0.816), while both AM1 and PM3 yield respective $R^2$ values around 0.65.

In Figure 6.7 to Figure 6.9, the data distributions of the pair-wise comparisons of AM1, PM3 and PM5 with B3LYP are shown for PPSA-1, PPSA-3, PNSA-1, and PNSA-3 as well as for PPSA-1Z and PPSA-3Z. Considering again B3LYP (which represents the highest level of theory) as reference method, it can be seen from the left column with 3 plots in Figure 6.7 that the overall outlier patterns of AM1 (top left) and PM5 (bottom left) are more similar to each other than to that of PM3 (middle left). All three semiempirical methods show greater similarities with regard to the positive outliers (compounds providing overestimations as compared B3LYP results).

Among the compounds with very large AM1, PM3, and PM5 overestimations of PPSA-1 are the brominated aromatics pentabromophenol, 2,4,6-tribromophenol and 2,4,5-tribromoimidazole, and also for AM1 $\alpha,\alpha,\alpha',\alpha'$-tetrabromo-*o*-xylene (cf. Figure 6.1). These overestimations can be traced back to the systematic difference in bromine net atomic charges between the semiempirical and *ab initio* schemes: AM1, PM3, and PM5 also yield positively charged bromine, while with HF and B3LYP bromine is always negatively charged (cf. Table 6.6). As a consequence, the large positive PPSA-1 outliers are found as large negative outliers (providing large underestimations) in the corresponding PNSA-1 plots (Figure 6.8; AM1: top left, PM3: middle left, PM5: bottom left).

Interestingly, AM1 is the only method where the aliphatically bound bromine atoms of $\alpha,\alpha,\alpha',\alpha'$-tetrabromo-2-xylene carry positive net atomic charges (0.024 and 0.018 au), contrasting with B3LYP (–0.064 and –0.079 au) and with the other schemes. Moreover, B3LYP yields positive charges for the aromatic carbons to which the two –CHBr$_2$ groups are attached (0.115 au), as opposed to small negative charges (–0.063 au) provided by AM1. For the remaining two aliphatic and four aromatic carbon atoms, the net atomic charges are negative according to both B3LYP (from –0.082 to –0.217 au) and AM1 (from –0.105 to –0.206 au). As a consequence, PPSA-1 is only 60.0 Å$^2$ with B3LYP, but 208.5 Å$^2$ with AM1. When considering only the molecular geometries, the total surface areas are 243 (B3LYP) and 250.4 Å$^2$ (AM1), built from contributions from 4 bromine atoms (B3LYP: 151.4 Å$^2$, AM1: 154.8 Å$^2$), 8 carbon atoms (B3LYP: 41.9 Å$^2$, AM1: 41.9 Å$^2$) and 6 hydrogen atoms (B3LYP: 49.7 Å$^2$, AM1: 53.7 Å$^2$).

**Table 6.7  Correlation in Terms of $R^2$ between AM1, PM3, PM5, HF/6-31G**, and B3LYP/6-31G** for CPSA Descriptors[a]**

| Descriptor | No. Compounds[b] | AM1 vs. | | | | PM3 vs. | | | PM5 vs. | | HF vs. |
|---|---|---|---|---|---|---|---|---|---|---|---|
| | | PM3 | PM5 | HF | B3LYP | PM5 | HF | B3LYP | HF | B3LYP | B3LYP |
| PPSA-1 | 607 | 0.898 | 0.954 | 0.867 | 0.885 | 0.901 | 0.931 | 0.920 | 0.854 | 0.878 | 0.971 |
| PPSA-2 | 607 | 0.913 | 0.967 | 0.914 | 0.902 | 0.894 | 0.852 | 0.823 | 0.958 | 0.955 | 0.989 |
| PPSA-3 | 607 | 0.692 | 0.947 | 0.798 | 0.722 | 0.650 | 0.541 | 0.440 | 0.888 | 0.827 | 0.957 |
| PNSA-1 | 607 | 0.651 | 0.848 | 0.567 | 0.614 | 0.670 | 0.750 | 0.700 | 0.542 | 0.601 | 0.892 |
| PNSA-2 | 607 | 0.878 | 0.951 | 0.853 | 0.851 | 0.871 | 0.775 | 0.754 | 0.909 | 0.907 | 0.965 |
| PNSA-3 | 607 | 0.804 | 0.912 | 0.695 | 0.659 | 0.845 | 0.660 | 0.648 | 0.848 | 0.816 | 0.983 |
| PPSA-1Z | 436 | 0.690 | 0.879 | 0.515 | 0.546 | 0.641 | 0.765 | 0.717 | 0.445 | 0.492 | 0.888 |
| PPSA-2Z | 436 | 0.695 | 0.866 | 0.534 | 0.494 | 0.801 | 0.638 | 0.597 | 0.538 | 0.540 | 0.899 |
| PPSA-3Z | 436 | 0.684 | 0.937 | 0.584 | 0.524 | 0.758 | 0.599 | 0.567 | 0.660 | 0.616 | 0.938 |
| PPSA-1H | 602 | 0.999 | 0.999 | 0.999 | 0.999 | 0.998 | 0.998 | 0.998 | 0.999 | 0.999 | 1.000 |
| PPSA-2H | 602 | 0.976 | 0.998 | 0.994 | 0.974 | 0.973 | 0.959 | 0.920 | 0.994 | 0.977 | 0.991 |
| PPSA-3H | 602 | 0.904 | 0.990 | 0.969 | 0.860 | 0.893 | 0.816 | 0.646 | 0.972 | 0.880 | 0.950 |

[a] For the net atomic charges, the Coulson scheme (Equation 6.53) was used for AM1, PM3, and PM5, while for HF/6-31G** and B3LYP/6-31G**, the Mulliken population analysis (Equation 6.55 and Equation 6.57) was employed. The surface areas were calculated using the Molecular Volume and Surface Area Calculation (MOLSV, 1985) and the Pauling van der Waals radii: C: 1.5 Å, H: 1.20 Å, N: 1.5 Å, O: 1.4 Å, F: 1.35 Å, Cl: 1.80 Å, Br: 2.00, I: 2.15 Å, S: 1.80 Å, P: 1.80 Å.

[b] Five of the 607 compounds have no hydrogen, resulting in 602 compounds for the PPSA correlations confined to hydrogen atoms. For evaluating the PPSA correlations confined to heavy atoms, all compounds without any positively or at least zero-charged surface area contribution according to any of the five quantum chemical methods were excluded.

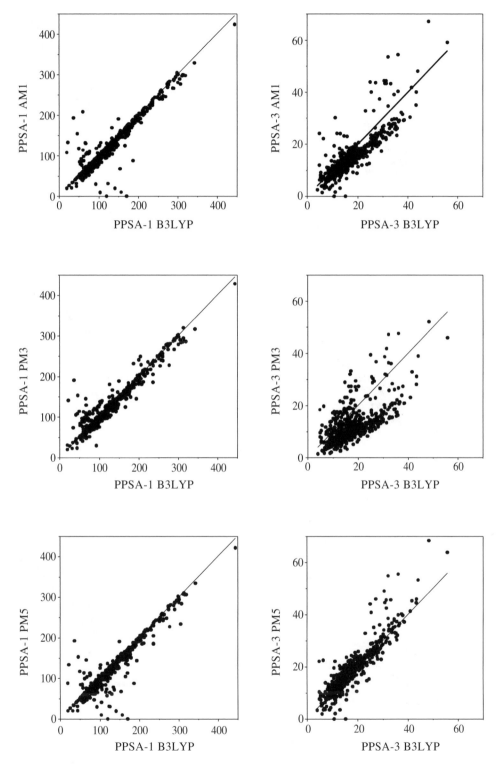

**Figure 6.7** PPSA-1 (left) and PPSA-3 (right) comparison between the semiempirical methods AM1 (top), PM3 (middle), and PM5 (bottom) and B3LYP/6-31G** for the set of 607 compounds.

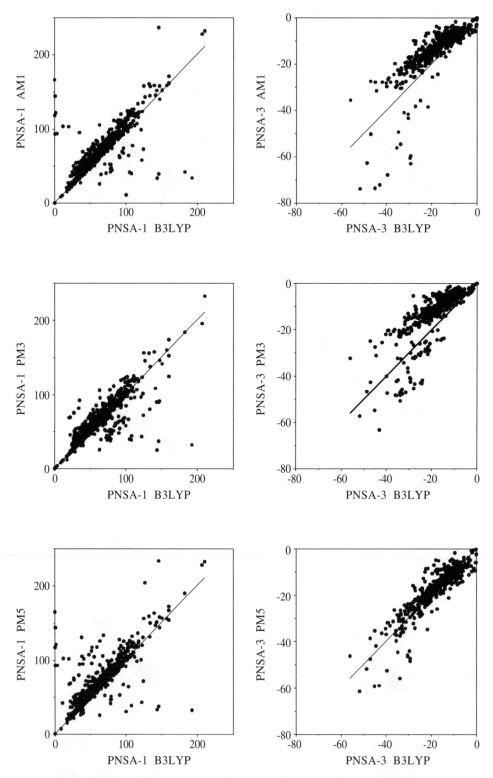

**Figure 6.8**    PNSA-1 (left) and PNSA-3 (right) comparison between the semiempirical methods AM1 (top), PM3 (middle), and PM5 (bottom) and B3LYP/6-31G** for the set of 607 compounds.

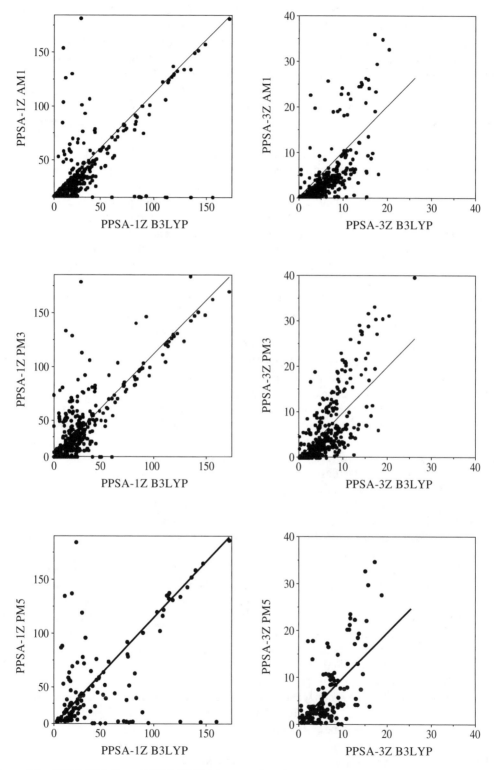

**Figure 6.9**    PPSA-1Z (left) and PPSA-3Z (right) comparison between the semiempirical methods AM1 (top), PM3 (middle), and PM5 (bottom) and B3LYP/6-31G** for the set of 607 compounds.

Note that PM3 and PM5 yield negatively charged bromine atoms for $\alpha,\alpha,\alpha',\alpha'$-tetrabromo-2-xylene (PM3: $-0.048$ au; PM5: $-0.058$ and $-0.037$ au), but positively charged bromine atoms in the case of pentabromophenol (PM3: from 0.075 to 0.097 au; PM5: from 0.078 to 0.120 au) and other brominated aromatics, indicating a corresponding difference in the PM3 and PM5 electronegativities between aliphatically and aromatically bound bromine. Accordingly, the PM3 and PM5 results of PPSA-1 for the latter compound (190.8 $\text{Å}^2$, 192.6 $\text{Å}^2$) are similar to the AM1 value (193.2 $\text{Å}^2$), but much larger than the B3LYP value (36.2 $\text{Å}^2$). In fact, pentabromophenol is the largest positive outlier (providing the largest PM3 and PM5 overestimation when compared with B3LYP) with respect to PPSA-1 (Figure 6.7, middle and bottom left), and the largest negative outlier of the corresponding PNSA-1 data distribution (Figure 6.8, middle and bottom left).

The greatest PPSA-1 underestimations of AM1 and PM5 when compared with B3LYP (Figure 6.7, top and bottom left) are found for hexachloroethane and carbon tetrachloride (both of which have PPSA-1 = 0 according to AM1 and PM5) as well as for pentachloroethane, 1,1,2,2-tetrachloroethane, chloroform, 1,1,1-trichloroethane, and 2,2,2-trichloroethanol (Figure 6.1). Again, the discrepancies are caused by a systematic difference in the net atomic charge calculation, which in this case relates to the aliphatically bound chlorine atom as mentioned above (with AM1 and PM5, chlorine attached to aliphatic carbon appears to be more electronegative than with HF and B3LYP; cf. Table 6.6).

The situation is illustrated with hexachloroethane, which has calculated total surface areas of 170.5 (B3LYP) and 166.2 $\text{Å}^2$ (AM1), respectively. In this case, the molecular surface is almost completely covered by chlorine atoms (B3LYP: 170.4 $\text{Å}^2$, AM1: 166.1 $\text{Å}^2$), with an essentially negligible contribution from the carbon skeleton (B3LYP: 0.1 $\text{Å}^2$; AM1: 0.1 $\text{Å}^2$). Because B3LYP yields a positive net atomic charge for chlorine (0.089 au) and a negative charge for carbon ($-0.268$ au), the resultant PPSA-1 is identical with the chlorine surface area (170.4 $\text{Å}^2$). It follows that with B3LYP the molecular surface of hexachloroethane appears to be positively charged over the whole area (except the very small carbon-associated region). In contrast, AM1 provides a very small but negative charge for chlorine ($-0.001$ au) and an again small but positive charge for carbon (0.002 au), such that now PPSA-1 is identical to the carbon surface area (0.1 $\text{Å}^2$). In this case, the molecular surface carries a small negative charge (again except for a negligible region associated with the carbon skeleton). These results illustrate that even with quite small absolute values of the net atomic charges, different charge signs may lead to very large differences for CPSA descriptor values.

In the right columns of Figure 6.7 and Figure 6.8, the AM1-B3LYP, PM3-B3LYP, and PM5-B3LYP data distributions are shown for the atomic charge weighted partial positive and negative surface areas, PPSA-3 and PNSA-3. With AM1 (top right in Figure 6.7), the greatest overestimations of PPSA-3 as compared to B3LYP are found for 1-benzylpyridinium-3-sulfonate (see Section V.B, for a method comparison of respective net atomic charges), fensulfothion (a phosphorothionate containing the central $[RO]_2P[=S]OR$ group; cf. Figure 6.1), and triphenyl phosphate, and the largest underestimations for hexachloroethane and carbon tetrachloride (see discussion above about the method-specific treatment of chlorine net atomic charges). Regarding PNSA-3, the greatest positive AM1 outlier is flucythrinate, and among the many negative outliers the six greatest underestimations are observed for the phosphorodithioates carbophenothion (mentioned already as $E_{HOMO}$ outlier, Section V.B), terbufos, disulfoton, and azinophos-methyl (which are also $E_{LUMO}$ outliers, Section V.B) for O-ethyl-O-(4-nitrophenyl)-phenyl-phosphonothioate (an organophosphorus compound with the central $(RO)_2P(=S)R$ group; cf. Figure 6.1) and for the above-mentioned fensulfothion.

Because PPSA-3 and PNSA-3 are affected by both the sign and extent of the net atomic charges in a local manner, different types of disagreement between the methods may lead to similar discrepancies in the descriptor values. An example of the PPSA-3 outliers where AM1 agrees with B3LYP in the signs of all net atomic charges is triphenyl phosphate. Again, the calculated total surface areas are quite similar (B3LYP: 183.8 $\text{Å}^2$ vs. AM1: 185.7 $\text{Å}^2$), which holds also true for the surface area contributions from the central phosphorus (B3LYP: 11.4 $\text{Å}^2$ vs. AM1: 11.6 $\text{Å}^2$), from the 4 oxygen atoms (B3LYP: 41.6 Å$^2$ vs. AM1: 42.3 Å$^2$), from the 18 carbon atoms (B3LYP: 134.6 Å$^2$ vs. AM1: 135.0 Å$^2$), and from the 15 hydrogen atoms (B3LYP: 153.6 Å$^2$ vs. AM1: 155.5 Å$^2$). By contrast, the net atomic charges show substantial differences in their absolute values, which is particularly striking for phosphorus (B3LYP: 1.285 au vs. AM1: 2.622), the P=O oxygen (B3LYP: –0.536 au vs. AM1: –1.042 au) and three positively charged carbon atoms (B3LYP: around 0.3 au vs. AM1: around 0.1 au), and to a smaller degree applies also for the negatively charged carbons (B3LYP: around –0.1 au vs. AM1: –0.13 au) and all hydrogen atoms (B3LYP: around 0.1 au vs. AM1: around 0.14 au). Overall, the generally greater absolute values of the AM1 net atomic charges led to a significantly greater PPSA-3 value according to AM1 when compared with B3LYP (54.4 vs. 36.1). The detailed analysis shows that this difference is because of greater PPSA-3 contributions provided by phosphorus (30.4 vs. 14.6) and all hydrogens (22.1 vs. 15.9), which largely overcompensate the smaller PPSA-3 contribution of the three positively charged carbon atoms (2.0 vs. 5.6).

In Figure 6.9, the semiempirical results for the partial positive surface area confined to heavy atoms, PPSA-1Z, and its atomic charge weighted counterpart, PPSA-3Z, are compared with B3LYP for all 607 compounds (including those with zero PPSA-1Z values, as opposed to the selection procedure applied for the R$^2$ statistics of Table 6.7). With regard to the PPSA-1Z comparison between AM1 and B3LYP (top left), the greatest AM1 underestimations are found for chlorinated alkanes (hexachloroethane, pentachloroethane, 1,1,2,2-tetrachloroethane, carbon tetrachloride, chloroform, 1,1,1-trichloroethane, 1,1,2-trichloroethane) and compounds with corresponding moieties (dicofol, 1,1,1-trichloroethanol, 1,1,1-trichloro-2-methyl-2-propanol). Again, this is caused by the fact that in contrast to the general B3LYP pattern, AM1 yields negative net atomic charges for aliphatically bound chlorine, making PPSA-1Z zero or very small in these cases. Accordingly, all of these compounds are also negative outliers in the PPSA-1 plot (Figure 6.7, left column) as discussed above. Interestingly, among the above-mentioned chlorinated alkanes 1,1,2-trichloroethane is the only compound with negative PM3 net atomic charges at the chlorine atoms (–0.001 au and –0.015 au), and at the same time the greatest negative PPSA-1Z outlier with PM3 (Figure 6.9, middle left). Note further that 1,1,2-trichloroethane is also exceptional with B3LYP, which yields a negative charge for 1 of the 3 chlorine atoms (–0.032 au).

A further large PPSA-1Z discrepancy relates to 1,2-bis(4-pyridyl)ethane (cf. Figure 6.1), where the B3LYP value of 41.6 Å$^2$ contrasts with 0 Å$^2$ according to AM1. Inspection of the net atomic charges reveals that apart from a large difference in the magnitude of the negative charge of the aromatic nitrogens (B3LYP: –0.431 au vs. AM1: –0.139 au), there is an important difference in the treatment of the aromatic carbons. Here, AM1 yields negative values throughout, while with B3LYP only the carbon atoms in the 3-position meta position relative to nitrogen are negatively charged (B3LYP: –0.127 au vs. AM1: –0.180 au), and the carbon atoms in the 2-position ortho position relative to nitrogen as well as the ones connected to the ethylene bridge carry positive B3LYP charges (B3LYP: 0.95 au, 0.140 au; AM1: –0.070 au, –0.036 au). Consequently, the AM1 surface area of all heavy atoms is negatively charged, and here PPSA-1 (124.9 Å$^2$) is identical to the hydrogen-associated part of the molecular surface area, PPSA-1H. By contrast, with B3LYP six aromatic carbon atoms contribute to the positive part of the molecular surface area (PPSA-1 = 166.2 Å$^2$), which is nonetheless still dominated by hydrogen atoms (PPSA-1H = 124.6 Å$^2$).

Large AM1 overestimations of PPSA-1Z when compared with B3LYP (Figure 6.9, top left and bottom left) are found for brominated aromatics (pentabromophenol, 2,4,6-tribromophenol, 2,4,5-tribromoimidazole, 1,2-dibromobenzene, 3,5-dibromo-4-hydroxybenzonitril), all of which also correspond to PPSA-1 outliers (Figure 6.7, top left) for α,α,α′,α′-tetrabromo-2-xylene with aliphatically bound bromine (which was already mentioned as PPSA-1 outlier for AM1 in contrast to PM3 and PM5) and for dibromoacetonitrile. Again, the reason is the systematically lower Br electronegativity with AM1 as compared to the *ab initio* methods, leading to positive AM1 net atomic charges that contrast with negatively charged bromine according to B3LYP. With regard to aromatically bound bromine, the π-electron donor strength overcompensates the negative inductive (σ electron) effect with all three semiempirical methods, which contrasts with an opposite pattern according to the *ab initio* methods.

A further large positive AM1 outlier with respect to PPSA-1Z is 2,4,6-trichlorophenol, despite identical values of B3LYP and AM1 for the total surface area (171.7 $Å^2$) and the carbon-associated part (36.9 $Å^2$), and very similar values for the surface area contributions from chlorine (95.4 vs. 95.2 $Å^2$), oxygen (11.1 $Å^2$ vs. 11.3 $Å^2$) and hydrogen (28.2 vs. 28.3 $Å^2$). In this case, AM1 provides small but positive net atomic charges for all chlorine atoms that are attached to aromatic carbon (ortho: 0.003 and 0.001 au vs. para: 0.027 au), while the two chlorine atoms in the 2-position ortho position attain a small but negative charge with B3LYP (ortho: –0.002 au vs. para: 0.029 au). As a consequence, the AM1 partial positive surface area confined to heavy atoms, PPSA-1Z, includes all chlorine atoms (101.9 $Å^2$), as opposed to a much smaller B3LYP result (38.5 $Å^2$) where only one of the three chlorine atoms is positively charged. These results indicate further that with AM1, aromatically bound chlorine is less electronegative than aliphatically bound chlorine.

Returning to the PPSA-1Z calculation with PM3 (Figure 6.9, middle left), the largest underestimations following 1,1,2-trichloroethane (s.a.) are observed for organic sulfides (R–S–R, R–S–S–R) and for 1,2-*bis*-4(pyridyl)ethane (that was already mentioned as a corresponding outlier to AM1). For the latter, PM3 yields negative net atomic charges for all heavy atoms as does AM1, resulting in a zero value for PPSA-1Z as opposed to 41.6 $Å^2$ according to B3LYP (s.a.). Interestingly, with PM5 the two ethylene bridge carbons are positively charged (0.017 au), resulting in a PPSA-1Z value above zero (12.0 $Å^2$), but still substantially smaller than with B3LYP.

Taking 3,6-dithiaoctane as an example of the above-mentioned sulfide group, the semiempirical and *ab initio* methods agree in negatively charged carbon atoms, but only B3LYP, HF and AM1 provide positive net atomic charges for sulfur (0.080, 0.122, and 0.049 au). With PM3 and PM5, the sulfide sulfur is apparently more electronegative as reflected by negative net atomic charges (–0.106 and –0.052 au). It follows that despite similar values for the B3LYP and PM3 molecular surface areas (214.5 vs. 213.5 $Å^2$) and the respective part associated with the two sulfur atoms (50.4 vs. 50.0 $Å^2$), the two methods differ significantly in their PPSA-1Z values; with B3LYP, the result is identical to the sulfur-associated surface area, while with PM3 there is no positively charged heavy atom, making PPSA-1Z zero for this molecule. Accordingly, the organic sulfides belong to the great positive PNSA-1 outliers of PM3 (Figure 6.8, middle left), as is illustrated by PNSA-1 values of 20.5 (B3LYP) vs. 68.7 $Å^2$ (PM3).

With regard to PPSA-3Z (Figure 6.9, right column), the outlier pattern is more complex because of the combined influence of local charge modus and charge extent. When compared with B3LYP, the greatest AM1 overestimations are observed for pentabromophenol, 2,4,6-triiodophenol, 1-benzyl-pyridinium-3-sulphonate, 2,4,5-tribromoimidazole, and saccharin. All of these compounds also belong to the group of large positive outliers with PM5, and most yield great positive overestimations with PM3.

Here, saccharin (Figure 6.1) is an example where all methods agree in the charge modus of all atoms, and nonetheless yield significantly different PPSA-3Z values. For the sulfur atom, the positive charge is large but still significantly smaller with B3LYP (1.198 au) than with AM1 (2.854 au), PM3 (2.310 au), and PM5 (2.846 au). At the same time, the negative B3LYP nitrogen charge (–0.681 au) is in between the values according to PM3 (–0.512 au) and AM1 (–0.859) as well as PM5 (–0.953 au). Only B3LYP yields similar charges for the carbonyl oxygen and the oxygen atoms attached to sulfur (–0.467 vs. –0.474 au), which contrasts with large differences for AM1 (–0.285 vs. –0.914 au), PM3 (–0.326 vs. –0.813 au), and PM5 (–0.374 vs. –0.896 au). Only one aromatic carbon atom has a positive (but very small) net atomic charge, but the carbonyl carbon carries a substantial positive charge that is greater with B3LYP than with the semiempirical methods (B3LYP: 0.568 au, AM1: 0.356 au, PM3: 0.346 au, PM5: 0.456 au). Because the positively charged surface portions are similar for all methods, the PPSA-1Z values vary only little (B3LYP: 25.3 $Å^2$, AM1: 26.1 $Å^2$, PM3: 27.0 $Å^2$, PM5: 25.8 $Å^2$). However, the substantial difference in the positive sulfur charge between B3LYP and the semiempirical schemes results in correspondingly different values for PPSA-3Z (B3LYP: 18.9, AM1: 34.6, PM3: 30.5, PM5: 35.2), which in turn are dominated by the respective sulfur contributions (B3LYP: 14.6, AM1: 32.0, PM3: 28.0, PM5: 31.6). Note further that the negatively charged part of the molecular surface area, PNSA-1, is similar across the methods (B3LYP: 89.5 $\text{Å}^2$; AM1: 87.2 $\text{Å}^2$, PM3: 88.4 $\text{Å}^2$, PM5: 87.0 $\text{Å}^2$), but that because of the above-mentioned differences in the extent of negative charge associated with oxygen attached to sulfur as well as with nitrogen, PNSA-3 varies significantly (B3LYP: –30.3, AM1: –43.4, PM3: –38.2, PM5: –46.4).

With regard to large underestimations of PPSA-3Z with respect to B3LYP predictions (Figure 6.9, right column), hexachloroethane is the greatest outlier from both AM1 and PM5 calculations (both of which yield negatively charged chlorine atoms); the second largest deviation is observed for tripropargylamine (cf. Figure 6.1), which is also the greatest negative outlier with PM3. Here, the negative charge of the central nitrogen varies considerably (B3LYP: –0.410 au, AM1: –0.264 au, PM3: –0.070 au, PM5: –0.307 au). More importantly, however, are the systematic differences in the charge modus for the non-terminal $sp^2$ carbons and the $sp^3$ carbons. For the former, a large positive net atomic charge according to B3LYP (around 0.34 au) contrasts with negative semiempirical charges, and for the latter a negative B3LYP charge is opposed to positive charges with AM1, PM3 and PM5. Because the three nonterminal $sp^2$ carbons have a much greater surface area contribution than the three $sp^3$ carbon atoms attached to the central nitrogen (e.g., B3LYP: 36.16 vs. 8.71 $\text{Å}^2$), both PPSA-1Z and PPSA-3Z are significantly greater for B3LYP (36.2 $\text{Å}^2$ and 12.4) than for the semiempirical methods (e.g., AM1: 8.7 $\text{Å}^2$ and 0.4). For the same reason, PNSA-1 is correspondingly smaller with B3LYP (58.4 $\text{Å}^2$) than with AM1 (84.3 $\text{Å}^2$), PM3 (85.3 $\text{Å}^2$) and PM5 (82.5 $\text{Å}^2$).

## D. Atom Type-Specific Subset Correlations with B3LYP

In Table 6.8, atom type-specific subset correlations of AM1, PM3, PM5, and HF with B3LYP are shown for frontier orbital energies and related descriptors, for parameters based on the charge distribution, and for the PPSA and PNSA descriptors. In contrast to Table 6.4 and Table 6.7, however, now all compounds providing a zero descriptor value by definition (e.g., PPSA-1Z = 0 if, for a given method, the compound has no positively charged heavy atom) were included for generating the statistics.

Taking $E_{HOMO}$ as an example, PM3 correlates much less with B3LYP ($R^2 = 0.500$) for the 39 halogenated compounds than the other three methods. As mentioned in Section V.B, above, major outliers of the corresponding PM3-B3LYP data distribution (Figure 6.2, top right) are polyfluorinated hydrocarbon derivatives and polychlorinated hydrocarbons.

Even within the two *ab initio* schemes there is only a moderate agreement with regard to the 107 nitrogen-containing compounds (without other heteroatoms except halogen; $R^2 = 0.732$), the 265 oxygen-containing compounds (without other heteroatoms except halogen; $R^2 = 0.701$) and the 133 compounds with both nitrogen and oxygen (but no other heteroatoms except halogen; $R^2 = 0.690$). There is also quite a poor correlation for the 42 compounds containing phosphorus or sulfur (without any restriction for other heteroatoms; $R^2 = 0.516$) this corrleation suggests that for the first vertical ionization potential, a more detailed investigation is needed to clarify the actual prediction quality of HF and B3LYP (and possibly other DFT method) frontier orbital energies and the respective performances of the semiempirical schemes.

With respect to $E_{LUMO}$, the trend of PM3 calculations is again much more different from B3LYP than all other methods for halogenated compounds ($R^2 = 0.567$), and all three semiempirical methods show significantly lower correlations with B3LYP than HF for the subset of 42 compounds with P or S (AM1: $R^2 = 0.519$, PM3: $R^2 = 0.499$, PM5: $R^2 = 0.297$, HF: $R^2 = 0.973$). As discussed above in the context of Figure 6.3 (Section V.B), all three semiempirical methods yield substantial deviations from the HF and B3LYP trends for the $E_{LUMO}$ energies of organophosphorus compounds

**Table 6.8** Correlation in Terms of $R^2$ between AM1, PM3, PM5, HF/6-31G**, and B3LYP/6-31G** Descriptors for Atom Type-Specific Subsets[a]

| Descriptor | B3LYP vs.: | C, H | Halogen | N | O | N, O | P, S | Non-aromatic | Aromatic | All |
|---|---|---|---|---|---|---|---|---|---|---|
| No. compounds | | 21 | 39 | 107 | 265 | 133 | 42 | 276 | 331 | 607 |
| $E_{HOMO}$ | AM1 | 0.977 | 0.899 | 0.911 | 0.832 | 0.878 | 0.593 | 0.818 | 0.886 | 0.830 |
| | PM3 | 0.953 | 0.500 | 0.911 | 0.727 | 0.862 | 0.687 | 0.638 | 0.826 | 0.726 |
| | PM5 | 0.902 | 0.909 | 0.912 | 0.835 | 0.878 | 0.623 | 0.819 | 0.827 | 0.829 |
| | HF | 0.955 | 0.960 | 0.732 | 0.710 | 0.690 | 0.516 | 0.713 | 0.914 | 0.689 |
| $E_{LUMO}$ | AM1 | 0.940 | 0.805 | 0.928 | 0.888 | 0.943 | 0.519 | 0.796 | 0.865 | 0.860 |
| | PM3 | 0.904 | 0.576 | 0.931 | 0.870 | 0.936 | 0.499 | 0.737 | 0.819 | 0.822 |
| | PM5 | 0.894 | 0.788 | 0.914 | 0.806 | 0.882 | 0.297 | 0.655 | 0.799 | 0.765 |
| | HF | 0.958 | 0.873 | 0.966 | 0.905 | 0.971 | 0.973 | 0.920 | 0.978 | 0.939 |
| EN | AM1 | 0.610 | 0.878 | 0.933 | 0.859 | 0.956 | 0.567 | 0.872 | 0.926 | 0.891 |
| | PM3 | 0.604 | 0.192 | 0.898 | 0.785 | 0.951 | 0.547 | 0.776 | 0.897 | 0.830 |
| | PM5 | 0.672 | 0.821 | 0.898 | 0.647 | 0.890 | 0.462 | 0.760 | 0.870 | 0.774 |
| | HF | 0.053 | 0.954 | 0.849 | 0.703 | 0.952 | 0.847 | 0.854 | 0.973 | 0.856 |
| HD | AM1 | 0.972 | 0.798 | 0.957 | 0.889 | 0.852 | 0.447 | 0.725 | 0.515 | 0.844 |
| | PM3 | 0.939 | 0.925 | 0.973 | 0.868 | 0.829 | 0.512 | 0.658 | 0.458 | 0.809 |
| | PM5 | 0.912 | 0.899 | 0.941 | 0.857 | 0.805 | 0.207 | 0.600 | 0.519 | 0.779 |
| | HF | 0.975 | 0.923 | 0.972 | 0.858 | 0.902 | 0.894 | 0.786 | 0.843 | 0.887 |
| $\mu$ (dip.m.) | AM1 | 0.946 | 0.968 | 0.910 | 0.776 | 0.828 | 0.863 | 0.763 | 0.859 | 0.847 |
| | PM3 | 0.777 | 0.911 | 0.821 | 0.701 | 0.725 | 0.873 | 0.731 | 0.801 | 0.795 |
| | PM5 | 0.686 | 0.942 | 0.840 | 0.726 | 0.769 | 0.716 | 0.782 | 0.781 | 0.792 |
| | HF | 0.966 | 0.992 | 0.980 | 0.976 | 0.961 | 0.990 | 0.982 | 0.975 | 0.978 |
| $Q_{Cmax}^{+}$ | AM1 | 0.000 | 0.250 | 0.571 | 0.771 | 0.748 | 0.905 | 0.886 | 0.680 | 0.776 |
| | PM3 | 0.054 | 0.425 | 0.409 | 0.602 | 0.384 | 0.799 | 0.793 | 0.440 | 0.579 |
| | PM5 | 0.067 | 0.163 | 0.519 | 0.757 | 0.693 | 0.922 | 0.865 | 0.651 | 0.752 |
| | HF | 0.234 | 0.540 | 0.927 | 0.934 | 0.947 | 0.988 | 0.963 | 0.944 | 0.950 |
| $Q_{Cavp}^{+}$ | AM1 | 0.000 | 0.095 | 0.327 | 0.317 | 0.308 | 0.779 | 0.656 | 0.176 | 0.441 |
| | PM3 | 0.056 | 0.189 | 0.286 | 0.188 | 0.110 | 0.681 | 0.619 | 0.075 | 0.319 |
| | PM5 | 0.058 | 0.048 | 0.242 | 0.376 | 0.360 | 0.749 | 0.681 | 0.174 | 0.457 |
| | HF | 0.176 | 0.292 | 0.800 | 0.573 | 0.592 | 0.903 | 0.890 | 0.496 | 0.718 |
| $Q_{Cmax}^{-}$ | AM1 | 0.420 | 0.108 | 0.208 | 0.232 | 0.234 | 0.020 | 0.416 | 0.060 | 0.113 |
| | PM3 | 0.027 | 0.000 | 0.000 | 0.005 | 0.086 | 0.001 | 0.293 | 0.016 | 0.012 |
| | PM5 | 0.544 | 0.112 | 0.325 | 0.339 | 0.258 | 0.067 | 0.449 | 0.142 | 0.213 |
| | HF | 0.938 | 0.880 | 0.856 | 0.866 | 0.867 | 0.838 | 0.956 | 0.823 | 0.864 |
| $Q_{Cavn}^{-}$ | AM1 | 0.188 | 0.133 | 0.250 | 0.299 | 0.199 | 0.385 | 0.261 | 0.072 | 0.245 |
| | PM3 | 0.002 | 0.104 | 0.075 | 0.044 | 0.033 | 0.202 | 0.295 | 0.019 | 0.011 |
| | PM5 | 0.287 | 0.052 | 0.198 | 0.347 | 0.099 | 0.428 | 0.250 | 0.077 | 0.231 |
| | HF | 0.683 | 0.936 | 0.840 | 0.844 | 0.851 | 0.939 | 0.910 | 0.637 | 0.867 |
| $Q_{Hmax}^{+}$ | AM1 | 0.108 | 0.673 | 0.452 | 0.822 | 0.684 | 0.654 | 0.826 | 0.785 | 0.722 |

| Descriptor | Method | | | | | | | | | |
|---|---|---|---|---|---|---|---|---|---|---|
| | PM3 | 0.585 | 0.573 | 0.682 | 0.092 | 0.477 | 0.865 | 0.001 | 0.504 | 0.116 |
| | PM5 | 0.767 | 0.767 | 0.857 | 0.658 | 0.758 | 0.900 | 0.264 | 0.721 | 0.140 |
| | HF | 0.959 | 0.966 | 0.987 | 0.860 | 0.950 | 0.972 | 0.949 | 0.906 | 0.034 |
| $Q_{Havp}^{+}$ | AM1 | 0.597 | 0.686 | 0.846 | 0.051 | 0.645 | 0.614 | 0.592 | 0.665 | 0.062 |
| | PM3 | 0.372 | 0.372 | 0.602 | 0.000 | 0.357 | 0.589 | 0.047 | 0.510 | 0.117 |
| | PM5 | 0.646 | 0.717 | 0.848 | 0.066 | 0.663 | 0.697 | 0.562 | 0.727 | 0.025 |
| | HF | 0.841 | 0.891 | 0.971 | 0.235 | 0.832 | 0.854 | 0.869 | 0.892 | 0.005 |
| PPSA-1 | AM1 | 0.885 | 0.852 | 0.906 | 0.979 | 0.951 | 0.920 | 0.905 | 0.083 | 0.951 |
| | PM3 | 0.920 | 0.850 | 0.972 | 0.908 | 0.934 | 0.920 | 0.915 | 0.786 | 0.949 |
| | PM5 | 0.878 | 0.843 | 0.899 | 0.931 | 0.924 | 0.911 | 0.916 | 0.124 | 0.963 |
| | HF | 0.971 | 0.944 | 0.991 | 0.967 | 0.974 | 0.972 | 0.986 | 0.832 | 0.969 |
| PPSA-2 | AM1 | 0.902 | 0.850 | 0.947 | 0.885 | 0.964 | 0.972 | 0.961 | 0.605 | 0.836 |
| | PM3 | 0.823 | 0.758 | 0.923 | 0.813 | 0.664 | 0.950 | 0.916 | 0.576 | 0.756 |
| | PM5 | 0.955 | 0.929 | 0.978 | 0.914 | 0.958 | 0.977 | 0.965 | 0.802 | 0.868 |
| | HF | 0.989 | 0.983 | 0.996 | 0.989 | 0.989 | 0.990 | 0.988 | 0.929 | 0.914 |
| PPSA-3 | AM1 | 0.722 | 0.661 | 0.871 | 0.607 | 0.803 | 0.812 | 0.607 | 0.172 | 0.589 |
| | PM3 | 0.440 | 0.411 | 0.772 | 0.507 | 0.101 | 0.544 | 0.186 | 0.107 | 0.254 |
| | PM5 | 0.827 | 0.808 | 0.906 | 0.714 | 0.859 | 0.892 | 0.657 | 0.222 | 0.654 |
| | HF | 0.957 | 0.962 | 0.988 | 0.944 | 0.962 | 0.948 | 0.916 | 0.890 | 0.801 |
| PPSA-1Z | AM1 | 0.525 | 0.598 | 0.353 | 0.761 | 0.713 | 0.632 | 0.516 | 0.229 | 0.000 |
| | PM3 | 0.703 | 0.615 | 0.828 | 0.262 | 0.635 | 0.707 | 0.569 | 0.887 | 0.015 |
| | PM5 | 0.463 | 0.525 | 0.298 | 0.349 | 0.527 | 0.568 | 0.531 | 0.291 | 0.327 |
| | HF | 0.878 | 0.839 | 0.939 | 0.649 | 0.829 | 0.873 | 0.905 | 0.909 | 0.364 |
| PPSA-2Z | AM1 | 0.522 | 0.503 | 0.459 | 0.869 | 0.667 | 0.358 | 0.384 | 0.278 | 0.000 |
| | PM3 | 0.626 | 0.585 | 0.674 | 0.796 | 0.561 | 0.666 | 0.447 | 0.808 | 0.000 |
| | PM5 | 0.572 | 0.551 | 0.496 | 0.854 | 0.599 | 0.425 | 0.482 | 0.384 | 0.022 |
| | HF | 0.911 | 0.900 | 0.920 | 0.949 | 0.904 | 0.889 | 0.863 | 0.989 | 0.452 |
| PPSA-3Z | AM1 | 0.539 | 0.539 | 0.529 | 0.842 | 0.571 | 0.299 | 0.110 | 0.266 | 0.000 |
| | PM3 | 0.596 | 0.574 | 0.601 | 0.823 | 0.375 | 0.466 | 0.188 | 0.637 | 0.000 |
| | PM5 | 0.638 | 0.658 | 0.592 | 0.877 | 0.631 | 0.504 | 0.139 | 0.373 | 0.055 |
| | HF | 0.948 | 0.948 | 0.937 | 0.981 | 0.957 | 0.917 | 0.848 | 0.973 | 0.325 |
| PPSA-1H | AM1 | 1.000 | 1.000 | 1.000 | 0.999 | 0.999 | 1.000 | 1.000 | 1.000 | 0.998 |
| | PM3 | 0.998 | 0.999 | 0.998 | 0.996 | 0.998 | 0.998 | 0.999 | 1.000 | 0.995 |
| | PM5 | 1.000 | 0.999 | 1.000 | 1.000 | 0.999 | 1.000 | 1.000 | 1.000 | 0.998 |
| | HF | 1.000 | 1.000 | 1.000 | 1.000 | 1.000 | 1.000 | 1.000 | 1.000 | 1.000 |
| PPSA-2H | AM1 | 0.974 | 0.968 | 0.994 | 0.966 | 0.959 | 0.984 | 0.973 | 0.979 | 0.932 |
| | PM3 | 0.920 | 0.915 | 0.961 | 0.884 | 0.885 | 0.958 | 0.916 | 0.955 | 0.839 |
| | PM5 | 0.977 | 0.974 | 0.994 | 0.972 | 0.971 | 0.986 | 0.975 | 0.980 | 0.938 |
| | HF | 0.991 | 0.989 | 0.997 | 0.988 | 0.984 | 0.994 | 0.993 | 0.993 | 0.967 |

Table 6.8 (continued) Correlation in Terms of $R^2$ between AM1, PM3, PM5, HF/6-31G**, and B3LYP/6-31G** Descriptors for Atom Type-Specific Subsets[a]

| Descriptor | B3LYP vs.: | C, H | Halogen | N | O | N, O | P, S | Non-aromatic | Aromatic | All |
|---|---|---|---|---|---|---|---|---|---|---|
| PPSA-3H | AM1 | 0.633 | 0.874 | 0.824 | 0.892 | 0.851 | 0.763 | 0.972 | 0.876 | 0.865 |
| | PM3 | 0.269 | 0.755 | 0.549 | 0.725 | 0.623 | 0.382 | 0.880 | 0.674 | 0.658 |
| | PM5 | 0.671 | 0.888 | 0.837 | 0.907 | 0.888 | 0.811 | 0.970 | 0.897 | 0.884 |
| | HF | 0.833 | 0.958 | 0.956 | 0.957 | 0.945 | 0.913 | 0.991 | 0.953 | 0.952 |
| PNSA-1 | AM1 | 0.858 | 0.001 | 0.436 | 0.684 | 0.846 | 0.949 | 0.399 | 0.589 | 0.614 |
| | PM3 | 0.853 | 0.788 | 0.470 | 0.631 | 0.787 | 0.724 | 0.807 | 0.534 | 0.700 |
| | PM5 | 0.880 | 0.046 | 0.494 | 0.660 | 0.767 | 0.797 | 0.382 | 0.573 | 0.601 |
| | HF | 0.904 | 0.812 | 0.903 | 0.875 | 0.913 | 0.909 | 0.923 | 0.825 | 0.892 |
| PNSA-2 | AM1 | 0.851 | 0.603 | 0.693 | 0.921 | 0.927 | 0.871 | 0.901 | 0.824 | 0.851 |
| | PM3 | 0.741 | 0.877 | 0.538 | 0.883 | 0.570 | 0.801 | 0.859 | 0.720 | 0.754 |
| | PM5 | 0.878 | 0.778 | 0.711 | 0.911 | 0.927 | 0.900 | 0.921 | 0.898 | 0.907 |
| | HF | 0.871 | 0.929 | 0.912 | 0.945 | 0.979 | 0.965 | 0.986 | 0.957 | 0.965 |
| PNSA-3 | AM1 | 0.588 | 0.318 | 0.379 | 0.879 | 0.785 | 0.813 | 0.674 | 0.625 | 0.659 |
| | PM3 | 0.477 | 0.867 | 0.348 | 0.880 | 0.399 | 0.834 | 0.763 | 0.598 | 0.648 |
| | PM5 | 0.649 | 0.468 | 0.445 | 0.902 | 0.824 | 0.868 | 0.827 | 0.798 | 0.816 |
| | HF | 0.787 | 0.984 | 0.918 | 0.976 | 0.983 | 0.991 | 0.990 | 0.979 | 0.983 |

[a] The compound subsets (21 hydrocarbons, 39 halogenated hydrocarbons, etc.) are selected according to the structural characteristics listed in Table 6.2. In contrast to the atom type specific statistics in Table 6.4 and Table 6.7, all compounds belonging to a certain subset defined through structural features were included for the corresponding statistics (e.g., compounds without positively charged heavy atoms yield zero values for PPSA-1Z etc., which are included in the statistical evaluation of this table).

and other hypervalent structures, and in this respect there appears to be no major difference between AM1 on the one hand, and PM3 as well as PM5 on the other hand. Surprisingly, low $R^2$ values are found for EN when considering the 21 hydrocarbons, where the trend with HF calculations is almost orthogonal to the one achieved with B3LYP ($R^2 = 0.053$), and a similarly low value is obtained for the PM3-B3LYP correlation with respect to the 39 halogenated hydrocarbons ($R^2 = 0.192$). With regard to molecular hardness, particularly low $R^2$ values are obtained for the 42 phosphorous and sulfur compounds except for HF (AM1: 0.447, PM3: 0.512, PM5: 0.207, HF: 0.894), and interestingly the overall statistics are superior those confined to aromatic and non-aromatic compounds, respectively.

The subset-specific trend comparison with respect to the dipole moment reveals significantly lower correlations with B3LYP for PM3 ($R^2 = 0.686$) and PM5 ($R^2 = 0.777$) in the case of hydrocarbons than for AM1 ($R^2 = 0.946$) and HF ($R^2 = 0.966$). At the same time, however, AM1 is the only method where all C atoms of the 21 hydrocarbons are negative, while according to B3LYP, ca. half of these compounds contain $sp^3$ carbon atoms with a (small) positive charge, with an again quite different trend as compared to HF (AM1: $R^2 = 0.0$, HF: $R^2 = 0.234$). These results demonstrate that a good agreement with respect to the dipole moment does not necessarily imply a good agreement with regard to the individual net atomic charges. It follows that net atomic charge fitting to dipole moments (cf. Equation 6.49) does not guarantee that the resultant individual charges reflect local chemical reactivity patterns, an aspect that appears to have not yet been analysed in detail with the so-called class-IV charges (cf. Section IV.D) according to the mapping schemes CM1 (Storer et al., 1995), CM2 (Li et al., 1998), and CM3 (Thompson et al., 2003).

Returning to the statistics for dipole moment in Table 6.8, the group of 265 compounds containing oxygen (but no other heteroatom except halogen) shows the lowest correlations between the semiempirical methods and B3LYP (AM1: $R^2 = 0.776$, PM3: $R^2 = 0.701$, PM5: $R^2 = 0.726$) except for PM5 with a still slightly lower $R^2$ value for the 42 compounds with phosphorus or sulfur atoms or both (0.716). Moreover, with both AM1 and PM3 the 276 non-aromatic compounds yield lower correlation coefficients than the 331 aromatic compounds (AM1: 0.763 vs. 0.859, PM3: 0.731 vs. 0.801).

With respect to the maximum and average positive and negative charges at carbon atoms, $Q_{Cmax}^+$, $Q_{Cavp}^+$ (considering only positively charged carbons) and $Q_{Cmax}^-$ and $Q_{Cavn}^-$ (considering only negatively charged carbons), Table 6.8 shows generally moderate to very low correlations between the 3 semiempirical methods and B3LYP. Interestingly, the agreement in trends is higher for non-aromatic than for aromatic compounds (e.g., $Q_{Cmax}^+$, AM1: $R^2 = 0.886$ vs. 0.680), and here (as well as generally) higher for maximum positive charges than for average positive charges ($Q_{Cavp}^+$, AM1: $R^2 = 0.656$ vs. 0.176). With regard to HF, suprisingly low $R^2$ values are found for $Q_{Cmax}^+$ and $Q_{Cavp}^+$ with hydrocarbons (0.234 and 0.176) and halogenated hydrocarbons (0.540 and 0.292), for $Q_{Cavn}^-$ with hydrocarbons (0.683), and for both $Q_{Cavp}^+$ and $Q_{Cavn}^-$ with respect to the whole subset of 331 aromatic compounds (0.496 and 0.637).

With regard to non-polar compounds, the B3LYP results for $Q_{Hmax}^+$ and $Q_{Havp}^+$ do not correlate at all with the ones of any of the other four methods ($R^2$ values below 0.15 throughout). For the MNDO-type methods, hydrogen atoms attached to oxygen yield significantly greater $Q_{Hmax}^+$ inter-correlations than those attached to nitrogen (e.g., AM1, 265 oxygen compounds vs. 107 nitrogen compounds: $R^2 = 0.822$ vs. 0.452). As a further general trend, Table 6.8 shows greater $R^2$ values for both $Q_{Hmax}^+$ and $Q_{Havp}^+$ in the case of non-aromatic compounds as compared to aromatic compounds (e.g., $Q_{Hmax}^+$, AM1: 0.846 vs. 0.686).

With regard to the CPSA descriptors, PPSA-1 (cf. Table 6.1 and Equation 6.66) yields generally high $R^2$ values except for halogenated compounds, where the very low correlations of AM1 and PM5 with B3LYP ($R^2 = 0.083$ and 0.124) are caused by systematic differences in the net atomic charge calculation of halogen and bromine atoms as discussed in the Sections V.B and C. Note further the quite low $R^2$ values for all three PPSA parameters as well as for the three corresponding descriptors confined to heavy atoms (cf. Table 6.1) when considering the group of 21 hydrocarbons, which reflects the fact that B3LYP differs from all other methods with respect to the assignment of positive net atomic charges to individual $sp^3$ carbon atoms in such compounds as discussed above. This concerns also the HF vs. B3LYP comparison (e.g., PPSA-3Z: $R^2 = 0.325$), as was already seen with very low $R^2$ values for $Q_{Cmax}^+$ (0.234) and $Q_{Cavp}^+$ (0.176).

A somewhat similar situation is observed when comparing the correlations with B3LYP for PPSA-1H and PPSA-3H. The very high $R^2$ values for the former reflect overall similar geometries with regard to the hydrogen-associated part of the molecular surface. The much lower $R^2$ values of the latter, however, reflect significant differences in the net atomic charge trends of hydrogen atoms, with almost negligible correlations between B3LYP and all other methods for $Q_{Hmax}^+$ and $Q_{Havp}^+$ as mentioned above.

The bottom part of Table 6.8 summarizes the comparative analysis for the negatively charged surface area descriptors PNSA-1, PNSA-2, and PNSA-3 (cf. Table 6.1 and Equations 6.67, 6.71, and 6.73). In agreement with the previously discussed systematic differences between B3LYP and both AM1 and PM5 with regard to net atomic charges of chlorine atoms attached to aliphatic carbon and aromatically bound bromine (see Section V.C), there is no correlation for PNSA-1 when considering the group of 21 halogenated compounds (AM1: $R^2 = 0.001$, PM5: $R^2 = 0.046$), and still quite low values for the corresponding PNSA-3 statistics (AM1: $R^2 = 0.318$, PM5: $R^2 = 0.468$). For the latter descriptor, the oxygen compounds show a much greater agreement with B3LYP than the nitrogen compounds (e.g., AM1: $R^2 = 0.879$ vs. 0.379), and for compounds containing both N and O atoms PM3 differs significantly more from B3LYP ($R^2 = 0.399$) than AM1, PM5, and HF ($R^2$ values: 0.785, 0.824, 0.983). Interestingly, the overall HF-B3LYP correlation of PNSA-1 is significantly lower for aromatic compounds as compared to non-aromatic compounds ($R^2 = 0.825$ vs. 0.923), and lower for halogenated compounds ($R^2 = 0.812$) than for any other subset. By contrast, PNSA-3 yields by far the lowest HF-B3LYP correlation for the 21 hydrocarbons ($R^2 = 0.787$).

## E.  Delocalizabilities

The delocalizability descriptors characterizing molecular reactivities toward the attack of a nucleophile ($D^N$; cf. Table 6.1 and Equation 6.58) and electrophile ($D^E$; cf. Table 6.1 and Equation 6.59) have been calculated and analyzed comparatively only for the three semiempirical methods AM1, PM3, and PM5. The resultant statistics are summarized in Table 6.9 and show generally high intercorrelations, but also some cases with only moderate to low $R^2$ values.

In the majority of cases, PM5 correlates more with AM1 than with PM3. Moreover, the average delocalizabilities ($D_{av}^N$, $D_{av}^E$, $D_{C-av}^N$, etc.) yield higher $R^2$ values than the maximum counterparts ($D_{max}^N$, $D_{max}^E$, $D_{C-max}^N$, etc.), except for the subset of parameters confined to oxygen atoms (last four rows in Table 6.9). Here, the AM1 and PM3 trends agree more with each other than with PM5 when considering the donor delocalizabilities (e.g., $D_{O-max}^E$: AM1 vs. PM3: $R^2 = 0.939$, AM1 vs. PM5: $R^2 = 0.805$) and the corresponding acceptor delocalizabilities (that are probably less useful for practical applications, but reflect the energy-weighted contributions of unoccupied MO wavefunctions to oxygen atoms according to the LCAO-MO scheme; cf. Equation 6.18 and Equation 6.58) yield the by far lowest correlations for AM1 vs. PM3 ($D_{O-max}^N$: $R^2 = 0.683$, $D_{O-av}^N$: $R^2 = 0.496$) as well as for PM3 vs. PM5 ($D_{O-max}^N$: $R^2 = 0.458$; $D_{O-av}^N$: $R^2 = 0.183$).

A further interesting difference concerns the intercorrelations for acceptor and donor delocalizabilities confined to carbon atoms. For the former, the three comparisons AM1 vs. PM3, AM1 vs. PM5, and PM3 vs. PM5 yield similar $R^2$ values for both $D_{C-max}^N$ (0.887, 0.869, 0.861) and $D_{C-av}^N$ (0.928, 0.944, 0.936). By contrast, AM1 agrees much more with PM5 than with PM3 with regard to $D_{C-max}^E$ and $D_{C-av}^E$ ($R^2 = 0.916$ and 0.970 vs. 0.562 and 0.799).

The greatest $R^2$ values are obtained for delocalizabilities confined to nitrogen atoms except for the maximum and average acceptor delocalizabilities, $D_{N-max}^N$ and $D_{N-av}^N$, with respect to AM1 vs. PM3. Moreover, $D_{H-max}^N$ provides relatively low intercorrelations as compared to most other delocalizability descriptors, and at the same time a still significantly greater similarity of PM5 with

**Table 6.9  Correlation in Terms of $R^2$ between AM1, PM3, PM5, HF/6-31G\*\*, and B3LYP/6-31G\*\* Delocalizabilities**

| Descriptor | No. Compounds[a] | AM1 vs. PM3 | PM5 | PM3 vs. PM5 |
|---|---|---|---|---|
| $D_{max}^N$ | 607 | 0.930 | 0.890 | 0.901 |
| $D_{av}^N$ | 607 | 0.995 | 0.985 | 0.976 |
| $D_{max}^E$ | 607 | 0.890 | 0.899 | 0.932 |
| $D_{av}^E$ | 607 | 0.994 | 0.994 | 0.996 |
| $D_{C-max}^N$ | 607 | 0.887 | 0.869 | 0.861 |
| $D_{C-av}^N$ | 607 | 0.928 | 0.944 | 0.936 |
| $D_{C-max}^E$ | 607 | 0.709 | 0.916 | 0.562 |
| $D_{C-av}^E$ | 607 | 0.842 | 0.970 | 0.799 |
| $D_{H-max}^N$ | 602 | 0.806 | 0.626 | 0.734 |
| $D_{H-av}^N$ | 602 | 0.941 | 0.929 | 0.883 |
| $D_{N-max}^N$ | 255 | 0.884 | 0.956 | 0.965 |
| $D_{N-av}^N$ | 255 | 0.884 | 0.954 | 0.969 |
| $D_{N-max}^E$ | 255 | 0.959 | 0.992 | 0.969 |
| $D_{N-av}^E$ | 255 | 0.962 | 0.992 | 0.971 |
| $D_{O-max}^N$ | 424 | 0.683 | 0.750 | 0.458 |
| $D_{O-av}^N$ | 424 | 0.496 | 0.729 | 0.183 |
| $D_{O-max}^E$ | 424 | 0.939 | 0.826 | 0.805 |
| $D_{O-av}^E$ | 424 | 0.921 | 0.850 | 0.829 |

[a] The total set of 607 compounds contains 602 molecules with at least 1 hydrogen atom, 255 molecules with at least 1 nitrogen atom, and 424 molecules with at least 1 oxygen atom.

PM3 than with AM1 ($R^2 = 0.734$ vs. 0.626), while greater $R^2$ values are achieved for $D_{H-av}^N$ with the lowest correlation between PM3 and PM5 ($R^2 = 0.883$).

It would be interesting to evaluate the performance of delocalizabilites on the *ab initio* level, which appears to have not yet been undertaken so far. Note that in contrast to semiempirical valence-electron minimum basis set approaches, *ab initio* calculations typically involve a much greater number of LCAO-MO wavefunctions (Equation 6.18), which results in correspondingly larger ranges of the delocalizability sums (Equations 6.58 and 6.59). This holds true in particular for the manifold of unoccupied molecular orbitals that increases with increasing basis set size. The implementation and analysis of *ab initio* delocalizabilities would probably include the consideration of reasonable thresholds for both occupied and unoccupied MO levels, which goes beyond the scope of the present investigation. At the same time, it should be kept in mind that according to recent findings as mentioned above in Section IV.F, the delocalizability parameter appears to show some instability with increasing molecular size (Hosoya and Iwata, 1999), which needs further investigation.

## VI. CONCLUDING REMARKS

In the context of computational toxicology, quantum chemical descriptors provide distinct probes to unravel mechanistic causes for the hazardous effects of chemical substances. At the same time, the level of theory employed may be crucial for the molecular property under analysis, which is particularly true for descriptors based on net atomic charges (that, in turn, are not physically observable, despite their intuitive meaning for charge-controlled intermolecular interactions).

A systematic difference between semiempirical and *ab initio* results relates to the charge distribution around halogen atoms. Aliphatically bound chlorine atoms are more electronegative in semiempirical NDDO-based schemes such as AM1 than according to *ab initio* calculations; the reverse is true for bromine attached to aromatic carbon. Overall, the semiempirical net atomic charge trends differ substantially from the ones calculated by *ab initio* methods, which is particularly striking for oxygen atoms, and even between HF and B3LYP there are significant discrepancies with regard to the net atomic charge trend of hydrogen atoms. Moreover, PM3 yields systematically greater positive nitrogen charges than all other methods.

Within the semiempirical methods, AM1 generally correlates more with PM5 than with PM3, which holds true for descriptors based on the charge distribution and on frontier orbital energies, and somewhat less pronounced also for delocalizabilities. At the same time, AM1 shows an overall greater degree of agreement with HF and B3LYP than PM3 and PM5 with regard to $E_{HOMO}$ and $E_{LUMO}$ energy, molecular EN, molecular HD, and dipole moment. Here, larger deviations occur for organophosphorus compounds and other hypervalent structures, and surprisingly PM3 and PM5 do not appear to be superior to AM1 in this respect when taking B3LYP as the reference method.

As a general trend, B3LYP correlates significantly more with HF than with all semiempirical methods. A notable exception is the $E_{HOMO}$ energy with HF-B3LYP that had $R^2$ values around 0.7 for non-aromatic compounds as well as for molecules containing nitrogen or oxygen or both, and around 0.5 for compounds with phosphorus or sulfur or both. It suggests that a more detailed analysis, including experimental data, will be required to assess the actual predictive quality of the vertical ionization potential based on $E_{HOMO}$ energies.

So far, there has been little use of *ab initio* methods for the derivation of QSARs to predict the toxic potency and mode of action of chemicals. The present results indicate that with regard to net atomic charges and charge partial surface area descriptors, *ab initio* methods may indeed provide a different type of information, which should be kept in mind for future investigations, particularly when focusing on the charge-controlled impact on bioreactivity of drugs and toxicants.

## ACKNOWLEDGMENTS

The author thanks Ralph Kühne and Aynur O. Aptula (both UFZ Leipzig) for their technical support to generate the statistics and figures, and Walter Thiel (Max Planck-Institut Mülheim) and Joachim Reinhold (University of Leipzig) for critically reading the manuscript. The set of 607 compounds and their 3D geometry generation were performed as part of the European Union IMAGETOX Research Training Network (HPRN-CT-1999-00015) under the coordination of Emilio Benfenati (Istituto di Ricerche Farmacologiche "Mario Negri," Milano, Italy), which also provided the fellowship for the research stay of Aynur Aptula at Leipzig.

## REFERENCES

Alemán, C., Luque, F.J. and Orozco, M., Suitability of the PM3-derived molecular electrostatic potentials, *J. Computational Chem.*, 14, 799–808, 1993.

Aptula, A.O., Kühne, R., Ebert, R.-U., Cronin, M.T.D., Netzeva, T.I., and Schüürmann G., Modeling discrimination between antibacterial and non-antibacterial activity based on 3D molecular descriptors, *QSAR Combinatorial Sci.*, 22, 113–128, 2003.

Atkins, P.W., *Molecular Quantum Mechanics*, 2nd ed., Oxford University Press, Oxford, 1983.

Ayers, P.W. and Parr, R.G., Variational principles for describing chemical reactions: the Fukui function and chemical hardness revisited, *J. Am. Chem. Soc.*, 122, 2010–2018, 2000.

Baerends, E.J., Perspective on "self-consistent equations including exchange and correlation effects", *Theor. Chem. Acc.*, 103, 265–269, 2000.

Baerends, E.J. and Gritsenko, O.V., A quantum chemical view of density functional theory, *J. Phys. Chem. A*, 101, 5383–5403, 1997.

Bakowies, D. and Thiel, W., Semiempirical treatment of electrostatic potentials and partial charges in combined quantum mechanical and molecular mechanical approaches, *J. Computational Chem.*, 17, 87–108, 1996.

Bauschlicher, C.W., Jr., A comparison of the accuracy of different functionals, *Chem. Phys. Lett.*, 246, 40–44, 1995.

Beck, B., Horn, A., Carpenter, J.E., and Clark, T., Enhanced 3D-databases: a fully electrostatic database of AM1-optimized structures, *J. Chem. Inf. Comput. Sci.*, 38, 1214–1217, 1998.

Becke, A.D., Density-functional thermochemistry. III. The role of exact exchange, *J. Chem. Phys.*, 98, 5648–5652, 1993.

Bergmann, D. and Hinze, J., Elektronegativität und Moleküleigenschaften, *Angewandte Chemie*, 108, 162–176, 1996.

Bergström, C.A.S., Strafford, M., Lazorova, L., Avdeef, A., Lutzman, K., and Artursson, P., Absorption classification of oral drugs based on molecular surface properties, *J. Med. Chem.*, 46, 558–570, 2003.

Berkowitz, M. and Parr, R.G., Molecular hardness and softness, local hardness and softness, hardness and softness kernels, and relations among these quantities, *J. Chem. Phys.*, 88, 2554–2557, 1988.

Besler, B.H., Merz. K.M. and Kollman, P.A., Atomic charges derived from semiempirical methods, *J. Computational Chem.*, 11, 431–439, 1990.

Bickelhaupt, F.M. and Baerends, E.J., Kohn-Sham density functional theory: predicting and understanding chemistry, in *Reviews in Computational Chemistry*, Lipkowitz, K.B. and Boyd, D.B., Eds., Vol. 15, Wiley-VCH, New York, 2000, pp. 1–86.

Bingham, R.C., Dewar, M.J.S., and Lo, D.H., Ground states of molecules. XXV. MINDO/3. An improved version of the MINDO semiempirical SCF-MO method, *J. Am. Chem. Soc.*, 97: 1285-1293.

Breneman, C.M. and Wiberg, K.B., Determining atom-centered monopoles from molecular electrostatic potentials: the need for high sampling density in formamide conformational analysis, *J. Computational Chem.*, 11, 361–373, 1990.

Brüstle, M., Beck, B, Schindler, T., King, W., Mitchell, T., and Clark, T., Descriptors, physical properties, and drug-likeness, *J. Med. Chem.*, 45, 3345–3355, 2002.

Bultinck, P., Langenaeker, W., Carbó-Dorca, R. and Tollenaere, J.P., Fast calculation of quantum chemical molecular descriptors from the electronegativity equalization method, *J. Chem. Inf. Comput. Sci.*, 43, 422–428, 2003.

Bultinck, P., Langenaeker, W., Lahorte, P., de Proft, F., Geerlings, P., Waroquier, M., and Tollenaere, J.P. The electronegativity equalization method I: Parametrization and validation for atomic charge calculations, *J. Phys. Chem. A*, 106, 7887–7894, 2002a.

Bultinck, P., Langenaeker, W., Lahorte, P., de Proft, F., Geerlings, P., van Alsenoy, C., and Tollenaere, J.P., The electronegativity equalization method. I. Applicability of different atomic charge schemes, *J. Phys. Chem. A*, 106, 7895–7901, 2002b.

Cammarata, A., Some electronic factors in drug-receptor interactions, *J. Med. Chem.*, 11, 1111–1115, 1968.

Cammarata, A. and Rogers, K.S., Electronic representation of the lipophilic parameter $\pi$, *J. Med. Chem.*, 14, 269–274, 1971.

Chandrakumar, K.R.S., and Pal, S., DFT and local reactivity descriptor studies on the nitrogen sorption selectivity from air by sodium and calcium exchanged zeolite-A, *Colloids Surf. A – Physicochemical Eng. Aspects*, 205, 127–138, 2002a.

Chandrakumar, K.R.S. and Pal, S., Study of local hard-soft acid-base principle to multiple-site interactions, *J. Phys. Chem. A*, 106: 5737–5744, 2002b.

Chandrakumar, K.R.S. and Pal, S., A systematic study on the reactivity of Lewis acid-base complexes through the local hard-soft acid-base principle, *J. Phys. Chem. A*, 106, 11775–11781, 20002c.

Chandrakumar, K.R.S. and Pal, S., Study of local hard-soft acid-base principle: effects of basis set, electron correlation, and the electron partitioning method, *J. Phys. Chem. A*, 107, 5755–5762, 2003.

Chattarai, P.K., Maiti, B. and Sarkar, U., Philicity: a unified treatment of chemical reactivity and selectivity, *J. Phys. Chem. A*, 107, 4973–4975, 2003.

Chermette, H., Chemical reactivity indexes in density functional theory, *J. Computational Chem.*, 20, 129–154, 1999.

Chirgwin, B.H. and Coulson, C.A., The electronic structure of conjugated systems. VI, *Proc. R. Soc. (London) A*, 201, 196–209, 1950.

Clark, T., Quo vadis semiempirical MO-theory?, *J. Mol. Struct. (Theochem)*, 530, 1–10, 2000.

Cohen, M.H., Strengthening the foundations of chemical reactivity theory, in *Theory of Chemical Reactivity: Topics in Current Chemistry*, Vol. 183, Nalewajski, R.F., Ed., Springer-Verlag, Heidelberg, Germany, 1996, pp. 143–173.

CORINA, Version 2.6, (2001) 3D Structure Generator, Sadowski, J., Schwab, C.H., and Gasteiger, J., Molecular Networks GmbH Computerchemie, Erlangen, Germany, 2001.

Coulson, C.A. and Longuett-Higgins, H.C., The electronic structure of conjugated systems. I. General theory, *Proc. R. Soc. (London) A*, 191, 39–60, 1947.

Cramer, C.J., Famini, G.R., and Lowrey, A.H., Use of calculated quantum chemical properties as surrogates for solvatochromic parameters in structure-activity relationships, *Acc. Chem. Res.*, 26, 599–605, 1993.

Cramer, C.J. and Truhlar, D.G., Implicit solvation models: Equilibria, structure, spectra, and dynamics, *Chem. Rev.*, 94, 2027-2094, 1999.

Cronin, M.T.D., Gregory, B.W., and Schultz, T.W., Quantitative structure-activity analyses of nitrobenzene toxicity to *Tetrahymena pyriformis*, *Chem. Res. Toxicol.*, 11, 902–908, 1998.

Cronin, M.T.D., Manga, N., Seward, J.R., Sinks, G.D., and Schultz, T.W., Parametrization of electrophilicity for the prediction of the toxicity of aromatic compounds, *Chem. Res. Toxicol.*, 14, 1498–1505, 2001.

Csonka, G.I., Analysis of the core-repulsion functions used in AM1 and PM3 semiempirical calculations: conformational analysis of ring systems, *J. Computational Chem.*, 14, 895–898, 1993.

Csonka, G.I. and Ángyán, J.G., The origin of the problems with the PM3 repulsion function, *J. Mol. Struct. (Theochem)*, 393, 31–38, 1997.

Curtiss, L.A., Redfern, P.C., and Pople, J.A., Assessment of Gaussian-2 and density functional theories for the computation of ionization potentials and electron affinities, *J. Chem. Phys.*, 109, 42–55, 1998.

Dannenberg, J.J., Hydrogen bonds: a comparison of semiempirical and *ab initio* treatments, *J. Mol. Struct. (Theochem)*, 401, 279–286, 1997.

Dearden, J. and Schüürmann, G., Quantitative structure-property relationships for predicting Henry's law constant from molecular structure, *Environ. Toxicol. Chem.*, 22, 1755–1770, 2003.

de Proft, F. and Geerlings, P., Calculation of ionization energies, electron affinities, electronegativities, and hardnesses using density functional methods, *J. Chem. Phys.*, 106, 3270–3279, 1997.

de Proft, F., Martin, J.M.L., and Geerlings, P., On the performance of density functional methods for describing atomic populations, dipole moments and infrared intensities, *Chem. Phys. Lett.*, 250. 393–401, 1996.

de Proft, F., Tielens, F., and Geerlings, P., Performance and basis set dependence of density functional theory dipole and quadrupole moments, *J. Mol. Struct. (Theochem)*, 506, 1–8, 2000.

Dewar, M.J.S. and Thiel, W., Ground states of molecules. 38. The MNDO method. Approximations and parameters, *J. Am. Chem. Soc.*, 99, 4899–4907, 1977.

Dewar, M.J.S., Zoebisch, E.G., Healy, E.F., and Stewart, J.J.P., AM1: a new general purpose quantum mechanical molecular model, *J. Am. Chem. Soc.*, 107, 3902–3909, 1985.

Famini, G.R., Penski, C.A., and Wilson, L.Y., Using theoretical descriptors in quantitative structure activity relationships: some physicochemical properties, *J. Phys. Org. Chem.*, 5, 395–408, 1992.

Famini, G.R. and Wilson, L.Y., Linear free energy relationships using quantum mechanical descriptors, in *Reviews in Computational Chemistry*, Vol. 18, Lipkowitz, K.B. and Boyd, D.B., Eds., Wiley-VCH, New York, 2002, pp. 211–255.

Fischer-Hjalmars, I., Deduction of the zero differential overlap approximation from an orthogonal atomic orbital basis, *J. Chem. Phys.*, 42, 1962–1972, 1965.

Francl, M.M. and Chirlian, L.E., The pluses and minuses of mapping atomic charges to electrostatic potentials, in *Reviews in Computational Chemistry*, Vol. 15, Lipkowitz, K.B. and Boyd, D.B., Eds., Wiley-VCH, New York, 2000, pp. 1–31.

Fukui, K., Yonezawa, T., and Nagata, C., Theory of substitution in conjugated molecules, *Bull. Chem. Soc. Jap.*, 27, 423–427, 1954.

Fukui, K., Yonezawa, T., and Nagata C., Interrelations of quantum-mechanical quantities concerning chemical reactivity of conjugated molecules, *J. Chem. Phys.*, 26, 831–841, 1957.

Fukui, K., Kato, H., and Yonezawa, T., A new quantum-mechanical reactivity index for saturated compounds, *Bull. Chem. Soc. Jap.*, 34, 1111–1115, 1961.

Fukui, K., Theory of orientation and stereoselection, *Top. Curr. Chem.*, 15, 1–85, 1970.

Gasteiger, J. and Marsili, M., Iterative partial equalization of orbital electronegativity: a rapid access to atomic charges in molecules, *Tetrahedron*, 36, 3219–3228, 1990.

Gasteiger, J., Rudolph, C., and Sadowski, J., Automatic generation of 3D-atomic coordinates for organic molecules, *Tetrahedron Comput. Methodol.*, 3, 537–547, 1990.

Gaussian 98, Revision A.7, Frisch, M.J., Trucks, G.W., Schlegel, H.B., Scuseria, G.E., Robb, M.A., Cheeseman, J.R., Zakrzewski, V.G., Montgomery, J.A., Stratmann, R.E., Burant, J.C., Dapprich, V., Millam, J.M., Daniels, A.D., Kudin, K.N., Strain, M.C., Farkas, O., Tomasi, J., Barone, V., Cossi, M., Cammi, R., Mennucci, B., Pomelli, C., Adamo, C., Clifford, S., Ochterski, J., Petersson, G.A., Ayala, P.Y., Cui, Q., Morokuma, K., Malick, D.K., Rabuck, A.D., Raghavachari, K., Foresman, J.B., Ciolowski, J., Ortiz, J.V., Stefanov, B.B., Liu, G., Liashenko, A., Piskorz, P., Komaromi, I., Gomperts, R., Martin, R.L., Fox, D.J., Keith, T., Al-Laham, M.A., Peng, C.Y., Nanayakkara, A., Gonzalez, C., Challacombe, M., Gill, P.M.W., Johnson, B.G., Chen, W., Wong, M.W., Andres, J.L., Head-Gordon, M., Replogle, E.S., and Pople, J.A., Gaussian Inc., Pittsburgh, PA, 1999.

Geerlings, P., de Proft, F., and Langenaeker, W., Conceptual density functional theory, *Chem. Rev.*, 103, 1793–1873, 2003.

Gough, K.M. and Kaiser, K.L.E., QSAR of the acute toxicity of para-substituted nitrobenzene and aniline derivatives to *Photobacterium phosphoreum*, in *QSAR 88. Proceedings of the Third International Workshop on Quantitative Structure-Activity Relationships in Environmental Toxicology*, Turner, J.E., England, M.W., Schultz, T.W., and Kwaak, N.J., Eds., National Technical Information Service, U.S. Department of Commerce, Springfield, VA, 1988, pp. 111–121.

Grigoras, S., A structural approach to calculate physical properties of pure organic substances: The critical temperature, critical volume and related properties, *J. Computational Chem.*, 11, 493–510, 1990.

Gross, K.C., Seybold, P.G., Peralta-Inga, Z., Murray, J.S., and Politzer, P., Comparison of quantum chemical parameters and Hammett constants in correlating pKa values of substituted anilines, *J. Org. Chem.*, 66, 6919–6925, 2001.

Hansch, C., Leo, A., and Taft, R.W., A survey of Hammett substituent constants and resonance and field parameters, *Chem. Rev.*, 91, 165–195, 1991.

Hariharan, P.C. and Pople, J.A., The influence of polarization functions on molecular orbital hydrogenation energies, *Theoretica Chimica Acta*, 28, 213–222, 1973.

Hehre, W.J., Radom, L., Schleyer, P.V.R., and Pople, J.A., *Ab initio Molecular Orbital Theory*, John Wiley, New York, 1986.

Hosoya, H. and Iwata, S., Revisiting superdelocalizablity: mathematical stability of reactivity indices, *Theor. Chem. Acc.*, 102, 293–299, 1999.

Iczkowski, R.P. and Margrave, J.L., Electronegativity, *J. Am. Chem. Soc.*, 83, 3547–3551, 1961.

Ivanov, J.M., Karabunarliev, S.H., and Mekenyan, O.G., 3DGEN: a system for exhaustive 3D molecular design, *J. Chem. Inf. Comput. Sci.*, 34, 234–243, 1994.

Jensen, F., *Introduction to Computational Chemistry*, John Wiley and Sons, Chichester, England, 1998.

Karelson, M., Lobanov, V.S., and Katritzky, A.R., Quantum-chemical descriptors in QSAR/QSPR studies, *Chem. Rev.*, 96, 1027–1043, 1996.

Klessinger, M., *Elektronenstruktur Organischer Moleküle*, Verlag Chemie, Weinheim, Germany, 1982.

Klopman, G., Chemical reactivity and the concept of charge- and frontier-controlled reactions, *J. Am. Chem. Soc.*, 90, 223–234, 1968.

Koch, W. and Holthausen, M.C., *A Chemist's Guide to Density Functional Theory*, 2nd ed., Wiley-VCH, Weinheim, Germany, 2001.

Kohn, W., Nobel lecture: Electronic structure of matter – wave functions and density functionals, *Rev. Modern Phys.*, 71, 1253–1266, 1999.

Kolb, M. and Thiel, W., Beyond the MNDO model: methodical considerations and numerical results, *J. Computational Chem.*, 14, 775–789, 1993.

Lee, C., Yang, W., and Parr, R.G., Development of the Colle-Salvetti correlation-energy formula into a functional of the electron density, *Phys. Rev. B*, 37, 785–789, 1988.

Levine, I.N., *Quantum Chemistry*, 4th ed., Prentice-Hall, Upper Saddle River, NJ, 1991.

Lewis, D.F.V., Molecular orbital calculations on tumor-inhibitory nitrosoureas: QSARs, *Int. J. Quantum Chem.*, 33, 305–321, 1988.

Lewis, D.F.V., Molecular orbital calculations on tumor-inhibitory aniline mustards: QSARs, *Xenobiotica*, 19, 243–251, 1989.

Li, J., Zhu, T., Cramer, C.J., and Truhlar, D.G., A new class IV charge model for extracting accurate partial charges from wave functions, *J. Phys. Chem. A*, 102, 1820–1831, 1998.

Lipkowitz, K.B., Baker, B., and Larter, R., Dynamic molecular surface areas, *J. Am. Chem. Soc.*, 111, 7750–7753, 1989.

Löwdin, P.O., The non-orthogonality problem connected with the use of atomic wave functions in the theory of molecules and crystals, *J. Chem. Phys.*, 18, 365–375, 1950.

Löwdin, P.O., Nonorthogonality problem, *Adv. Quantum Chem.*, 5, 185–199, 1970.

Lopez, X. and York, D.M., Parameterization of semiempirical methods to treat nucleophilic attacks to biological phosphates: AM1/d parameters for phosphorus, *Theor. Chem. Acc.*, 109, 149–159, 2003.

Mattioni, B.E., Kauffman, G.W., and Jurs, P.C., Predicting the genotoxicity of seconddary and aromatic amines using data subsetting to generate a model ensemble, *J. Chem. Inf. Comput. Sci.*, 43, 949–963, 2003.

Mekapati, S.B. and Hansch, C., On the parametrization of the toxicity of organic chemicals to *Tetrahymena pyriformis*: the problem of establishing a uniform activity, *J. Chem. Inf. Comput. Sci.*, 42, 956–961, 2002.

Mekenyan, O.G., Dimitrov, D., Nikolova, N., and Karabunarliev, S., Conformational coverage by a genetic algorithm, *J. Chem. Inf. Comput. Sci.*, 39, 997–1016, 1999.

Mekenyan, O.G., Ivanov, J.M., Veith, G.D., and Bradbury, S.P., DYNAMIC QSAR: a search for active conformations and significant stereoelectronic indices, *Quant. Struct.-Act. Relat.*, 13, 302–307, 1994.

Mekenyan, O.G., Nikolova, N., and Schmieder, P., Dynamic 3D QSAR techniques: applications in toxicology, *J. Mol. Struct. (Theochem)*, 622, 147–165, 2003.

Mekenyan, O.G., Roberts, D.W., and Karcher, W., Molecular orbital parameters as predictors of skin sensitization of halo- and pseudohalobenzenes acting as $S_NAr$ electrophiles, *Chem. Res. Toxicol.*, 10, 994–1000, 1997.

Mekenyan, O.G., Veith, G.D., Bradbury, S.P., and Russom, C.L., Structure-toxicity relationships for $\alpha,\beta$-unsaturated alcohols in fish, *Quant. Struct.-Act. Relat.*, 12, 132–136, 1993.

MOLSV (Calculation of Molecular Volume & Surface Area), Smith, G.M., QCPE program No. 509, 1985.

MOPAC 93, Revision 2, Fujitsu Limited, 9-3, Nakase 1-Chome, Mihama-ku, Chiba-city, Chiba 261, Japan, and Stewart Computational Chemistry, 15210 Paddington Circle, Colorado Springs, CO, 1994.

Mortier, W.J., van Gnechten, K., and Gasteiger, J., Electronegativity equalization: application and parametrization, *J. Am. Chem. Soc.*, 107, 829–835, 1985.

Mosier, P.D., Jurs, P.C., Custer, L.L., Durham, S.K., and Pearl, G.M., Predicting the genotoxicity of thiophene derivatives from molecular structure, *Chem. Res. Toxicol.*, 16, 721–732, 2003.

Mujika, J.I., Mercero, J.M., and Lopez, X., A theoretical evaluation of the pKa for twisted amides using density functional theory and dielectric continuum methods, *J. Phys. Chem. A*, 107, 6099–6107, 2003.

Mulliken, R.S., A new electroaffinity scale, together with data on valence states and on valence ionization potentials and electron affinities, *J. Chem. Phys.*, 2, 782–793, 1934.

Mulliken, R.S., Electronic population analysis on LCAO-MO [linear combination of atomic orbital-molecular orbital] molecular wave functions. I, *J. Chem. Phys.*, 23, 1833–1840, 1955.

Mulliken, R.S., Criteria for the construction of good self-consistent-field molecular orbital wave functions, and the significance of L.A.A.O.M.O. population analysis, *J. Chem. Phys.*, 36, 3428–3439, 1962.

Pal, S. and Chandrakumar, K.R.S., Critical study of local reactivity descriptors for weak interactions: qualitative and quantitative analysis of adsorption of molecules in the zeolite lattice, *J. Am. Chem. Soc.*, 122, 4145–4153, 2000.

Palm, K., Luthman, K., Ungell, A.-L., Strandlund, G., and Artursson, P., Correlation of drug absorption with molecular surface properties, *J. Pharm. Sci.*, 85, 32–39, 1996.

Parr, R.G., Donnelly, R.A., Levy, M., and Palke, W.E., Electronegativity: the density functional viewpoint, *J. Chem. Phys.*, 68, 3801–3807, 1978.

Parr, R.G. and Yang, W., Density-functional theory of the electronic structure of molecules, *Annu. Rev. Phys. Chem.*, 46, 701–728, 1995.

Pearson, R.G., Absolute electronegativity and hardness correlated with molecular orbital theory, *Proc. Natl. Acad. Sci.*, 83, 8440–8441, 1986.

Pearson, R.G., Absolute electronegativity and hardness, *Chem. Br.*, 27, 444–447, 1991.

Perks, H.M. and Liebman, J.F., Paradigms and paradoxes: electronegativity and bond energies – living legacies of Linus and Lee, *Struct. Chem.*, 11, 375–378, 2000.

Pople, J., Nobel lecture: quantum chemical models, *Rev. Mod. Phys.*, 71, 1267–1274, 1999.

Purdy, R., Quantitative structure-activity relationships for predicting toxicity of nitrobenzenes, phenols, anilines, and alkylamines to fathead minnows, in *QSAR 88. Proceedings of the Third International Workshop on Quantitative Structure-Activity Relationships in Environmental Toxicology*, Turner, J.E., England, M.W., Schultz, T.W., and Kwaak, N.J., Eds., National Technical Information Service, U.S. Department of Commerce, Springfield, VA, 1988, pp. 99–110.

Purdy, R., The utility of computed superdelocalizabilities for predicting LC50 values of epoxides to guppies, *Sci. Total Environ.*, 109/110, 553–556, 1991.

Raghavachari, K. and Anderson, J.B., Electron correlation effects in molecules, *J. Phys. Chem.*, 100, 12960–12973, 1996.

Rappé, A.K. and Goddard, W.A., Charge equilibration for molecular dynamics simulation, *J. Phys. Chem.*, 95, 3358–3363, 1991.

Repasky, M.P., Chandrasehar, J., and Jorgensen, W.L., Improved semiempirical heats of formation through the use of bond and group equivalents, *J. Computational Chem.*, 23, 498–510, 2002a.

Repasky, M.P., Chandrasehar, J., and Jorgensen, W.L., PDDG/PM3 and PDDG/MNDO: improved semiempirical methods, *J. Computational Chem.*, 23, 1601–1622, 2002b.

Romero, M. de L. and Méndez, F., Is the hydrogen atomic charge representative of the acidity of para-substituted phenols?, *J. Phys. Chem. A*, 107, 4526–4530, 2003a.

Romero, M. de L. and Méndez, F., The local HSAB principle and bond dissociation energy of p-substituted phenols, *J. Phys. Chem. A*, 107, 5874–5875, 2003b.

Russom, C.L., Bradbury, S.P., Broderius, S.J., Hammermeister, D.E., and Drummond, R.A., Predicting modes of toxic action from chemical structure: acute toxicity in the fathead minnow (*Pimephales promelas*), *Environ. Toxicol. Chem.*, 16, 948–967, 1997.

Sadowski, J. and Gasteiger, J., From atoms and bonds to three-dimensional atomic coordinates: automatic model builders, *Chem. Rev.*, 93, 2567–2581, 1993.

Sanderson, R.T., An interpretation of bond lengths and a classification of bonds, *Science*, 114, 670–672, 1951.

Sanderson, R.T., Principles of electronegativity. I. General nature, *J. Chem. Educ.*, 65, 112–118, 1988.

Scheiner, A.C., Baker, J., and Anzelm, J.W., Molecular energies and properties from density functional theory: exploring basis set dependence of Kohn-Sham equation using several density functionals, *J. Computational Chem.*, 18, 775–795, 1997.

Schlegel, H.B., Exploring potential energy surfaces for chemical reactions: an overview of some practical methods, *J. Computational Chem.*, 24, 1514–1527, 2003.

Schmitt, H., Altenburger, R., Jastorff, B., and Schüürmann, G., Quantitative structure-activity analysis of the algae toxicity of nitroaromatic compounds, *Chem. Res. Toxicol.*, 13, 441–450, 2000.

Schüürmann, G., QSAR analysis of the acute toxicity of organic phosphorothionates using theoretically derived molecular descriptors, *Environ. Toxicol. Chem.*, 9, 417–428, 1990.

Schüürmann, G., Ecotoxicology and structure-activity studies of organophosphorus compounds, in *Rational Approaches to Structure, Activity and Ecotoxicology of Agrochemicals*, Fujita, T. and Draber, W., Eds., CRC Press, Boca Raton, FL, 1992, pp. 485–541.

Schüürmann, G., Quantum chemical approach to estimate physicochemical compound properties: application to substituted benzenes, *Environ. Toxicol. Chem.*, 14, 2067–2076, 1995.

Schüürmann, G., Modelling pKa of carboxylic acids and phenols, *Quant. Struct.-Act. Relat.*, 15, 121–132, 1996.

Schüürmann, G., Ecotoxic modes of action of chemical substances, in *Ecotoxicology*, Schüürmann, G. and Markert, B., Eds., John Wiley and Spektrum Akademischer Verlag, New York, 1998a, pp. 665–749.

Schüürmann, G., Quantum chemical analysis of the energy of proton transfer from phenol and chlorophenols to H2O in the gas phase and in aqueous solution, *J. Chem. Phys.*, 109, 9523–9528, 1998b.

Schüürmann, G., Cossi, M., Barone, V., and Tomasi, J., Prediction of the pKa of carboxylic acids using the *ab initio* continuum-solvation model PCM-UAHF, *J. Phys. Chem. A*, 102, 6706–6712, 1998.

Schüürmann, G., Aptula, A.O., Kühne, R. and Ebert, R.-U., Stepwise discrimination between four modes of toxic action of phenols in the *Tetrahymena pyriformis* assay, *Chem. Res. Toxicol.*, 16, 974–987, 2003.

Schüürmann, G. and Funar-Timofei, S., Multilinear regression and comparative molecular field analysis (CoMFA) of azo dye–fiber affinities. 2. Inclusion of solution-phase molecular orbital descriptors, *J. Chem. Inf. Comput. Sci.*, 43, 1502–1512, 2003.

Snyder, B.H. and Merril, C.R., A relationship between the hallucinogenic activity of drugs and their electronic configuration, *Proc. Nat. Aca. Sci.*, 54, 258–266, 1965.

Soffers, A.E.M.F., Boersma, M.G., Vaes, W.H.J., Vervoort, J., Tyrakowska, B., Hermens, J.L.M., and Rietjens, I.M.C.M., Computer-modeling-based QSARs for analysing experimental data on biotransformation and toxicity, *Toxicol. in Vitro*, 15, 539–551, 2001.

Springborg, M., *Methods of Electronic-Structure Calculations*, John Wiley and Sons, Chichester, England, 2000.

Stanton, D.T. and Jurs, P.C., Development and use of charged partial surface area structural descriptors in computer-assisted quantitative structure-property relationships studies, *Anal. Chem.*, 62, 2323–2329, 1990.

Stenberg, P., Luthman, K., Ellens, H., Lee, C.P., Smith, P.L., Lago, A., Elliott, J.D., and Artursson, P., Prediction of the intestinal absorption of endothelin receptor agonists using three theoretical models of increasing complexity, *Pharm. Res.*, 16, 1520–1526, 1999.

Stewart, J.J.P., MOPAC 2002, software package, 2002

Stewart, J.J.P., Optimization of parameters for semiempirical methods. I. Method, *J. Computational Chem.*, 10, 209–220, 1998a.

Stewart, J.J.P., Optimization of parameters for semiempirical methods. II. Application, *J. Computational Chem.*, 10, 221-264, 1998b.

Stewart, J.J.P., MOPAC: A semiempirical molecular orbital program, *J. Comput.-Aided Mol. Design*, 4, 1–105, 1990a.

Stewart, J.J.P., Semiempirical molecular orbital methods. in *Reviews in Computational Chemistry*, Vol. 1, Lipkowitz, K.B. and Boyd, D.B., Eds., Wiley-VCH, New York, 1990b, pp. 45–118.

Storer, J.W., Giesen, D.J., Cramer, C.J., and Truhlar, D.G., Class IV charge models: a new semiempirical approach in quantum chemistry, *J. Comput.-Aided Mol. Design*, 9, 87–110, 1995.

Stowasser, R. and Hoffmann, R., What do the Kohn-Sham orbitals and eigenvalues mean?, *J. Am. Chem. Soc.*, 121, 3414–3420, 1999.

Sutcliffe, B.T., The nuclear motion problem in molecular physics, *Adv. Quantum Chem.*, 28, 65–80, 1997.

SYBYL 6.8, Tripos Associates, St. Louis, MO, 2001.

Thanikaivelan, P., Padmanabhan, J., Subramanian, V., and Ramasami, T., Chemical reactivity and selectivity using Fukui functions: basis set and population scheme dependence in the framework of B3LYP theory, *Theor. Chem. Acc.*, 107, 326–335, 2002.

Thiel, W., Semiempirical methods: current status and perspectives, *Tetrahedron*, 44, 7393–7408, 1988.

Thiel, W., Perspectives on semiempirical molecular orbital theory, *Adv. Chem. Phys.*, 93, 703–757, 1996.

Thiel, W., Thermochemistry from semiempirical molecular orbital theory, in *Computational Thermochemistry*, ACS Symposium Series 677, Irikura, K.K. and Frurip, D.J., Eds., American Chemical Society, Washington, D.C., 1988, pp. 142–161.

Thiel, W., Semiempirical methods, *Modern Methods and Algorithms of Quantum Chemistry*, Vol. 1, NIC Series, John von Neumann Institute for Computing, Grotendorst, J., Ed., Jülich, Germany, 2000, pp. 233–255.

Thiel, W. and Voityuk, A.A., Extension of MNDO to d orbitals: parameters and results for the halogens, *Int. J. Quantum Chem.*, 44, 807–829, 1992.

Thiel, W. and Voityuk, A.A., Extension of MNDO to d orbitals: parameters and results for the second-row elements and for the zinc group, *J. Phys. Chem.*, 100, 616–626, 1996.

Thompson, J.D., Cramer, C.J. and Truhlar, D.G., Parameterization of charge model 3 for AM1, PM3, BLYP and B3LYP, *J. Computational Chem.*, 24, 1291–1304, 2003.

Veith, G.D. and Mekenyan, O.G., A QSAR approach for estimating the aquatic toxicity of soft electrophiles [QSAR for soft electrophiles], *Quant. Struct.-Act. Relat.*, 12, 349–356, 1993.

Wang, N.X. and Wilson, A.K., Effects of basis set choice upon the atomization energy of the second-row compounds SO2, CCl and ClO2 for B3LYP and B3PW91, *J. Phys. Chem. A*, 107, 6720–6724, 2003.

Weber, W. and Thiel, W., Orthogonalization corrections for semiempirical methods, *Theor. Chem. Acc.*, 103, 495–506, 2000.

Wiberg, K.B. and Rablen, P.R., Comparison of atomic charges derived via different procedures, *J. Computational Chem.*, 14, 1504–1518, 1993.

Williams, D.E., Net atomic charge and multipole models for the *ab initio* molecular electric potential, *Reviews in Computational Chemistry*, Vol. 2, Lipkowitz, K.B. and Boyd, D.B., Eds., Wiley-VCH, New York, 1991, pp. 219–271.

Winget, P., Thompson, J.D., Xidos, J.D., Cramer, C.J., and Truhlar, D.G., Charge model 3: a class IV charge model based on hybrid density functional theory with variable exchange, *J. Phys. Chem. A*, 106, 10707–10717, 2002.

Yoneda F. and Nitta Y., Electronic structure and antibacterial activity of nitrofuran derivatives, *Chem. Pharm. Bull.*, 12, 1264–1268, 1964.

Zerner, M.C., Semiempirical molecular orbital methods, in *Reviews in Computational Chemistry*, Vol. 2, Lipkowitz, K.B. and Boyd, D.B., Eds., Wiley-VCH, New York, 1991, pp. 313–365.

# CHAPTER 7

# Building QSAR Models: A Practical Guide

David J. Livingstone

## CONTENTS

## I. INTRODUCTION

The purpose of this chapter is to give a practical guide to the construction and interpretation of quantitative structure-activity relationship (QSAR) models and to illustrate, in particular, some common pitfalls. It is not possible in the space of a single book chapter to cover any of the topics in great depth; there are, for example, entire books devoted to single subjects such as regression analysis (Draper and Smith, 1998). Nevertheless, with the help of references for further reading, it is hoped that this chapter will prove useful to those who are unfamiliar with QSAR modeling. Examples of QSAR models may be found in many excellent books and reviews (e.g., Ford et al., 1996; Gundertofte and Jorgensen, 2000; Holtje and Sippl, 2001; Livingstone, 1996). Illustrations of the process of producing and interpreting QSAR models are more elusive, however. This author has produced a "how to do it" book (Livingstone, 1995) and Kubinyi (1993) nicely demonstrated many aspects of the QSAR modeling process. More advanced topics are described in detail in the two books edited by van de Waterbeemd (1994; 1995).

The chapter is set out in the same order in which QSAR models are generally constructed starting with the selection of compounds for modeling, collection of response data, assembly of physicochemical descriptor data, data reduction, or selection, and the construction and interpretation of the models.

## II. COMPOUND SELECTION

At first sight the selection of compounds may appear to be self-evident. If we are interested in the biological effects of substituted phenols, for example, then we simply collect or measure results for all the phenols that can be found. Unfortunately, this strategy is not the best or most efficient way to gather data and, in extreme cases, too many results for the wrong compounds may prevent the establishment of any instructive QSAR models. There is a series of substitutions well known to medicinal chemists that goes: methyl, ethyl, propyl, butyl, futile. If a reaction can be successfully used to make the methyl substituent, then often it can be easily modified to produce any higher alkane derivatives. Although this rapidly produces further compounds their information content is limited since the physicochemical properties of these compounds change in the same unhelpful way. In the terms of electronic effects all these alkyl groups are the same whereas as they get longer their bulk, as expressed by molecular weight or molar refraction, increases collinearly with their lipophilicity. This series of derivatives provides no information on electronic effects on activity and also does not differentiate between changes in size or hydrophobicity. This may be a fairly obvious series, but consider the substituents shown in Table 7.1.

At first sight these may appear to be reasonably diverse in terms of their chemistry even though there are three halogens in the set. The $\pi$ values are nicely spread out from hydrophilic ($-1.23$) to hydrophobic ($1.44$) but, unfortunately, so are the $\sigma$ values. In general there is no correlation between the hydrophobic substituent constant $\pi$ and the electronic substituent constant $\sigma$, but for this set the squared correlation coefficient between these two physicochemical properties is 0.95. In much the same way as the methyl-futile set is uninformative, so would this set be in terms of distinguishing electronic effects from hydrophobic effects in any QSAR model derived from it.

The challenge here is experimental design. Scientists from the more experimental disciplines will probably already be aware of the need for design when planning an experiment, but it may not be obvious that the successful construction of QSAR models also calls for experimental design. Each compound tested or included in a QSAR analysis corresponds to a design point; the experimental

**Table 7.1  Values of $\pi$ and $\sigma$ Substituent Constants for a Set of Common Substituents**

| Substituent | $\pi$ | $\sigma$ |
|---|---|---|
| NH$_2$ | −1.23 | −0.66 |
| OH | −0.67 | −0.37 |
| OCH$_3$ | −0.02 | −0.27 |
| H | 0.00 | 0.00 |
| F | 0.14 | 0.06 |
| Cl | 0.70 | 0.23 |
| Br | 0.86 | 0.23 |
| SCF$_3$ | 1.44 | 0.50 |

*Source:* From Livingstone, D.J., *Data Analysis for Chemists: Applications to QSAR and Chemical Product Design*, Oxford University Press, Oxford, 1995. Reproduced with permission of Oxford University Press.

factors that need to be varied in order to create the design are the physicochemical properties used to characterize the compounds. The biological endpoint, the result of the experiment, is also, unusually, an experimental factor; in this case we are eventually interested in creating a model that links the endpoint to the physicochemical descriptors. The design needs to include compounds that yield both high and low values of the endpoint as well as, preferably, a well-spread range of intermediate values. There is not sufficient space to elaborate on these concepts here; the interested reader may find help in texts on experimental design or specifically in the context of QSAR modeling Chapter 2 of Livingstone (1995) and Sections 3.1 to 3.4 of van de Waterbeemd (1995).

## III. RESPONSE DATA

The biological effects of compounds, their responses, or $y$ variables, can be assembled in one of two main ways: measured directly by the investigators or collected from literature sources. In the case of the latter it is important where possible to refer to the original literature. Transcription errors are surprisingly common and often propagate from paper to paper to review. Reviews, naturally, are a particularly attractive source of literature information since they can contain a lot of data in one handy source. Toxicity data sources and common errors encountered in literature data are discussed in detail in Chapter 2.

A response set may consist of a mixture of measurements from both of these sources but, whatever their origin, it is necessary to know the precision and range of the data. Some measurements, such as percentage inhibition, have a natural range, but others may cover many orders of magnitude. This may be deceptive, and it is often dangerous to take response data at face value. Examination of the distribution of the data is usually very instructive and may indicate where some transformation or other processing is required. Responses such as $LD_{50}$, for example, may be found to consist of just some very high and very low values with little or nothing in between. In this case it is better to treat the data as classified (i.e., toxic/non-toxic) and use an appropriate method for model construction (see later). Many of the standard statistical methods are based on the properties of data which are normally distributed. If the distribution of the response data, in particular, is markedly different from normal, then the assumptions used will be violated and the results will be misleading. Transformation of the data in order to make it more normal, a log transformation for example, can help in this situation.

Precision of the response data is another important property that needs to be considered since any model of the response in terms of physicochemical descriptors should have a standard error no better than the measurement errors. After all, with certain notable exceptions, it should not be possible to calculate something more precisely than it can be measured. A model standard error that is less than the experimental error is a very good sign that the model has been overfitted. In the case of an iteratively trained model such as a neural network (see later), such a situation may indicate a model that has been overtrained but not necessarily overfitted. The problem with overfitted or overtrained models is that they appear to fit the training set data well, indeed too well, but are unlikely to generalize to other sets, which is the usual purpose for fitting a model in the first place.

## IV. DESCRIPTOR DATA

Descriptors of chemical structure, the independent variables, or $x$ variables can come from a variety of sources and may be measured, estimated, or calculated as discussed in Chapters 3 to 6. It is quite common these days to use many descriptors in a QSAR study, often many more variables than there are compounds (cases) in the set. This leads to the need for dimension reduction, variable elimination and variable selection. In the early days of QSAR modeling, the 1960s and 1970s, there was little need for any of these methods since the descriptors used were generally tabulated physicochemical properties and the choice of these was very limited. There were exceptions, such as the topological descriptors, which found popularity quite early on in QSAR modeling. One reason for this is that they can be computed for any structure and do not need a common parent as with tabulated properties. They may be used for studies of heterogeneous collections of compounds. Many collections of data of environmental concern consist of quite diverse chemical structures and thus topological descriptors have often been used to characterize them. Today there are many different types of descriptors that may be computed for any structure. Todeschini and Consonni, for example, list over 3000 different chemical descriptors (Todeschini and Consonni, 2000), and descriptors of different types may be usefully combined as suggested in a recent review (Livingstone, 2000). So, what is meant by dimension reduction, variable elimination, and variable selection? These are all means by which we can reduce the complexity of a problem in order to be able to recognize useful and informative patterns in the data. Although they all have a common aim, the means by which these three different processes achieve this is quite different as can be seen in the following sections.

## A. Dimension Reduction

If a data set consists of just 20 compounds described by 3 variables, then there is little need for any dimension reduction; relationships among the variables can be observed in a simple correlation matrix and relationships between the compounds in three simple pair-wise variable plots. If the data set contains hundreds of compounds described by dozens of variables, perceiving these relationships is much more problematic. Cluster analysis is a useful tool for the visualization of both the variables and the compounds in a data matrix. Cluster analysis produces a diagram, called a dendrogram, in which similar objects (descriptors or cases) are grouped together in clusters. Similarity is defined by some measure of distance between the objects. In the case of variables this can be the simple pair-wise correlation between them, and in the case of compounds this can be the Euclidean distance in the descriptor space. Figure 7.1 shows the dendrogram for a set of 70 variables

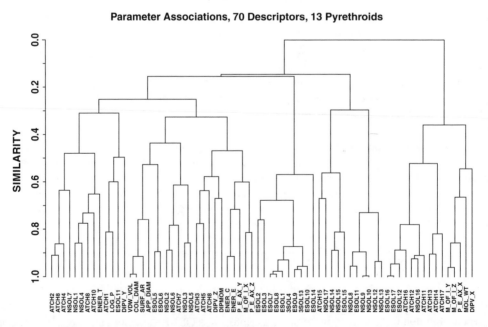

**Figure 7.1**    Dendrogram showing the association between 70 computed descriptors for a set of 13 pyrethroid insecticides. Distance on the vertical axis is a measure of similarity (correlation) between the variables. (From Ford, M.G. and Livingstone, D.J., *Quant. Struct.-Act. Relat.,* 9, 107–114, 1990. With permission.)

describing a set of pyrethroid analogs. The clusters contain groups of variables that are correlated or, in other words, share the same information. A diagram such as this allows the analyst to see which descriptors may be redundant; to choose variables to use in the construction of models; or, where the variables include biological responses, to see how biological mechanisms group together.

Principal component analysis (PCA) is a mathematical procedure (Jackson, 1991; Chapter 4 of Livingstone, 1995) in which new variables, called principal components (PCs), are created from linear combinations of the original variables in such a way that:

The first principal component explains the maximum variance (information) in the data set. Subsequent components describe the maximum part of the remaining variance subject to the condition that:
The principal components are orthogonal to one another.
The construction of the PCs is shown in Equation 7.1:

$$PC1 = a_{1,1}v_1 + a_{1,2}v_2 + \ldots a_{1,n}v_n$$

$$PC2 = a_{2,1}v_1 + a_{2,2}v_2 + \ldots a_{2,n}v_n \tag{7.1}$$

$$PCq = a_{q,1}v_1 + a_{q,2}v_2 + \ldots a_{q,n}v_n$$

where the subscripted term, $a_{ij}$, shows the contribution of the original variable, $v_j$, to the $i^{th}$ PC. Here the equation shows the generation of q PCs from a set of n variables. This is the third condition for PCA:

As many PCs may be extracted as the smaller of $p$ (data points) or $n$ (dimensions) for a $p \times n$ matrix (denoted by $q$ in Equation 7.1).

The contributions of the original variables (loadings) to the PC's give insight into the correlation structure of the variables in the original data set. The values of the new variables, the PC scores, may be plotted against one another to show the relationships between the compounds. Since the first two PCs explain the maximum variance in the data, a plot of PC2 vs. PC1 often gives a very good representation of the data set in just two dimensions. Figure 7.2 shows a PC plot for a set of compounds which have teratogenic activity. The principal components were calculated from computed molecular properties, and it can be seen that the teratogens are clearly distinguished from the non-teratogenic compounds. The PC scores can also be used as new variables for the construction of models relating a response to physicochemical properties (see later). Since the PCs are orthogonal to one another, the resulting models may have some useful properties such as stability when new compounds are added to the model.

## B.  Variable Elimination

Variable elimination is the name given to the process by which unhelpful or unnecessary variables are removed from a data set. One means by which a variable may be judged is from the information it contains. If the standard deviation of the variable is very small, then it does not contain much information and is thus not likely to be useful in the construction of models. Another common situation is that a variable may contain only a small number of different values, an extreme case being where the values are the same for all compounds in the set except one. If a variable such as this is used in the construction of a model then it may appear to be useful but is usually only serving to identify, and thus explain, the response value for that single compound. An example of this is shown in Table 7.2, which contains values of receptor binding and computed properties for a set of quinuclidine-based muscarinic receptor agonists (Saunders et al., 1990). For some of the compounds (6, 7, 8, and 12) the substitution pattern means that there is not, in fact, an

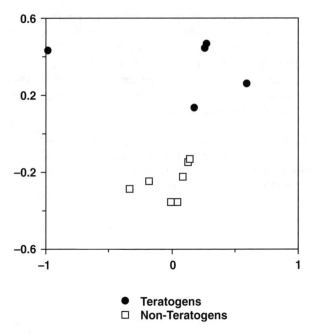

**Figure 7.2**   Principal component scores plot for a set of dopamine mimetics. Compounds with teratogenic activity are indicated by filled circles. (From Ridings, J.E., Manallack, D.T., Saunders, M.R., Baldwin, J.A., and Livingstone, D.J., *Toxicology*, 76, 209–217, 1992. With permission.)

**Table 7.2   Values of Binding Ratio and Computed Descriptors for Muscarinic Agonists**

| Compound No. | Log Binding Ratio[a] | V–2[b] | V–4[b] | V[c] |
|:---:|:---:|:---:|:---:|:---:|
| 5 | 2.69 | –78 | –72 | 63.5 |
| 6 | 1.72 | –88 | 0 | 64.7 |
| 7 | 1.59 | 0 | –80 | 63.5 |
| 8 | 1.18 | 0 | 0 | 65.8 |
| 9 | 2.11 | –24 | –87 | 62.4 |
| 10 | 1.73 | –78 | –37 | 59.0 |
| 12 | 1.28 | –50 | 0 | 66.9 |
| 18 | 3.08 | –78 | –82 | 59.0 |
| 23 | 2.60 | –67 | –61 | 66.1 |

[a] Log of the measured binding at two sites — a measure of biological activity.
[b] Values of the calculated electrostatic potential adjacent to atoms 2 and 4.
[c] Calculated van der Waals volume of the substituent.

electrostatic potential minimum at that position and the entries in the table are incorrectly listed as zero (they should in fact be listed as missing). Regression analysis of the binding data against the V-2 and V-4 variables gives an apparently significant 2-term equation with an $R^2$ of 0.83, but omission of the compounds with the false zero values for V-4 completely destroys this relationship. This shows that the model is not generally applicable and that the electrostatic potential minima at position 4 were simply acting as indicator variables for these 3 compounds.

Another reason why variables may be unnecessary is that they may be correlated with one or more of the other descriptors in the set. This is known as collinearity, as demonstrated in Table 7.1, or multicollinearity where one variable is more or less explained by a combination of the others.

**Parameter Associations for Retained Descriptors**

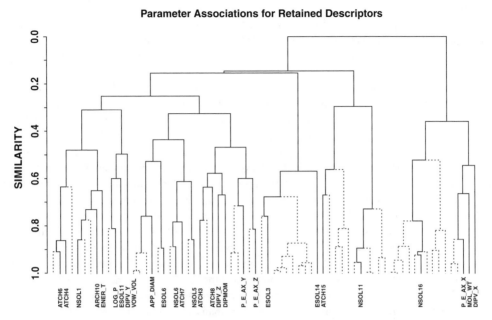

**Figure 7.3**    The dendrogram from Figure 7.1 with the retained variables shown in solid lines. (From Ford, M.G. and Livingstone, D.J., Q*uant. Struct.-Act. Relat.*, 9, 107–114, 1990. With permission.)

Examination and counting of the pair-wise correlations between variables is the basis of a method of variable elimination called Corchop (Livingstone and Rahr, 1989). Figure 7.3 shows the end result of applying this procedure to the data set used to generate the dendrogram shown in Figure 7.1. This data set consisted of 70 computed descriptors; the Corchop process reduced this to just 28 molecular properties which, as can be seen from the diagram, are quite well-spread with regards to their correlations. Another variable elimination procedure that takes account of multicollinearity is known as unsupervised forward selection (Whitley et al., 2000).

## C.  Variable Selection

Having eliminated unsuitable descriptor variables from a data set, the analyst may still be faced with the problem that there are many variables to choose from in order to construct models relating the response to chemical structure (via the physicochemical properties). There is quite a large variety of procedures for variable selection, some built in to the process of model construction, and thus it is only possible here to consider some of the general principles of variable selection. First, though, it is necessary to issue a warning — variable selection can be fraught with danger. The overall aim of variable selection is to choose descriptors that will be useful in some sort of mathematical model and will lead to a model that will generalize to other unseen compounds. It is always possible given enough variables to choose from in the beginning, that properties may be selected that appear to relate to the response but in fact are correlated by chance. The danger of chance correlations has been examined for regression analysis (Topliss and Edwards, 1979), partial least squares (PLS) (Cramer et al., 1988), and neural networks (Manallack and Livingstone, 1992; Livingstone and Manallack, 1993); the guidelines given in these papers may help to avoid chance effects.

How do we set about variable selection? One obvious approach is to examine the pair-wise correlations between the response and the physicochemical descriptors. One form of model building, forward stepping multiple regression, begins by choosing the descriptor that has the highest correlation with a response variable. If the response is a categorical variable such as toxic/non-toxic, then variable selection involves the choice of a descriptor that best separates the two (or more) classes. Forward stepwise regression analysis, or discriminant analysis for a classified response, continues to select variables on the basis of their fit, or discrimination, until some stopping rule is reached. This is most often the value of an F-to-enter statistic, although other measures may be used. For these two methods at least, some form of automatic variable selection is available. Unfortunately, these automated procedures rarely give the best result, and for any reasonable sized data set there can be many alternative models that give as good or better results. Perhaps an example will illustrate this. There is a well-known data set, which has become something of a standard for the testing of new data analysis methods (Comley et al., 1990). However, it has been pointed out that this set suffers from a number of problems (Whitley et al., 2000). This set consists of 31 compounds described by 53 computed molecular properties and the original study reported a 3-term regression model for the data. Many other models can be calculated for this set, and it was pointed out by Kubinyi (1994) that if each combination of the possible regression models (from all 1-term equations up to all possible 29-term equations: there are $7.16 \times 10^{15}$ possible combinations) was computed at a speed of 1 per second each, it would take 227 million years! Further Kubinyi (1994) reported the application of an evolutionary algorithm to explore possible regression models for this set, demonstrating quite impressive results. Other genetic algorithm (GA)-type approaches have been used on this set (Leardi, 1996) and applied to other problems of model selection. The GA method is generally a very useful approach for solving problems involving large solution spaces. Finally, one procedure that has proven its worth on many occasions is to examine the data in a variety of ways (PC plots, cluster analysis, factor analysis, etc.) and take the most important variables as identified by these methods to form the set that is then used to build mathematical models.

## V. BUILDING MODELS

There are very many methods available to build mathematical models that relate biological properties to chemical structure. Given the limitations of space here, this section will just focus on a few of the more commonly used techniques. To give a taste of the variety of methods available, Table 7.3 lists some of the better known techniques.

**Table 7.3   Some of the More Common Pattern Recognition Techniques**

| Supervised Learning | Unsupervised Learning |
| --- | --- |
| Multiple Linear Regression | Principal Component Analysis |
| Discriminant Analysis | Cluster Analysis |
| Partial Least Squares (PLS) | Nonlinear Mapping |
| Soft Independent Modelling of Class Analogy (SIMCA) | Factor Analysis |
| Canonical Correlation Analysis | k-Nearest Neighbours (KNN)[a] |
| Principal Component Regression[b] | Correspondence Analysis |
| Classification and Regression Trees (CART) | Kohonen Mapping or Self-Organizing Map (SOM) |
| Linear Learning Machine | |
| Neural Networks | |
| Adaptive Least Squares | |
| Genetic Programming | |
| Logistic Regression | |

[a] This method uses class membership to make predictions but does not use this information to train the analysis.
[b] The generation of principal components is unsupervised, but fitting the regression coefficients is supervised.

The procedures listed in Table 7.3 are referred to as pattern recognition methods. This is partly because this is a general term for the methods used to analyze the data that we are interested in (Livingstone, 1991), but also because the pattern recognition concept of learning is useful to bear in mind. So what does the term learning mean? Some pattern recognition methods have their origins in artificial intelligence research; attempts to devise algorithms that could learn to distinguish patterns in a data set. These algorithms can be classified as supervised and unsupervised, where supervision refers to the use made of the response data that we are trying to model. Unsupervised learning, or training without a teacher, makes no use of the response; the algorithms seek to recognize patterns in the descriptor data only. Many of the unsupervised learning techniques are display methods (dimension reduction) in which a higher dimensional space is shown in lower dimensions so that the patterns may be more easily discerned. The advantage of unsupervised learning is that since the algorithm is not trying to fit a model there should be a lower likelihood of chance effects. Since supervised learning, on the other hand, does use the response, care needs to be taken to avoid chance effects (see testing models). The ratio of compounds ($p$) to variables ($n$) in a data set is also important in the difference between supervised and unsupervised learning methods. When $n \geq p$, some supervised learning techniques may not work because of failure to invert a matrix, while others may give a false, apparently correct, classification. This is not a problem for unsupervised methods, although the presence of extra variables that do not contain useful information may obscure meaningful patterns (see variable elimination).

Another important feature of mathematical modeling techniques is the nature of the response data that they are capable of handling. Some methods are designed to work with data that are measured on a nominal or ordinal scale; this means the results are divided into two or more classes that may bear some relation to one another. Male and female, dead and alive, and aromatic and nonaromatic, are all classifications (dichotomous in this case) based on a nominal scale. Toxic, slightly toxic, and non-toxic are classifications based on an ordinal scale since they can be written as: toxic > slightly toxic > non-toxic. The rest of this section is divided into three parts; methods that deal with classified responses, methods that handle continuous data, and artificial neural networks that can be used for both.

## A.  Modeling Classified Data

Discriminant analysis is probably the best known technique for the analysis of classified response data. Discriminant analysis fits a line or surface between two classes of data as illustrated in Figure 7.4.

In this simple two-dimensional example the discriminant function is a straight line; in the case of a set of compounds described by $n$ physicochemical properties the discriminant function would be an $n$-dimensional hypersurface. The discriminant function can be represented as:

$$W = a_1 x_1 + a_2 x_2 + \dots a_n x_n \tag{7.2}$$

where the $x_i$s are the independent variables and the $a_i$s are their fitted coefficients. Most statistics programs will offer a variety of options for fitting discriminant functions (e.g., forward stepping, backward stepping, complete estimation, quadratic) and will produce a number of statistical measures of the fit. It is beyond the scope of this chapter to discuss these in detail; any good statistics textbook will explain these quite adequately, but it is worth examining the way that the results from a discriminant analysis (and some other types of classification methods) are usually reported. This is a classification matrix, which is commonly called a confusion matrix. Table 7.4 shows a confusion matrix for the classification of compounds as active or inactive against a gene family target, G-protein coupled receptor amine subset, as predicted by a committee of 100 trained neural networks (Manallack et al. 2002).

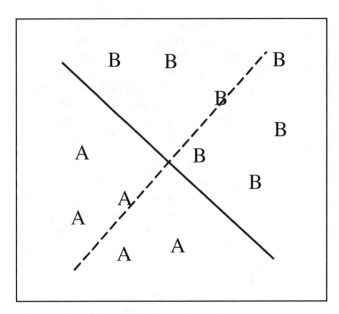

**Figure 7.4**   2D representation of discriminant analysis. The dotted line represents the discriminant function and the solid line represents a discriminant surface that separates the two classes of samples. (From Livingstone, D.J., *Data Analysis for Chemists: Applications to QSAR and Chemical Product Design*, Oxford University Press, Oxford, 1995. Reproduced with permission of Oxford University Press.)

**Table 7.4   Confusion Matrix from an Ensemble of 100 Trained Neural Networks Predicting Gene Family Target Activity**

| GPCR Amine Consensus Networks | | | |
| --- | --- | --- | --- |
| | **Active** | **Inactive** | **Total** | **% Correct** |
| Active | 1106 | 262 | 1368 | 80.85 |
| Inactive | 245 | 1123 | 1368 | 82.09 |
| Total | 1351 | 1385 | 2736 | |
| % Correct | 81.87 | 81.08 | | 81.47 |

*Source:* Manallack, D.T., Pitt, W.R., Gancia, E., Montana, J.G., Livingstone, D.J., Ford, M.G., and Whitley, D.C., *J. Chem. Inf. Comput. Sci.*, 42, 1256–1262, 2002. Reproduced with permission of the American Chemical Society.

In the confusion matrix the row entries are the observed classes and the columns are the predicted classes. The first row shows that 1106 compounds are correctly predicted as active with 262 incorrectly predicted to be inactive. The column on the right shows the percentage of active and inactive compounds that were correctly predicted, 80.85 and 82.09%, respectively. These quantities are also often called *sensitivity* (percentage of positives or actives correctly predicted) and specificity (percentage of negatives or inactives correct). The entry 81.47 in the table is an overall measure of the percentage correctly predicted. The two percentages reported in the bottom row are called the predictive power of a positive test and, unsurprisingly perhaps, the predictive power of a negative test. In other words, when the classification method calls a prediction positive (or negative), these terms describe how often the method makes a correct prediction. In this example the analysis is quite well balanced and the sensitivity and specificity values are very similar to the predictive powers. Unfortunately, this is not always the case. Sometimes a classification technique will work well for one class but not another and this will be reflected in differences between these quantities.

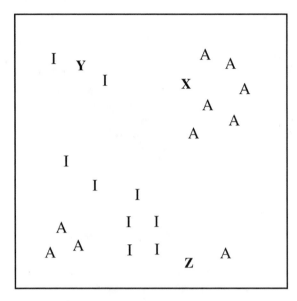

**Figure 7.5** 2D representation of the *k*-nearest neighbours method. (From Livingstone, D.J., *Data Analysis for Chemists: Applications to QSAR and Chemical Product Design*, Oxford University Press, Oxford, 1995. Reproduced with permission of Oxford University Press.)

*K*-nearest neighbors (KNN) is the name of a classification method that is unsupervised in the sense that class membership is not used to train the technique, but that makes predictions of class membership. The principle behind KNN is an appealingly common sense one; objects that are close together in the descriptor space are expected to show similar behavior in their response. Figure 7.5 shows a two-dimensional representation of the way that KNN operates.

The points on Figure 7.5 represent compounds in the *n*-dimensional hyperspace of the *n* descriptors used to characterize them. Active compounds are labeled A and inactive compounds I with three unknowns shown as X, Y, and Z. Prediction of activity for X and Y is unambiguous since they are surrounded by active and inactive compounds, respectively. Prediction for compound Z is more problematic since its two nearest neighbors are active and inactive; this is where the term *k* comes into play. The obvious prediction for compound Z is inactive since more of the neighbors are inactive, so if we choose three for the value of *k* then we will get what seems to be a sensible result. The only adjustable parameter in the KNN technique is the value of *k*, and this is most often chosen to be an odd number so that a majority vote will reach a conclusion rather than having a stalemate. Choice of the value for *k* is clearly critical for this method and is usually chosen on the results of predictions for the training set.

## B. Modeling Continuous Data

Multiple linear regression analysis (MLR) is the most commonly used, and abused, tool for the construction of models of continuous response data. There are a number of ways to construct models (e.g., forward stepwise, backward stepwise, stepwise with addition and elimination, best subsets, etc.). There are also a number of different types of linear regression models. Details of the fitting process and the statistics used to assess the goodness of fit are covered in many standard texts (Draper and Smith, 1981; Kubinyi, 1993; Livingstone, 1995) so here we will just summarize some of the more important features using an example. Equation 7.3 shows the results of fitting a multiple linear regression model to the toxicity ($-\log[IG_{50}]^{-1}$) of a set of substituted phenols to the protozoan *Tetrahymena pyriformis* (Cronin et al., 2002).

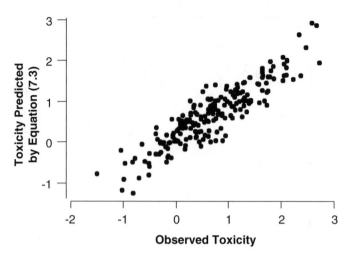

**Figure 7.6**   Plot of predicted (from Equation 7.3) against observed toxicity of phenols to *T. pyriformis*. (From Cronin, M.T.D., Aptula, A.O., Duffy, J.C., Netzeva, T.I., Rowe, P.H., Valkova, I.V., and Schultz, T.W., *Chemosphere,* 49, 1201–1221, 2002. With permission.)

$$\log\left(IGC_{50}\right)^{-1} = 0.38(0.024)\log D - 0.58(0.058)E_{LUMO} + 0.0047(0.0008)MW$$

$$- 0.018(0.0048)P_{NEG} + 0.05(0.008)SsOH - 0.61(0.11)ABSQon \qquad (7.3)$$

$$+ 2.69(1.15)MaxHp - 0.99(.29)$$

$$n = 185, \ R^2 = 0.83, \ R^2_{CV} = 0.82, \ s = 0.34, \ F = 128$$

The statistics reported for the fit are the number of compounds used in the model (n), the squared multiple correlation coefficient ($R^2$), the cross-validated multiple correlation coefficient ($R^2_{CV}$), the standard error of the fit (s), and the F statistic. The squared multiple correlation coefficient can take values between 0 (no fit at all) and 1 (a perfect fit) and when multiplied by 100 gives the percentage of variance in the response explained by the model (here 83%). This equation is actually quite a good fit to the data as can be seen by the plot of predicted against observed values shown in Figure 7.6.

When building or comparing regression models, it is tempting to use just $R^2$ as a measure of fit since it is a handy single number. This can be very misleading because any regression model will improve as extra terms are added to it. The extra terms allow greater flexibility in accommodating the structure of the response data. The number of data points also has an effect on $R^2$ since, for small data sets, the effect of extra terms is more pronounced. In order to judge the quality of regression models that contain different numbers of variables or different numbers of data points it is necessary to use a different type of $R^2$, one that is adjusted for the number of variables. This is known unsurprisingly as $R^2_{adjusted}$ or sometimes by the symbol $\bar{R}^2$ and is given by:

$$\bar{R}^2 = 1 - \left(1 - R^2\right)\frac{n-1}{n-p} \qquad (7.4)$$

where n is the number of data points and p is the number of terms in the equation. The value of $\bar{R}^2$ for Equation 7.3 is 0.82. This value is not very different from $R^2$, which is perhaps not unexpected since this is a reasonably large data set.

The cross validated multiple correlation coefficient is an attempt to give an idea of how well the model might predict by leaving out compounds one at a time and predicting their response value from a model fitted to the remainder of the set. This is discussed further in the section on testing models. The standard error of the fit is the quantity that should be compared with the measurement error of the biological test in order to avoid overfitting as discussed in the section on response data. The F statistic is the ratio of the variance explained by the regression model (explained mean square) to the residual variance (residual mean square), and this is a measure of the overall significance of the regression model. F statistics are used by consulting a look-up table of values. If the calculated value is greater than the tabulated value, then the model is significant at that level of confidence. F statistics have two associated degrees of freedom: $v_1$ and $v_2$. The first of these relates to the number of terms in the equation, and the second relates to the number of samples. F statistics are tabulated for various confidence levels (e.g., 1%, 5%, 10%). Commonly the F-tables are computerized by standard statistical packages.

The final statistical values that are reported for Equation 7.3 are the standard errors of the regression coefficients. These allow us to assess the significance of the individual terms by computing a statistic, called the t statistic, by dividing the regression coefficient by its standard error:

$$t = \left| \frac{b}{\text{SE of } b} \right| \qquad (7.5)$$

where b is the regression coefficient and SE refers to the standard error of the coefficient. The t statistics for the regression coefficients and the constant in Equation 7.3 are 15.7, 10.1, 5.9, 3.7, 6.0, 5.6, 2.4, and 3.4 respectively. These t statistic values are used, like the F statistic, by consulting a look-up table of values (again, now normally computerized). The tables contain entries for different confidence levels (e.g., 0.1, 1, 5, and 10%) for different degrees of freedom. The associated degrees of freedom for a t statistic is given by (n-p-1) where n is the number of data points and p is the number terms in the equation, including the constant term, as before. From the tables, the value of the t statistic for a reasonable number of degrees of freedom (10 or more) at the 5% confidence level is 2.2. This is equivalent to saying that the regression coefficient should be at least twice as big as its standard error if it is to be considered to be significant. This seems like good common sense. As can be seen from the reported values for Equation 7.3, all of the terms in the equation are significant.

Regression on principal components (PCR) is another from of regression modeling that may be used for continuous response data. Here, the independent variables (the x set) are computed from the descriptor variables using PCA as shown in Equation 7.1. These are the principal component scores and they have several advantages:

- There is a smaller number of PC scores than the original descriptor variables, sometimes a much smaller number.
- The PCs explain large amounts of variance in the descriptor set and thus may be expected to capture a large amount of the information in the physicochemical properties.
- Generation of the PCs is an unsupervised learning process that may help to reduce the possibility of chance effects.
- The PCs are orthogonal to one another which has important consequences for the resulting regression models (see Chapter 7 of Livingstone [1995]).

The PC regression model is constructed in exactly the same way as any other regression model:

$$y = a_1.PC1 + a_2.PC2 + ... a_n PC_n \qquad (7.6)$$

where the $a_i$ are regression coefficients computed by least squares and all the regular statistics that characterize the fit of a regression model can be calculated and used to assess how well the model explains the data. PLS regression is related to PCR, but here the equivalent of the PC scores, usually called latent variables (LVs), are calculated so that:

- The first LV accounts for the maximum variance in the descriptor set, and has the highest correlation with the dependent variable.
- Subsequent LVs, which are orthogonal to the first LV and one another, have the next highest correlations with the dependent variable and explain the next largest amounts of variance in the descriptor set.

PLS combines the two separate steps of PCA and regression modeling that are seen in PCR. In addition, it often results in more compact models (fewer terms) than PCR which may also perform better at fitting the data. Application of PLS to the data used to generate Equation 7.3 resulted in a 3-term (3 LVs) PLS model with an $R^2$ of 0.82 for 197 compounds (12 more than used to generate Equation 7.3). This PLS model used a selected set of 12 descriptors from the original physico-chemical descriptor set, compared with the 7 variables selected by the stepwise regression process used to generate Equation 7.3. Of these 12 variables, 7 corresponded to the descriptors used in the regression model with the additional descriptors making up indicator variables for specific molecular features (indicator variables take a value of 1 when a particular feature is present, 0 otherwise). The use of indicator variables may explain why the PLS model was able to accommodate an extra 12 compounds compared with Equation 7.3.

How can we judge the quality of these two models? The regression model has an $R^2$ value of 0.83, with an adjusted $R^2$ of 0.82 (and an $R^2_{CV}$ of 0.82, but see later), while the PLS model has an $R^2$ of 0.82 ($R^2_{CV} = 0.80$). Both models have a more or less equivalently good fit, although the cross-validated $R^2$ is lower for the PLS model. The PLS model accounts for a larger number of compounds. This may be an explanation of why the $R^2_{CV}$ is smaller since the extra compounds may be outliers from the main model, which are accounted for by the extra indicator variables. In terms of interpretation, the regression model is arguably simpler since it involves just seven variables with their associated regression coefficients. The PLS model, on the other hand, involves 12 variables with 3 sets of coefficients, one set for each of the 3 latent variables. Ease of interpretation is not necessarily a criterion by which quality is judged, so how else can this question be decided? The best test of any model, as discussed later, is to apply it to another set of compounds. Fortunately for us, this is exactly the procedure Cronin et al. (2002) undertook. Predictions were made using both models (and others) for a test set of 46 compounds. The statistics of the fit comparing predicted and measured values were $R^2 = 0.74$, $s = 0.44$ for the regression model and $R^2 = 0.82$, $s = 0.37$ for the PLS model. On the basis of their performance on this test set the PLS model appears to be better.

PLS is able to handle collinearity and multicollinearity in the data set. These terms refer to the case when descriptor variables are correlated in pairs or when one descriptor is explained by a combination of variables, as discussed in the section on variable elimination. Most implementations of PLS use an algorithm (e.g., Non-Linear Iterative Partial Least Squares [NIPALS]), which is able to handle data sets that contain more variables than cases (compounds). This is because the method operates in an iterative fashion by making guesses at the solution to the eigenvector/eigenvalue problem and then refining these guesses. Because of this capacity to handle oversquare matrices, PLS is used extensively to analyze the data resulting from the use of 3D QSAR methods such as Comparative Molecular Field Analysis (CoMFA), GRID, Comparative Molecular Similarity Analysis (COMSIA) and so on.

## C. Artificial Neural Networks

Artificial neural networks (ANNs) are attempts to mimic biological intelligence systems (brains) by copying some of their structure and functions. The basic building block of an ANN is an artificial

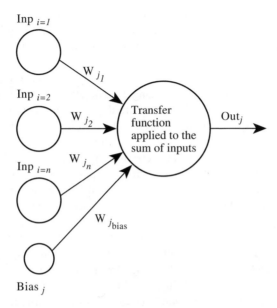

**Figure 7.7**   Diagram of an artificial neuron connected to three input neurons and a bias unit. (From Manallack, D.T. and Livingstone, D.J., *Med. Chem. Res.*, 2, 181–190, 1992. With permission.)

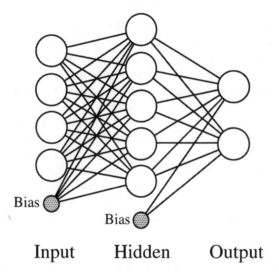

**Figure 7.8**   Illustration of how an ANN is built from layers of artificial neurons. (From Manallack, D.T. and Livingstone, D.J., *Med. Chem. Res.*, 2, 181–190, 1992. With permission.)

neuron or processing element as shown in Figure 7.7. This unit takes inputs from the units connected to it, sums up these inputs, applies a non-linear transfer function to the sum, and produces an output according to the transfer function (see Manallack and Livingstone, 1992; or Chapter 9 of Livingstone, 1995).

The processing elements are typically arranged in layers; one of the most commonly used arrangements is known as a back propagation, feed forward network as shown in Figure 7.8. In this network there is a layer of neurons for the input, one unit for each physicochemical descriptor. These neurons do no processing, but simply act as distributors of their inputs (the values of the variables for each compound) to the neurons in the next layer, the hidden layer. The input layer also includes a bias neuron that has a constant output of 1 and serves as a scaling device to ensure that the summed signals will reach a value on the transfer function which will produce an output. Each neuron in the input layer is connected to each of the units in the hidden layer. The hidden layer neurons process their inputs through the transfer function and pass their output to the output layer unit or units.

Training a neural network is achieved by first initializing the connection weights between the neurons, usually to random values, and then passing the data set through the network and comparing the output with the target values where the targets are the values of the biological response. ANNs are able to train to classified or continuous target values so they are suitable for handling most types of response data. Training continues by repeatedly passing the data through the network, evaluating the errors between the output and target, adjusting the weights of the connections between neurons, and reevaluating the outputs. Because the data pass forward through the network, it is known as feed-forward. Since the errors are evaluated at the output and then connections altered back toward the input layer, the process is known as back propagation of errors (hence the terminology back propagation, feed forward network). These networks also glory under a number of alternative names such as multilayer perceptron (MLP) and back propagation neural network (BNN). There are other types of network architecture that have found application in QSAR, and the following reviews and books may be helpful: Manallack and Livingstone (1994), Devillers (1996), Livingstone and Salt (1995), and Zupan and Gasteiger (1999).

How does one go about setting up a neural network to solve a particular problem? First of all, it is necessary to choose a computer program. There are a number of such dedicated packages, both commercial and public domain, as well as toolboxes for some of the well-known general purpose mathematics programs such as Matlab (The MathWorks, Inc., Natick, MA). Any Internet search engine will reveal a host of sources of such software. After obtaining the program and transforming the data into the appropriate format for input it is then necessary to choose the architecture of the network, although some programs even automate this step. Choice of input layer and output layer neurons is usually self-determined by the number of descriptor variables and the form of the response data. A single output neuron is sufficient for most purposes whether the response is classified or continuous, although multicategory classified data may call for more output neurons. The next step is to choose the number of hidden layer neurons (some papers have reported the use of more than one hidden layer, but most successful reported applications only use one). This needs to be considered with regards to the number of data points and the number of input neurons (descriptors). A sufficient number need to be used to allow the complexity of the data to be modeled, but not so many that there will be as many or more connection weights than compounds in the set. In this situation the neural network will overfit the data (Manallack and Livingstone, 1994). In practice, the smaller the number of hidden units the better as far as speed of training and generality of the model are concerned.

After setting up the network the next step is to choose a training algorithm (there are several) and values for the adjustable training parameters. This can be a complex business, although most computer packages will have detailed help for this step. After this we are nearly ready to begin training; all that remains is to choose some suitable stopping criterion to end the training. There is a variety of ways to choose this stopping method, such as the absolute error between output and targets, but one particularly useful method is known as early stopping (Tetko et al., 1995). In this procedure the data set is divided into three sets:

- A training set on which the network is trained
- A control set used to monitor the performance of the network as it trains
- A test set used to judge network performance at the end

This is quite demanding of data points and not all data sets contain sufficient examples. In such cases it may be necessary to just retain a control set. Once all this has been done then network training can begin. It is hoped a trained network will be able to make useful predictions of the data. It should be borne in mind that a trained network is just one solution. There is no way of knowing whether this network is anywhere near the global minimum, and thus if such a minimum exists, it is necessary to reinitialize and train the network several times. A popular method for using neural networks is to train a large number of networks on the same problem and then use them as a committee (see Table 7.4) to make predictions. Not only does this help to avoid rogue predictions from a particular network solution, but it also allows some error estimates of the predictions to be calculated. A recent area of application of ANN, which shows much promise, is in the field of 3D QSAR (Livingstone and Manallack, 2003).

## VI. TESTING MODELS

After building a mathematical model of whatever type, it is important to try to assess how well the model might be expected to work. Most statistics packages will generate a variety of statistical quantities for the more common modeling approaches. These will allow a judgment of significance and will give some guidance on whether the model may have arisen by chance. This of course is based on the assumption that the data conform to some statistical distribution, usually multivariate normal. Does this really give us any information on how well the model might work or does it just show how well the model fits the data within the modeling assumptions? The answer, unfortunately, is really the latter; the only way to learn how well a model may work is to try it out.

One common, albeit flawed, approach is known as leave-one-out cross validation (LOO or CV) or jack-knifing. This process involves, as the name suggests, leaving out one compound, fitting the model to the remainder of the set, making a prediction for the left out compound, and then repeating the process until every compound in the set has been left out. A variety of statistics can be generated using this procedure, which are the LOO equivalents of regular fit statistics. Examples include the LOO $R^2$, commonly called $Q^2$, and a predictive residual sum of squares (PRESS), the LOO equivalent of the residual sum of squares. A distinct advantage of LOO is that it is reproducible. The disadvantage of the LOO approach is that only a small part of the data set is omitted, particularly for large (>50) data sets, and if outliers occur in pairs or groups then this method will not identify them. A better technique is to leave out some larger portion of the set, say 10 or 20%, and to repeat this a number of times. This allows the generation of a set of predicted values for the compounds so that estimates may be made of the likely errors in prediction. The disadvantage of this approach is that it is computationally intensive and suffers from a combinatorial explosion as the sample size is increased. Bootstrapping, in which new data sets are generated by randomly sampling the original data set, is a related approach (Efron and Tibshirani, 1993).

What about chance effects? Do any of these procedures allow us to judge whether a relationship is real or if it can have happened by chance? Unfortunately, they do not although LOO, or better still leave-n-out (LNO), does give us some idea of confidence in the model. One way to check for chance effects is to scramble the response values and then try to build models using the scrambled data. This can be repeated a number of times and then some fit statistics, such as $R^2$, can be tabulated for the resulting models. Figure 7.9 shows a plot of the distribution of $R^2$ values for 7 variable regression equations fitted to the data used to generate Equation 7.3. The response data had been randomly scrambled 500 times and the same 7 variable regression equations generated. The $R^2$ value for the model of the unscrambled response is 0.83 which is convincingly higher than the $R^2$ for the scrambled sets. We can therefore be reasonably confident that this model is not a chance fit.

The best test of a model, of course, is to present it with unseen data. This may be achieved by holding back some of the original data to form a test set. A more convincing test is to synthesize or test some more compounds once the model has been built.

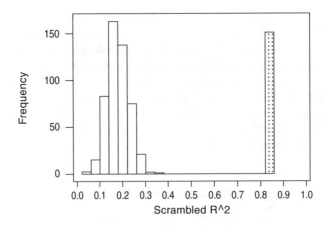

**Figure 7.9**   Plot of the distribution of $R^2$ values for equations fitted to scrambled response data. The $R^2$ value for the equation fitted to the unscrambled data is indicated by the spotted bar.

## VII. INTERPRETING MODELS

There is an argument that it is not necessary to try to interpret models. If the aim of modeling is simply prediction, then this can be true. There is also an argument that it is dangerous to attempt to interpret models, since correlation does not imply causality, and there is also an element of truth in this. Having said that, human nature makes us curious and it is always tempting to try to interpret the models we create. Some modeling approaches lend themselves more easily to attempts at interpretation. A multiple linear regression equation and a discriminant function are examples of models that may be easily interpreted. Here, the important variables are identified and the sign and magnitude of their influence on the response can be judged from their coefficients in the model. Only experiment will confirm the effects of these properties, but examination of the models can be helpful in suggesting the experiments to perform. Other modeling procedures are much more obscure; PCR and PLS models, for example, involve several coefficients for each variable and ANN models contain their relationship in an often very complex set of weights. There are ways in which the importance of variables in the ANN model can be assessed (Tetko et al., 1996), but this still does not make for very easy interpretation.

## VIII. CONCLUSIONS AND RECOMMENDATIONS

The methods and their requirements to build models relating biological activities to their physicochemical and/or structural properties must be borne in mind from the outset of a (Q)SAR study. The following are a small number of simple recommendations that should assist in successful modeling:

- If possible select chemicals for testing to provide as much physicochemical information as possible.
- Ensure that the biological response data are appropriate for modeling.
- If a large number of descriptor variables are utilized, reduce the number by variable elimination prior to modeling.
- Use a variable selection technique appropriate to the problem.
- Use a modeling technique appropriate to the data and data set being modeled. Preference should be given to simple modeling techniques.

- Models should be assessed not only in terms of their goodness of fit (i.e., statistical quality), but also in terms of their predictive power. The predictive power of a model can only be assessed by estimating the activity of a set of compounds not included in the original model.
- In all modeling techniques, and neural networks in particular, care must be taken not to overtrain or overfit the model.
- If possible, models should be interpreted in terms of their mechanistic meaning.

## REFERENCES

Comley, J.C.W., O'Dowd, A.B., Hudson, A.T., Jackson, P., Jandu, K.S., Livingstone, D.J., Rose, V.S., Selwood, D.L., and Stables, J.N., Structure-activity relationships of antifilarial antimycin analogues, a multivariate pattern recognition study, *J. Med. Chem.*, 33:136-142, 1990.

Cramer, R.D., Bunce, J.D., Patterson, D.E. and Frank, I.E., Cross-validation, bootstrapping, and partial least squares compared with multiple regression in conventional QSAR studies, *Quant. Struct.-Act. Relat.*, 7, 18–25, 1988.

Cronin, M.T.D., Aptula, A.O., Duffy, J.C., Netzeva, T.I., Rowe, P.H., Valkova, I.V., and Schultz, T.W., Comparative assessment of methods to develop QSARs for the prediction of toxicity of phenols to *Tetrahymena pyriformis*, *Chemosphere*, 49, 1201–1221, 2002.

Devillers, J., Ed., *Neural Networks in QSAR and Drug Design*, Academic Press, London, 1996.

Draper, N.R. and Smith, H., *Applied Regression Analysis*, John Wiley and Sons, New York, 1991.

Efron, B. and Tibshirani, R., *An Introduction to the Bootstrap*, Chapman and Hall, London, 1993.

Ford, M.G. and Livingstone, D.J., Multivariate techniques for parameter selection and data analysis exemplified by a study of pyrethroid neurotoxicity, *Quant. Struct.-Act. Relat.*, 9, 107–114, 1990.

Ford, M.G., Greenwood, R., Brooks, G.T., and Franke, R., Eds., *Bioactive Compound Design: Possibilities for Industrial Use*, Bios, Oxford, 1996.

Gundertofte, K. and Jorgensen, F.S., Eds., *Molecular Modeling and Prediction of Bioactivity*, Kluwer Academic/Plenum Publishers, New York, 2000.

Holtje, H.-D. and Sippl, W., Eds., *Rational Approaches to Drug Design*, Prous, Barcelona, 2001.

Jackson, J.E., *A User's Guide to Principal Components*, John Wiley and Sons, New York, 1991.

Kubinyi, H., *QSAR: Hansch Analysis and Related Approaches*, VCH, Weinheim, 1993.

Kubinyi, H., Variable selection in QSAR studies, *Quantitative Structure-Activity Relationships* 13, 285–294, 1994.

Leardi, R., Genetic algorithms in feature selection, in J. Devillers (ed.) *Genetic Algorithms in Molecular Modeling*, Academic Press, London, 1996.

Livingstone, D.J., Pattern recognition methods for use in rational drug design, in *Molecular Design and Modeling: Concepts and Applications. Vol. 203 of Methods in Enzymology*, Langone, J.J., Ed., Academic Press, San Diego, 1991, pp. 613–638.

Livingstone, D.J., *Data Analysis for Chemists: Applications to QSAR and Chemical Product Design*, Oxford University Press, Oxford, 1995.

Livingstone, D.J., Structure-property correlations in molecular design, in *Structure-Property Correlations in Drug Research*, van de Waterbeemd, H., Ed., R.G. Landes, Austin, TX, 1996, pp. 81–110.

Livingstone, D.J., The characterization of chemical structures using molecular properties. A survey, *J. Chem. Inf. Compu. Sci.*, 40, 195–209, 2000.

Livingstone, D.J. and Manallack, D.T., Statistics using neural networks: chance effects, *J. Med. Chem.*, 36, 1295–1297, 1993.

Livingstone, D.J. and Manallack, D.T., Neural networks in 3D QSAR, *QSAR Combinatorial Sci.*, 22, 510–518, 2003.

Livingstone, D.J. and Rahr, E., Corchop: an interactive routine for the dimension reduction of large QSAR data sets, *Quant. Struct.-Act. Relat.*, 8, 103–108, 1989.

Livingstone, D.J. and Salt, D.W., Neural networks in the search for similarity and structure-activity, in P.M. Dean (ed.) *Molecular Similarity in Drug Design*, Blackie Academic and Professional, Glasgow, 1995, pp. 187–214.

Manallack, D.T. and Livingstone, D.J., Artificial neural networks: application and chance effects for QSAR data analysis, *Med. Chem. Res.,* 2, 181–190, 1992.

Manallack, D.T. and Livingstone, D.J., Neural networks: a tool for drug design, in *Advanced Computer-Assisted Techniques in Drug Discovery,* van de Waterbeemd, H., Ed., VCH, Weinheim, 1993, pp. 293–319.

Manallack, D.T., Pitt, W.R., Gancia, E., Montana, J.G., Livingstone, D.J., Ford, M.G., and Whitley, D.C., Selecting screening candidates for kinase and G protein-coupled receptor targets using neural networks, *J. Chem. Inf. Comput. Sci.,* 42, 1256–1262, 2002.

Ridings, J.E., Manallack, D.T., Saunders, M.R., Baldwin, J.A., and Livingstone, D.J., Multivariate quantitative structure-toxicity relationships in a series of dopamine mimetics, *Toxicology,* 76, 209–217, 1992.

Saunders, J., Cassidy, M., Freedman, S.B., Harley, E.A., Iversen, L.L., Kneen, C., Macleod, A.M., Merchant, K.J., Snow, R.J. and Baker, R., Novel quinuclidine-based ligands for the muscarinic cholinergic receptor, *J. Med. Chem.,* 33, 1128–1138, 1990.

Tetko, I.V., Livingstone, D.J., and Luik, A.I., Neural network studies. 1. Comparison of overfitting and overtraining, *J. Chem. Inf. Comput. Sci.,* 35, 826–833, 1995.

Tetko, I.V., Villa, A.E.P., and Livingstone, D.J., Neural network studies. 2. Variable selection, *J. Chem. Inf. Comput. Sci.,* 36, 794–803, 1996.

Todeschini, R. and Consonni, V., *Handbook of Molecular Descriptors,* Wiley-VCH, Mannheim, 2000.

Topliss, J.G. and Edwards, R.P., Chance factors in studies of quantitative structure-activity relationships, *J. Med. Chem.,* 22, 1238–1244, 1979.

van de Waterbeemd, H., Ed., *Advanced Computer-Assisted Techniques in Drug Discovery,* VCH, Weinheim, 1994.

van de Waterbeemd, H., Ed., *Chemometric Methods in Molecular Design,* VCH, Weinheim, 1995.

Whitley, D.C., Ford, M.G., and Livingstone, D.J., Unsupervised forward selection: a method for eliminating redundant variables, *J. Chem. Inf. Comput. Sci.,* 40, 1160–1168, 2000.

Zupan, J. and Gasteiger, J., *Neural Networks in Chemistry and Drug Design,* Wiley-VCH, Weinheim, 1999.

# QSARs for Human Health Endpoints

# CHAPTER 8

# Prediction of Human Health Endpoints: Mutagenicity and Carcinogenicity

**Romualdo Benigni**

## CONTENTS

## I. INTRODUCTION

One of the most ambitious goals of structure-activity relationship/quantitative structure-activity relationship (SAR/QSAR) applications to toxicology is the modeling of the chemical carcinogenicity; this because of the severity of its consequences on the quality of life and because of the enormous investment in time, financial resources, and animal lives required to test chemicals adequately. Mutagenicity is another important toxicological endpoint: chemical mutagens provoke heritable — mostly deleterious — changes to the genetic material.

From a mechanistic point of view, carcinogens can be classified as: (1) genotoxic and (2) epigenetic carcinogens (Woo and Lai, 2003). Genotoxic carcinogens, also known as deoxyribonucleic acid- (DNA) reactive carcinogens, interact directly with DNA either as parent chemicals or as reactive metabolites (Miller and Miller, 1977). Genotoxic, or mutagenic, carcinogens are thought to work by inducing mutations; the first step in several carcinogenic processes often consists of one or more mutations (the somatic mutation theory of cancer) (Arcos and Argus, 1995). The major classes of genotoxic carcinogens are: direct-acting carcinogens (such as epoxides, aziridines, nitrogen and sulfur mustards, β-halo-ethers, and lactones); aromatic amines and nitroaromatics; nitrosoamines and nitrosoamides; hydroazo and azoxy compounds; carbamates; organophosphates; aflatoxin-type furocoumarins; and homocyclic, heterocyclic, and polycyclic aromatic hydrocarbons (Ashby, 1995; Woo et al., 1995; 2002).

Epigenetic carcinogens act through mechanisms that do not involve direct DNA damage. In reality, there is seldom an absolute demarcation, and a better definition would be that of carcinogens that are predominantly genotoxic and predominantly epigenetic (Woo and Lai, 2003). Epigenetic carcinogens include cytotoxic chemicals that induce compensatory regenerative hyperplasia, agents that act via receptors, agents that cause indirect DNA damage via reactive oxygen species, and agents that regulate gene expression. For an updated review on this explosively growing literature see Woo and Lai (2003). As opposed to genotoxic carcinogens, no unifying mechanistic theory exists for the action of epigenetic carcinogens, and each class has to be studied separately.

## II. DATA FOR QSAR MODELING

A direct way to assess the mutagenic and carcinogenic potential of a chemical is to test it in an appropriate experimental system. At the same time, collation of these experimental data constitutes the necessary database to build SAR/QSAR models that can be used for predicting the activity of untested chemicals.

A wide range of experimental systems have been generated to determine mutagenicity (Zeiger, 1987; 1994). In their classical form, the mutagenicity tests (e.g., those based on the various *Salmonella typhimurium* bacterial strains) provide both a yes/no (mutagenic/nonmutagenic) outcome, and a quantitative definition of potency of the mutagenic compounds (e.g., increase of mutants per dose). Since chemical mutagenicity can be studied in model experimental systems more quickly and easily than carcinogenicity, the study of the mutagenicity has contributed remarkably to the study of cancer, and several short-term mutagenicity tests are used as surrogates and predictors of carcinogenicity.

The main tool to assess the carcinogenic potential of a chemical is the rodent bioassay. Because of its central role in the regulation of chemicals, the rodent bioassay has been under intense scrutiny. The overall evidence points to the validity of the bioassay as a basis for human risk assessment (Fung et al., 1995; Haseman et al., 2001; Huff, 1999; Huff, 2002; Tomatis et al., 1997).

The bioassay provides three types of information:

1. Yes/no response (if a chemical has to be considered as being carcinogenic or not) in the various experimental groups. These individual group yes/no responses allow for the generation of the overall carcinogenicity score.
2. Potency of the carcinogenic compounds. For each tumor type induced, a potency index can be calculated (e.g., $TD_{50}$, which is the dose required to halve the probability of the animals remaining tumorless). A measure often used is the geometrical mean of the $TD_{50}$'s, averaged over the whole range of tumors (Gold et al., 1991).
3. The profile of tumors (e.g., target organs) induced by the chemical.

These three endpoints are quite different in terms of relevance for the human health, and in terms of suitability for modeling.

The yes/no response is highly relevant and predictive for the human health. All human carcinogens are also animal carcinogens, and several human carcinogens were first discovered in animals (Huff, 1993; 1999).

Carcinogenic potency is also very relevant. There is evidence of a strong correlation between the ranking of potency in rats and mice (Benigni and Giuliani, 1999). Moreover, a strong correlation between carcinogenic potency estimated from epidemiologic data and that estimated from animal carcinogenesis bioassays, hence between human and rodent carcinogenic potency, has been demonstrated (Allen et al., 1988; Goodman and Wilson, 1991). The similar ranking of carcinogenic potency in different species suggests that potency is an intrinsic property of a chemical carcinogen that is derived to a greater extent from its chemical reactivity.

As opposed to the generality of carcinogenic potency, the tumor types induced appear to be highly variable from species to species. Not only do they depend on the species, but also may vary with the condition of use (e.g., age of the host, as well as dose and route of administration) of the carcinogen. Differences in the tumor profiles may result not only from differences in targeted reactions of the ultimate carcinogens, but also in the myriad of events that mediate and surround these reactions (Bucci, 1985). In this sense, the information on tumor profiles is of limited importance for the extrapolation of the risk to humans.

An important issue concerns the quality and reproducibility of the rodent carcinogenicity bioassay data. The bioassay, as performed by the standard protocols, is very costly and time consuming, and full replicate experiments are seldom performed. Based on analyses of a relatively small set of 38 replicate experiments, Gold et al. (1987) have estimated the overall reproducibility of the rat bioassay to be 85% and the mouse bioassay to be 80%; both values are quite satisfactory. Gottmann et al. (2001) analyzed a larger set of 121 chemicals for which replicate rodent bioassay results for the same chemicals, but tested under different protocols, were available. The estimated overall concordance in rodent carcinogenicity classification was only 57%. Since the results analyzed by Gold et al. were all generated under the strict protocols adopted by the U.S. National Toxicology Program (NTP), and the results considered by Gottmann et al. were of more varied origin, the conclusion that can be made is that adherence to strict experimental protocols is an essential requirement. At the same time, attention to the origin of the data, and to the protocols used for their generation, is equally necessary for those who want to build QSAR models.

An important issue for QSAR modeling is an appreciation of the manner in which data are reported. For example, the typical outcome of the *Salmonella typhimurium* mutagenicity assay consists of results in a number of tester strains. The results from each strain are a measure of potency (number of mutants) and a yes/no score (positive/negative; mutagenic/nonmutagenic). In addition, a summary score is defined (*Salmonella* positive if the chemical is positive in at least one of the tester strains, otherwise it is *Salmonella* negative). The data selected have important implications for the QSAR modeling. If the goal of the study is to provide mechanistic insight into the activity under consideration, then the experimental data have to provide a clear measure of the chemical-induced biological activity. In this case, models will be built for each strain, and separately for the yes/no and potency responses. If the goal is to create a model for use in hazard assessment, then the summary scores can be used for a coarser grain analysis. A similar issue applies to the rodent carcinogenicity bioassay. The NTP protocol for this assay consists of four experimental groups (rat and mouse, male and female); each group produces both a yes/no and a potency measure. These outcomes can be summarized into an overall carcinogenicity score (Huff, 1999).

## A. Public Sources of Carcinogenicity and Mutagenicity Data

The issues related to the availability of data from public sources, and how this influences the practical feasibility of QSAR modeling, are discussed in detail by Richard (2003). Richard also discussed the ongoing attempts to develop databases that combine toxicological and chemical structure information (Richard, 2003; Richard and Williams, 2002). Table 8.1 provides a listing and brief description of websites that are the most prominent public sources of chemical mutagenicity and carcinogenicity data. Further information on sources of toxicological information is given in Chapter 2.

**Table 8.1    Selection of the Main Public Databases of Mutagenicity and Carcinogenicity Data**

| Website | Program/Database |
|---|---|
| ntp-server.niehs.nih.gov | National Cancer Institute (NCI)/National Toxicology Program (NTP) |
| toxnet.nlm.nih.gov/ | National Library of Medicine (NLM)/TOXNET |
| | Environmental Protection Agency (EPA)/Gene-Tox |
| | NCI/Chemical Carcinogenesis Research Information System (CCRIS) |
| | EPA/Integrated Risk Information System (IRIS) |
| | National Institute for Occupational Safety and Health(NIOSH)/Registry of Toxic Effects of Chemical Substances (RTECS) |
| potency.berkeley.edu/cpdb.html | University of California – Berkeley/Carcinogenic Potency Database (CPDB) Project |
| www.epa.gov/gap-db | EPA/Genetic Activity Profiles (GAP) |
| monographs.iarc.fr | World Health Organization (WHO)/International Agency for Research on Cancer (IARC) |

*Note:*   The Gene-Tox and GAP databases specifically focus on mutagenicity data; all the other databases contain both mutagenicity and carcinogenicity data. For a detailed discussion of the databases, see Richard (2003) and Richard and Williams (2002).

## III. QSAR MODELING OF MUTAGENICITY AND CARCINOGENICITY

As for other biological activities, QSAR modeling is a powerful tool for understanding the determinants (substructures, chemical physical forces, etc.) of a chemical's action (Hansch and Leo, 1995). It thus contributes remarkably to the comprehension of the mechanisms of chemical mutagenicity and carcinogenicity. Another important goal of the use of QSAR analyses is risk assessment: QSARs can be employed to estimate the activity of other chemicals not tested experimentally. Thousands of chemicals currently in commerce and in the environment have not undergone carcinogenicity testing (e.g., the European INventory of Existing Commercial Substances (EINECS) compilation of 101,000 chemicals existing before toxicity testing was required for new chemicals imported or produced in the European Union [McCutcheon, 1994]). Moreover, the accelerating pace of chemical discovery and synthesis has heightened the need for efficient prioritization and toxicity screening methods.

The most informative QSAR analyses are those performed on individual classes of congeneric chemicals: chemicals that share a basically similar structure, act by the same mechanism of action, and share the same rate-limiting step in their mechanism. For the best modeling results, the chemicals in the set should induce the same well-defined biological effect. A well-defined biological effect is, for example, the induction of mutations in one specific *Salmonella typhimurium* strain, rather than a summary score of mutagenicity based on the entire profile of responses in the various strains. In this case, only one mechanism is (usually) acting, and this can be modeled quite efficiently; the resulting QSAR points to the chemical determinants and can be used to predict the effect of other chemicals possessing chemical features in the same range of the modeled set of chemicals.

### A.  QSARs for Individual Chemical Classes

QSARs have been generated for a number of individual chemical classes, (Benigni and Giuliani, 1996; Cronin and Dearden, 1995a; Debnath et al., 1994; Hansch, 1991; Passerini, 2003). The majority relate to *in vitro* mutagenicity, but a number of QSAR models for animal carcinogenicity exist as well. Among the carcinogens, the QSARs refer almost exclusively to genotoxic carcinogens. Overall, they provide a consistent picture of the genotoxic mechanisms of toxicity of the chemical mutagens and carcinogens.

**Table 8.2   Selected QSARs for Individual Classes of Mutagens and Carcinogens**

| Chemical Class | Reference |
|---|---|
| **Mutagenicity** | |
| Aromatic amines | Debnath et al. (1992a) |
| Nitroaromatics | Debnath et al. (1992b) |
| Quinolines | Debnath et al. (1992c); Smith (1997) |
| Thiazoles | Biagi et al. (1986) |
| Carbazoles | Andre et al. (1995) |
| Triazenes | Shusterman et al. (1989) |
| Furanones | Tuppurainen (1999) |
| Halogenated methanes | Benigni et al. (1993) |
| Propylene oxides | Hooberman et al. (1993) |
| Styrene oxides | Tamura et al. (1982) |
| Nitrofurans | Debnath et al. (1993) |
| **Carcinogenicity** | |
| Aromatic amines | Benigni et al. (2000); Franke et al. (2001) |
| *N*-Nitroso compounds | Dunn III and Wold (1981) |
| Polycyclic aromatics | Norden et al. (1978); Richard and Woo (1990); Zhang et al. (1992) |
| Miscellaneous | Loew et al. (1985) |

The most important conclusion to be made from these studies is the great importance of hydrophobicity in the modulation of the potential for mutagenicity and carcinogenicity. Hansch and coworkers have showed that compounds that require S9 activation to become mutagenic in bacteria all have log $K_{ow}$ terms with coefficients near 1.0 (Debnath et al., 1994). Other QSARs show that where a direct chemical reaction with DNA appears to occur, without metabolic activation, no hydrophobic term enters into the equation (Hansch et al., 2001). In these cases, usually only the electronic (reactivity) properties are important. Notable examples of QSARs based on electronic terms and without a hydrophobic term relate to the mutagenicity to *Salmonella* of aniloacridines, *cis*-platinum analogs, lactones, and epoxides. All of these examples are for chemicals that do not require activation (Hansch et al., 2001).

Table 8.2 provides a list of representative QSARs for individual classes of mutagens and carcinogens. The QSARs in the original references can be used in two ways. First, the equations can be used to estimate the activity of untested chemicals belonging to the same chemical class. It is intended that interpolation, and not extrapolation, should be performed; the untested chemicals should have parameters in the same range of the original set. Second, the inspection of the published QSARs may suggest parameters and methods for new QSAR analyses of sets of chemicals similar to those already considered in the literature.

The next section presents in detail the results of QSAR analyses of the most studied chemical class: the aromatic amines.

## B.   An Example: QSARs for the Aromatic Amines

The aromatic amines are chemicals with a great environmental and industrial importance, so a large database of experimental results has been generated (Woo and Lai, 2001). The availability of such a large quantity of data has stimulated several investigators to develop QSARs for the aromatic amines. From a practical point of view, this gives the opportunity to estimate the mutagenicity and carcinogenicity of untested amines. This is of great importance, since new amines are produced continuously by the chemical industry; the QSAR predictions can help to lead production toward safer aromatic amines. From a methodological point of view, the availability of several

QSAR models that relate to different organisms permits an interesting discussion of the issues related to the modeling of carcinogenicity and mutagenicity data.

In our first QSAR analysis of the carcinogenicity of the aromatic amines, we considered only the carcinogenic aromatic amines, and we investigated the structural factors that influence the gradation of carcinogenic potency in rodents (Benigni et al., 2000). The study focused on an homogeneous class of nonheterocyclic amines. The following are the QSAR models that emerged from the analysis of the bioassay data (BRM = carcinogenic potency in mice; BRR = carcinogenic potency in rats):

$$BRM = 0.88(\pm0.27)\log K_{ow} * I(monoNH_2) + 0.29(\pm0.20)\log K_{ow} * I(diNH_2)$$
$$+ 1.38(\pm0.76)\, E_{HOMO} - 1.28(\pm0.54)\, E_{LUMO} - 1.06(\pm0.34)\sum MR_{2,6}$$
$$- 1.10(\pm0.80)\, MR_3 - 0.20(\pm0.16)\, E_S(R) + 0.75(\pm0.75)\, I(diNH_2)$$
$$+ 11.16(\pm6.68)$$

$$n = 37, r = 0.907, r2 = 0.823, s = 0.381, F = 16.3, p < 0.001$$

(8.1)

$$BRR = 0.35(\pm0.18)\log K_{ow} + 1.93(\pm0.48)\, I(Bi) + 1.15(\pm0.60)\, I(F) - 1.06(\pm0.53)\, I(BiBr)$$
$$+ 2.75(\pm0.64)\, I(RNNO) - 0.48(\pm0.30)$$

$$n = 41, r = 0.933, r^2 = 0.871\ s = 0.398, F = 47.4, p < 0.001$$

(8.2)

where BRM = log $(MW/TD_{50})_{mouse}$ and BRR = log $(MW/TD_{50})_{rat}$.

$TD_{50}$ is the daily dose required to halve the probability of an experimental animal to remain tumorless to the end of its standard life span (Gold et al., 1991). The chemical parameters in the equations are: log $K_{ow}$, which is a measure of hydrophobicity; $E_{HOMO}$, energy of the highest occupied molecular orbital; $E_{LUMO}$, energy of the lowest unoccupied molecular orbital; $\sum MR_{2,6}$, sum of molar refractivity of substituents in the ortho-positions of the aniline ring; $MR_3$, molar refractivity of substituents in the meta-position of the aniline ring; Es(R), Charton's substituent constant for substituents at the functional amino group; $I(monoNH_2)$ = 1 for compounds with only one amino group; $I(diNH_2)$ = 1 for compounds with more than one amino group; $I(Bi)$ = 1 for biphenyls; $I(I(BiBr)$ = 1 for biphenyls with a bridge between the phenyl rings; $I(RNNO)$ = 1 for compounds with the group N(Me)NO; and $I(F)$ = 1 for fluoroamines. N(Me)NO is a nitroso group, with a methyl substitution at the amino nitrogen. $E_{HOMO}$ and $E_{LUMO}$ were calculated by the SYBYL software (Tripos) after optimization with the Austin Model 1 (AM1) Hamiltonian; log $K_{ow}$ was calculated from the TSAR software (Oxford Molecular, now Accelrys).

The key factor for carcinogenic potency is hydrophobicity (log $K_{ow}$). Both BRM and BRR increase with increasing hydrophobicity. In the case of BRM (mice) the influence of hydrophobicity is stronger for compounds with one amino group (characterized by the indicator variable $I[monoNH_2]$) in comparison with compounds with more than one amino group (characterized by the indicator variable $I[diNH_2]$). For BRM, electronic factors also play a role: potency increases with the increasing $E_{HOMO}$ and with the decreasing $E_{LUMO}$. Such effects seem to be less important for BRR (rats); no electronic terms occur in Equation 8.2. Carcinogenic potency also depends on the type of the ring system. Aminobiphenyls (indicator variable $I[Bi]$) and, in the case of BRR, fluorenamines (indicator variable $I[F]$) are intrinsically more active than anilines or naphthylamines. The bridge between the rings in the biphenyls decreases potency ($I[BiBr]$). Steric factors are involved in the case of BRM, but cannot be detected in the case of BRR. BRM strongly decreases with the addition of bulky substituents adjacent to the functional amino group, on the nitrogen itself and in position 3. The latter effects are not so important. In the case of BRR, R = (Me)NO strongly enhances potency (compounds with this substituent have no measured value for BRM).

Equation 8.1 and Equation 8.2 were derived from the analysis of carcinogenic aromatic amines only, and are very powerful to help explain the gradation of their carcinogenic potency. However, when we applied the equations to the noncarcinogenic amines, we found that the equations did not predict the lack of carcinogenic effects well (the non-carcinogens were predicted as having some, albeit low, degree of activity). This means that the molecular determinants that rule the gradation of carcinogenic potency are not the same as those that determine the difference between carcinogens and noncarcinogens. In a subsequent report we studied the differences in molecular properties between the two classes of carcinogenic and noncarcinogenic aromatic amines specifically (Franke et al., 2001). Four equations were derived, one for each of the experimental groups (rat and mouse, male and female). The 2 classes were coded as: 1= inactive and 2 = active compounds.

The following discriminant function achieves a highly significant separation of classes for female rat carcinogenicity:

$$w = 0.65 \ L(R) + 0.79 \ E_{HOMO} - 1.54 \ E_{LUMO} + 0.76 \ MR_2 - 0.50 \ MR_5 + 1.32 \ I(An)$$
$$- 0.53 \ I(o\text{-}NH_2) + 0.99 \ I(BiBr) + 0.99 \ I(diNH_2) - 1.08 \ \log K_{ow} * I(diNH_2) \tag{8.3}$$

$$w_{(mean,class1)} = 1.05, \ N_1 = 30$$

$$w_{(mean,class2)} = -1.21, \ N_2 = 26$$

where L(R) is the length of the substituent at the amino group, I(An) = 1 for anilines, and I(o-NH$_2$) = 1 if nonsubstituted amino group occurs in the ortho-position to the functional amino group. $w_{(mean,class1)}$ is the mean of the w values of the Class 1 chemicals, and $w_{(mean,class2)}$ is the mean of the w values of the Class 2 chemicals. Chemicals with calculated w values closer to 1.05 are reclassified (predicted) as inactives; chemicals with calculated w values closer to –1.21 are reclassified as actives.

The correct reclassification rate of discriminant function (Equation 8.3) amounts to 91.1% (Class 1: 93.3%; Class 2: 88.5%) with a fairly stable cross validation (all compounds: 80.4%; Class 1: 76.7%; Class 2: 84.6%). Cross validation is a tool to assess the robustness of the model, and is performed by constructing a model on two thirds of the compounds, and checking the ability of the model to predict the activity of the remaining one third correctly.

For male rat carcinogenicity a good separation of classes is achieved by the following discriminant function:

$$w = 0.48 \ L(R) + 0.90 \ E_{HOMO} - 1.43 \ E_{LUMO} + 0.72 \ MR_2 + 1.13 \ I(An)$$
$$- 0.54 \ I(o\text{-}NH_2) - 0.45 \ MR_5 + 0.70 \ I(diNH_2) - 0.80 \ \log K_{ow} * I(diNH_2) \tag{8.4}$$
$$+ 0.65 \ I(BiBr)$$

$$w_{(mean,class1)} = 1.15, \ N_1 = 28$$

$$w_{(mean,class2)} = -1.01, \ N_2 = 32$$

The correct reclassification rate amounts to 91.7% (Class 1: 92.9%; Class 2: 90.6%) with a good result for cross-validation (all compounds: 83.3%; Class 1: 82.1%; Class 2: 84.4%).

The results obtained for male and female rats resemble each other. Of key importance for class separation are the electronic properties as expressed by $E_{HOMO}$ and $E_{LUMO}$, the type of ring system, and substitution in the ortho-position as well as at the amino nitrogen. The probability of a compound being assigned to the active class increases with increasing values of $E_{LUMO}$, decreasing values of $E_{HOMO}$, decreasing bulk of substituents in position 2 (ortho-position), decreasing length (or bulk) of substituents at the amino nitrogen, and increasing number of aromatic rings (anilines have a distinctively lower probability to be active than biphenyls, fluorenes, or naphthalenes). Another important feature promoting carcinogenicity is the occurrence of an amino group in ortho-position to the functional amino group. Of lesser importance are the variables $I(diNH_2)$, $I(BiBr)$, $MR_5$, and the cross product log $K_{ow}*I(diNH_2)$. It appears that the key factors differentiating active and inactive compounds on the one hand and governing potency within the group of active compounds are different. The most pronounced differences are with respect to the importance of hydrophobicity and the directionality of electronic effects.

For female mouse carcinogenicity, the following discriminant function reclassifies 85.7% of the compounds correctly (Class 1: 87.9%; Class 2: 83.3%) and has acceptable cross-validation (all compounds: 81.0%; Class 1: 84.8%; Class 2: 76.7%):

$$w = -0.47 \ I(NR) + 1.38 \ \log K_{ow} * I(monoNH_2) + 1.68 \ \log K_{ow} * I(diNH_2)$$
$$- 0.37 \ I(An)(4.3.1) + 0.33 \ I(o\text{-}NH2) - 0.55 \ MR_5 - 0.45 \ I(BiBr)$$

(8.5)

$$w_{(mean,class1)} = -0.92, \ N_1 = 33$$
$$w_{(mean,class2)} = 1.01, \ N_2 = 30$$

where $I(NR) = 1$ if the amino nitrogen is substituted.

For male mouse carcinogenicity, the following discriminant function is obtained:

$$w = -1.96 \ L(R) + 1.69 \ B_5(R) - 0.83 \ E_{HOMO} + 0.97 \ E_{LUMO} - 1.22 \ I(An)$$
$$+ 0.73 \ I(o\text{-}NH2) + 0.59 \ MR_3 + 0.69 \ MR_5 + 0.77 \ MR_6 - 0.76 \ I(diNH_2)$$

(8.6)

$$+ 1.09 \ \log K_{ow} * I(diNH2) - 0.79 \ I(BiBr)$$

$$w_{(mean,class1)} = -1.11, \ N_1 = 25$$
$$w_{(mean,class2)} = 1.16, \ N_2 = 24$$

where $B_5$ is the maximal width of the substituent at the amino group.

It should be noted that the difference in sign of the average w values for the 2 classes in Equations 8.3 to 8.6 is only formal, and does not have any relevance on mechanisms.

The discriminant function in Equation 8.6 shows a good reclassification rate (all compounds: 89.8%; Class 1: 96.0%; Class 2: 83.3%) and stability in cross validation (all compounds: 83.7%; Class 1: 96.0%; Class 2: 70.8%).

The results for mice were similar to those found for rats. Hydrophobicity is a key factor determining the gradation of the carcinogenic potency (Equations 8.1 and 8.2), but only of small importance for yes/no activity (Equations 8.3 to 8.6). The reverse is true for electronic properties ($E_{HOMO}$, $E_{LUMO}$), which show a minor effect for the gradation of potency, but a pronounced effect for yes/no activity. Equations 8.4 to 8.6 also demonstrate the importance of steric (shape, size) factors for yes/no activity. For example, in all four equations the first term indicates that the probability of being noncarcinogenic increases with increasing length of the substituent (L[R]) or simply with the presence of a substituent (I[NR]) on the amino nitrogen. Hydrophobicity is a force involved in the absorption and transport of the drugs in the cells and organisms, as well as in the interaction between drugs and metabolizing enzymes. The electronic parameters are measures of chemical reactivity, and hence of the ability to undergo metabolic transformations. It should be noted that the results of the QSAR analyses agree with the notion that the aromatic amines require metabolic activation to become carcinogenic (Woo and Lai, 2001). For amines and amides, this typically involves an initial oxidation to $N$-hydroxylamine and $N$-hydroxylamide. In particular, $E_{HOMO}$ is a parameter for oxidation reactions.

The successful QSARs obtained in modeling the rodent carcinogenicity of the aromatic amines contradict the view that carcinogenicity is difficult to model and predict. This argument is largely based on the recognition that chemical carcinogenesis is a multistage, multifactorial process that involves exogenous and endogenous factors that are often intertwined in an interrelated network. Moreover, the carcinogenesis process has three operational stages: initiation, promotion, and pro-gression. The ideal QSAR model of carcinogenicity should consider all the different stages and factors. Fortunately, this is not the case. As a generality, it should be remembered that no model is a complete representation of reality, but only a description of a sufficient number of elements that are relevant for the problem under consideration. In particular, QSAR modeling attempts to discover the rate-limiting factors of the (often complex) interaction between chemicals and biolog-ical systems. The same applies to the QSARs of physical and chemical reactions, where the concept of the rate-limiting step is even more familiar. Thus, complex and multi-step processes are often modeled, with a very good fit, by just one or a few parameters (Hansch and Leo, 1995). The example of the aromatic amines demonstrates that rodent carcinogenicity can be modeled success-fully, with 80 to 90% accuracy; the critical requirement is that sufficient data are available to build the QSAR model. As a matter of fact, the large industrial and environmental impact of the aromatic amines has been instrumental in the testing of a large number of these chemicals: a total of 200 aromatic amines were found in a database of about 800 chemicals bioassayed (unpublished results).

In addition to the rodent bioassay, the aromatic amines have been studied in the shorter term test *Salmonella typhimurium* mutagenicity as well as in a variety of acute toxicity assays. A number of QSARs have been generated from such data. The work of Hansch in recent years has demon-strated that the comparison of the QSAR models obtained in different systems, by putting them in a wider perspective, can provide useful clues in the study of the mechanisms of action of individual chemical classes, and can give precious hints on how appropriate the specific models and parameters selected are (Hansch, 2001; Hansch et al., 2002). An exercise of the mechanistic comparison of QSARs has been performed on aromatic amines (Benigni and Passerini, 2002). The results are detailed below.

Debnath et al. (1992a) collected a large database of chemicals with various different basic structures (e.g., aniline, biphenyl, anthracene, pyrene, quinoline, carbazole, etc.). The experimental data referred to *Salmonella* TA98 and TA100 strains, with S9 metabolic activation. The mutagenic potency was expressed as log (revertants/nmol). The AM1 molecular orbital energies are given in electron volts. The mutagenic potency in TA98 + S9 was modeled by:

$$\log TA98 = 1.08(\pm 0.26) \log K_{ow} + 1.28(\pm 0.64) \ E_{HOMO} - 0.73(\pm 0.41) \ E_{LUMO}$$

$$+ 1.46(\pm 0.56) \ I_L + 7.20(\pm 5.4)$$

(8.7)

$$n = 88, \ r = 0.898, \ s = 0.860$$

where $I_L$ is an indicator variable that assumes a value of 1 for compounds with 3 or more fused rings. The electronic terms $E_{HOMO}$ and $E_{LUMO}$, though statistically significant, accounted for only 4% of the variance of the biological data, whereas log $K_{ow}$ alone accounted for almost 50%. The most hydrophilic amines ($n = 11$) could not be treated by Equation 8.7, and were modeled by a separate equation containing only log $K_{ow}$, suggesting that these amines may act by a different mechanism. The mutagenic potency in the *Salmonella* strain TA100 + S9 was expressed by:

$$\log TA100 = 0.92\,(\pm0.23)\,\log K_{ow} + 1.17\,(\pm0.83)\,E_{HOMO} - 1.18\,(\pm0.44)\,E_{LUMO}$$
$$+ 7.35\,(\pm6.9)$$

(8.8)

$$n = 67, \, r = 0.877, \, s = 0.708$$

Also in this case, a different equation was necessary for the most hydrophilic amines (n = 6).

The QSARs show that for both bacterial mutagenicity and rodent carcinogenicity, the gradation of the potency of the active aromatic amines first depends on hydrophobicity, and then depends on electronic and steric properties. This confirms the existence of a common first step in the mutagenicity and carcinogenicity of these compounds, and adds confidence to the models obtained.

The QSARs for the potency of the active aromatic amines in *Salmonella typhimurium* were not suitable to differentiate the inactives from the actives, and appropriate QSAR models were derived. The QSAR models specific for the separation were based on electronic and steric terms, and hydrophobicity was not found to be significant (Benigni et al., 1998). This result is analogous to that obtained for the rodent carcinogenicity of amines (see above), and provides further evidence of the similarity of the mechanisms of action in *Salmonella* and rodents. This similarity obviously applies only to the first steps of the process by which the aromatic amines provoke cancer on one side (rodents), and mutations on the other side (bacteria). Once the initial insult to the cells has occurred, the process follows different pathways in the two biological systems; the strength of the QSAR model is that it is able to describe quantitatively the first step, which appears to be rate limiting in both systems.

The QSARs for the acute toxicity of the aromatic amines in a variety of other experimental systems (e.g., fathead minnow, guppy, etc.) were also considered. They were much simpler than those for bacterial mutagenicity and rodent carcinogenicity, and usually relied only on hydrophobicity; these findings point to a specific mechanism of action, different from the mechanisms of genotoxic carcinogenicity (Benigni and Passerini, 2002).

## C. QSAR Models for Noncongeneric Chemicals

The application of QSAR modeling to individual classes of chemicals, acting through distinct mechanisms of action, constitutes the area where the classical QSAR approaches give their best contribution, in terms of both the understanding of chemical mutagenicity and carcinogenicity and of predictive ability. Unfortunately there are a number of drawbacks: (1) each QSAR model is specific for one individual chemical class; (2) the database of experimental results is not sufficiently populated with representative chemicals to provide a sufficiently representative basis to model carcinogenicity or mutagenicity of each chemical class of interest; and (3) the chemicals of practical interest change with the time, since new chemical entities (pharmaceuticals, dyes, etc.) are produced and marketed every year. As a consequence, for the most part it is not possible to retrieve from the historical carcinogenicity and mutagenicity databases sufficiently chemically similar molecules from which to derive QSAR models. The above problems have been instrumental in stimulating attempts to develop SARs and QSARs for noncongeneric sets of chemicals (i.e., general prediction models with the hope of being able to predict the activity of any type of chemical). A scientific aspect of this is the interest in exploring nonapparent associations that cross traditional boundaries of chemical class.

General models to predict mutagenicity and carcinogenicity have ranged from those based on purely heuristic human expert judgement, those that require prior hypothesis and human intervention in choosing relevant parameters, and those that use automated rule-based approaches derived from human expert knowledge, to automated statistical, machine learning and pattern recognition approaches that derive algorithms independently for the prediction of existing data. Many of these approaches are driven by the recognition of structural alerts (SAs) in the molecules. SAs are chemical substructures, or functional groups, that have been mechanistically or statistically associated with the induction of mutations or cancer. At a more sophisticated level, the prediction methods contain information on both known SAs and modulating factors. Some approaches for the prediction of the long-term carcinogenicity bioassay also include the knowledge on short-term test results, such as *Salmonella typhimurium*, as basis for the prediction. Some of these approaches have been implemented into commercial software programs. Commercial systems for toxicity prediction are described in more detail in Chapter 9. The following citations also provide full and detailed reviews of such systems: Benigni and Richard (1998), Cronin and Dearden (1995b), Dearden et al. (1997), and Richard and Benigni (2002).

## IV. THE ASSESSMENT OF THE PREDICTION ABILITY

It is crucial to assess the real predictive ability of the different approaches. Most often, authors and developers report some measure of the accuracy of system performance, which varies according to the type of method. A more stringent criterion is prospective prediction: the predictions are performed on compounds whose experimental results did not exist at the time the model was generated. Because of their unbiased character, the results of such exercises constitute the most important source of information in the field. Two important prospective prediction exercises were performed in the past decade under the aegis of the NTP; the exercises invited the modeling community to submit predictions on the rodent carcinogenicity of chemicals that were in the process of being bioassayed by the NTP. The results of the two NTP exercises have been analyzed in various papers (Benigni, 1997; Benigni and Zito, 2004). Only a very short summary is given below.

### A. The First NTP Comparative Exercise on the Prediction of Rodent Carcinogenicity

In the first comparative exercise, 44 compounds from different chemical classes were selected (Tennant et al., 1990). An analysis of the results of the comparative exercise is provided by (Benigni, 1997). Table 8.3 reports the main features of the participating approaches. Most of the systems that participated in the comparative exercise were SAR or QSAR approaches; other approaches searched for relationships between carcinogenesis and shorter-term biological events (activity-activity relationships [AARs]).

Table 8.4 reports the overall accuracy of the predictions. A wider range of measures of the predictive ability can be calculated (e.g., sensitivity, specificity, etc.), but the overall accuracy index

**Table 8.3    First NTP Comparative Exercise on the Prediction of Rodent Carcinogenicity: Prediction Systems**

| System | Approach | Reference |
|---|---|---|
| Tennant et al. | Toxicity + SAs, human expert | Tennant et al. (1990) |
| RASH (Rapid Screening of Hazard) | Toxicity + SAs, human expert | Jones and Easterly (1991) |
| Weisburger | Chemistry, human expert | Anon. (1993) |
| Lijinsky | Chemistry, human expert | Anon. (1993) |
| DEREK | Expert system, rules | Sanderson and Earnshaw (1991) |
| DEREK Hybrid | Expert system, rules | Sanderson and Earnshaw (1991) |
| COMPACT | Metabolism estimation | Lewis et al. (1990) |
| MultiCASE | Automatic SAs generation | Rosenkranz and Klopman (1990) |
| TOPKAT | QSAR (noncongeneric) | Enslein et al. (1990) |
| Benigni | SAs + Electrophilicity (estimated) | Benigni (1991) |
| Ke | Electrophilicity, experimental | Bakale and McCreary (1992) |
| *Salmonella typhimurium* | Mutagenicity, experimental | Anon. (1993) |

**Table 8.4    First NTP Comparative
Exercise on the Prediction
of Rodent Carcinogenicity:
Prediction Accuracy**

| System | Accuracy |
| --- | --- |
| Tennant et al. | 0.75 |
| RASH | 0.68 |
| Weisburger | 0.65 |
| Ke | 0.65 |
| DEREK | 0.59 |
| TOPKAT | 0.57 |
| Benigni | 0.57 |
| Salmonella typhimurium | 0.57 |
| DEREK Hybrid | 0.56 |
| Lijnsky | 0.55 |
| COMPACT | 0.53 |
| MultiCASE | 0.49 |

provides sufficient information in this context. It appears that for the approaches that relied solely on information derived from chemical structure, the overall accuracy in terms of positive or negative predictions was in the range of 50 to 65%. The more biologically based approaches, which incorporated knowledge of short-term mutagenicity tests or subchronic bioassay results together with the recognition of chemical substructural alerts and used the subjective human expert judgement to combine the various pieces of information, attained 75% accuracy (namely, the Tennant and Ashby approach [Tennant et al. 1990]).

A problem common to most of the participating approaches was that quite a number of noncarcinogens were predicted to be positive by different systems. The difficult chemicals all contained SAs, and the prediction approaches were not able to consider the presence of modulating factors (e.g., detoxifying functionalities). In other terms, the various QSAR (and, to a different degree, the AAR) approaches essentially acted as gross class-identifiers. They pointed to the presence or absence of alerting chemical functionalities, but were not able to make gradations within each potentially harmful class.

## B.  The Second NTP Comparative Prediction Exercise on the Prediction of Rodent Carcinogenicity

The second comparative exercise devised by the NTP involved 30 chemicals in the progress of being bioassayed (Bristol et al., 1996). Table 8.5 presents the participating approaches, and Table 8.6 reports the accuracy of the predictions. An analysis of the results of this exercise is reported by Benigni (2000). It appears that the human expert-based predictions performed best overall, especially those methods that incorporated the most information. In addition, the Syrian Hamster Embryo (SHE) assay, an experimental system specifically designed to incorporate the key elements of the transformation process that a cell can undergo in becoming malignant, was among the best performing methods. The highest overall accuracy in this second exercise was in the range 65 to 70%. As in the first comparative exercise (Benigni, 1997), several noncarcinogens were predicted as carcinogens. The predictive approaches were almost invariably unable to make gradations between potential and actual carcinogenicity.

## C.  Lessons from the Comparative Exercises on the Prediction of Carcinogenicity

The results of the two NTP comparative exercises are very important for judging the real capability of predicting the carcinogenicity of untested chemicals. The general predictive models

**Table 8.5 Second NTP Comparative Exercise on the Prediction of Rodent Carcinogenicity: Prediction Systems**

| System | Approach | Reference |
|---|---|---|
| Huff et al. | SAs + toxicity, human expert | Huff et al. (1996) |
| OncoLogic | Expert system + human expert | Woo et al. (1997) |
| Bootman | Human expert | Bootman (1996) |
| FALS | QSAR, noncongeneric | Moriguchi et al. (1996) |
| Ashby | SAs + toxicity, human expert | Ashby (1996) |
| Tennant et al. | SAs + toxicity, human expert | Tennant and Spalding (1996) |
| Benigni et al. | Chemistry, human expert | Benigni et al. (1996) |
| RASH | SAs + toxicity, human expert | Jones and Easterly (1996) |
| R1, R2, Lee et al. | Activity-activity model, computerized | Lee et al. (1996) |
| SHE transformation | Experimental | Kerckaert et al. (1996) |
| Benigni, old | SAs + electrophilicity (estimated) | Benigni (1991) |
| Progol | Automatic SAs generation + mutagenicity | King and Srinivasan (1996) |
| Purdy | Human expert | Purdy (1996) |
| MultiCASE | Automatic SAs generation | Zhang et al. (1996) |
| DEREK | Expert system, rules + human expert | Marchant (1996) |
| *Salmonella typhimurium* | Experimental | Bristol et al. (1996) |
| COMPACT | Metabolism estimation + human expert | Lewis et al. (1996) |
| KeE | Estimated electrophilicity | Benigni (1991) |

**Table 8.6 Second NTP Comparative Exercise on the Prediction of Rodent Carcinogenicity: Prediction Accuracy**

| System | Accuracy |
|---|---|
| Oncologic | 0.67 |
| SHE transformation | 0.67 |
| R2 | 0.65 |
| Benigni et al. | 0.64 |
| Huff et al. | 0.63 |
| R1 | 0.63 |
| Tennant et al. | 0.58 |
| Bootman | 0.58 |
| Ashby | 0.54 |
| Fuzzy Adaptive Least Squares (FALS) | 0.54 |
| Benigni (Old) | 0.52 |
| RASH | 0.48 |
| COMPACT | 0.44 |
| KeE | 0.39 |
| DEREK | 0.38 |
| Purdy | 0.35 |
| *Salmonella typhimurium* | 0.33 |
| Prolog | 0.32 |
| CASE/MULTICASE | 0.18 |

for noncongeneric chemicals attained a maximum accuracy of around 65 to 70%; this maximum was not attained easily and many approaches, including several commercially available ones, showed quite poor performance. In both exercises, the best performance was attained by approaches that relied largely on human expert judgement. Regardless of this, classical QSAR methods can generate quite satisfactory models for individual classes of mutagens and carcinogens, with accuracy in the range of 80 to 90%. This is also true for rodent carcinogenicity, indicating that there is no inherent difficulty in modeling this biological endpoint (Benigni et al. 2000; Franke et al., 2001; Zhang et al., 1992).

The reasons for the difficulties encountered by the general approaches have been discussed in several publications (Benigni and Zito, 2004; Richard, 1994; Richard and Benigni, 2002). In summary, some of the arguments are that: (1) whereas the classical QSAR models describe one class and one main mechanism at a time, the general approaches aim to model several mechanisms at the same time, which is a far more difficult task; (2) the database of rodent experiments (around 1500 chemicals bioassayed) is not sufficiently representative in terms of number of chemicals per mechanism/chemical class; and (3) the evolution and changes of the chemical types of interest (new drugs, new pesticides, etc.) continuously poses new challenges that cannot be faced based on the information provided by the historical carcinogenicity database. To some extent, the human expert approaches were able to overcome these difficulties by their ability to access several sources of information in a flexible way.

## V. HOW SHOULD A USER APPROACH THE PREDICTION OF MUTAGENICITY AND CARCINOGENICITY?

The above discussion had the role of setting the scene for the next crucial step of this chapter: predicting the activity of chemicals based on SAR criteria. To explore this step, different situations have to be considered.

Generally speaking, one wants to make predictions to narrow down health research priorities. Mutagenicity and carcinogenicity are separate toxicological endpoints, but a mutagenic chemical has a high probability of being also a genotoxic carcinogen. The short-term mutagenicity tests (e.g., *Salmonella typhimurium*) can be used to prescreen for carcinogenicity; its results can also add or subtract value to an estimated carcinogenicity result predicted by SAR. The mutagenicity tests are quite inexpensive. Instead of attempting a prediction of mutagenicity, in most cases it may be preferable to perform a mutagenicity test rather than to predict its result.

There is a fundamental difference between cases in which biological results for chemicals congeneric to the query exist, and for cases in which the query chemical does not have congeneric chemicals with known biological activity. The former case simply requires that a QSAR model is built on the congeneric chemicals with known biological activity by applying an appropriate approach. The QSAR model can then be used to predict the activity of the untested chemical. This procedure does not pose any particular challenge specifically related to carcinogenicity and mutagenicity endpoints, except that the knowledge of, or hypothesis on the mechanisms of action, can assist in the selection of the relevant chemical parameters. As usual with QSAR practice, it is essential that chemicals used to build the model (the training set) are available in a statistically sufficient number, and that they have sufficient diversity in terms of chemical characteristics.

Unfortunately, most often one has to predict the activity of chemicals for which no or few similar chemicals with known activity are available. Such a conundrum has no simple solution. Table 8.3 and Table 8.5 report a wide range of approaches that have been designed for assisting in such an enterprise; Tables 8.4 and Table 8.6 show that the performances of the approaches are not sufficiently satisfactory to allow for their predictions for individual chemicals to be taken at face value. In addition, it appears that the poorest approaches were those implemented in completely automated computer programs. In spite of these difficulties, there is room enough for the use of structure-activity criteria in the prediction of toxic activity to set health research priorities. An excellent example of such a practice is provided by the NTP selection process of chemicals to be tested with the rodent bioassay. Because of obvious resource limitations, the selection process had to be directed toward chemicals for which there was a high degree of suspicion of carcinogenicity. A major factor was the consideration of structure-activity relationships. Out of the ~400 chemicals tested by National Cancer Institute/NTP, two thirds were selected as suspect carcinogens; and one third was selected on production/exposure considerations. In the first class, 68% were demonstrated to be rodent carcinogens, whereas only 20% of the second class were positive (and only 7% were positive in two species) (Fung et al., 1995). This suggests that the general level of our knowledge on QSARs is already adequate to usefully prioritize chemicals for testing and to direct the development of safer chemicals.

## A. The Human Expert Approach

Both the process of priority setting at NTP and the approaches that performed best in the comparative exercises have a human expert component. An in-depth discussion of the practical process of performing a human expert prediction is presented by Woo et al. (2002). Since the paper gives a very informative view of the various steps of the process, we will follow his presentation and comment on it.

The paper presents an analysis aimed at establishing a ranking of carcinogenic potential ranking for drinking water disinfection byproducts (DBPs). The analysis was performed at the Environmental Protection Agency (EPA) to prioritize research efforts. Disinfection is necessary to destroy pathogenic organisms. Some of the DBPs may present undesirable side effects; when disinfectants such as chlorine react with natural inorganic and organic matter in the water, chlorine-derivative DBPs are formed, and a number of them have been shown to be carcinogenic in animal studies. Several hundred DBPs have been identified, and no toxicological information is available for the large majority of them.

The first step of the process required the expertise of chemists. These experts examined a list of more than 600 DBPs identified and cataloged by the EPA. DBPs with a low probability of being formed in actual drinking water were excluded, and a second list of 239 actual or probable DBPs remained for research prioritization. Thirty DBPs were found to have sufficient toxicological data, thus 209 chemicals were subjected to a detailed SAR analysis based on expert judgement. Mechanism-based SAR analysis has been used effectively by the EPA for many years to assess the potential carcinogenic hazard of new chemicals, for which there are no or scanty data, under the Premanufacture Substances Control Act. Essentially, mechanism-based SAR analysis involves comparison of an untested chemical with structurally related compounds for which carcinogenic activity is known. According to Woo et al. (2002),

> All available knowledge and data relevant to evaluation of carcinogenic potential of the untested chemical are considered. These include a) SAR knowledge base of the related chemicals; b) toxicokinetics and toxicodynamics parameters (including physicochemical properties, route of potential exposure, and mode of activation or detoxification) that affect the delivery of biologically active intermediates to target tissue(s) for interaction with cellular macromolecules or receptors; c) supportive non-cancer screening or predictive data known to correlate to carcinogenic activity.

This step is quite complicated and requires skill and expertise in chemistry, biochemistry, and toxicology. This is performed by a highly interdisciplinary team. The experts look for the presence of structural features known, or reasonably supposed, to be involved in carcinogenesis, together with the modifying features that can enhance or negate the effect of the main structural features. The impact of the route of exposure is taken into account as well. The outcome of this step is the construction of a semiquantitative concern rating scale of low, marginal, low- or high-moderate classification. Based on the resulting concern scale, appropriate experiments are decided on and resources are allocated.

It should be noted that the most important tool in the process described above is the skill of the human experts. At the same time, great support is provided by the access to all the relevant literature. The ability to search relationally across public toxicity databases using both biological and chemical criteria represents a potentially powerful approach for SAR analysis. The databases in Table 8.1 constitute the first source of publicly available data for retrieving toxicological information. Searches in chemical abstracts can provide a wealth of chemical and biochemical data on individual chemicals. Whilst large pharmaceutical and chemical companies have invested heavily in relational database platforms aimed at performing chemical analog searching, the public database initiatives in this field are still in early stages of development, and are urgently needed (Richard and Williams, 2002).

Within this perspective, the role of the automated general approaches for the prediction of carcinogenicity should be remembered as well. The results of the prediction challenges indicate that, unfortunately, their predictions cannot be taken at face value. They can be very useful if accompanied by the supporting reasons for the prediction. Such reasons can have different forms: the SAs identified, together with the associated probability of carcinogenicity; a list of similar chemicals from the historical carcinogenicity database, together with a measure of similarity with the query; and the expert system rules encompassing the query. All this information can provide precious support to the ultimate decision of the human expert.

## VI. RECOMMENDATIONS: A SUMMARY

1. QSAR modeling is more crucial to the prediction of carcinogenicity than of mutagenicity. Because of its ease and speed, experimental assessment of mutagenicity through short-term tests is preferable to QSAR predictions.
2. A mutagenic chemical has a high probability of also being a genotoxic carcinogen. Experimental mutagenicity results can be used as support to QSAR predictions of carcinogenicity.
3. In spite of the intrinsic complexity of the carcinogenesis process, QSAR models can be generated successfully for individual classes of congeneric chemicals. Crucial requirements are (a) the class is well defined (truly congeneric, with the same mechanism of action), and (b) a statistically sufficient number of compounds with measured biological activity exists.
4. A number of QSAR models for individual classes of congeneric chemicals (e.g., aromatic amines) are available in the literature (Table 8.2). These models can be used to predict the biological activity of untested chemicals belonging to the same class.
5. Often one wants to predict the activity of chemicals for which no, or few, similar chemicals with known activity are available. QSAR models aimed at predicting the carcinogenic activity of compounds of any class have been generated. Evidence exists that their predictions for the individual chemicals cannot be taken at face value.
6. Among the predictive approaches for non-congeneric chemicals, the best performance was attained by human experts that combined, in a non-formalized manner, several lines of evidence and information.
7. Among others, crucial to the human expert judgement is (a) access to all the relevant literature, and (b) the ability to search across toxicity databases using both biological and chemical criteria.
8. Commercially available prediction software packages can be a useful tool to support expert judgement, provided that they offer transparent predictions and not black-box responses.
9. For large numbers of chemicals, QSAR theory and practice are of great value to prioritize chemicals for testing, and to direct the development of safer chemicals.

## REFERENCES

Allen, B.C., Crump, K.S., and Shipp, A.M., Correlation between carcinogenic potency of chemicals in animals and humans, *Risk Analysis*, 8, 531–544, 1988.

Andre, V., Boissart, C., Sichel, F., Gauduchon, P., Letalaer, J.Y., Lancelot, J.C., Robba, M., Mercier, C., Chemtob, S., Raoult, E., and Tallec, A., Mutagenicity of nitro-and amino-substituted carbazoles in *Salmonella typhimurium*. 2. ortho-aminonitro derivatives of 9H-carbazole, *Mutation Research*, 345: 11-25.

Anon., *Predicting Chemical Carcinogenesis in Rodents: An International Workshop*, National Institute of Environmental Health Sciences, Research Triangle Park, NC, 1993.

Arcos, J.C. and Argus, M.F., Multifactor interaction network of carcinogenesis: a 'Tour Guide', *Chemical Induction of Cancer: Modulation and Combination Effects*, in Arcos, J.C., Argus, M.F., and Woo, Y.T., Eds., Birkhauser, Boston, 1995 pp. 1–20.

Ashby, J., Fundamental structural alerts to potential carcinogenicity or non-carcinogenicity, *Environ. Mutagenesis*, 7, 919–921, 1995.

Ashby, J., Prediction of rodent carcinogenicity for 30 Chemicals, *Environ. Health Perspect.*, 104 (Suppl. 5), 1101–1104, 1996.

Bakale, G. and McCreary, R.D., Prospective Ke screening of potential carcinogens being tested in rodent bioassay by the US National Toxicology Program, *Mutagenesis*, 7, 91–94, 1992.

Benigni, R., QSAR prediction of rodent carcinogenicity for a set of chemicals currently bioassayed by the US National Toxicology Program, *Mutagenesis*, 19, 83–89, 1991.

Benigni, R., The first US National Toxicology Program exercise on the prediction of rodent carcinogenicity: definitive results, *Mutation Res.*, 387, 35–45, 1997.

Benigni, R., Andreoli, C., Conti, L., Tafani, P., Cotta-Ramusino, M., Carere, A., and Crebelli, R., Quantitative structure-activity relationship models correctly predict the toxic and aneuploidizing properties of six halogenated methanes in Aspergillus nidulans, *Mutagenesis*, 8, 301–305, 1993.

Benigni, R., Andreoli, C., and Zito, R., Prediction of the carcinogenicity of further 30 chemicals biossayed by the U.S. National Toxicology *Program, Environ. Health Perspect.,* 104 (Suppl. 5), 1041–1044, 1996.

Benigni, R. and Giuliani, A., Quantitative structure-activity relationship (QSAR) studies of mutagens and carcinogens, *Med. Res. Rev.*, 16, 267–284, 1996.

Benigni, R. and Giuliani, A., Tumor profiles and carcinogenic potency in rodents and humans: value for cancer risk assessment, *Environ. Carcinogenesis Ecotoxicology Rev.*, C17, 45–67, 1999.

Benigni, R., Giuliani, A., Franke, R., and Gruska, A., Quantitative structure-activity relationships of mutagenic and carcinogenic aromatic amines, *Chem. Rev.*, 100, 3697–3714, 2000.

Benigni, R. and Passerini, L., Carcinogenicity of the aromatic amines: from structure-activity relationships to mechanisms of action and risk assessment, *Mutation Res. Rev.*, 511, 191–206, 2002.

Benigni, R., Passerini, L., Gallo, G., Giorgi, F., and Cotta-Ramusino, M., QSAR models for discriminating between mutagenic and nonmutagenic aromatic and heteroaromatic amines, *Environ. Mol. Mutagenesis*, 32, 75–83, 1998.

Benigni, R. and Richard, A.M., Quantitative structure-based modeling applied to characterization and prediction of chemical toxicity, *Methods*, 14, 264–276, 1998.

Benigni, R. and Zito, R., The second National Toxicology Program comparative exercise on the prediction of rodent carcinogenicity: definitive results, *Mutation Res. Rev.,* 566, 49–63, 2004.

Biagi, L.G., Hrelia, P., Gerra, M.G., Paolini, M., Barbaro, A.M. and Cantelli-Forti, G., Structure-activity relationships of nitroimidazo (2,1-b) thiazoles in the salmonella mutagenicity assay, *Arch. Toxicol.*, Suppl. 9, 425–429, 1986.

Bootman, J., Speculations on the rodent carcinogenicity of 30 chemicals currently under evaluation in rat and mouse bioassays organised by the US National Toxicology Program, *Environ. Mol. Mutagenesis*, 27, 237–243, 1996.

Bristol, D.W., Wachsman, J.T., and Greenwell, A., The NIEHS predictive-toxicology evaluation project: chemcarcinogenicity biossay, *Environ. Health Perspect.*, 104 (Suppl. 5), 1001–1010, 1996.

Bucci, T.J., Profiles of induced tumors in animals, *Toxicol. Pathol.*, 13, 105–109, 1985.

Cronin, M.T D. and Dearden, J.C., QSAR in toxicology. 3. Prediction of chronic toxicities, *Quant. Struct.-Act. Relat.*, 14, 329–334, 1995a.

Cronin, M.T.D. and Dearden, J.C., QSAR in toxicology. 4. Prediction of non-lethal mammalian toxicological end points, and expert systems for toxicity prediction, *Quant. Struct.-Act. Relat.*, 14, 518–523, 1995b.

Dearden, J.C., Barratt, M.D., Benigni, R., Bristol, D.W., Combes, R.D., Cronin, M.T.D., Judson, P.N., Payne, M.P., Richard, A.M., Tichy, M., Worth, A.P., and Yourick, J.J., The development and validation of expert systems for predicting toxicity, *Alt. Lab. Anim. (ATLA)*, 25, 223–252, 1997.

Debnath, A.K., Debnath, G., Shusterman, A.J. and Hansch, C., A QSAR investigation of the role of hydrophobicity in regulating mutagenicity in the Ames test. 1. Mutagenicity of aromatic and heteroaromatic amines in *Salmonella typhimurium* TA98 and TA100, *Environ. Mol. Mutagenesis,* 19, 37–52, 1992a.

Debnath, A.K., Lopez de Compadre, R.L., Shusterman, A.J. and Hansch, C., Quantitative structure-activity relationship investigation of the role of hydrophobicity in regulating mutagenicity in the Ames test. 2. Mutagenicity of aromatic and heteroaromatic nitro compounds in *Salmonella typhimurium* TA100, *Environ. Mol. Mutagenesis*, 19, 53–70, 1992b.

Debnath, A.K., de Compadre, R.L.L., and Hansch, C., Mutagenicity of quinolines in *Salmonella typhimurium* TA100: a QSAR study based on hydrophobicity and molecular orbital determinants, *Mutation Res.*, 280, 55–65, 1992c.

Debnath, A.K., Hansch, C., Kim, K.H., and Martin, Y.C., Mechanistic interpretation of the genotoxicity of nitrofurans (antibacterial agents) using quantitative structure-activity relationships (QSAR) and comparative molecular field analysis (CoMFA), *J. Med. Chem.*, 36, 1009–1116, 1993.

Debnath, A.K., Shusterman, A., Lopez de Compadre, R.L., and Hansch, C., The importance of the hydrophobic interaction in the mutagenicity of organic compounds, *Mutation Res.*, 305, 63–72, 1994.

Dunn III, W.J. and Wold, S., An assessment of carcinogenicity of N-nitroso compounds by the SIMCA method of pattern recognition, *J. Chem. Inf. Comput, Sci.*, 21, 8–13, 1981.

Enslein, K., Blake, B.W., and Borgstedt, H.H., Prediction of probability of carcinogenicity for a set of ongoing NTP bioassays, *Mutagenesis*, 5, 305–306, 1990.

Franke, R., Gruska, A., Giuliani, A., and Benigni, R., Prediction of rodent carcinogenicity of aromatic amines: a quantitative structure-activity relationships model, *Carcinogenesis*, 22, 1561–1571, 2001.

Fung, V.A., Barrett, J.C., and Huff, J., The carcinogenesis biossay in perspective: application in identifying human cancer hazards, *Environ. Health Perspect.*, 103, 680–683, 1995.

Gold, L.S., Slone, T.H., Manley, N.B., Garfinkel, G.B., Hudes, E.S., Rohrbach, L., and Ames, B.N., The carcinogenic potency database: analyses of 4000 chronic animal cancer experiments published in the general literature and by the U.S. National Cancer Institute/National Toxicology Program, *Environ. Health Perspect.*, 96, 11–15, 1991.

Gold, L.S., Wright, C., Bernstein, L. and de Veciana, M., Reproducibility of results in 'near-replicate' carcinogenesis bioassays, *J. Nat. Cancer Inst.*, 78, 1149–1158, 1987.

Goodman, G. and Wilson, R., Quantitative prediction of human cancer risk from rodent carcinogenic potencies: a closer look at the epidemiological evidence for some chemicals not definitively carcinogenic in humans, *Regul. Pharmacol. Toxicol.*, 14, 118–146, 1991.

Gottmann, E., Kramer, S., Pfahringer, B. and Helma, C., Data quality in predictive toxicology: reproducibility of rodent carcinogenicity experiments, *Environ. Health Perspect.*, 109, 509–514, 2001.

Hansch, C., Structure-activity relationships of chemical mutagens and carcinogens, *Sci. Total Environ.*, 109/110, 17–29, 1991.

Hansch, C., Hoekman, D., Leo, A., Weininger, D., and Selassie, C.D., Chem-bioinformatics: comparative QSAR at the interface between chemistry and biology, *Chem. Rev.*, 102, 783–812, 2002.

Hansch, C., Kurup, A., Garg, R., and Gao, H., Chem-bioinformatics and QSAR: a review of QSAR lacking positive hydrophobic terms, *Chem. Rev.*, 101, 619–672, 2001.

Hansch, C. and Leo, A., *Exploring QSAR. 1. Fundamentals and Applications in Chemistry and Biology*, American Chemical Society, Washington, D.C., 1995.

Haseman, J.K., Melnick, R.L., Tomatis, L., and Huff, J., Carcinogenesis bioassays: study duration and biological relevance, *Food Chem. Toxicol.*, 39, 739–744, 2001.

Hooberman, B.H., Chakraborty, P.K., and Sinsheimer, J.E., Quantitative structure-activity relationships for the mutagenicity of propylene oxides with *Salmonella*, *Mutation Res.*, 299, 85–93, 1993.

Huff, J., Chemicals and cancer in humans: first evidence in experimental animals, *Environ. Health Perspect.*, 100, 201–210, 1993.

Huff, J., Value, validity, and historical development of carcinogenesis studies for predicting and confirming carcinogenic risks to humans, in *Carcinogenicity. Testing, Predicting, and Interpreting Chemical Effects*, Kitchin, K.T., Ed., Marcel Dekker Inc., New York, 1999, pp. 21–123.

Huff, J., IARC monographs, industry influence, and upgrading, downgrading, and under-grading chemicals. A personal point of view, *Int. J. Occup. Environ. Health*, 8, 249–270, 2002.

Huff, J., Weisburger, E., and Fung, V.A., Multicomponent criteria for predicting carcinogenicity: dataset of 30 NTP chemicals, *Environ. Health Perspect.*, 104 (Suppl. 5), 1105–1112, 1996.

Jones, J.D. and Easterly, C.E., On the rodent bioassays currently being conducted on 44 chemicals: a RASH analysis to predict test results from the National Toxicology Program, *Mutagenesis*, 6, 507–514, 1991.

Jones, T.D. and Easterly, C.E., A RASH analysis of National Toxicity Program data: predictions for 30 compounds to be tested in rodent carcinogenesis experiments, *Environ. Health Perspect.*, 104 (Suppl. 5), 1017–1030, 1996.

Kerckaert, G.A., Brauninger, R., LeBoeuf, R.A., and Isfort, R.J., Use of the Syrian hamster embryo cell transformation assay for carcinogenicity prediction of chemicals currently being tested by the National Toxicology Program in rodent biossays, *Environ. Health Perspect.*, 104 (Suppl. 5), 1075–1084, 1996.

King, R.D. and Srinivasan, A., Prediction of rodent carcinogenicity biossays from molecular structure using inductive logic programming, *Environ. Health Perspect.*, 104 (Suppl. 5), 1031–1040, 1996.

Lee, Y., Buchanan, B.G., and Rosenkranz, H.S., Carcinogenicity predictions for a group of 30 chemicals undergoing rodent cancer biossays based on rules derived from subchronic organic toxicities, *Environ. Health Perspect.*, 104 (Suppl. 5), 1059–1064, 1996.

Lewis, D.F.V., Ioannides, C., and Parke, D.V., COMPACT and molecular structure in toxicity assessment: a prospective evaluation of 30 chemicals currently being tested for rodent carcinogenicity by the NCI/NTP, *Environ. Health Perspect.*, 104 (Suppl. 5), 1011–1016, 1996.

Lewis, D.F.V., Ionnides, C., and Parke, D.V., A prospective toxicity evaluation (COMPACT) on 40 chemicals currently being tested by the National Toxicology Program, *Mutagenesis*, 5, 433–435, 1990.

Loew, G.H., Poulsen, M., Kirkjian, E., Ferrel, J., Sudhindra, B.S., and Rebagliati, M., Computer-assisted mechanistic structure-activity: application to diverse classes of chemical carcinogens, *Environ. Health Perspect.*, 61, 69–96, 1985.

Marchant, C.A., Prediction of rodent carcinogenicity using the DEREK system for 30 chemicals currently being tested by the National Toxicology Program, *Environ. Health Perspect.*, 104 (Suppl. 5), 1065–1074, 1996.

McCutcheon, P., Implications of Council Regulation 793/93 on the evaluation and control of existing substances, *Annali dell' Istituto Superiore di Sanità*, 30, 367–372, 1994.

Miller, J.A. and Miller, E.C. Ultimate chemical carcinogens as reactive mutagenic electrophiles, in *Origins of Human Cancer*, Hiatt, H.H., Watson, J.D., and Winstein, J.A., Eds., Cold Spring Harbor Laboratory, Cold Spring Harbor, NY, 1977, pp. 605–628.

Moriguchi, I., Hirano, H., and Hirono, S., Prediction of the rodent carcinogenicity of organic compounds from their chemical structures using the FALS method, *Environ. Health Perspect.*, 104 (Suppl. 5), 1051–1058, 1996.

Norden, B., Edlund, U., and Wold, S., Carcinogenicity of polycyclic aromatic hydrocarbons, studied by SIMCA pattern recognition, *Acta Chemica Scandinavica Series B*, 32, 602–608, 1978.

Passerini, L., QSARs for classes of mutagens and carcinogens, in *Quantitative Structure-Activity Relationship (QSAR) Models of Mutagens and Carcinogens*, Benigni, R., Ed., CRC Press, Boca Raton, FL 2003, pp. 81–123.

Purdy, R., A mechanism-mediated model for carcinogenicity: model content and prediction of the outcome of rodent carcinogenicity biossays currently being conducted on 25 organic chemicals, *Environ. Health Perspect.*, 104 (Suppl. 5), 1085–1094, 1996.

Richard, A.M., Application of SAR methods to non-congeneric data bases associated with carcinogenicity and mutagenicity: issues and approaches, *Mutation Res.*, 305: 73–97, 1994.

Richard, A.M. Public sources of mutagenicity and carcinogenicity data: use in structure-activity relationship models, in *Quantitative Structure-Activity Relationship (QSAR) Models of Mutagens and Carcinogens*, Benigni, R., Ed., CRC Press, Boca Raton, FL 2003, pp. 145–173.

Richard, A.M. and Benigni, R., AI and SAR approaches for predicting chemical carcinogenicity: survey and status report, *SAR QSAR Environ. Res.*, 13: 1-19, 2002.

Richard, A.M. and Williams, C.R., Distributed structure-searchable toxicity (DSSTox) public database network: a proposal, *Mutation Res.*, 499, 27–52, 2002.

Richard, A.M. and Woo, Y.T., A CASE-SAR analysis of polycyclic aromatic hydrocarbon carcinogenicity, *Mutation Res.*, 242, 285–303, 1990.

Rosenkranz, H.S. and Klopman, G., Prediction of the carcinogenicity in rodents of chemicals currently being tested by the US National Toxicology Program: structure-activity correlations, *Mutagenesis*, 5, 425–432, 1990.

Sanderson, D.M. and Earnshaw, C.G., Computer prediction of possible toxic action from chemical structure: the DEREK system, *Hum. Exp. Toxicol.*, 10, 261–271, 1991.

Shusterman, A.J., Debnath, A.K., Hansch, C., Horn, G.W., Fronczek, F.R., Greene, A.C., and Watkins, S.F., Mutagenicity of dimethyl heteroaromatic triazenes in the Ames test: the role of hydrophobicity and electronic effects, *Mol. Pharmacol.*, 12, 939–944, 1989.

Smith, C.J., Hansch, C., and Morton, M.J., QSAR treatment of multiple toxicities: the mutagenicity and cytotoxicity of quinolines, *Mutation Res.*, 379, 167–175, 1997.

Tamura, N., Takahashi, K., Shirai, N., and Kawazoe, Y., Studies on chemical carcinogens. XXI. Quantitative structure-mutagenicity relationships among substituted styrene oxides, *Chem. Pharmacol. Bull.* 30, 1393–1400, 1982.

Tennant, R.W. and Spalding, J., Predictions for the outcome of rodent carcinogenicity bioassays: identification of trans-species carcinogens and noncarcinogens, *Environ. Health Perspect.*, 104 (Suppl. 5), 1095–1100, 1996.

Tennant, R.W., Spalding, J., Stasiewicz, S., and Ashby, J., Prediction of the outcome of rodent carcinogenicity bioassays currently being conducted on 44 chemicals by the National Toxicology Program, *Mutagenesis*, 5, 3–14, 1990.

Tomatis, L., Huff, J., Hertz-Picciotto, I., Sandler, D.P., Bucher, J., Boffetta, P., Axelson, O., Blair, A., Taylor, J., Stayner, L., and Barrett, J.C., Avoided and avoidable risks of cancer, *Carcinogenesis*, 18, 97–105, 1997.

Tuppurainen, K., Frontier orbital energies, hydrophobicity and steric factors as physical QSAR descriptors of molecular mutagenicity. A review with a case study: MX compounds, *Chemosphere*, 38, 3015–3030, 1999.

Woo, Y.T. and Lai, D.Y., Aromatic amino and nitro-amino compounds and their halogenated derivatives in *Patty's Toxicology*, 5th ed., Vol. 4, Bingham, E., Cohrssen, B., and Powell, C.H., Eds., John Wiley and Sons Inc., New York, 2001, pp. 969–1105.

Woo, Y.T. and Lai, D.Y., Mechanisms of action of chemical carcinogens and their role in structure-activity relationships (SARs) analysis and risk assessment in *Quantitative Structure-Activity Relationship (QSAR) Models of Mutagens and Carcinogens*, Benigni, R., Ed., CRC Press, Boca Raton, FL 2003, pp. 41–80.

Woo, Y.T., Lai, D.Y., Arcos, J.C., Argus, M.F., Cimino, M.C., DeVito, S., and Keifer, L., Mechanism-based structure-activity relationship (SAR) analysis of carcinogenic potential of 30 NTP test chemicals, *Environ. Carcinogenesis Ecotoxicology Rev.*, C15, 139–160, 1997.

Woo, Y.T., Lai, D.Y., Argus, M.F., and Arcos, J.C., Development of structure-activity relationship rules for predicting carcinogenic potential of chemicals, *Toxicol. Lett.*, 79, 219–228, 1995.

Woo, Y.T., Lai, D.Y., McLain, J.L., Ko Manibusan, M., and Dellarco, V., Use of mechanism-based structure-activity relationships analysis in carcinogenic potential ranking for drinking water disinfection by-products, *Environ. Health Perspect.*, 110 (Suppl. 1), 75–87, 2002.

Zeiger, E., Carcinogenicity of mutagens: predictive capability of the *Salmonella* mutagenesis assay for rodent carcinogenicity, *Cancer Res,*, 47, 1287–1296, 1987.

Zeiger, E., Strategies and philosophies of genotoxicity testing what is the question?, *Mutation Res.*, 304, 309–314, 1994.

Zhang, L., Sannes, K., Shusterman, A.J., and Hansch, C., The structure-activity relationships of skin carcinogenicity of aromatic hydrocarbons and heterocycles, *Chem. Biol. Interaction*, 81, 149–180, 1992.

Zhang, Y.P., Sussman, N., Macina, O.T., Rosenkranz, H.S., and Klopman, G., Prediction of the carcinogenicity of a second group of organic chemicals undergoing carcinogenicity testing, *Environ. Health Perspect.*, 104 (Suppl. 5), 1045-1050, 1996.

CHAPTER **9**

# The Use of Expert Systems for Toxicity Prediction: Illustrated with Reference to the DEREK Program

**Robert D. Combes and Rosemary A. Rodford**

## CONTENTS

# I. INTRODUCTION

## A. The Nature of Expert Systems

An expert system is any formalized system that is often, but not necessarily, computer based that can be used to make predictions on the basis of prior information (Combes and Judson, 1995; Dearden et al., 1997). Expert systems are designed to emulate the way in which a group of human experts solve problems. They are intended to help users make decisions, rather than make decisions for them. While expert systems are applicable to any discipline, the use considered here is for toxicity (and metabolism) prediction. There are two main types of computerized expert system: automated rule-induction (ARI) and knowledge-based systems (KBS). The two types differ fundamentally in the way they operate. ARI systems make predictions by learning from and discovering patterns in existing data, whereas KBS predict by reasoning on the basis of existing human knowledge.

## B. The Basis for Using Expert Systems

The use of expert systems for toxicity prediction is based upon the premise that the activity of a molecule in any particular biological system is determined by its physicochemical properties, in particular its molecular structure (Barratt, 2000; Barratt and Rodford, 2001; Richard et al., 2000). From a knowledge of the latter, structural alerts — structural parts of molecules that are responsible for or can modulate biological activity, can be identified.

Structural features that promote biological activity are sometimes called biophores. They are divisible into pharmacophores and toxicophores. Pharmacophores impart desirable properties on a molecule (e.g., pharmacological activity or a particular fragrance). Toxicophores are responsible for undesirable effects such as toxicity (e.g., mutagenicity and skin sensitization). The same molecule can have more than one descriptor that can act as both a pharmacophore and a toxicophore in the same or different biological systems. Examples here are the toxic side effects of anti-cancer drugs and the use of Warfarin, a commercially available rat poison, to help reduce the formation of blood clots in human heart disease.

Other structural features can reduce biological activity, and these may be termed biophobes or modifiers. An example of a biophobe is a bulky substituent that reduces the effects of an adjacent biophore by steric hindrance.

## C. Biological Activities Predicted by Expert Systems

A wide range of biological activities is predicted by the main available expert systems. The main commercially available expert systems are listed in Table 9.1. Examples include pharmacological activity such as receptor binding and a variety of information on toxicological activity relevant to toxic hazard assessment. The latter includes: acute toxicity, mutagenicity/carcinogenicity, eye/skin irritation, skin/respiratory sensitization, target organ toxicity, teratogenicity, reproductive toxicity, endocrine disruption, and neurotoxicity. There is little information on systemic toxicity, although the prediction of this would be possible if the data were made available. The prediction of pharmacological activity and toxic hazard assessment can be combined in the process of computer-aided drug design (CADD).

# II. TYPES OF EXPERT SYSTEMS

## A. Automated Rule-Induction (ARI) Systems

### 1. The Nature of ARI Systems

ARI systems analyze information relating to chemical structures for associations between active and inactive molecules. Molecules from a training set of chemicals of known activity for a particular

**Table 9.1    Main Commercially Available Expert Systems for the Prediction of Toxicity**

| Name | Supplier | Website | Main Endpoints Predicted |
|------|----------|---------|--------------------------|
| **Knowledge-Based Systems** | | | |
| DEREK for Windows | LHASA Ltd. | www.lhasalimited.org | Mutagenicity, carcinogenicity, skin sensitization, acute toxicity, and many other effcts |
| OncoLogic® | LogiChem Inc. | www.logichem.com | Carcinogenicity |
| HazardExpert | CompuDrug | www.compudrug.com/hazard.html | Carcinogenicity, mutagenicity, teratogenicity, membrane irritation, neurotoxicity, and other effects |
| **Automated Rule-Induction Systems** | | | |
| TOPKAT | Accelrys Ltd. | www.accelrys.com/products/topkat/ | Carcinogenicity, mutagenicity, various mammalian acute and chronic toxicities, developmental toxicity, and many other effects |
| MCASE, CASE, CASETOX, etc. | MultiCASE Inc. | www.multicase.com/ | Carcinogenicity, mutagenicity, teratogenicity, various mammalian acute and chronic toxicities, and many other effects |
| PASS | Laboratory of Structure Function Based Drug Design, V.N. Orekhovich Institute of Biochemical Chemistry | www.ibmh.msk.su/PASS/ | Mutagenicity, carcinogenicity, teratogenicity, embryotoxicity, and many other effects |
| ToxScope | LeadScope Inc. | www.leadscope.com/products/txs.htm | Carcinogenicity and many other mammalian toxicological endpoints |
| ToxFilter | Pharma Algorithms Inc. | ap-algorithms.com/tox_filter.htm | Mammalian acute toxicity |
| ECOSAR | U.S. Environmental Protection Agency | www.epa.gov/opptintr/exposure/docs/episuit edl.htm | Many ecological effect and environmental fate endpoints |

biological endpoint are fragmented into all possible atom pairs and other associations. Pattern recognition techniques are then used, together with other statistical analyses, to compare the frequency of occurrence of specific structural features in sets of active and inactive molecules. In this way, the most important features determining or modifying activity are identified. After it has been trained, the system can then be used to search for the presence of biophores and biophobes in novel molecules. Some ARI systems also utilize methods such as quantitative structure-activity relationships (QSAR) and molecular modeling of 3D structures. ARI systems appear very much as black boxes because they are not transparent in the explanation of the basis of their predictions to the user.

## 2.  Types of ARI Systems

ARI systems make quantitative predictions, for example, by providing a probability value of carcinogenicity being induced by a molecule or a quantitative prediction of an acute toxicity. Two widely used ARI systems are Toxicity Prediction by Komputer-Assisted Technology (TOPKAT) (Enslein, 1988; Enslein et al., 1994) and Multiple Computer Automated Structure Evaluation

(MultiCASE) (Dearden et al., 1997; Klopman and Rosenkranz, 1994). A further system is Computerized Optimized Parametric Analysis of Chemical Toxicity (COMPACT) (Lewis et al., 1994). The latter analyses the ability of a molecule to fit into the active site of the CYP1A1 isozyme of cytochrome P450 (CYP) (and some other CYP isozymes), by modeling molecular shape (planarity or area/depth) and chemical reactivity (covalent bond formation). The use of COMPACT is limited to molecules that are activated by these CYP enzymes.

### 3.  Examples of the Use of an ARI System

#### a.  Estrogenicity

The MultiCASE system has been used to identify a common 6-Å unit biophore on a range of hormonally active chemicals with estrogenic activity that act as endocrine disruptors. This structural feature is a spacer biophore that is thought to be involved in the molecules binding to the estrogen receptor and is found on the standard estrogenic chemical, 17-beta-estradiol (see Combes, 2000). Other examples of molecules possessing this biophore include 4-hydroxytamoxifen, 2-chloro-4-hydroxybiphenyl, 3,4-dihydroxyfluorene, and 2,2-(*bis*-4-hydroxyphenyl–1,1,1-trichloroethane).

#### b.  Tubulin Inhibition

Biophores of some tubulin inhibitors were identified by CASE on molecules such as colchicine, podophyllotoxin, and dihydrocombrestatin (see Combes, 2000). These chemicals might act as nongenotoxic carcinogens by being able to bind to tubulin, inducing phenomena such as aneuploidy.

#### c.  Draize Eye Irritation

Rosenkranz et al. have also identified a total of 13 different CASE biophores and 7 biophobes for Draize eye irritation (Rosenkranz et al., 1998). These included structural features on 2,4-dihydroxybenzoic acid and sodium lauryl sulphate. In the case of a quantitative prediction of the eye irritation potential of 2-methylbutyric acid, the ocular irritation potential of the chemical was predicted to be high and was assigned a value of 49 CASE units.

### B.  Knowledge-Based Systems (KBS)

### 1.  The Nature of KBS

KBS use structural alerts to develop rules devised by experts based on a database of previous information, for example, on different endpoints in toxicity. All the information is stored in the software's rule base for later recall. The rules describe toxicophores/pharmacophores in molecules of known activity. These structural features can then be identified in novel molecules drawn on the computer screen using commercially available chemical drawing packages, such as ISISDRAW and CHEMDRAW, or imported as standard mol files. Examples of KBS for toxicity are HAZARDEXPERT (Smithing and Darvas, 1992) and Deductive Estimation of Risk from Existing Knowledge (DEREK) (Judson, 2002; Langowski, 1993; Sanderson and Earnshaw, 1991). The development of KBS is crucially dependent on the availability of experts to identify the relevant structural alerts and use them to write rules for each toxicity endpoint and the availability of good quality toxicity data. In the case of DEREK, which is supplied by LHASA Ltd., a small not-for-profit organization based in the chemistry department of Leeds University in the U.K., the users are all part of a collaborative group that meets regularly, at least three times a year, to discuss the development of the knowledge base and to share experiences on the use of the software. Sometimes, users' companies will be able to provide test data to LHASA, which can then be used to derive the information and knowledge needed to construct a new rule, without revealing either the original or its source. In the case where a company provides data or resources, LHASA will offset the value of this against the cost of the software. This is not only a very cost-effective method of obtaining the software, but it is also an excellent way to become involved in the development and the future direction of the software. Generally, KBS are transparent in that they explain the basis of their predictions to the user, providing literature sources and references to original data where possible. It should be noted that currently KBS only make qualitative predictions.

## 2. Examples of the Use of DEREK as a Toxicity Predictor

### a. Skin Sensitization

DEREK, which has now been developed into a user-friendly Microsoft Windows format, has an extensive rule base for skin sensitization (Barratt and Basketter, 1994; Barratt et al., 1994; Payne and Walsh, 1994). An example of the toxicophore identified for the skin sensitization of citronellal using DEREK for Windows v. 6.0.0 is shown in Figure 9.1. This shows that once the structure has been processed against the skin sensitization rule base, the toxicophore is highlighted. The number of occurrences of the toxicophore in the molecule is also recorded, which in this case is one. Also displayed are the rule number and description that has been fired. In this case, the rule is 419 — skin sensitization (aldehyde). An alert overview gives the basis of the rule, and why the program has identified a toxicophore alerting to skin sensitization. This information also includes citations to relevant publications in the literature, as well as further comments facilitating interpretation of the rules by the user. Examples are also provided of known compounds expressing the alert, and 2 of these for rule 419 are shown in Figure 9.2. If the reasoning engine is invoked, further information is provided on the likelihood of the hazard alerted being expressed in the chosen species. In fact, the information provided by the program is totally transparent, allowing the user to understand the basis of the prediction; the user can either accept or reject the prediction, based on extra information that might have become available since the generation of the rule base.

Locations:

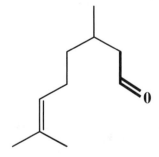

List of alerts found:

419 Aldehyde / Skin sensitization, number of matches = 1

**Figure 9.1**  Toxicophore identified for the skin sensitization of citronellal.

<u>**Examples:  (419 Aldehyde / Skin sensitization)**</u>

<u>**Example 1.**</u>        **3-(4-tert-butylphenyl)propanal**
**AS Number:**        18127-01-0

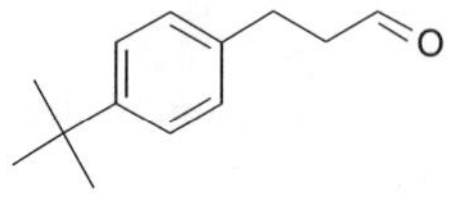

<u>**Test Data:  (3-(4-tert-butylphenyl)propanal)**</u>

**1.**
**Species:**  guinea pig
**Assay:**    maximization test
**Result:**   strong

<u>**References:**</u>

**Title:**    Multivariate QSAR analysis of a skin sensitization database.
**Author:**   Cronin MTD and Basketter DA.
**Source:**   SAR and QSAR in Environmental Research, 1994, 2, 159-179.

<u>**Example 2.**</u>        **3-(4-isopropyl-phenyl)-2-methyl-propionaldehyde**
**CAS Number:**       103-95-7

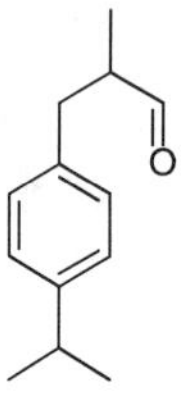

<u>**Test Data:  (3-(4-isopropyl-phenyl)-2-methyl-propionaldehyde)**</u>

**1.**
**Species:**  guinea pig
**Assay:**    maximization test
**Result:**   strong

<u>**References:**</u>

**Title:**    Multivariate QSAR analysis of a skin sensitization database.
**Author:**   Cronin MTD and Basketter DA.
**Source:**   SAR and QSAR in Environmental Research, 1994, 2, 159-179.

**Figure 9.2**   Two examples of compounds expressing the skin sensitization alert 419.

## b.   Skin Permeability

For a chemical to act as a skin sensitizer, it has to be absorbed by the skin, traverse the dermal barrier, and then pass to responding cells in the epithelial layers. There it has to react with immunological proteins via a process involving covalent binding (Kimber et al., 2001). If a chemical is either unable to penetrate the skin or it lacks the ability to react with such proteins, then it cannot act as a skin sensitizer.

To help establish the relevance of a skin sensitization alert from DEREK for Windows, the system incorporates reasoning that allows the determination of skin permeability coefficients (log Kp values) according to the algorithm developed by Potts and Guy (1990). DEREK for Windows derives log Kp from the logarithm of the octanol/water partition coefficient (log $K_{ow}$) using the following equation (Moriguchi et al., 1992):

$$\log K_{ow} = 0.246 \ CX - 0.386 \ NO + 0.466 \tag{9.1}$$

where CX is the sum of the empirical weighted numbers of carbon and halogen atoms and NO is the total number of oxygen and nitrogen atoms.

DEREK for Windows calculates log Kp in cm/h using a modified Potts and Guy equation:

$$\text{Log Kp (cm/h)} = -2.72 + 0.71 \log K_{ow} - 0.0061 \ MW \tag{9.2}$$

where MW is the molecular weight.

These additional parameters enable DEREK for Windows to indicate to the user whether a skin sensitization hazard alert given is likely to be expressed. In the example of citronellal given above, DEREK for Windows indicates that the hazard of skin sensitization is plausible. In another example, diacetyl-diperoxyadipic acid, alert 406 — skin sensitization (diacyl peroxide) is fired as highlighted in Figure 9.3.

The skin permeability of this chemical is low; the log Kp value from the reasoning algorithm is –6.856 and the chemical is known to give a weak response in the guinea pig maximization test, which indicates an equivocal response in mammals. This information leads to the prediction from the software that skin sensitization in humans is doubtful.

### c.  Predicting Mutagenicity and Carcinogenicity

Some 18 different structural alerts (toxicophores) have been identified as being present on chemicals that had been shown previously to be rodent carcinogens in 2-year rodent lifetime bioassays, as well as possessing genotoxicity in one or more short-term genotoxicity assays, as part of the U.S. National Toxicology Program (NTP) collaborative toxicity testing studies (Ashby and Paton, 1993). This information was used to generate mutagenicity rules for DEREK, and the program is being continuously updated and improved for this important toxicity endpoint (Ridings et al., 1996). Some mutagenicity and carcinogenicity rules were written for chemicals found in foods (Long and Combes, 1995) and an example of these, the bisfuranoid mycotoxin substructure, is shown in Figure 9.4.

### 3.  Rule-Base Development in DEREK for Windows

The need for new rules for certain classes of chemicals, or improvements to existing rules, may be identified by LHASA or any DEREK users. The DEREK Collaborative Group prioritizes the work. The first stage of rule development is to carry out a thorough literature search to establish whether there is sufficient published information to support the proposed rule. Unpublished test results may also be used instead of, or in addition to, data from the open literature. The next stage is to propose a mechanism of action and identify any structural conditions that could lead to exclusion from the rule. Finally, it is necessary to make sure that the new rule fires for all appropriate chemicals in the class. Having completed the support for the rule, this information may be sent to LHASA; the organization will check the quality of the work and incorporate the rule in the knowledge base. However, after training, or by carefully following the detailed user guide, companies can enter their own in-house alerts using the DEREK for Windows knowledge-base editor.

Locations:

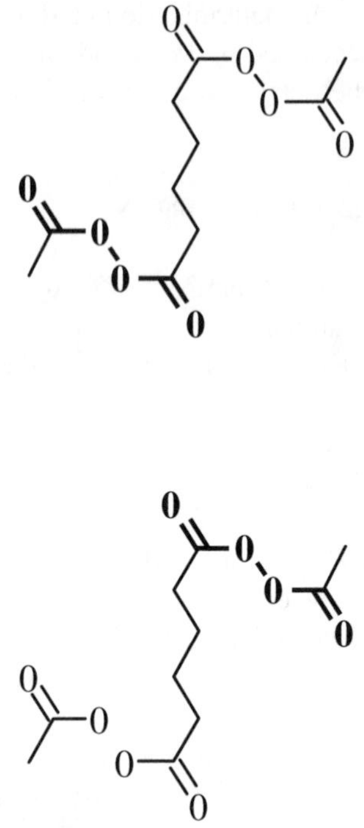

**List or alerts found:**

406 Diacyl peroxide / Skin sensitization, number of matches = 2

**Figure 9.3**    Toxicophore identified for the skin sensitization of diacetyl-diperoxy adipic acid.

Rule 201

Toxic End-Point:                    **Mutagenicity and Carcinogenicity**

Compound Type:                  **Bisfuranoid Mycotoxin Substructure/Analogue**

Toxicophore (in boldface):

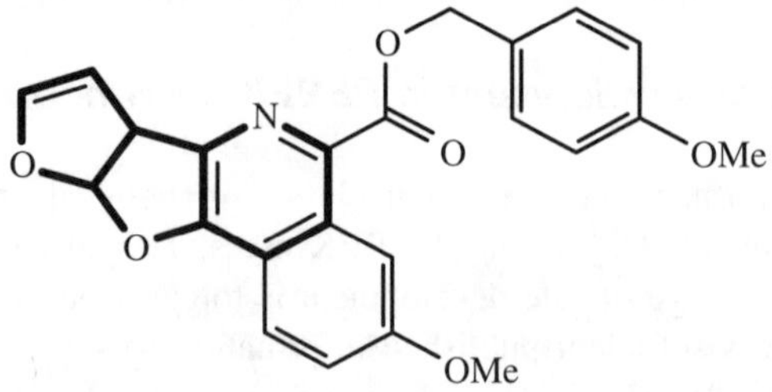

**Figure 9.4**    Illustration of the mutagenicity toxicophore for bisfuranoids in DEREK.

This feature is especially useful if a company has proprietary data that cannot be shared with the collaborative group, but that would be very helpful for in-house predictions. It should be remembered that if private new alerts have been created or other local changes have been made to the knowledge base, it is important for each company to consider merging these novel rules with a new release of DEREK for Windows when it receives it.

## III. DISCUSSION

### A. Data Quality

One of the most serious problems encountered in the development of expert systems is gaining access to high-quality test data. To help in overcoming this difficulty, LHASA have been working with the International Life Sciences Institute/Health and Environmental Sciences Institute (ILSI/HESI) in Washington, D.C., on a project aimed at producing an International Toxicology Information System, the so-called Structure-Activity Relationships (SAR) Database. This is intended to provide a structure-searchable database of toxicological information by chemical — more information is available from www.ilsi.org. The first phase of this project has been running since July 2000 with the goal being to build a pilot database containing a limited amount of data on just four toxicological endpoints (skin sensitization, mutagenicity, carcinogenicity, and hepato-toxicity). This goal was met, although the database needs to be populated further so that effective evaluations can be carried out by the project sponsors. This will provide an opportunity to ensure that all sponsors' requirements have been met and to identify other improvements to the database. The resulting product will also provide a better demonstration tool with which to gain further sponsorship for the next phase of the project. Some ten institutions have been sponsoring the first phase of the project. Greater sponsorship, especially from companies willing to provide non-confidential in-house data, is required to ensure the long-term success of this important initiative.

### B. Validation of Expert Systems

There have been relatively few studies in which expert systems have been compared for their ability to correctly predict the same biological activity, except in the case of rodent carcinogenicity, by using the NTP database (Parry, 1994). In these studies, several systems (including MultiCASE, DEREK, and TOPKAT) showed overall accuracies in correctly identifying rodent carcinogens varying from 60 to 90%, depending on the system and the database. Optimal levels of performance were obtained using combinations of the systems. It is concluded from these kinds of studies that expert systems should be used as screens in conjunction with each other and with *in vitro* tests. Expert systems are proving especially useful for high throughput screening of drugs.

It is most important that expert systems should be properly validated, just like any other test method. This is especially so at the present time, because (Q)SAR and expert systems potentially offer the most realistic and practical way to address the requirements of the recent European Union White Paper on chemicals testing (Anon., 2001; Worth and Balls, 2003). Unfortunately, this is not happening and in some situations, such as COMPACT and CASE, the systems are being used only by one research group. The validation of these approaches will require special considerations as outlined in Chapter 20 as well as Worth et al. (1998) and Worth and Cronin (2004).

As stated earlier, expert systems are developed using training sets of chemicals. It is important that such training sets of chemicals are of varying structures containing differing biological activities. Training sets should consist of chemicals acting via as wide a range of mechanisms as possible causing the toxic endpoint of interest. It is also crucial that systems are developed that can correctly predict not only the activities of the chemicals in the training set, but also those of novel untested chemicals, which are structurally related to those in the training set. This is so that the utility of the expert system for predicting the unknown is as comprehensive as possible.

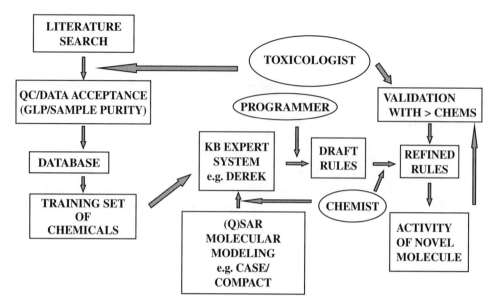

**Figure 9.5** Development and integrated use of expert systems for toxicity prediction.

## C. Limitations of Expert Systems

Expert systems have several noteworthy limitations. First, their development depends crucially on the availability of accurate, relevant, and high-quality biological data on individual chemical entities with well-defined structures. Unfortunately, no test sample is completely pure, and it is very important that information used to construct rules for expert systems is derived from studies using test samples of high purity. Also, the identity of a chemical responsible for toxicity might not be known due to metabolism, which can also often result in the generation of several active molecules that might act antagonistically or synergistically together. Several expert systems for predicting metabolism have been developed (Darvas et al., 1999, 2002; Greene et al., 1999; Langowski and Long, 2002). The METEOR program (also from LHASA Ltd.) is designed to be used in conjunction with DEREK to facilitate the processing of predicted metabolites for structural alerts (Long, 2002). This topic is discussed in Chapter 10.

Strict criteria should be applied to the process of data acceptance for rule development, and ideally, the effects of metabolism should be known or be predictable. It is also important that toxicological information is correctly interpreted and this can be achieved by a close cooperation among toxicologists, chemists, and experts in computer programming (Figure 9.5).

## IV. CONCLUSIONS

There is little doubt that expert systems will continue to improve and provide a very useful tool for the toxicologist in the early prediction of adverse biological effects. The current main areas of application are for compound prioritization for the safety assessment of existing chemicals and for the early screening of candidate chemicals, particularly in the agrochemical and pharmaceutical industries, in the process of high throughput screening.

One of the major limitations of expert systems is the availability of high quality toxicity data for a wide range of endpoints. It is only once this problem has been overcome that the true potential of expert systems for predicting all forms of toxicity will be realized. It is most important that both newly generated and existing data are fed into the process where rules can be written for such computer prediction methods to assist with new initiatives such as those involving endocrine disruptor testing and for existing and high production volume chemicals testing (Figure 9.5). It is hoped that more emphasis will be placed on validating expert systems (Barratt and Langowski, 1999; Worth and Cronin, 2004) according to the criteria adopted for other new test methods.

## ACKNOWLEDGMENTS

The authors thank Carol Marchant for critically reviewing the manuscript.

## REFERENCES

Anon., white paper, *Strategy for a Future Chemicals Policy(COM(2001)88* final, 2001, www.europa.eu.int/comm/environment/chemicals/whitepaper.htm

Ashby, J. and Paton, D., The influence of chemical structure on the extent and sites of carcinogenesis for 522 rodent carcinogens and 55 different human carcinogen exposures, *Mutation Res.*, 286, 3–74, 1993.

Barratt, M.D., Principles of toxicity prediction from chemical structure, in *Progress in the Reduction, Refinement and Replacement of Animal Experimentation,* Balls, M., Zeller, A.-M., and Halder, M., Eds., Elsevier Science B.V., Amsterdam, 2000, pp. 449–456.

Barratt, M.D. and Basketter, D.A., Structure-activity relationships for skin sensitization: an expert system, in *In vitro Toxicology,* : Rougier, A., Goldberg, A.M., and Mailbach, H.I., Eds., Mary Ann Liebert Inc., New York, 1994, pp. 293–301.

Barratt, M.D., Basketter, D.A., Chamberlain, M., Admans, G.D., and Langowski, J.J., An expert system rulebase for identifying contact allergens, *Toxicol. in Vitro,* 8, 1053–1060, 1994.

Barratt, M.D. and Langowski, J.J., Validation and subsequent development of the DEREK skin sensitization rulebase by analysis of the BgVV list of contact allergens, *J. Chem. Inf. Comput. Sci.,* 9, 294–298, 1999.

Barratt, M.D. and Rodford, R.A., The computational prediction of toxicity, *Curr. Opinion Chem. Biol.,* 5, 383–388, 2001.

Combes, R.D., The use of structure-activity relationships and markers of cell toxicity to detect non-genotoxic carcinogens, *Toxicol. In Vitro* 14, 387–399, 2000.

Combes, R.D. and Judson, P., The use of artificial intelligence systems for predicting toxicity, *Pesticide Sci.,* 45, 179–194, 1995.

Darvas, F., Papp, A., Allardyce, A., Benfenati, E., Gini, G., Tichy, M., Sobb, N., and Citti, A., Overview of different artificial intelligence approaches combined with a deductive logic-based expert system for predicting chemical toxicity, in *AAAI Spring Symposium: Predictive Toxicology of Chemicals*, Stanford, CT, pp. 94–99, 1999.

Darvas, F., Szabó, I., Karancsi, T., Slégel, P., Dormán, G., László, Ü., and Krajcsi, P., Dual uses of *in silico* and *in vitro* metabolism data in lead discovery, ComGenex' MAID is a metabolism alerting system for the early phase of drug discovery, *Genet. Eng. News,* B22B, 32–34, 2002.

Dearden, J.C., Barratt, M.D., Benigni, R., Bristol, D.W., Combes, R.D., Cronin, M.T.D., Judson, P.M., Payne, M.P., Richard, A.M., Tichy, M., Worth, A.P., and Yourick, J.J., The development and validation of expert systems for predicting toxicity: the report and recommendations of an ECVAM/ECB workshop (ECVAM workshop 24), *Alternatives Lab. Anim. (ATLA),* 25, 223–252, 1997.

Enslein, K., An overview of structure-activity relationships as an alternative to testing in animals for carcinogenicity, mutagenicity, dermal and eye irritation, and acute oral toxicity, *Toxicol. Ind. Health,* 4, 479–498, 1988.

Enslein, K., Gombar, V.K., and Blake, B.W., Use of SAR in computer-assisted prediction of carcinogenicity and mutagenicity of chemicals by the TOPKAT program, *Mutation Res.,* 305, 47–62, 1994.

Greene, N., Judson, P.N., Langowski, J.J., and Marchant, C.A., Knowledge based expert systems for toxicity and metabolism predictions: DEREK, StAR and METEOR, *SAR QSAR Environ. Res.,* 10, 299–314, 1999.

Judson, P.N., Prediction of toxicity using DEREK for Windows, *J. Toxicol. Sci.,* 27, 278, 2002.

Kimber, I., Pichowski, J.S., Betts, C.J., Cumberbatch, M., Basketter, D.A., and Dearman, R.J., Alternative approaches to the identification and characterisation of chemical allergens, *Toxicol. In Vitro,* 15, 307–312, 2001.

Klopman, G. and Rosenkranz, H.S., Approaches to SAR in carcinogenesis and mutagenesis. Prediction of carcinogenicity/mutagenicity using MULTI-CASE, *Mutation Res.,* 305, 33–46, 1994.

Langowski, J., Computer prediction of toxicity, *Pharm. Manuf. Int. (Int. Rev. Pharm. Technol., Res. Dev.)* 1993, pp. 77–80.

Langowski, J. and Long, A., Computer systems for the prediction of xenobiotic metabolism, *Adv. Drug Delivery Rev.,* 54, 407–415, 2002.

Lewis, D.V.F., Moereels, H., Lake, B.G., Ioannides, C., and Parke, D.V., Molecular modelling of enzymes and receptors involved in carcinogenesis: QSARs and COMPACT-3D, *Drug Metab. Rev.,* 26, 261–285, 1994.

Long, A., Rule-based prioritisation of metabolites: some recent developments in METEOR, *Drug Metab. Rev.,* 34 (Suppl. 1), 71, 2002.

Long, A. and Combes, R.D., Using DEREK to predict the activity of some carcinogens/mutagens found in foods, *Toxicol. In Vitro,* 9, 563–569, 1995.

Moriguchi, I., Hirono, S., Liu, Q., Nakagome, I. and Matsushita, Y., Simple method of calculating octanol-water partition coefficient, *Chem. Pharm. Bul.,* 40. 127–130, 1992.

Parry, J.M., Detecting and predicting the activity of rodent carcinogens, *Mutagenesis,* 9, 3–5, 1994.

Payne, M.P. and Walsh, P.T., Structure-activity relationships for skin sensitization potential: development of structural alerts for use in knowledge-based toxicity prediction systems, *J. Chem. Inf. Comput. Sci.,* 34, 154–161, 1994.

Potts, R.O. and Guy, R.H., Predicting skin permeability, *Pharm. Res.,* 9, 663–669, 1990.

Richard, A.M., Bruce, R., and Greenberg, M., Structure activity relationship methods applied to chemical toxicity: making better use of what we have, in *Progress in the Reduction, Refinement and Replacement of Animal Experimentation,* Balls, M., Zeller, A.M., and Halder, M., Eds., Elsevier Science B.V., Amsterdam, 2000, pp. 457–468.

Ridings, J.E., Barratt, M.D., Cary, R., Earnshaw, C.G., Eggington, C.E., Ellis, M.K., Judson, P.N., Langowski, J.J., Marchant, C.A., Payne, M.P., Watson, W.P. and Yih, T.D., Computer prediction of possible toxic action from chemical structure: an update on the DEREK system, *Toxicology,* 106, 267-279, 1996.

Rosenkranz, H.S., Zhang, Y.P., and Klopman, G., The development and characterisation of a structure-activity relationship model of the Draize eye irritation test, *Alternatives Lab. Anim. (ATLA),* 26, 779–809, 1998.

Sanderson, D.M. and Earnshaw, C.G., Computer prediction of possible toxic action from chemical structure: the DEREK system, *Hum. Exp. Toxicol.,* 10, 261–273, 1991.

Smithing, M.P. and Darvas, F., Hazard expert: an expert system for predicting chemical toxicity, in *Food Safety Assessment,* Finlay, J.W., Robinson, S.F., Armstrong, D.J., Eds., American Chemical Society, Washington, D.C., 1992, pp. 191–200.

Worth, A.P. and Balls, M., Eds., Alternative (non-animal) methods for chemicals testing: current status and future prospects. A report prepared by ECVAM and the ECVAM Working Group on Chemicals, *Alternatives Lab. Anim. (ATLA),* 30 (Suppl. 1) 2002, 125 pp.

Worth, AP., Barratt, M.D., and Houston, J.B., The validation of computational prediction techniques, *Alternatives Lab. Anim. (ATLA),* 26, 241–247, 1998.

Worth, A.P. and Cronin, M.T.D., Report of the workshop on the validation of (Q)SARs and other computational prediction models, *Proceedings of the Fourth World Congress on Alternatives to Animal Use in Life Sciences, Alterantives Lab. Anim.,* 2004.

# ethods for the Prediction of
# lism and Biotransformation
# vithin Biological Organisms

## I. INTRODUCTION

The biotransformation of chemicals within biological organisms is of great significance in determining their biological effects (e.g., their effectiveness or toxicity as agents such as pharmaceuticals or agrochemicals). Increasingly, *in vitro* methods have been used for the assessment of chemical metabolism, additionally motivated by the need for the efficient screening of large numbers of chemicals and the concern to reduce the use of live animals. Today, the pressure in the pharmaceutical or agrochemical industry is increasingly to screen "earlier, faster, and smarter" to avoid failure of candidate chemicals (Ekins, 2000) and pharmacokinetic aspects (absorption, distribution, metabolism, and excretion [ADME]) are being considered earlier. Computer-based methods can increasingly be used as one component in this process.

Advances in computer technology, computational chemistry, and theoretical understanding have yielded a battery of different tools for rationalizing and predicting chemical metabolism. This chapter briefly surveys some available methodologies that have been applied to answer a variety of different questions concerned with chemical metabolism, primarily for mammalian biotransformations, although biotransformations within plants and other organisms follow similar principles. The desirable or undesirable metabolic characteristics of a chemical vary with its intended application. For human pharmaceuticals it is important that therapeutically active concentrations of the active component are maintained in the blood circulation for several, preferably 24, hours. Rates of biotransformation are particularly important in this context. Knowledge of the enzymes mediating the reactions or those induced or inhibited may allow drug-drug interactions, or problems arising from genetic variation such as enzymic polymorphism, to be anticipated or explained. Understanding the metabolic route of a drug candidate at an early stage may also assist in the design of related drugs with improved pharmacokinetic profiles. It will reduce the incidence of troublesome interactions with other co-administered drugs. Additionally, using structure-toxicity relationships, the early prediction of potential toxic effects that may originate from the metabolites becomes possible.

Chemical transformations in animals or man (biotransformations) may be spontaneous or enzyme-mediated reactions. They may involve the administered compound or its metabolites. Since the administered compound is usually, but not invariably, stable under physiological conditions (e.g., 37°C, pH 7.4, aqueous medium), or the reactivity is well known and drug-drug interactions and absorption properties may be enzyme-dependent, emphasis in work on predicting biotransformation is frequently placed on the enzyme-mediated aspects. This is particularly the case because frequently this has repercussions for development described above. However, some chemicals are significantly reactive *in vivo* and many are metabolized to reactive intermediates; consequently, the reaction chemistry may be very significant in determining the overall fate and the concentrations of biologically important species.

In the last 20 years advances in computer technology and computational chemistry have resulted in the availability of a variety of different commercially available software packages or methodologies that can assist in the prediction or rationalization of various aspects of chemical metabolism in mammals. Such tools broadly divide into those that may be used to predict whether a particular biotransformation is likely to be catalyzed by a particular class or type of enzyme and those (usually expert systems) that give predictions of the full range of likely, initial, and subsequent biotransformations for a chemical in a particular biological system.

Chemical metabolism can be described qualitatively or quantitatively. Many scientists can make qualitative predictions of the likely excretion products or blood plasma metabolites in mammals, or a particular animal including man, based on accumulated knowledge and experience. Such knowledge, in its raw form, generally consists of structure-metabolism relationships that are frequently expressible as qualitative structure-based rules that may be encoded into computer-based expert systems (see Chapter 9 for a full definition). Examples of such systems, in their more fully developed commercial forms, are discussed toward the end of this chapter.

Increasing structural knowledge of the active sites of important enzymes has allowed predictions of biotransformations to be made based on the fit of substrates to active sites and the energetics of the molecular interactions within them. With such computational tools the prediction of metabolic fate is becoming increasingly possible. The quantitative prediction of the complete pattern of metabolites in blood plasma, bile, urine, or feces in a range of species (rat, mouse, dog, and man) requires a detailed consideration of the interplay of all possible metabolic and non-metabolic spontaneous chemical reactions, which is presently not achievable. This chapter presents an overview of the main approaches and achievements of computer-based methods for predicting metabolic transformation and complete metabolic profiles. Detailed descriptions of the methodologies are not given; they are amply described elsewhere.

## II. STRUCTURE-METABOLISM RELATIONSHIPS

### A. Modeling Specific Enzyme-Mediated Biotransformation Reactions

Much published work has derived structure-activity relationships (SARs) for a particular enzyme and range of substrates. Such results may be used for predictive purposes provided the chemical in question is sufficiently similar to the chemicals used to develop the SAR (the training set) and the computational technology and skills to use them are available.

There are two components determining chemical metabolism by an isolated enzyme. These are the extent of binding, characterized by the binding constant $K_m$, and the reactivity of the substrate with respect to the specific biotransformation in the enzyme site, commonly characterized by the kinetic parameter $V_{max}$. The latter property reflects the innate reactivity of the substrate and the structural and physicochemical requirements of the enzyme. In addition, the extent of biotransformation *in vivo* is determined by the physiological route to the enzyme and the concentration within tissues. The reactivity of the substrate is determined by its electronic structure (i.e., electron or charge distribution and molecular orbital energies) and the ease of effective interaction with the enzyme active site. Interactions with enzyme active sites depend on electrostatic, hydrophobic, and steric interaction forces. While some of these requirements may be expressed empirically as a substrate pharmacophore, described below, the influence of detailed aspects of shape and size are not usually discernible without reference to a structural model of the enzyme site. The effects of additional hydrogen bonding capabilities, hydrophobic regions, or substituent variation may be difficult to assess without more precise information on the active site. The use of homology modeling of enzymes and substrate-active site docking procedures and the associated calculations of the energies of interactions has assisted in giving a better understanding of the structural requirements.

### B. Modeling Substrate Reactivity: Application of Quantum Mechanical Calculations to Metabolism Prediction

Many workers have used calculated molecular orbital (MO) energies as indicators of chemical reactivity. In much work the values of MO-based parameters related to the intrinsic reactivity may assist in highlighting the most intrinsically reactive positions in the molecule or the relative reactivity of a series of molecules. The idea behind this is that enzyme-driven Phase I reactions generally occur at positions of highest nucleophilic or electrophilic reactivity or are determined by the relative stability of free radical intermediates. This approximation only applies adequately for enzymes with few steric requirements or lacking specific binding sites on the substrate (such as basic nitrogen atoms or anionic sites). The properties of the enzyme and the dynamics of the enzyme-substrate complex frequently also need to be considered if regiospecificity of reactions are to be predicted reliably. This enzymic component, added to the electronic component of the substrate reactivity to give the overall site specific reactivity, is usually referred to as an accessibility correction factor.

Methods for estimating these corrections for cytochrome P450 (CYP) isoforms have been patented by the Camitro Corporation. The use of quantum mechanical (QM) methods in metabolism prediction thus varies in complexity. The simplest use is from the examination of empirical correlations of enzymic specific activity (μmol/min mg enzyme) with a single parameter such as the energy of the lowest unoccupied molecular orbital ($E_{LUMO}$).

A good example of the use of simple semiempirical methods is reported by Jolivette and Anders (2002) for haloalkane glutathione conjugation. More elaborate is the calculation of energy barriers to the metabolic reactions in the enzyme-substrate complex using quantum mechanical calculations in combination with molecular mechanical methods, such as exemplified by the work of Ridder et al. (1999). These authors examined the 3-hydroxylation of 4-hydroxybenzoate by microbial flavoprotein monooxygenase hydroxybenzoate-3-hydroxylase. Quantum mechanical calculations were used to describe the part of the system that reacted (i.e., the substrate, cofactor, and catalytic amino acids). Molecular mechanics were used to describe the surrounding protein and solvent molecules. It was found that the activation energy of this reaction was strongly dependent on the existing hydroxy group. This group was found to deprotonate and increase the nucleophilicity at the 3-position. The study used the crystal structure of the C4a-hydroperoxyflavin intermediate from the enzyme-substrate complex. It assumed a reaction profile that involves the cleavage of the peroxide oxygen-oxygen bond and the formation of the carbon-oxygen bond between the C3 atom of the substrate and the distal oxygen of the peroxide moiety of the cofactor. Changes in the conformation of amino acids and water molecules in the active site, involving hydrogen bond interactions that stabilize the transfer of oxygen to the substrate, are indicated by molecular mechanics calculations. Replacing the hydroxy group with fluorine increased the activation barrier by about 15 kcal/mol, corresponding to a rate factor of over $10^{10}$.

In a related study, the effect on the activation energies of fluorine substituents in other positions of the ring was examined (Ridder et al., 2000). Results showed a good correlation with the logarithm of experimental rate constants. This supported the proposal that the electrophilic attack is rate limiting under physiological conditions. Such studies provide an insight into the required structural properties of substrates and the mechanism of reduction of activation energy barriers in enzymatic reactions. However, they also require considerable expertise in computational chemistry and detailed knowledge of the active site structure. As such they are not practical for most enzymatic reactions. A more empirical approach is usually required.

Despite the drawbacks described above, much simpler calculations on the substrate alone, with empirical steric corrections, may give useful indications of regiospecificity for biotransformations that do not have a strict substrate orientation preference (usually for substrates devoid of hydrogen-bonding donor/acceptor groups). Koerts et al. (1997) employed substrate site relative reactivities, calculated using frontier orbital interaction theory. A function (f) was used, which is an estimate of the effective π-electron population at a particular atom site calculated using a combination of the electron density contributions of the highest ($c_1^2$) and second highest ($c_2^2$) molecular orbitals, with a weighting based on the energy difference ($\Delta\lambda$) between them (Fleming, 1976). $c_1$ and $c_2$ are the atomic orbital coefficients for the molecular orbitals:

$$f = 2\left(c_1^2 + c_2^2 \exp(-D\Delta\lambda)\right)\Big/\left(1 - \exp[-D\Delta\lambda]\right) \tag{10.1}$$

where D is a constant (3 is used). The electron densities were calculated using semi-empirical MO calculations with the Austin Model 1 (AM1) Hamiltonian, in this instance from the AMPAC program (Quantum Chemistry Program Exchange, Indiana University, Bloomington, Indianapolis), but other programs could be used. For ortho-hydroxylation to bromine and iodine, it was found necessary to introduce correction factors, 0.4 and 0.06, respectively, described as an allowance for steric factors. The presence of hydrogen bonding groups (e.g., an aromatic hydroxyl group) can greatly affect the regioselectivity of hydroxylation reactions. For instance, CYP isoforms vary in their regioselectivities for a given substrate. In the case of quinoline, CYP2E1 produces 3-hydroxyquinoline and CYP2A6, quinoline N-oxide, and probably the 5,6-diol via the epoxide (Reigh, 1996).

## C. Empirical Modeling of Substrate Requirements: QSARs without 3D Molecular Modeling

Hansch type quantitative structure-activity relationships (QSARs) using lipophilicity measures as the only parameter have been derived for congeneric series (e.g., Gao and Hansch, 1996). For data sets of greater structural diversity more descriptors are required. Most such work has concentrated on P450 enzymes, relating the binding constants ($K_m$) or inhibition constants ($K_i$) to measures of lipophilicity (log $K_{ow}$). QSARs are applicable only to compounds in the structural space defined by the training set, limiting their usefulness as a predictive tool.

Gao and Hansch (1996) reported examples of P450 metabolism, specifically N-demethylation, where the overall rate of the reaction for the isolated enzyme, increased with increasing lipophilicity (as measured by log $K_{ow}$). Further, it was shown to be independent of the electron donating or withdrawing effects of substituents, which appeared to have approximately equal and opposite effects on the two components, substrate binding, and reaction rate. For microsomes *in vitro* the lipophilicity was a particularly significant factor.

Correlations of enzymatic parameters for series of chemicals with descriptors related to reactivity have also been published. Soffers et al. (1996) looked at the correlation of the rate of conjugation of a small series of fluoronitrobenzenes with several classes of glutathione S-transferases (GST). The reaction involves the nucleophilic substitution of fluorine activated by ortho- or para-nitro groups with glutathione. Correlations were found with both the $E_{LUMO}$ and calculated relative heats of formation of the Meisenheimer complex intermediates ($\Delta\Delta HF$). The $V_{max}$ values of purified enzymes, or human cytosol containing it, increased with the number of fluorine substituents and with decreasing calculated $E_{LUMO}$. The enzyme binding affinity as measured by $1/K_m$ increased with increasing lipophilicity (or increasing number of fluorine substituents). The turnover rate of the enzyme (measured by $V_{max}/K_m$) also increases with the number of fluorine substituents. Substitution of the fluorine at the ortho-position was favored over substitution of a fluorine at a para-position even for 2,3,4,6-tetrafluoronitrobenzene, which has a substituent ortho to the fluorine. Weak steric effects of adjacent fluorines are also seen in aromatic hydroxylations (Koerts et al., 1997).

Neural network methods for predicting whether screened molecules are CYP450 2D6 substrates have been developed by GlaxoSmithKline researchers. A 20% false positive and a 10% false negative rate were stated (Ekins and Rose, 2002).

A recent study of *in vivo* structure-metabolism relationships for substituted anilines used a principal components (PCs) analysis of quantum mechanical and other calculated physicochemical descriptors (Scarfe et al., 2002). Predictions were made using two PCs, for whether N-acetylation and subsequent oxanilic acid formation would occur or not. Subjectively defined regions on a simplified plot of PC1 vs. PC2, using the eight most significant descriptors offered a predictive capability. Only a relatively small data set of 15 compounds was analyzed. The predictive approach utilized a combination of the PC analysis and a consideration of close analogs and the influence of competing metabolic pathways. For example, 2-methyl-4-trifluoromethyl fell in the region of PC plot where oxanilic acid formation occurred. In this case it was considered more likely that the aromatic methyl would be oxidized instead of the acetyl methyl, as found for 3-methyl-4-trifluoromethylaniline. Caution is clearly required, particularly in making predictions from such analyses on a limited data set. It is important to consider substituent effects and competing pathways very carefully, rather than taking the results of a limited statistical analysis at face value. A similar pattern-recognition approach was used earlier by the same group to derive structure-metabolism relationships for Phase II conjugation reactions of benzoic acids in the rabbit and rat (Cupid et al., 1996; 1999).

## D.  3D QSAR: Molecular Field Analysis

Molecules are characterized by potential hydrogen bonding, polar, hydrophobic, and electrostatic interactions in 3D space, using 3D molecular fields. Techniques such as Comparative Molecular Field Analysis (CoMFA), which considers the 3D distribution of electrostatic and steric fields, have been applied to congeneric series of enzyme substrates or inhibitors generating 3D QSAR equations. Most examples of such applications are to modeling CYP substrate and inhibitor specificity and these have been extensively reviewed in the literature (Ekins et al., 2000; 2001; Ter Laak and Vermeulen, 2001; Ter Laak et al., 2002).

Quantitative measures of enzymatic activity such as enzyme affinities ($K_m$) or $K_i$ for inhibitors have also been correlated with the field parameters. The results may provide a useful insight to the steric and electrostatic field requirements for the effective interaction with the enzyme for a particular series of substrates or inhibitors. In common with most QSAR methods such results are not applicable to non-congeneric molecules unless they can be effectively aligned with the training set of molecules, since the position within the active site is not known. The appropriate alignment of molecules is a critical factor for CoMFA and assumes a common orientation of the substrate within the active site. Properties other than steric and electrostatic properties have been used, including reactivity based fields of $E_{LUMO}$ and energy of the highest unoccupied molecular orbital ($E_{HOMO}$). An example of a CoMFA application is the prediction of the affinity of molecules for CYP2C9 by Rao et al. (2000), specifically the inhibition constant, $K_i$.

Other measures of properties in 3D, such as Molecular Lipophilicity Potential (MLPot) and Molecular Hydrogen Bond Potential (MHBP), have been used to characterize 3D properties. They are defined for points on a molecular surface created around the molecule and calculated from the summation of contributions from the substructural fragments making up the molecule weighted by the distance function. The hydrogen bond potentials include an angle-dependent function.

An alternative approach is the use of the GRID field, a widely used computational tool to map molecular surface of drugs and macromolecules. The GRID force field includes steric, electrostatic, and H-bonding effects using different probes to estimate each of these effects. Other measures of substrate molecular interaction fields with enzymes have also been developed. The VolSurf procedure employing a GRID force field has become popular. In this approach the properties of the enzyme site is modeled by various probe molecules that represent putative polar and hydrophobic interaction sites. The VolSurf method converts the information in the 3D molecular force fields into molecular descriptors related to the size and shape to the balance of hydrophilic and hydrophobic regions. The VolSurf descriptors have been explained elsewhere (Cruciani et al., 2001). They have the advantage of computational efficiency and have been recommended for quantitative structure-property relationship studies, particularly when large numbers of compounds are involved.

Correlations of enzymatic parameters such as $K_m$ and $V_{max}$ with more complex descriptors relating to molecular surface properties or other properties in 3D have been used for QSAR models. Unfortunately, although such equations may have predictive utility they are difficult to interpret mechanistically because the descriptors selected have rather obscure meanings. For example, in a QSAR study of a series of 4-substituted phenols $V_{max}$ and $K_m$ values for O-glucuronidation were measured for expressed human UGT1A6 and UGT1A9. QSAR models derived using Molecular Surface-Weighted Holistic Invariant Molecular (MS-WHIM) descriptors (Ethell et al., 2002). Multivariate analysis of $K_m$ values using a statistically determined subset of 102 MS-WHIM descriptors resulted in linear functions of 4 of these descriptors for each of the 2 enzymes. The descriptors have obscure meanings and it is not possible to derive useful mechanistic information from the equations. Consequently, their application is confined to phenols similar to those in the data set used to derive them. For the generation of more useful QSARs, the need to increase the structural diversity of the training set was noted by the authors. The use of simpler descriptors can be more informative. In the same paper, a simple plot of UGT1A6 $V_{max}$ (but not $K_m$) values (nmol/min/mg) representing turnover rate against molecular volume showed a negative correlation. Their analysis suggested that phenols with very bulky substituents would be bound, but not glucuronidated and inhibit the glucuronidation of other phenols with smaller substituents, as observed experimentally. In another study, of O-glucuronidation of 19 indolocarbazole analogs, absence of glucuronidation observed in 9 of the compounds, was related to a molecular dimension in one direction exceeding a certain limiting value (Takenaga et al., 2002).

## III. PHARMACOPHORE MODELING

Pharmacophores are descriptions of the structural requirements of small molecules to fit into enzymes or receptors. In the present context we are concerned with pharmacophores for enzyme substrates. The term has tended to be used generally regardless of the presence or absence of pharmacological activity. Pharmacophore modeling in metabolism has largely concentrated on CYP isoforms and has been described and reviewed recently, for example, by Ter Laak and Vermeulen (2001) and by De Groot and Ekins (2002). In the simplest treatments it is assumed that all substrates will be oriented in a similar manner in the active site of the enzyme. Pharmacophore models are built up by examining the common geometric conditions that are required for a satisfactory superposition of substrates onto a template substrate, usually chosen by virtue of its large size and relative rigidity. For substrates the sites of known metabolism are usually superimposed at the beginning of pharmacophore construction, together with chosen common features of the structures. The superimposed structures may then suggest features such as hydrogen bond donors, hydrogen bond acceptors, hydrophobic centers, or the presence of centers of negative or positive charge. They may allow definition of distance or angle constraints between them. It is usually assumed that the geometry in the active site of the enzyme is the same as that obtained from theoretical energy-minimization procedures, or alternatively several conformations that are employed within a fixed energy of the theoretical minimum energy conformation are considered (Ter Laak and Vermeulen, 2001; Ter Laak et al., 2002). When several conformations for each substrate are allowed several pharmacophore models or hypotheses may be generated, that with the lowest energy cost is usually selected. The resultant models require validation. Recently pharmacophores have been defined using a 3D QSAR approach using binding or inhibition constants, $K_m$ or $K_i$, as measures of activity. Computer programs such as Catalyst (Molecular Simulations Ltd.) are available to derive pharmacophore hypotheses in this way (Ekins et al., 2000; 2001).

Several different pharmacophore models of particular P450 isoforms have been proposed, and are reviewed by De Groot and Ekins (2002), Ekins et al. (2001), and Ter Laak and coworkers (2001, 2002). These generally show an evolution of revision with increasing complexity. The simplest methods employ a manual overlay of chemical structures using molecular modeling programs and minimum energy conformations. More recent analyses have employed several con-formations and a 3D QSAR approach using $K_m$ or $K_i$ values. For example, for CYP3A4 the program Catalyst generated a pharmacophore including a hydrophobic area, a hydrogen bond donor, and two hydrogen bond acceptor features with defined positions in 3D space (Ekins et al., 1999).

Features of protein structure have been incorporated into pharmacophore models (e.g., for CYP2D6), and site-specific mutagenesis has indicated a frequent role of binding to the Asp 301 group. The aspartate group is attached to the I-helix of the enzyme positioned over one half of the heme moiety. Accordingly, the aspartate group attached to the I-helix positioned over a heme ring was used as positional and steric constraints (De Groot et al., 1999a; 1999b).

## IV. EMPIRICAL 3D QSAR APPROACHES

The Computer-Optimized Molecular Parameterized Analysis of Chemical Toxicity (COMPACT) methodology is based on electronic and molecular shape parameters. It predicts whether specific

P450 isoforms would be induced by particular substrates, leading on to toxicity, notably carcino-genicity. The COMPACT method has been used primarily to identify the carcinogenicity potential of molecules. This is based upon the recognition by COMPACT of potential metabolic activation by, or induction of, P450 1A and to a lesser degree P450 2E1 (Lewis et al., 2002a). Lewis characterized CYP1A substrate/inducers by low values for the difference between $E_{HOMO}$ and $E_{LUMO}$ ($\Delta E$) and as being flat planar molecules as defined by large values of molecular area (molecular depth)$^2$ (Lewis et al., 1989). In later work, combinations of three descriptors (area/depth$^2$, $\Delta E$, and diameter) were reported to show discrimination among substrates of CYP2E, CYP1A, and CYP2B/C (Lewis, 1992). The work has been extended to include other parameters such as molecular length, length/width$^2$, and log $K_{ow}$ in QSARs for CYP induction potency (Lewis et al., 2002b).

Recent work by Lewis has also looked at interactions in the enzyme-substrate complex via homology modeling discussed below (Lewis et al., 1999). Hydrogen bond and aromatic ring-ring interactions between substrate functional groups and amino acid components around the active site are identified for orientations of the substrate giving the observed position of oxidation. Differences between the activity of expressed enzymes between different species are attributable to differences in the enzyme structures. For example, the metabolism of coumarin shows extensive differences between species. Coumarin 7-hydroxylation occurs extensively in Old World primates such as the baboon and cynomulgous monkey and also in man, but to a very small extent occurs in New World primates, such as the marmoset and squirrel monkey. This is partly attributable to differences in the amount of expressed coumarin 7-hydroxylase activity between various species of primate.

The lipophilicity parameter, log $K_{ow}$, is generally greater than zero for CYP substrates or inducers. This is attributable, at least in part, to the fact that in most eukaryotic cells P450 is embedded in a phospholipid membrane across which a substrate has to travel (Lewis et al., 2001).

Other parameters, such as numbers of hydrogen-bonding groups and MO energies, have been introduced into QSARs for binding and chemical reactivity (Lewis et al., 2002b). Lewis et al. (2001) presented QSARs for the free energies of binding of substrates to human CYP isoforms and P450 isoform induction potencies. These indicate quantitative dependencies on, for example, hydrogen bond donor and acceptor ability, molecular mass and dipole moment, molecular length/width ratios, or number of potential pi-pi stacking interactions in the enzyme/substrate complex. However, the ratio of the number of data points to variables (1.75 to 4) is very low, lower than the recommended minimum of five. The relationships must be considered in need of further validation. The descriptors were chosen using insights from mechanistic knowledge. Because the overall enzymic conversion rate is also dependent on substrate reaction rates, the relationships are of limited utility for predicting actual rates of metabolism. The quantitative significance of such processes on biotransformation is also dependent on the relative concentrations of the different isoforms in the particular tissues, which varies between species and individuals and is dependent on factors such as diet and exposure to other chemicals.

Factors that contribute to binding to P450 enzymes have been further considered by Lewis and Dickens (2002). Distinguishing characteristics of P450 isoform substrates (molecular weight range, chemical classes, shape, etc.) have been given in several such papers (Lewis, 2001; 2003). The use of parameters from protein-substrate homology modeling such as the number of hydrogen bonds and the number of pi-pi stacking interactions were found to be useful parameters with estimated contributions of –2.0 and –0.9 kcal/mol, respectively. Molecular shape and electronic properties, primarily the energy difference between occupied and unoccupied molecular orbitals as well as acidity, and basicity measures have been found to be important factors determining P450 specificity. A decision tree for evaluating specificity for substrates based on these factors has been presented by Lewis and Dickens (2002).

Equations for the potency of P450 inhibitors were also presented by Lewis et al. (2001). Such relationships relate to metabolism in *in vitro* systems rather than whole organisms. They represent one component that needs to be integrated with absorption and distribution factors.

## V. USE OF MOLECULAR MODELS OF PROTEINS: HOMOLOGY MODELS AND DOCKING

MO calculations on substrates can indicate the most susceptible site for a particular transformation. When located in the active site of an enzyme, the properties of the enzyme are also important. Specifically these include the steric constraints of the docking site, which influence the proximity of the reactive groups to the various potential sites of reaction on the substrate. Also important are the electronic characteristics of the enzyme reactive groups, which influence the extent of discrimination between different sites. It is frequently necessary, particularly when the enzyme site is small or has strong directional binding, to consider both reactivity and the interactions in the active site. Although these may be implied from substrate pharmacophores, a more detailed understanding and a potentially broader predictivity is in principle achievable from active site modeling.

Because details of the active site of many enzymes are frequently not known, use has been made of the structures of bacterial P450 proteins such as CYP102, CYP101, CYP105, CYP107A, and CYP108, as well as structures of eukaryotic CYP55 from *Fusarium oxysporum* and CYP119 from the Archaeon *Sulfolobus solfataicus*. For example, a model of the active site of CYP2D6 has been formed from the structures of CYP101, CYP102 and CYP108, while 2C9 was formed from the structure of CYP2C5/3 and the B′-helical region of CYP102. Specialised homology modeling programs such as Modeler (e.g., as implemented by Quanta97) are used for this purpose. Although sequence alignments are done automatically, adjustments can be made to take into account the results of site-directed mutagenesis experiments and clear errors in the structures. The methods are described in detail in the literature (Sali and Blundell, 1993; Ter Laak et al., 2001; 2002).

Homology modeling evidently requires considerable expertise to be successful. Several slightly different models of the protein may be produced by these methods. The structures are energy minimized by molecular mechanics methods. Once a satisfactory homologos structure has been achieved, experiments are conducted in which substrates are docked into the binding site. These allow important interactions between amino acid residues and substrates to be discerned. For example, protein homology models for CYP2B1, 2B4, and 2B5 were constructed from the crystal structure of mammalian CYP2C5. Molecular modeling and site-directed mutagenesis experiments showed the importance of particular residues, Ile114, Ser294, Ile363, and Val367, in the active site of CYP2B1 (Spratzenegger, 2001).

Recently, homology modeling has been combined successfully with the use of molecular orbital calculations of reactivity. For example, De Groot et al. (1999a, 1999b) combined the results of MO calculations and molecular modeling in rationalizing the positions of hydroxylation of debrisoquine and related molecules by CYP2D6.

Debrisoquine is hydroxylated by CYP2D6 at the alicyclic C4-position (see Figure 10.1). Phenolic metabolites of debrisoquine have also been reported. Lightfoot et al. (2000), using a molecular

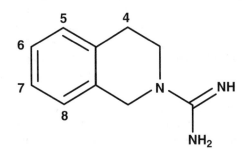

**Figure 10.1**   Structure of debrisoquine.

model of the active site of CYP2D6 and calculated energies of interaction, showed that while the adoption of different binding orientations was required for hydroxylation of the aromatic ring; the energetics of the direct mechanism of aromatic hydroxylations were unfavorable. A mechanism of hydroxylation based on benzylic radical formation with spin delocalisation on to the aromatic rings and subsequent reaction was proposed. In another study, Ter Laak et al. (2002) described the disadvantages and limitations of pharmacophore modeling, protein homology, and substrate docking techniques used to predict drug metabolism.

## VI. COMBINED APPLICATIONS OF PHARMACOPHORE, MOLECULAR MODELING AND MO CALCULATIONS

Analysis of the structures of substrates for a given enzyme may allow a pharmacophore to be derived. This describes general structural features of substrates and distances from key atoms. Using this pharmacophore different possible sites for metabolism of a potential substrate may be identified. For some metabolic processes, using quantum mechanical techniques, it has been possible to calculate relative energies of formation of the key reactive intermediates. This may be used to predict a favored metabolite. An alternative is to use rules based on generic structure-metabolism relationships, such as for N-dealkylations which become less favorable the greater the steric bulk of the alkyl group (Upthagrove and Nelson, 2001). For P450-mediated metabolic transformation, the recent availability of crystal structures of non-mammalian soluble CYP enzymes, starting with P450cam, has facilitated homology modeling of CYP enzymes. Using information on the orientation of model substrates as a template for other substrates, it is possible to examine potential favorable and unfavorable interactions for alternative metabolic reactions.

De Groot et al. (1999a) reported that MO calculations alone could not predict the metabolic site of oxidation in their data set because oxidation next to a nitrogen or oxygen atom was most favored. However, when combined with the distance constraints imposed by CYP2D6 (a distance of 5, 7, or 10 Å) between the site of oxidation and the basic nitrogen atom, it was possible to identify the metabolic products correctly based on the relative stability of the various hydroxylated species. De Groot et al. (1999a; 1999b) combined the use of simple 3D pharmacophore descriptions, MO calculations to determine reactive sites and molecular modeling of the substrate within the active site, to provide qualitative predictions of the likely metabolism of various CYP isozymes such as CYP2D6. For that isoform, the predicted metabolism was largely supported by the experimental results, although there were differences in some instances ascribable in part to alternative mechanisms or new pathways not considered in the modeling. The pharmacophore model describes the structural requirements of the substrates in general terms (e.g., distance constraints, the presence of hydrophobic groups, or particular binding groups in particular locations relative to the substrate's site of metabolism). Reduction in binding and turnover due to unfavorable steric interactions at the active site, or electronic effects of groups are not included. 3D modeling of a potential substrate in the enzyme active site, guided by an orientation based on a model substrate, particularly when coupled with calculations of the free energies of interactions, allows identification of unfavorable intermolecular interactions.

MO calculations of the substrate and products of metabolism, or transition states, can identify the most likely intermediates and products. When intermolecular effects result in preferred orientations within the active site or a poor fit, the predictions from MO calculations may not be observed experimentally.

Recently, pharmacophores have been defined from homology models of CYP isoforms in the ALMOND program (Pastor et al., 2000). This method is based on an assessment of complementarity between binding properties of substrates and the enzymic binding sites. It assumes that within the enzyme-substrate complex, the site of reaction in the enzyme (the reactive oxygen atom) coincides in space, to a useful approximation, with the site of oxidation on the substrate in the substrate. The distribution of protein hydrogen bond donor or acceptor groups and hydrophobic regions within the active site relative to the site of reaction are defined; this is then compared with the potential substrate's distribution of such groups relative to the various potential reactive sites to find the most complementary match. The ALMOND technique, is well described and illustrated by Boyer and Zamora (2002) for oxidation of two alternative sites on a steroid molecule. Their assessment of the method showed that in more than 50% of all CYP2C9 catalyzed reactions of 43 substrates the methodology gave a correct prediction of the most important observed site of oxidation, and in 90% of cases the predicted favored oxidation site was included in the three most important observed oxidation sites. This method appears to show some promise but is probably best used in combination with other predictive methods to achieve a consensus prediction. Such approaches are relatively complex computationally and depend on good models of enzyme active site structures.

## VII. PREDICTING RATES OF METABOLISM FOR SPECIFIC ENZYMES

When determined kinetically, the relative rates of metabolism are expected to be controlled by the activation energy of the reaction, which in turn is related to the relative energies of transition state and reactants. When under thermodynamic control, the relative energy of reactants and products determine the balance of products. In the simplest approach, factors such as the relative electron density on aromatic rings are assumed to determine the relative energies of transition states, the reaction rates and the site of reaction (e.g., the work of Koerts et al. [1997] described above). In the most accurate approach calculations are performed using a model of the reaction within the enzyme active site such as in the work of Ridder et al. (1999). Quantum mechanical calculations of the complexity of the intermediate of the substrate itself have also been employed. De Groot et al. (1999a, 1999b) used MO calculations to estimate the stabilities of reactants, intermediate radicals, and hydroxylation products. This allowed the prediction of the thermodynamically and kinetically favored products of CYP2D6 oxidation. Such approaches have been developed commercially by the Camitro Group (Ewing et al., 2002).

Bursi et al. (2001) reported two methods to calculate the stability of testosterone-like steroids. These were the use of a decision tree and molecular descriptors or quantum mechanical methods. For satisfactory accuracy, Bursi and colleagues (2001) had to use a 3-21G basis set with spin correction and equilibrium geometries. This required 12 hours of computation for reactants and products. Optimization of transition state geometry was also required. The simpler decision tree analysis approach indicated that descriptors such as the volumes and, to a lesser degree, the shape were important. Correlations of calculated and experimental rates of metabolism were reported.

Jones et al. (2002) have shown that the readiness of functional group oxidation generally follows the relative stability of the radical that is formed (e.g. N-dealkylation > O-dealkylation > secondary carbon oxidation > primary carbon oxidation). They also demonstrated that useful correlations with the rates of product formation can be obtained from MO calculations. Jones and colleagues (2002) used semiempirical MO calculations ("in combination with experimental data") to predict activation energies for aromatic and aliphatic hydroxylations. Correlations of activation energies with structural descriptors have been used to develop models to predict metabolic stability and regioselectivity of drug molecules and such methods have been patented (Ewing et al., 2002). Such calculations appear to be most useful when steric effects due to the substrate or enzyme site are not significant.

A good example of the influence of structure on metabolism and the application of quantum mechanical/molecular mechanics methods is provided by the immunosuppressive drugs sirolimus and everolimus, the structures of which differ only by the presence of the 2-hydroxyethyl group in the latter (see Figure 10.2). Sirolimus has a disadvantage of low and highly variable oral

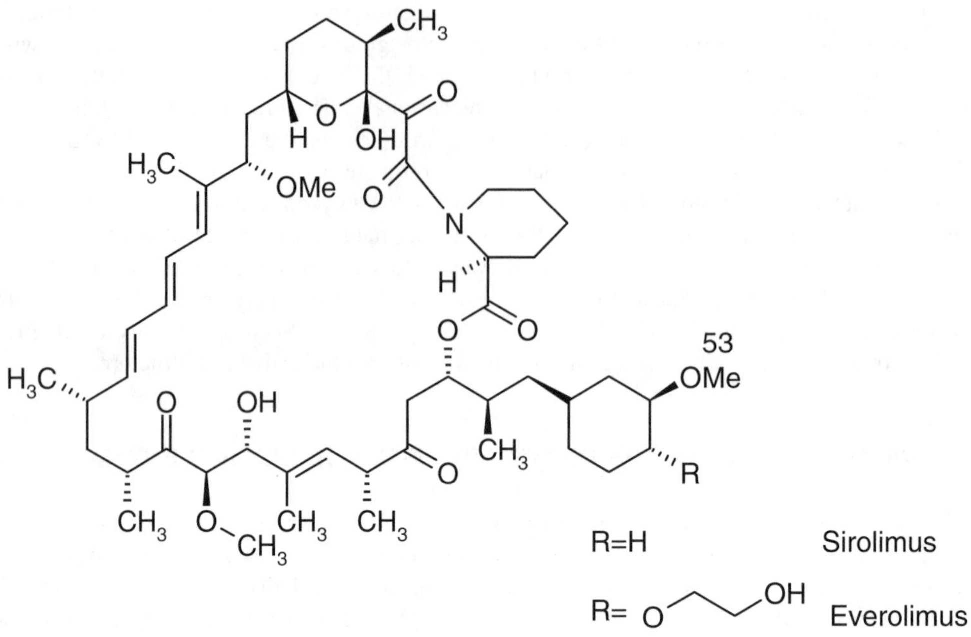

**Figure 10.2**  Structures of the immunosuppressive drugs sirolimus and everolimus.

bioavailability attributable to CYP3A4/5-mediated intestinal and hepatic first-pass metabolism. Hydrocarbon hydroxylation and O-dealkylation are the main CYP-dependent metabolic processes. The total clearance rate of the everolimus metabolites is threefold lower than for sirolimus. The major metabolism reaction in sirolimus, 53-O-demethylation, is slower by a factor of 15 in everolimus.

Using quantum mechanical calculations of the energies of the free radicals and the unreacted substrate and active site docking experiments and molecular dynamics, Kuhn et al. (2001) estimated the relative activation energies for the hydrogen abstraction reactions resulting in hydroxylations and O-dealkylations of the macrolide sirolimus. The molecular dynamics simulations were based on a homology model of CYP3A4, which included the heme structure and the iron attached oxygen atom (O*). The influence of the C-H...O* bond angle was considered. Qualitative prediction of the regiospecificity was in good agreement with experimental results. The much slower rate of O-demethylation of everolimus is attributable to a less favorable mean C-H...O* bond angle and to a stabilizing hydrogen bond interaction in the CYP3A4-substrate complex in sirolimus, absent in everolimus. The latter involves the hydroxyl group adjacent to the methoxy group, with the negatively charged ferryl oxygen atom of the CYP3A4 heme group.

## VIII. PREDICTION OF METABOLIC PROFILES *IN VIVO*

Factors affecting metabolism and disposition of drugs or xenobiotics *in vivo* have been amply described elsewhere (Timbrell, 1991). They may be divided into biological factors and molecular factors.

Regarding biological factors, chemical metabolism is dependent on a wide range of genetic and environmental factors that may be considered as sources of intraindividual, interindividual, and interspecies variation. Intraindividual factors include diet, coexposure to xenobiotics or drugs, health status, or disease. Interindividual factors include age, sex, and genetic makeup, including race.

Such factors determine the distribution and type of enzymes. For chemical metabolism taken as a whole, the extent of variation decreases in the order:

Interspecies > Interindividual > Intraindividual

The nature of the exposure, route (e.g., skin, oral, intraperitoneal injection), and duration (exposure vs. time profile) may also be important. The overall fate may be primarily determined by the stability and reactivity of the drug within the biological environments encountered.

Since chemicals are primarily metabolized in enzyme-rich tissues such as the liver and kidney, the extent of metabolism depends on the exposure of these various tissues to the chemical (i.e., for an oral route on the extent of intestinal absorption). The intestinal absorption into the systemic circulation and absorption into or out of other tissues may be influenced by carrier or reflux enzymes (e.g., P-glycoprotein enzymes in the intestinal wall that capture and return chemicals to the intestinal lumen). Similarly, there are carrier enzymes involved in transporting chemicals into the bile that have a marked specificity. The extent of metabolism and proportion of second and third generation metabolites increases with the time in contact with important organs such as the liver. This may be enhanced by reabsorption from the gut after excretion in the bile, which may be subsequent to metabolism (e.g., hydrolysis of glucuronides in the gut).

Testa and Cruciani (2001) divided molecular factors into the effects of global properties of the molecule as a whole, such as solubility characteristics (the octanol-water partition coefficient) or molecular size. These determine distribution within an organism and proximal molecular factors such as substituents or chemical functionality that modify the rate of functional group metabolism (e.g., changes in local electron densities or MO energies).

Many studies predicting metabolism have concentrated on quantitative aspects (e.g., relating the rate characteristics or proportion of a particular metabolic reaction to the experimental or calculated properties of a molecule), or the characterization of the structural requirements (i.e., structure-metabolism relationship for a particular enzyme or class of enzyme).

The sequences of biotransformations of a chemical can be described as a metabolic tree diagram. Chemical biotransformations may occur consecutively on a molecule, producing new generations of molecules. The proportions of subsequent generations of metabolites diminish after each step due to competition from other reactions or processes. The proportions of second generation metabolites may be undetectably small if the preceding biotransformation has a low fractional yield. For a highly efficient first generation biotransformation with a high fractional yield, the next generation of metabolites will be more important. Phase II reactions, involving the conjugation with hydrophilic biomolecules, result in increased water solubility and increased likelihood of excretion as urinary metabolites. Further biotransformations of Phase II conjugates are not frequently observed, although they are occasionally significant and may have toxicological consequences (e.g., some acyl glucuronides may covalently bind to proteins directly or indirectly via ring opening, which may contribute to hypersensitivity reactions (Wang and Dickinson, 1998). Metabolic biotransformations are generally facilitated by some degree of hydrophobicity, important in enzyme binding.

## IX. PREDICTING CHEMICAL BIOTRANSFORMATION PROFILES: COMPETING BIOTRANSFORMATIONS

Many biotransformations are simple functional group conversions, with rates dependent on a range of properties of the molecule and the organism. The various possible biotransformations, including spontaneous reactions, can be viewed as competing reactions; kinetically slow biotransformations are frequently only apparent in the absence of alternative rapid biotransformations. Noncovalent protein binding of chemicals may reduce the availability for enzymic metabolism and

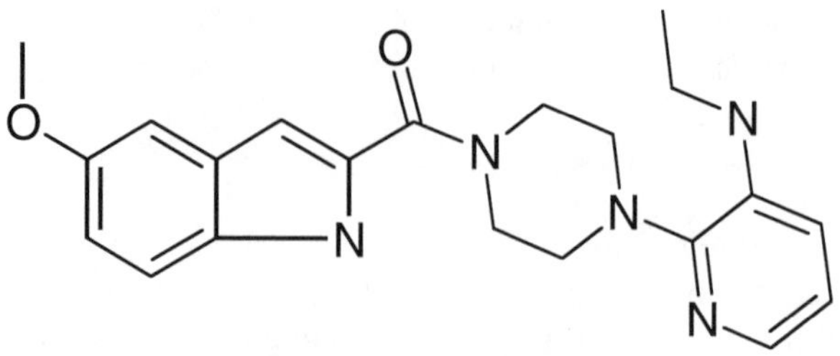

**Figure 10.3** Structure of atevirdine.

the absolute rates. When metabolic biotransformations can occur rapidly and in high yield, the observed products in the urine or feces may reflect several consecutive biotransformations.

## A. A Well-Documented Example of Metabolism: Atevirdine

Atevirdine mesylate is a non-nucleoside inhibitor of human immunodeficiency virus-1 (HIV-1) reverse transcriptase (see Figure 10.3). Its metabolism in the rat was described by Chang et al. (1998). The main metabolites were identified to result from N-dealkylation of the amine nitrogen or O-dealkylation of the alkyl ether followed in the latter case by hydroxylation of the indole ring in the 6-position. N-dealkylation and O-dealkylation, preceded by hydroxylation at the adjacent aliphatic carbon atom, are generally rapid and very common metabolic reactions (Testa, 1995). The metabolites are excreted as glucuronic acid conjugates. Hydroxylation of the pyridine ring occurs to a minor extent. A sex difference in the metabolism was apparent; N-desethyl 6-O-glucuronic acid atevirdine was the main metabolite in male rats, whereas O-desmethyl 5-O-glucuronic acid atevirdine was predominant in females. This gender difference was cited as being consistent with the involvement of the male-dominant CYP3A in the N-de-ethylation of atevirdine.

Prediction of the major metabolites of atevirdine requires the correct prioritization of the alternative metabolic processes. Differences in the distribution of enzymes between sexes (or equivalent information) need to be taken into account. In addition, the ability to assess the importance of metabolism following the first metabolic step, second-generation metabolites is important.

Delavirdine is related to atevirdine and has a methyl sulfonamide substituent on the indole ring, rather than the methoxy group present in atevirdine. In delavirdine, hydroxylation is observed on the pyridine ring, at C-4' and C-6', and not on the indole ring. This is attributable to the relative electron's withdrawing nature of the sulphonamide group compared with a methoxy group (Chang et al., 1997, 1998). Prediction of metabolism requires substituent effects to be taken into account (e.g., by empirical rules or by computational means).

A particular problem for prediction arises when reactive metabolites are produced. This is because they are short-lived and undergo rapid spontaneous reactions, in addition to enzyme-catalyzed reactions. Reactions of reactive metabolites with proteins or other biomolecules may make a very important quantitative contribution to the fate of the compound; however, such products, or their reactive precursor, are not revealed by conventional chemical analysis of urine or feces. Intramolecular reactions following generation of reactive sites in part of the molecule are not infrequently observed. For example, prinomide is an anti-inflammatory agent that contains a reactive pyrrole ring easily oxidized to reactive intermediates (see Figure 10.4). Major metabolites identified in the urine of experimental animals were the phenyl ring 4-hydroxy metabolite and a bicyclic spiro derivative that has been suggested to be formed via autooxidation of the pyrrolin-2-one metabolite (Dalvie et al., 2002). In addition, a fused succinamide metabolite was also detected in the urine, the mechanism of formation for which is unclear. The computer-based prediction of such metabolites would require a system with considerable knowledge of chemical reactivity.

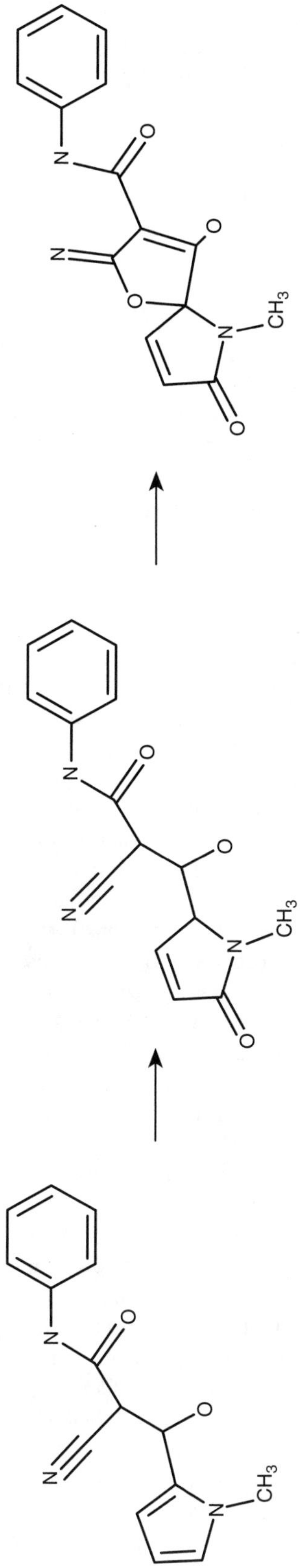

**Figure 10.4** Oxidation of prinomide to reactive intermediates.

A further example of the importance of non-enzymatic reactions is the metabolism of cyclo-phosphamides in humans (Joqueviel, 1998). Ring-opening reactions with spontaneous elimination of acrolein and a range of hydrolysis reactions at the phosphorus center occur.

## X. PREDICTION OF METABOLIC PROFILES AND CHEMICAL FATE WITHIN ORGANISMS: EXPERT SYSTEMS AND DATABASES

### A.  Expert (Knowledge-Based) systems

Several expert systems have been developed for the prediction of chemical fate within organisms, or primarily metabolism. Three of the best known are META (Klopman et al., 1994; Talafous et al., 1994), MetabolExpert (Darvas et al., 1999) initiated as a research project by CompuDrug in 1985, and a more recent program METEOR (LHASA Ltd., University of Leeds, U.K.). A formal comparison of the predictions of these systems does not appear to have been performed and published, although the predictions of META, MetabolExpert, and several human experts for a drug were presented for comparative purposes (Wilbury, 1999).

### 1.  MetabolExpert

MetabolExpert was originally the product of a research project initiated by CompuDrug in 1985. The project aimed to model chemical transformations in a living system, or in the environment, using an expert system approach. The expert system was released commercially in 1987. Several descriptions of the MetabolExpert system have been published (Darvas et al., 1987; 1988; 1999) and only a summary is given here. In its most recent form MetabolExpert (Version 10.1) comprises a family of expert system programs, each composed of one or more databases and prediction tools. Database modules are available for mammalian metabolism, plant metabolism, photodegradation chemistry, and soil degradation chemistry. For mammalian metabolism, the biotransformation database is reported to contain 179 biotransformations (Darvas et al., 1999); 112 are derived from Testa and Jenner (1976) and 67 are based on frequently occurring metabolic pathways (Pfeiffer and Borchert, 1963). The basic biotransformations are divided into Phase I and Phase II. The input to the system is the structure of the compound drawn graphically and the output is a predicted metabolic tree. Chromatographic (high performance liquid chromatography [HPLC]) retention times of the predicted metabolites can be calculated.

The biotransformations are implemented as rules based on the presence and absence of substructures. The rules act on the original query molecule to produce a first generation of metabolites and the process is repeated on the generated metabolites to produce a second generation and so on. No further metabolites are generated from a Phase II metabolite because this is regarded as hydrophilic and excretable. The overall process can be terminated when a set number of metabolites have been produced. No indication of the relative importance of the metabolites appears to be given.

Darvas et al. (1999) and Tarjanyi et al. 1998 presented an illustration of the predictions of MetabolExpert for the drug (-)-diprenyl. A metabolic tree was shown including 4 generations and 36 different metabolites. For 25 of these metabolites experimental retention times were reported. It is not clear whether all these 25 metabolites were observed in experimental metabolism studies. Most metabolic studies report relatively few metabolites; minor ones may not be detected or identified and their observation may be dependent on the experimental conditions used in the analysis. An important predicted metabolite of deprenyl, benzyl methyl ketone, was reported to be observed only when an appropriate HPLC extraction system was used (Darvas et al., 1999). Bencze et al. (2000) have illustrated the use of the system in conjunction with the related HazardExpert system in a hazard assessment of ethylene oxide.

## 2.  META

The META system was developed by Klopman and coworkers at Case Western Reserve University, Cleveland, OH. It has been described by the developers and by others (Klopman et al., 1994; 1997; 1999; Klopman and Tu, 1999; Talafous et al., 1994; Langowski and Long, 2002). Molecules can be drawn in graphically, entered as a Simplified Molecular Line Entry System (SMILES) or Klopman Line Notation (KLN) code, or supplied as a .MOL file. Similar to MetabolExpert it has a dictionary of metabolic transformations. In a description published in 1999 (Klopman and Tu, 1999), the metabolism dictionary was described as containing 1118 transforms covering the activity of 30 different enzyme systems and a stability module containing 326 transforms. To take into account different molecular environments of a given chemical functionality, several transforms cover a particular type of metabolic transformation. For example, 28 transforms model O-dealkylation of aromatic and aliphatic ethers (e.g., the conversion of codeine to morphine and phenacetin to acetaminophen), 16 transforms model S-dealkylation, and 80 transforms model N-dealkylation. The system also assesses whether the products contain unstable chemical functionalities and generates decomposition products using a stability module.

Klopman and Tu (1999) report that simple quantum mechanical calculations are used to detect unstable products. Only the stable molecules are displayed unless the user requests to see the intermediates. Unlike MetabolExpert, each biotransformation receives a numerical priority value based on its prevalence compared with other biotransformations. The priority number scale is 1 to 9, where 1 stands for the fastest reaction and 9 indicates the slowest. Once the primary metabolites have been evaluated, the program can be used to proceed to evaluate the second generation of metabolites by treating each of the first generation metabolites as parents and generating a metabolic tree. When the program identifies that a metabolite is very soluble in water no further metabolites are generated from it. Klopman and Tu (1999) report that information on the transformation used to generate a predicted metabolite, including literature references and the enzyme responsible, are provided by the program.

The program is reported to carry out simple Hückel molecular orbital calculations to determine the relative sensitivity of aromatic carbon atoms to oxidation and the relative stability of keto and enol tautomers. Klopman et al. (1999) have reported that for polycyclic aromatic hydrocarbons, adequate reactivity is an essential but not sufficient condition for enzyme catalyzed reaction. The accessibility of the reactive site (i.e., the absence of steric hindrance) was also found to be important. Genetic algorithms have been used to optimize the performance of the biotransformation dictionary by treating the initial priority scores set by "expert" assessment as adjustable parameters (Klopman et al., 1997).

For the metabolism of the molecule X (see Figure 10.5) used as in a recent comparison of several structure-metabolism relationship database systems (Wilbury, 1999), META produced the following pathways and rate priority numbers (in brackets):

1.  O-dealkylation of the methoxy group (6)
2.  Formation of N-methyl (5), sulphate (5), hydroxyl (6) and glucuronide (6) conjugates of the aromatic amino group (6)
3.  Reduction of the aromatic nitro group to a nitroso group (5)
4.  Aromatic hydroxylation (7)
5.  Hydrolysis and opening of the pyridone ring (6)

## 3.  METEOR

The METEOR expert system, first marketed in 1999, is a newer development than MetabolExpert or META. It provides qualitative predictions of metabolic fate in mammals and the likelihood of individual biotransformations. The system, developed and marketed by LHASA Ltd., evolved out

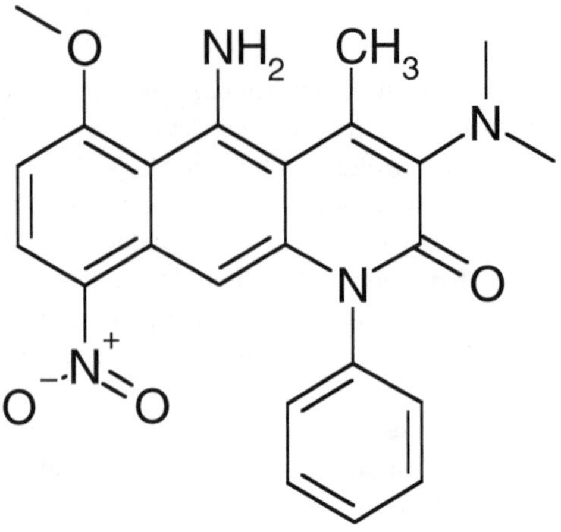

**Figure 10.5**   Structure of molecule X.

of the Deductive Estimation of Risk from Existing Knowledge (DEREK) system for toxicity prediction (Sanderson and Earnshaw, 1990). It uses a knowledge base of structure-metabolism rules describing the biotransformations. However, each of METEOR's biotransformation rules are carefully defined generic reaction descriptions rather than simple functional group transformations. For example, a given bond in an allowed structure may be defined as a single bond in a ring of size five, six, or seven which is not fused to another ring.

If all theoretically possible biotransformations are shown for each consecutive product a combinatorial explosion of products occurs. MetabolExpert and META terminate a sequence when a Phase II metabolite was produced or when the product was predicted to be very water soluble, respectively. METEOR has facilities to apply a combination of several prioritization methods to identify the most likely metabolites. First, the likelihood of the biotransformations are qualitatively described, for example, as probable, plausible, equivocal, or doubted. A filter is applied to display only those at or above a user-selected level of likelihood (plausible is the default level). Second, the likelihood of a biotransformation is defined differently according to the octanol-water partition coefficient of the substrate, which is divided into a value of high, medium or low. Log $K_{ow}$ is estimated for every submitted structure or predicted (Phase I) metabolite using either a call to an external program (ClogP for Windows, BioByte Corp.) or a less accurate internal calculation using the Moriguchi method (Moriguchi, 1992). Generally enzymic biotransformations become less likely with increasing hydrophilicity (Testa et al. 2000a; 2000b; Testa and Cruciani, 2001), so a biotransformation is ranked plausible for a lipophilic substrate (log $K_{ow}$ >2) and ranked doubted for a hydrophilic one (log $K_{ow}$ < 0). Metabolism generally proceeds to make more water soluble metabolites with a resultant reduction in rates of metabolism for many subsequent biotransformations.

Third, using the default settings, biotransformations are not produced from predicted Phase II metabolites because they are generally too hydrophilic and tend to be excreted. Fourth, rules exist defining the relative likelihood of pairs of biotransformations. For example, N-dealkylation is more likely than O-dealkylation, and in monosubstituted benzenes, para-hydroxylation > ortho-hydroxylation and ortho-hydroxylation > meta-hydroxylation. When two or more biotransformations in such a defined relationship are possible for a given substrate, the biotransformations are assigned to different levels or tiers reflecting this ordering. By only accepting for display the biotransformations at the highest, most likely, level, less probable metabolites are removed from the results display.

In a recent version of the program the likelihood of many Phase I biotransformations is reduced after the first generation. This action reflects the lower probability of second and subsequent generation metabolites being produced in significant quantities. The assigned likelihoods are assigned according to the context of the biotransformation. Species variation or other physicochemical parameters may also influence the likelihoods of biotransformations and in some instances these have been introduced as modulators of the predicted likelihood.

The derivation of likelihood levels for each metabolic step employs a form of reasoning. METEOR uses a system of non-numerical argumentation as a basis for its assignments of likelihoods combined with the rules describing the absolute and relative likelihoods. Two publications describe the underlying theory and its application within METEOR (Judson et al., 2003; Button et al., 2003). In Version 6.0 released in 2002, METEOR has over 300 relative likelihood (relative reasoning) rules, over 1100 absolute likelihood (absolute reasoning) rules, and over 250 biotransformations that include comments and references. The system is in the process of expansion, testing (validation), and refinement, and new releases are produced annually.

## B. The Use of Databases for Predicting Metabolism

Several databases of published biotransformations are commercially available, such as Molecular Design Ltd's Metabolite and the Accelerys' Metabolism Database (formerly produced by Synopsis). The former is quite extensive, and contains *in vivo* and *in vitro* biotransformation summaries from the literature, while the latter has as its core information based on the U.K. Royal Society of Chemistry's Biotransformations series (Hawkins, 1988–1996) supplemented by additional data from the literature. Both systems are searchable by reaction type. The intelligent use of such databases provides much valuable information on likely metabolic profile.

By searching for a particular biotransformation and dividing the number of occurrences with the number of times the relevant reactant structural fragment occurs for all its biotransformations, an occurrence ratio is obtained that is a crude measure of the relative likelihood of that biotransformation. This approach applied to the MDL Metabolite program and its implementation has been described (Boyer and Zamora, 2002 and Boyer, 2002). The program produces biotransformation occurrence ratios for each site in a molecule that is present (based on a local substructure). This allows an assessment of the regiospecificity and comparison with alternative biotransformations. Boyer (2002) advocated that a consensus approach, combining the results from all the different available methodologies, quantum mechanics, pharmacophore matching, etc. should be used to give the most reliable prediction of metabolism.

## XI. CONCLUSIONS AND RECOMMENDATIONS

As has been made amply clear, depending on the use and stage of development of a chemical agent a variety of questions are commonly asked about chemical metabolism and biotransformation. Increasingly powerful tools have been developed to assist in the answering of these questions. Some require little expertise in computational chemistry (e.g., the expert systems), while others require considerable experience, good judgment, and considerable computing power to be usefully applied (e.g., homology modeling). Obviously the choice of tool depends on what the important issues are, (i.e., what questions are being asked) and the available resources and expertise of scientists. In drug development, prediction of a complete metabolic profile by an expert system for a particular species without reference to the likely enzymology is useful to assist in the identification of metabolites and to suggest the most likely reactions and potential structural modifications influencing metabolism. However, it does not presently give a good indication of whether a drug is likely to exhibit drug-drug interactions. The prediction also does not show whether a drug will have wide interindividual variation due to an enzymic polymorphism or definitely exhibit too short a half-life to be viable.

To answer such questions, the appropriate methods, described briefly above, can potentially be employed when the relevant enzymic mechanisms are well characterized. CYP-mediated oxidations are relatively well characterized, but less information exists on metabolism by other enzymes. For a wide range of chemical classes the intelligent use of metabolism databases may also assist in predicting metabolism and in chemical design. There is not one system or one approach that will answer all questions in this area, although advances in technology will probably alleviate this situation in future.

Expert systems may potentially be extended to consider 3D structure and provide greater information on the enzymology of reactions and possibly species dependence. Reaction chemistry of intermediates or reactive starting materials, producing covalent binding to tissues and toxicity, also needs to be considered for many applications and can be incorporated into future expert systems. The intelligent integration of the results from available techniques and their use by appropriate experienced and trained scientists is required to reap the greatest benefits from available technology.

## ACKNOWLEDGMENTS

I thank Dr. Anthony Long and my other colleagues at LHASA Ltd for their advice and assistance with this review. Thanks are also due to Dr. Mark Cronin, Liverpool John Moores University, for reviewing the manuscript and giving encouragement throughout.

## REFERENCES

Bencze, L., Toth, G., and Kurdi, R., Predicting the environmental hazards of xenobiotics by QSPR methods I. Ethylene oxide, *Hungarian J. Ind. Chem.*, 28,187–193, 2000.

Boyer, S., The Advantages of Multiple Modeling Methods for Metabolism and Toxicity Prediction, oral presentation, Royal Society of Chemistry – Drug Metabolism Group 2002: New Technologies in Drug Discovery, Heriott-Watt University, Edinburgh, December 12–13, 2002.

Boyer, S. and Zamora, I., New methods in predictive metabolism, *J. Comput.-Aided Mol. Design*, 16, 403–413, 2002.

Bursi, R., de Gooyer, M.E., Grootenhuis, A., Jacobs, P.L., van de Louw, J., and Leysen, D., (Q)SAR study on the metabolic stability of steroidal androgens, *J. Mol. Graphics Modeling*, 19, 552–556, 2001.

Button W.G., Judson P.N., Long A., and Vessey J.D., Using absolute and relative reasoning in the prediction of the potential metabolism of xenobiotics, *J. Chem. Inf. Comput. Sci.,* 43, 1371–1377, 2003.

Chang, S., Sood, V.K., Wilson, G.J., Kloosterman, D.A., Sanders, P.E., Hauer, M.J., Vrbanac J.J., and Fagerness, P.E., Identification of the metabolites of the HIV-1 reverse transcriptase inhibitor delarvirdine in monkeys, *Drug Metab. Disposition*, 25, 228–242, 1997.

Chang, S., Sood V.K., Wilson, G.J., Kloosterman, D.A., Sanders P.E., Schuette M.R., Judy, R.W., Voorman R.L., Maio, S.M., and Slatter J.G., Absorption, distribution, metabolism, and excretion of atevirdine in the rat, *Drug Metab. Disposition,*, 26, 1008–1018, 1998.

Cruciani, G., Clementi, S., Crivori, P., Carrupt, P-A and Testa, B., VolSurf and its application in structure-disposition relationships, in *Pharmacokinetic Optimization in Drug Research*, Testa, B., van de Waterbeemd, H., Folkers, G., and Guy, R., Eds., Wiley-VCH, Weinheim, 2001, pp. 539–550.

Cupid, B.C., Beddell, C.R., Lindon, J.C., Wilson, I.D. and Nicholson, J.K., Quantitative structure-metabolism relationships for substituted benzoic acids in the rabbit: prediction of urinary excretion of glycine and glucuronide conjugates, *Xenobiotica*, 26, 157–176, 1996.

Cupid, B.C., Holmes, E., Wilson, I.D., Lindon, J.C. and Nicholson, J.K., Quantitative structure-metabolism relationships for substituted benzoic acids in the rabbit: prediction of urinary excretion of glycine and glucuronide conjugates, *Xenobiotica*, 29, 27–42, 1998.

Dalvie, D.K., Kalgutkar, A.S., Khojaasteh-Bakht S.C., Obach, R.S., and O'Donnell, J.P., Biotransformation reactions of five-membered aromatic heterocyclic rings, *Chem. Res. Toxicol.*, 15:269–299, 2002.

Darvas, F., MetabolExpert, an expert system for predicting metabolism of substances, in *QSAR in Environmental Toxicology*, Kaiser, K.L.E., Ed., Riedel, Dordrecht, 1987, pp. 71–81.

Darvas, F., Predicting metabolic pathways by logic programming, *J. Mol. Graphics*, 6, 80–86, 1988.

Darvas, F., Marokhazi, S., Kormos, P. Kulkarni G., Kalasz, H., and Papp, A., MetabolExpert: its use in metabolism research and in combinatorial chemistry, in *Drug Metabolism: Databases and High-Throughput Testing During Drug Design and Development*, Erhardt, P.W., Ed., Blackwell Science, Oxford, 1999, pp. 237–271.

De Groot, M.J., Ackland, M.J., Horne, V.A., Alex, A.A., and Jones B.C., Novel approach to predicting P450-mediated drug metabolism: development of a combined protein and pharmacophore model for CYP2D6, *J. Med. Chem.*, 42, 1515–1524, 1999a.

De Groot, M.J., Ackland, M.J., Horne, V.A., Alex, A.A., and Jones B.C., A novel approach to predicting P450-mediated drug metabolism: CYP2D catalysed N-dealkylation reactions and qualitative metabolite predictions using a combined protein and pharmacophore model for CYP2D6, *J. Med. Chem.*, 42, 4062–4070, 1999b.

De Groot, M.J. and Ekins, S., Pharmacophore modeling of cytochromes P450, *Adv. Drug Delivery Rev.*, 54, 367–383, 2002.

Ekins, S., Bravi, G., Wikel, J.H., and Wrighton, S.A., Three dimensional-quantitative structure-activity relationship analysis of cytochrome P-450 3A4 substrates, *J. Pharmacol. Exp. Ther.*, 291, 424–433, 1999.

Ekins, S., De Groot, M.J., and Jones, J.P., Minireview: Pharmacophore and three-dimensional quantitative structure activity relationship methods for modeling cytochrome P450 active sites, *Drug Metab. Disposition*, 29, 936–944, 2001.

Ekins, S. and Rose, J., *In silico* ADME/Tox: the state of the art, *J. Mol. Graphics Modeling*, 20, 305–309, 2002.

Ekins S., Waller, C.L., Swann, P.W., Cruciani, G., Wrighton, S.A., and Wikel, J.H., Progress in predicting ADME parameters *in silico*, *J. Pharmacol. Toxicol. Methods*, 44, 251–272, 2000.

Ethell, B.T., Ekins, S., Wang, J., and Burchell, B., Quantitative structure activity relationships for glucuronidation of simple phenols by expressed UGT1A6 and UGT1A9, *Drug Metab. Disposition*, 30, 734–738, 2002.

Ewing, T.J.A., Kocher, J-P., Tieu, H. and Korzekwa, K.R., Predicting susceptibility of reactive sites on molecules to metabolism and accessibility correction factors for electronic models of cytochrome P450 metabolism, *U.S. Patent Application Publication*, 2002.

Fleming, I., *Frontier Orbitals and Organic Chemical Reactions*, John Wiley, Chichester, 1976.

Gao, H. and Hansch, C., QSAR of P450 oxidation: on the value of kcat and Km with kcat/Km, *Drug Metab. Rev.*, 28, 513–526, 1996.

Hawkins D.R., *Biotransformations*, Vols. 1–7, Royal Society of Chemistry, Cambridge, U.K., 1988–1996.

Jolivette, L.J., and Anders, M.W., Structure-activity relationship for the biotransformation of haloalkenes by rat liver microsomal glutathione transferase 1, *Chem. Res. Toxicol.*, 15, 1036–1041, 2002.

Jones, P.J., Mysinger M., and Korzekwa, K.R., Computational models for cytochrome P450: A predictive electronic model for aromatic oxidation and hydrogen atom abstraction, *Drug Metab. Disposition*, 30, 7–12, 2002.

Joqueviel, C., Martini, R., Gilard, V., Malet-Martino, M., Canal, P., and Niemeyer, U., Urinary excretion of cyclophosphamide in humans, determined by phosphorus-31 nuclear magnetic resonance spectroscopy, *Drug Metab. Disposition*, 26, 418–428, 1998.

Judson P.N. and Vessey J.D., A comprehensive approach to argumentation, *J. Chem. Inf. Comput. Sci.*, 43, 1356–1363, 2003.

Klopman, G., Dimayuga, M. and Talafous, J., META. 1. A program for the evaluation of metabolic transformation of chemicals, *J. Chem. Inf. Comput. Sci.*, 34, 1320–1325, 1994.

Klopman, G. and Tu, M., META: A program for the prediction of the products of mammal metabolism of xenobiotics, in *Drug Metabolism: Databases and High Throughput Testing During Drug Design and Development*, Erhardt, P.W., Ed., Blackwell Science, Oxford, 1999, pp. 271–276.

Klopman, G., Tu, M., and Fan, B.T., META. 4. Prediction of the metabolism of polycyclic aromatic hydrocarbons, *Theor. Chem. Acc.*, 102, 33–38, 1999.

Klopman, G, Tu, M., and Talafous, J., META. 3. A genetic algorithm for metabolic transform priorities optimization, *J. Chem. Inf. Comput. Sci.*, 37, 329–334, 1997.

Koerts, J., Velraeds, M.M.C., Soffers, A.E.M.F., Vervoort, J., and Rietjens, I.M.C.M., Influence of substituents in fluorobenzene derivatives on the cytochrome P450-catalysed hydroxylation at the adjacent ortho aromatic carbon center, *Chem. Res. Toxicol.*, 10, 279–288, 1997.

Kuhn, B., Jacobsen, W., Christians, U., Benet, L.Z., and Kollman, P.A., Metabolism of sirolimus and its derivative everolimus by cytochrome P450 3A4: insights from docking, molecular dynamics, and quantum chemical calculations, *J. Med., Chem.*, 44, 2027–2034, 2001.

Langowski, J. and Long A., Computer systems for the prediction of xenobiotic metabolism, *Adv. Drug Delivery Rev.*, 54, 407–415, 2002.

Lewis, D.F.V., Computer-assisted methods in the evaluation of chemical toxicity, *Rev. Computational Chem.*, 3, 173–222, 1992.

Lewis, D.F.V., COMPACT: a structural approach to the modeling of cytochromes P450 and their interactions with xenobiotics, *J. Chem. Technol. Biotechnol.*, 76, 237–244, 2001.

Lewis, D.F.V., Essential requirements for substrate binding affinity and selectivity toward human CYP2 family enzymes, *Arch. Biochem. Biophys.*, 409, 32–44, 2003.

Lewis, D.F.V., Bird, M.G., and Jacobs, M.N., Human carcinogens: an evaluation study via the COMPACT and HazardExpert procedures, *Human Exp. Toxicol.*, 21, 115–122, 2002a.

Lewis, D.F.V., Modi S., and Dickins, M., Structure-activity relationships for human cytochrome P450 substrates and inhibitors, *Drug Metab. Rev.*, 34, 69–82, 2002b.

Lewis, D.F.V. and Dickens, M., Substrate SARs in human P450s, *Drug Discovery Today*, 7, 918–925, 2002.

Lewis, D.F.V., Dickins, M., Lake, B.G., Eddershaw, P.J., Tarbit, M.H., and Goldfarb, P.S., Molecular modeling of the cytochrome P450 isoform CYP2A6 and investigations of CYP2A substrate selectivity, *Toxicology*, 133, 1–33, 1999.

Lewis, D.F.V., Ioannides, C., and Parke, D.V., Prediction of chemical carcinogenicity from molecular structure: a comparison of MINDO/3 and CNDO/2 molecular orbital methods, *Toxicol. Lett.*, 45, 1–13, 1989.

Lewis, D.F.V., Modi, S., and Dickins, M., Quantitative structure-activity relationships (QSARs) within substrates of human cytochromes P450 involved in drug metabolism, *Drug Metab. Drug Interactions*, 18, 221–242, 2001.

Lightfoot, T., Ellis, S.W., Mahling, J., Ackland, M.J., Bijloo, G.J., De Groot, M.J., Vermeulen, N.P.E., Blackburn, G.M., Lennard, M.S., and Tucker, G.T., Regioselective hydroxylation of debrisoquine by cytochrome P450 2D6: implications for active site modeling, *Xenobiotica*, 30, 219–233, 2000.

Moriguchi, I., Hirono, S., Liu, Q., Nakagome, I., and Matsushita, Y., Simple method of calculating octanol/water partition coefficient, *Chem. Pharm. Bull.*, 40, 127–130, 1992.

Parke, D.V., Ioannides, C., and Lewis, D.F.V., Computer modeling and in vitro tests in the safety evaluation of chemicals: strategic applications, *Toxicol. in Vitro*, 4, 680–685, 1990.

Pastor, M., Cruciani G., McLay, I., Pickett, S., and Clementi S., Grid-Independent Descriptors (GRIND): a novel class of alignment-independent three-dimensional molecular descriptors, *J. Med. Chem.*, 43, 3233–3243, 2000.

Pfeifer, S. and Borchert, H., *Biotransformation von Arzneimitteln*, VEB Verlag Volk und Gesundheit, Berlin, 1963.

Rao, S., Aoyama, R., Schrag, M., Trager, W.F., Rettie, A. and Jones, J.P., A refined 3-dimensional QSAR of cytochrome P450 2C9: computational predictions of drug interactions, *J. Med. Chem.*, 43, 2789–2796, 2000.

Reigh, G., McMahon, H., Ishizaki, M. Ohara, T., Shimae, K., Esumi, Y, Green, C., Tyson, C., and Ninomiya, S., Cytochrome P450 species involved in the metabolism of quinoline, *Carcinogenesis*, 17, 1989–1996, 1996.

Ridder, L., Mulholland, A.J., Rietjens, I.M.C.M., and Vervoort J., Combined quantum mechanical and molecular mechanical reaction pathway calculation for aromatic hydroxylation by p-hydroxybenzoate-3-hydroxylase, *J. Mol. Graphics Modeling*, 17, 163–175, 1999.

Ridder, L., Mulholland, A.J., Rietjens, I.M.C.M., and Vervoot, J., A quantum mechanical/molecular mechanical study of the hydroxylation of phenol and halogenated derivatives by phenol hydroxylase, *J. Am. Chem. Soc.*, 122, 8728–8738, 2000.

Sali, A. and Blundell, T.L., Comparative protein modeling by satisfaction of spatial restraints, *Journal of Molecular Biology*, 234, 779–815, 1993.

Sanderson, D.M. and Earnshaw C.G., Computer prediction of possible toxic action from chemical structure; the DEREK system, *Human and Experimental Toxicology*, 10, 261–273, 1991.

Scarfe, G.B., Wilson, I.D., Warne, M.A., Holmes, E., Nicholson, J.K., and Lindon, J.C., Structure-metabolism relationships of substituted anilines: prediction of N-acetylation and N-oxanilic acid formation using computational chemistry, *Xenobiotica*, 32, 267–277, 2002.

Soffers, A.E.M.F., Ploemen, J.H.T.M., Moonen, M.J.H., Wobbes, T., van Ommen B., Vervoort, J., van Bladeren, P.J., and Rietjens, I.M.C.M., Regioselectivity and quantitative structure-activity relationships for the conjugation of a series of fluoronitrobenzenes by purified glutathione S-transferase enzymes from rat and man, *Chem. Res. Toxicol.*, 9, 638–646, 1996.

Spratzenegger M., Wang Q.M., He Y.Q., Wester M.R., Johnson E.F., and Halpert J.R., Amino acid residues critical for differential inhibition of CYP2B4, CYP2B5 and CYP2B1 by phenylimidazoles, *Mol. Pharmacol.*, 59, 475–484, 2001.

Takenaga, N., Ishii, M., and Kamei, T., Structure-activity relationship in O-glucuronidation of indolocarbazole analogs, *Drug Metab. Disposition*, 30, 494–497, 2002.

Talafous, J., Sayre, L.M., Mieyal, J.J., and Klopman, G., META. 2. A dictionary model of xenobiotic metabolism, *J. Chem. Inf. Comput. Sci.*, 34, 1326–1333, 1994.

Tarjanyi, Z., Kalasz, H., Hollosi, I., Bathori, M., Bartok, T., Lengyel, J., Maguar, K., and Fuerst, S., Chromatographic investigation and computer simulation of L-deprenyl metabolism, *Eur. J. Drug Metab. Pharmacokinetics*, 23, 324–328, 1998.

Ter Laak, A.M. and Vermeulen, N.P.E., Molecular-modeling approaches to predict metabolism and toxicity, in *Pharmacokinetic Optimization in Drug Research*, Testa, B., van de Waterbeemd, H., Folkers, G., and Guy, R., Eds., Wiley-VCH, Weinheim, 2001, pp. 551–588.

Ter Laak, A.M., Vermeulen, N.P.E. and de Groot, M.J., Molecular-modeling approaches to predict metabolism and toxicity, *Drugs Pharm. Sci.*, 116, 505–548, 2002.

Testa, B., *The Metabolism of Drugs and Other Xenobiotics: Biochemistry of Redox Reactions*, Academic Press, London, 1995.

Testa, B, Caron, G., Crivori, P. Rey, S., Reist, M., and Carrupt, P.A., Lipophilicity and related properties as determinants of pharmacokinetic behavior, *Chimia*, 54, 672–677, 2000a.

Testa, B, Crivori, P., Reist, M., and Carrupt, P.A., The influence of lipophilicity on the pharamacokinetic behaviour of drugs: concepts and examples, *Perspect. Drug Discovery Design,* 19, 179–211, 2000b.

Testa, B. and Cruciani, G., Structure-metabolism relations and the challenge of predicting biotransformation, in *Pharmacokinetic Optimization in Drug Research*, Testa, B., van de Waterbeemd, H., Folkers, G., and Guy, R., Eds., Wiley-VCH, Weinheim, 2001, pp. 65–84.

Testa, B. and Jenner, P., *Drug Metabolism: Chemical and Biochemical Aspects*, Marcel Dekker, New York, 1976.

Timbrell, J.A., *Principles of Biochemical Toxicology*, Taylor and Francis, London, 1991.

Upthagrove, A.L. and Nelson, W.L., Importance of amine pKa and distribution coefficient in the metabolism of fluorinated propanolol analogs: metabolism by CYP1A2, *Drug Metab. Disposition*, 29, 1389–1395, 2001.

Wang, M. and Dickinson, R.G., Disposition and covalent binding of diflunisal and diflunisal acyl glucuronide in the isolated perfused rat liver, *Drug Metab. Disposition*, 26, 98–104, 1998.

Wilbury T.T., Comparison of commercially available metabolism databases during the design of prodrugs and codrugs, in *Drug Metabolism: Databases and High Throughput Testing During Drug Design and Development*, Erhardt, P.W., Ed., Blackwell Science, Oxford, 1999, p. 208.

# CHAPTER 11

# Prediction of Pharmacokinetic Parameters in Drug Design and Toxicology

Judith C. Duffy

## CONTENTS

# I. INTRODUCTION

Over the last 40 years, the scientific literature has abounded with examples of the application of quantitative structure-activity relationships (QSARs) and molecular modeling techniques to the problem of predicting biological activity. The application of these techniques to the prediction of pharmacokinetic or toxicokinetic parameters has been, until recently, less intensely researched.

The increasing use of high throughput screening technologies has led to the generation of a prolific amount of pharmacokinetic data. These data need to be collated, organized, and rationally applied to the construction of more informative and accurate models that can be used to predict pharmacokinetic parameters. Previously the generation of hit and lead compounds was a rate-limiting factor in drug development, but the focus has now shifted more to the optimization of leads generated, particularly in relation to their pharmacokinetic properties. Ideally, we would wish to predict pharmacokinetic profiles using simple, easily calculable, or readily generated interpretable physicochemical parameters. Unfortunately, the complexity of the processes *in vivo*, has meant that few global models have been published, although there has been some success in the production of local models (Testa et al., 2000).

In terms of studying the time course of xenobiotics within the body there are two major areas of interest, drug design and toxicology. For both of these disciplines there are many important reasons for trying to predict kinetic parameters *a priori*. In drug development it would greatly reduce the economic cost and time associated with initial research of compounds that later fail because of adverse pharmacokinetics and would reduce the number of animals used in testing. In terms of toxicology it would allow rapid screening of large numbers of chemicals in primary hazard assessment.

One major difficulty in making these predictions is the complexity inherent in pharmacokinetic or toxicokinetic data. Consider, for example, the oral ingestion of a xenobiotic. The kinetic parameters of importance here are the same whether this is a therapeutic drug taken intentionally, or if the compound was a toxicant accidentally ingested. Overall, we wish to be able to predict the concentration-time profile of the compound in the organs of the body. This involves a multitude of interacting systems responsible for absorption, distribution, metabolism, and elimination.

The compound must first dissolve in the aqueous environment of the stomach and be absorbed across the membranes lining the gastrointestinal tract. It may be broken down by microflora or enzymes in the gut, or be degraded by first pass metabolism in the liver, prior to entering the general circulation. Once in the blood, binding to plasma proteins, affinity for different tissues and the ability of the compound to cross biological membranes will affect its distribution and storage within the body. The rate at which the body can eliminate the compound through renal clearance, hepatic metabolism, or other mechanisms will also, in part, determine its concentration-time profile. Each of these individual processes is, in itself, multifactorial. For example, hepatic metabolism requires the ability to cross the relevant biological membranes to allow entry to the liver, productive binding to one or more of the myriad of metabolic enzymes, and successful elimination of the metabolites.

To date, several excellent review articles have been published on the prediction of various pharmacokinetic parameters. Consequently, it is not intended that this chapter will provide a full review of the current literature in this area. This chapter provides an introduction to key pharmacokinetic parameters and an overview of references concerning their prediction. A discussion of how the data, available from diverse sources, can be structured and used to address specific questions in pharmacokinetics and toxicokinetics will then be presented.

## II. PHARMACOKINETIC PARAMETERS IN DRUG DEVELOPMENT

When developing a new drug product, a tablet, taken orally with a once-daily dosing regimen, is usually the most desirable formulation. The drug needs to reach its site of action in adequate concentration without accumulating in the body or producing toxic side-effects at other sites. The pharmacokinetic properties must be optimized such that the drug will be readily absorbed, transported to the appropriate site, and eliminated from the body in a timely manner. Figure 11.1 shows the host of interrelated pharmacokinetic properties that control the time course of a xenobiotic in the body.

It has been estimated that approximately 40% of drug candidates fail during the preclinical testing stage of development, due to unacceptable pharmacokinetics (Thompson, 2001). Clearly, the economic cost of failure at this late stage is high. The ability to screen out the likely failures at a much earlier stage would be more beneficial — the so-called fail fast, fail cheap philosophy. Conversely, there would be a more rapid screening-in of the candidates that appear promising from the results of high throughput or *in silico* screens.

The key pharmacokinetic parameters to be considered in drug development are discussed here, *vide infra*. Although this is not an exhaustive list, successful modeling of these key parameters would provide a useful preliminary screening tool. The properties are presented in the order in which a xenobiotic entering the body via the oral route would be exposed to them, rather than in order of priority.

### A. Solubility

On entering the stomach an oral dose must first dissolve in the aqueous environment, prior to being absorbed across the walls of the gastrointestinal tract. The rate and extent of dissolution is therefore an important parameter. Solubility issues are discussed elsewhere in this book (see Chapter 4), but it is important to note that any new drug must possess the correct balance between aqueous solubility, allowing it to dissolve in the stomach and lipophilicity, to permit transfer across biological membranes. Poor solubility can result in a low oral bioavailability, although this problem may be resolved in some instances by judicious choice of formulation.

### B. Fraction of Dose Absorbed

A fraction of the drug that has dissolved will then partition across the lipid interface of the gastrointestinal tract into the blood. The various biological barriers encountered by xenobiotics in getting to their site of action are discussed in a review article by Pagliara et al. (1999). This article also discusses the many techniques used as models for investigating absorption properties.

Cell membranes make up a bilayer of amphiprotic lipids, with the outer layer predominantly composed of zwitterionic lipids and the inner layer containing more negatively charged lipids. Adjacent cells are linked by an arrangement of membrane proteins referred to as tight junctions,

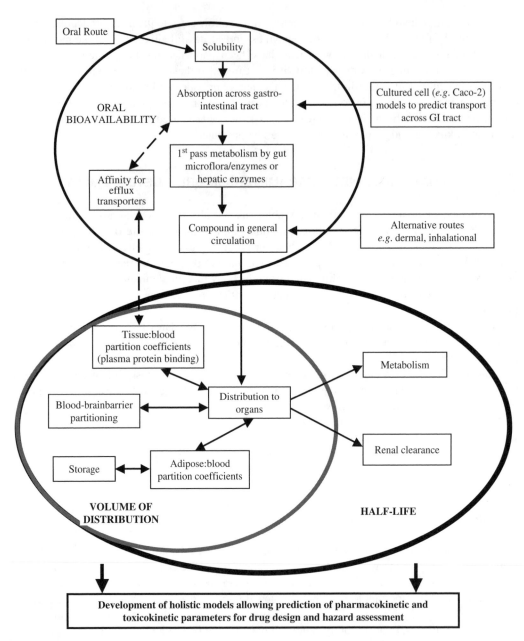

**Figure 11.1** Overview of factors controling the time course of xenobiotics in the body.

containing water-filled pores. In the small intestine these pores represent approximately 0.1% of the total surface area. Drug uptake mainly occurs via transcellular absorption through the lipoidal cell membrane, either by passive diffusion or via a specific carrier-mediated transport mechanism. Alternatively, drugs may use the paracellular route, passing through the aqueous pores of the tight junctions between cells. These channels generally only permit the passive diffusion of small, hydrophilic molecules (Pagliara et al., 1999).

Even if the drug possesses suitable properties for solubility and absorption, it may still be prevented from entering the blood if it is susceptible to degradation by microorganisms or enzymes residing in the gut wall. Drugs that have entered the intestinal epithelia may be transported back out of the cells by efflux transporters, discussed later. The P-glycoprotein efflux pump is highly expressed in intestinal epithelia. It is possible this pump acts in concert with the metabolizing enzymes, as drugs with affinity for this transporter would be pumped back out of the cell and again presented to the enzyme system (Veber et al., 2002).

## C.  Metabolism

To eliminate a xenobiotic the body must be able to either excrete the unchanged form via the kidneys, or metabolize the parent to a compound that can be readily excreted. Once successfully absorbed from the gastrointestinal tract, compounds enter the hepatic portal vein and are transported directly to the liver — the major organ of metabolism. Here, susceptible groups may undergo extensive metabolism, referred to as the first-pass effect, resulting in low concentrations of the drug entering the general blood circulation.

Metabolism is primarily the conversion of lipophilic compounds to more water-soluble compounds that can be readily excreted in urine. In Phase I-type metabolism, susceptible functional groups undergo oxidation reactions catalyzed by the cytochrome P450 (CYP) superfamily of enzymes. Research carried out on these enzymes has indicated that three forms, namely, CYP2D6, CYP2C9, and CYP3A4 are responsible for the metabolism of over 90% of drugs in man (Langowski and Long, 2002). In Phase II-type metabolism a specific transferase enzyme conjugates the parent drug, or its Phase I metabolite, with another group such as sulfate or glucuronide to produce, typically, a more polar molecule. This can then be eliminated from the body in the urine. Hepatic metabolism for highly metabolized drugs is dependent on the rate at which the compound is presented to the liver. Conditions that alter hepatic blood flow can affect the extent of metabolism.

Much research has been carried out on metabolic processes and is discussed in detail in Chapter 10. Not only must the rate and extent of metabolism be considered in drug development, but also the possibility of the drug being converted to an undesirable toxic metabolite must be taken into account.

## D.  Affinity for Efflux Transporters

The P-glycoprotein efflux transporter actively prevents certain drugs from accumulating within cells or being transported across them. The molecule is present in several organ systems, including the gastrointestinal epithelia (Osterberg and Norinder, 2000). Once a drug molecule has been absorbed into a cell lining the gut wall, it may be pumped back into the gastric lumen by the action of this transporter, before it has a chance to cross the cell. This will result in an overall reduction in transport of susceptible drugs into the blood. Although the P-glycoprotein transporter has been the subject of much research recently, it is possible that other transporters may play a similar role (Pagliara et al., 1999). The role of P-glycoprotein in preventing drugs from crossing the blood-brain barrier is discussed in a later section.

## E.  Bioavailability

Bioavailability is defined as "the fraction of unchanged drug reaching the systemic circulation following administration by any route" (Benet, 1987). In the case of oral bioavailability, it is the fraction of the oral dose that reaches the general circulation. Oral bioavailability is a composite parameter dependent on the extent of dissolution of the dosage form, absorption across the gut wall, and evasion of gut wall or first-pass hepatic metabolism. Enterohepatic recycling, where the drug may not be completely reabsorbed following excretion into bile, may also be a factor.

For the purposes of modeling and subsequent optimization of candidate drug structures, it is generally easier to consider each of the constituent parameters individually. However, there have been several successful studies that have considered overall oral bioavailability data.

## F.  Alternative Routes of Administration

Although the oral route is generally the preferred route of administration, there are instances where it may be more convenient to administer the drug via an alternative route. For example, in treatments requiring long-term maintenance doses, the use of transdermal patches may be more convenient for the patient. Dermal permeability and bioavailability values may also be important pharmacokinetic parameters. The rate and extent of absorption of drugs across the lining of the lungs are also significant for drugs given via the inhalational route.

There are many alternative routes and methods of administration that will not be listed here. Whatever the route, the factors governing absorption or distribution of drug from the site of administration to the general blood circulation would be the first consideration. Once in the general circulation, factors determining distribution to all organs, storage metabolism, and elimination will be the same, regardless of the route of administration.

## G.  Binding to Plasma Proteins

The percentage of drug bound to plasma proteins will significantly affect the overall distribution of the drug in the body, as it is generally only the free drug that is able to cross biological membranes and so enter tissues. Human serum albumin is the most abundant drug-binding protein in blood. It possesses several binding sites with varying affinity for different drugs, generally binding lipophilic, acidic drugs. $\alpha$-1 acid glycoprotein is more important in binding basic and neutral drugs, although it has recently also been shown to bind some acidic drugs (Saiakhov et al., 2000).

## H.  Tissue:Blood Partition Coefficients

The ratio of drug partitioning between blood and tissues determines distribution and storage properties of drugs. It is governed by the solubility of the drug in the two different phases as well as the ability to bind to macromolecules constitutive of tissues and blood (Poulin and Theil, 2000). As individual tissues make up different fractions of water, neutral lipids, phospholipids, and proteins, each tissue shows a different affinity for specific drugs. Affinity for adipose tissue is significant for long-term storage of lipophilic xenobiotics, particularly neutral and acidic drugs; basic drugs tend to accumulate more in lean tissue, as cations are more likely to bind to macromolecules here.

## I.  Blood-Brain Barrier Partition Coefficient

The blood-brain barrier (BBB) partition coefficient (PC) is the ratio of the amount of drug found in the brain to that found in blood. Of all tissues, partitioning into the brain is of particular significance as this is the site of action for centrally acting drugs and a site where xenobiotics may exert significant toxicity. The BBB is difficult to cross, as the junctions between cells are particularly tight. It is generally not possible for drugs to pass through the paracellular route; they must pass through the cells. The P-glycoprotein efflux transporter is highly expressed in cells forming the BBB and plays an important role in preventing compounds from entering the central nervous system. It has been suggested that this transporter may have evolved as a protection mechanism to prevent the entry of plant toxins into the body and the brain in particular (Hosoya et al., 2002). Drugs that exhibit lower than anticipated brain partitioning may be acting as substrates for this transporter. In designing drugs it may be necessary to optimize properties ensuring high brain penetration, if this is the site of action, or to screen out compounds that would enter the brain and cause central toxicity.

## of Distribution

ʌe of distribution ($V_d$) of a drug is given by the following equation:

$$V_d = \frac{\text{Total amount of drug in body}}{C} \tag{11.1}$$

ʒentration in blood or plasma.

; the whole body as one compartment, it represents the apparent volume into which ites. For an average 70 Kg man, approximate fluid volumes are as follows: plasma, ʟ; extracellular fluid, 12 L; total body water, 42 L (Benet, 1987). Drugs that have a or tissues consequently show relatively low blood or plasma concentrations (C). The ne of distribution may be far in excess of the given fluid volumes. Conversely, a drug ound to plasma proteins will be retained primarily in the blood giving high blood ʌtration values, resulting in much a lower value for the volume of distribution. The ʒortant as it reflects the overall distribution of a drug and determines its persistence it is the only drug in blood that can be presented to the organs of elimination.

ʒe

clearance is normally defined as the volume of blood or plasma from which drug tely removed in a finite time, typically measured in ml min$^{-1}$. For drugs that are ʒ, clearance may be limited by the rate of blood flow to the eliminating organ. ʒ from the blood generally occurs via the kidneys or liver. Elimination of drug by ʒion to metabolites by the liver has been discussed previously. Drugs may also ʌnged in the urine, referred to as renal clearance. The fraction of drug excreted ʒ urine can be used as an indicator of the relative contributions of hepatic or renal e elimination of the drug. Generally, drugs that are extensively metabolized are ʌ the body faster than those relying on excretion via the kidneys, as metabolic ore efficient than renal filtration. There are other routes available for the elimination ʌs excretion into expired air, saliva, sweat, or milk. These routes are generally less will not be considered here, although it must be noted that total clearance of a drug ʌ of clearance values via all routes.

ʒt ½) of a drug is the time taken for the amount of a drug in the body to fall to ʒginal value. It is useful as it determines the amount of time taken for the body ʒg and also determines the time required to reach steady-state concentrations of ʒntinuous administration. It is calculated from the following equation:

$$t_{\frac{1}{2}} = \frac{0.693 V_d}{Cl} \tag{11.2}$$

ʒ ne of distribution and Cl = total body clearance.

a composite parameter dependent on the volume of distribution and the rate of drug. This follows from the fact that it is only the drug presented to the eliminating be removed from the body. Optimization of the half-life of a new drug candidate too long a half-life may result in bioaccumulation and toxic side effects, whereas a short half-life may result in therapeutic levels not being maintained or inconvenient dosage regimens.

Although there are several other pharmacokinetic parameters that are significant in drug development, it is only these key parameters that will be discussed in this chapter.

## III. TOXICOKINETIC PARAMETERS IN HAZARD ASSESSMENT

All of the pharmacokinetic parameters listed above are not only important in drug design, but they are also significant in determining overall toxicity of hazardous substances. Being able to predict the above parameters would also be beneficial in performing preliminary hazard assessment of chemicals. The toxicokinetic parameters considered here refer only to those affecting man or other animals; environmental toxicology is discussed elsewhere in this book.

When estimating the timescale for the persistence of a toxicant in the body, although the same parameters are important, the perspective and the emphasis may be different. For example, oral absorption may be relevant in cases of poisoning, but for hazard assessment predicting uptake through alternative routes of exposure may be more important. Prediction of dermal penetration of hazardous materials, or inhalation of volatile organic compounds in industrial settings, may be more significant in determining overall exposure and toxicity levels. The rate at which a toxicant can be eliminated from the body is determined by the rate and extent of metabolism, renal clearance and clearance via other routes, such as expiration. The routes of metabolism may be of additional significance if the chemical is broken down to metabolites that also produce toxic effects. Parameters that affect the distribution of the toxicants such as affinity for blood proteins or tissue:blood PCs may indicate potential sites of toxic action. In terms of toxicity the ability to cross the BBB is particularly important because this may lead to serious central side-effects. In terms of accumulation, adipose represents a more significant site of storage for solvents and pollutants than it does for drugs (Poulin et al., 2001). Being able to predict the half-life of a toxicant is particularly important as this could indicate potential for bioaccumulation of the compound to dangerous levels.

Table 11.1 summarizes key pharmacokinetic parameters and their significance in both drug design and toxicology.

Depending on whether an individual study is undertaken for purposes of drug develo hazard assessment, identical physiological parameters in man are variously referred to pharmacokinetic or toxicokinetic. When discussing the modeling of these parameters more generic expression such as chemicokinetic could be used to encompass both pharmacokinetic will be used in this chapter to describe the physiological parameters c whether these be for the purposes of toxicology or drug development.

## IV. EXAMPLES OF MODELS AVAILABLE IN THE LITERATURE

With increasing interest in the modeling and prediction of pharmacokinetic pro has recently been a large increase in the number of published models. It is not int give a full review of all of the models published for each parameter. An overview is the key parameters listed above, and a few examples of models are given with refer review articles.

### A.  Solubility

Lipinski et al. (1997) published a review article on computational and experimental methods of estimating aqueous solubility, serving as a useful reference of available methods. The article discusses poor solubility as a potential limiting factor to overall drug absorption across the gut wall.

**Table 11.1    Significance of Pharmacokinetic Parameters in Drug Design and Toxicology**

| Parameter | Significance to Drug Design | Significance to Toxicology |
|---|---|---|
| Solubility | Affects rate and extent of dissolution and subsequent absorption from gastrointestinal tract following oral dose | May affect the extent of uptake of toxicants; dependent on route of exposure |
| Fraction of dose absorbed | Oral route is generally preferred for drug delivery; extent of oral absorption needs to be optimised | Determination of risk following oral ingestion of toxicant |
| Metabolism | May affect the overall bioavailability and the time course of a drug within the body | Important for assessing time course of exposure and the likelihood of producing toxic metabolites |
| Bioavailability | Composite parameter reflecting solubility, absorption, and 1st pass metabolism of compounds, the significance of which is given above for both drug design and toxicology | |
| Affinity for efflux transporters | May reduce therapeutic efficacy of drug, in particular for those designed to act on central nervous system | May provide degree of protection against entry of a toxicant into the central nervous system |
| Alternative routes of administration/exposure | Optimization of dermal penetration or absorption via lungs may be preferable for the administration of certain compounds formulated as patches, inhalers, etc. | Dermal penetration and inhalation are common routes of exposure for environmental and industrial pollutants, particularly volatile organic compounds |
| Binding to plasma proteins | Will affect overall distribution of drug to sites of action or storage and presentation to eliminating organs | Will affect overall distribution of toxicants to sites of action or storage and presentation to eliminating organs |
| Tissue:blood partition coefficients | Will determine the extent to which the target tissue will take up the drug, in addition to affecting overall distribution and storage properties | May indicate affinity of a chemical for a tissue where it exerts a toxic effect, in addition to affecting storage and distribution within the body |
| Blood-brain barrier partition coefficient | A specific case of tissue:blood partition coefficient of particular significance as a site of drug activity and toxic effects | |
| Volume of distribution | Indicates distribution of drug in the body; affects the overall half-life of the drug | Indicates distribution of a toxicant in the body; affects its overall half-life |
| Clearance | Indicates the rate at which the drug would be eliminated from the body; affects the overall half-life of the drug | Indicates the rate at which a toxicant would be eliminated from the body; affects its overall half-life |
| Half-life | Essential for determining useful and effective dosage regimens for drugs; ideally would allow for once-daily dosing | Indicates for how long a compound would persist in the body and its potential for bioaccumulation, resulting in long-term adverse effects |

In a later paper by Huuskonen et al. (2000), the authors advocated the use of large, diverse training sets to produce more robust, open models for prediction of aqueous solubility. Their study involved the prediction of solubility for 734 diverse organic compounds. Thirty-seven atom type electrotopological state (E-state) indices were calculated. These indices are generated using the compound's molecular graph and encode the electronic state of the atom type as it is influenced by surrounding atoms in their specific topological arrangement. Huuskonen and colleagues used 675 compounds in the training set, 38 in a validation set, and 21 in a test set. Multiple linear regression using 34 structural parameters produced $R^2$ values of 0.94 (s = 0.58) for 675 compounds in the training set and 0.80 (s = 0.87) for 21 compounds of the test set. Incorporation of melting point into the regression analysis produced only a marginal improvement for the training set and no improvement for the test set. An artificial neural network was used on 34 structural parameters, and the statistics were $R^2 = 0.96$ (s = 0.51) and $R^2 = 0.84$ (s = 0.75) for the training and test sets, respectively. Regression of aqueous solubility against log $K_{ow}$ and melting point only, produced, as would be anticipated, a negative correlation with both ($R^2 = 0.91$; s = 0.73) for 675 compounds of the training set. The findings were discussed in the context of similar studies.

The approach taken by Liu and So (2001) to predicting aqueous solubility is also useful, as it does not require the incorporation of any experimentally determined parameters or computationally expensive, calculated 3D parameters. In this paper 7 1D and 2D parameters were used in an artificial neural network model to predict the solubility of 1312 diverse organic compounds. The model included an estimated log $K_{ow}$ value (to indicate hydrophobicity), the calculated topological polar surface area (TPSA; as a measure of hydrophilicity), molecular weight, Wiener index, Balaban's relative electronegativity index, bonding information content, and molecular flexibility indices. The resulting model had a correlation coefficient of 0.92 and the standard deviation between the measured and estimated solubility values was 0.72 log units. This was considered reasonable as the experimental error is estimated to be in the region of 0.5 log units. The authors recommend this approach as being useful the in preliminary screening of drug candidates, even prior to their synthesis.

The method described by Kobayashi et al. (2001) uses a system where a solid drug is initially dissolved in a chamber at pH 1, representing stomach pH, and is then transferred to a vessel at pH 6, representing the small intestine. The permeation of the drug so dissolved is then determined across a Caco-2 cell monolayer. The method attempts to account for the extent to which a drug would dissolve in the varying pHs of the intestinal tract and correlate this with its overall absorption profile. Although the results are interesting, this is a relatively lengthy process and for preliminary screening more rapid methods may be more appropriate.

Programs such as QMPRPlus™ (SimulationPlus, Inc.) and WSKOWIN (EPISUITE available from the Environmental Protection Agency website, Syracuse Research Corporation, Inc.) can be used to predict aqueous solubility directly (see Chapters 3 and 4). As log $K_{ow}$ inversely correlates reasonably well with aqueous solubility for many compounds, programs that are available to predict log $K_{ow}$ may also be useful in preliminary screens for aqueous solubility.

## B. Fraction of Dose Absorbed

### 1. In Vivo *Studies*

In determining the amount of orally administered drug entering the systemic circulation, the fraction of dose absorbed is one of the most important parameters and has been studied extensively both *in vivo* and using *in vitro* models.

In 1997, Lipinski et al. published their "rule of 5" method after analyzing the absorption properties of 2287 compounds. The results of their study indicated that a drug would be more likely to show poor absorption if:

1. Molecular weight exceeded 500
2. The sum of OH and NH hydrogen bond donors exceeded 5
3. The log $K_{ow}$ exceeded 5
4. The sum of N and O atoms acting as hydrogen bond acceptors exceeded 10

A compound exceeding two of the above criteria would be highly likely to demonstrate absorption problems. The rule of five method for drug screening is significant as it provides a very simple screening tool, using readily calculable and easily interpretable parameters. It is also easy to recognize which modifications to drug structure may have a beneficial effect on absorption properties. Hence the method proved to be a widely applicable tool. Since 1997 more studies of this type have been carried out and the limitations to the use of the rule of 5 have been recognized. Despite the need for some refinement to the model it still provides a useful early screening tool from which other models to predict absorption have begun to evolve.

The significance of hydrogen bonding in intestinal absorption has in particular been recognized by other authors. In 1998, Wessel et al. used a neural network model to predict absorption of 86 drugs. Of the six significant parameters used in the model, three related to hydrogen bonding ability; two further parameters reflected overall size and shape of the molecule; and the final parameter was the number of single bonds, believed to reflect overall flexibility of the molecule. Interestingly, the work of Veber et al. (2002), discussed below, also refers to the role of overall flexibility/rigidity of molecules in determining oral bioavailability.

Raevsky et al. (2000) studied human intestinal absorption of a relatively small data set of 32 compounds. These authors also found a clear correlation indicating that hydrogen bonding ability had a significant negative effect on absorption; this effect was possibly due to the entropic cost of breaking multiple hydrogen bonds between the drug and water molecules or between drug and functional groups present on the surface of the membrane. Although polarizability and partial atomic charge had a limited effect on absorption, log D showed no correlation in this data set.

Egan et al. (2000) reviewed several published methods of predicting human intestinal absorption. Additionally, the authors undertook an extensive study of drugs and drug-like molecules to derive a model to discriminate between the well- and poorly absorbed compounds. The parameters used to generate the relationship were polar surface area (PSA) and a calculated log $K_{ow}$ value. Molecular weight as an independent parameter did not improve correlation as both log $K_{ow}$ and PSA are themselves influenced by molecular weight. A pattern recognition model was generated in which a 95% confidence ellipse for the well-absorbed compounds was computed. This utilized a training set of 199 well and 35 poorly absorbed compounds. Removal of outliers gave boundaries to the ellipse where a PSA above 148.1$A^2$ and log $K_{ow}$ above 5.88 would be predicted to be poorly absorbed. Outlier detection recognized many compounds that were actively transported, and the paper includes a useful compilation of references for compounds that are known to be actively transported and so often appear as outliers in these data sets.

When tested on external datasets comprising hundreds of compounds the vast majority of orally administered drugs were found to lie within the ellipse for good absorption, confirming the usefulness of this model to predict good absorption properties. The authors also comment that although historically it has been accepted that only the uncharged form of an ionizable drug crosses biological membranes, increasing evidence suggests that the contribution to partitioning made by the ionized form of the drug may be significant and warrants further investigation.

In 2001, Zhao et al. performed a study where human intestinal absorption data and the methods for its prediction were evaluated in detail. The authors reported that the most significant descriptors for absorption were hydrogen bonding descriptors, supporting the findings of previous studies. The authors reviewed absorption values, available in the literature for a range of compounds, concluding that for 169 drugs the reported values were reliable.

These 169 drugs were analyzed in more detail in a subsequent study by Abraham et al. (2002). Here the authors performed an investigation comparing rate constants for human intestinal absorption to those for diffusion and partition processes to determine whether it is diffusion or partition that is the dominant step in absorption. The results indicated that diffusion across the stagnant mucous layer and transfer across the mucous-membrane interface are the major processes. For 127 of the compounds, the percentage absorption in humans was found to correlate with excess molar refractivity, dipolarity/polarizability, McGowan characteristic volume, and the sums of hydrogen bond acidity and basicity. The constants in the generated QSAR equation are not similar to those previously generated for water/solvent PCs or log $K_{ow}$. The authors suggest that membrane systems, characterized by partitioning or rate of transfer, may not be the most appropriate models for predicting intestinal absorption. The rate constant for absorption of the ionized species of strongly acidic or basic compounds approximated that for the neutral species.

In a recent review article by Egan and Lauri (2002), many computational approaches available for the prediction of passive intestinal absorption are reviewed and assessed in terms of their relative advantages and disadvantages. Overall, lipophilicity, hydrophilicity, size, and charge are determined as being most significant in determining passive absorption.

## 2.  In Vitro *Studies*

There are several *in vitro* methods available that are used as predictors of gastrointestinal absorption. Culturing intestinal cells directly is difficult, as the cells do not retain their physiological properties in culture; transformed cell lines are therefore used. The most common of these is the Caco-2 cell line, grown from human colon carcinoma cells onto a membrane. The ability to cross this membrane is indicative of the ability to be absorbed across the gastrointestinal tract; many studies have been undertaken to measure absorption in this system. The review article by Pagliara et al. (1999) discusses many *in vitro* techniques used to predict absorption in humans with the relative advantages and disadvantages of each.

Several papers have also been published in which a correlation has been sought between permeation across Caco-2 cells and physicochemical properties of the compounds. The review article by Ekins et al. (2000) discusses several studies to predict Caco-2 cell permeation. Correlations have been found with polar surface area, hydrogen bond descriptors, VolSurf, and other parameters. Van de Waterbeemd et al. (2001a) also discuss models for predicting oral absorption of compounds, including the use of Caco-2 cell lines. This paper also provides much useful information on the optimization of pharmacokinetic parameters in drug development.

While Caco-2 cells typically require 21 days to become suitable for use, the Madin-Darby Canine Kidney (MDCK) cell line, derived from the dog kidney, typically only requires 3 days. This cell-line may prove to be a useful alternative, as the two cell lines show many common features. More investigation is required into how comparable the two systems are.

In the study by Fujiwara et al. (2002), 3D structures were determined by molecular orbital calculation for 87 compounds. The following descriptors were obtained: dipole moment, polarizability, and sum of charges on nitrogen atoms, oxygen atoms, and the hydrogen atoms bonded to them. Log $K_{ow}$ was also included as a parameter. Use of a neural network to determine the correlation of these parameters with Caco-2 permeability was reported to give good results.

Kulkarni et al. (2002) discuss the development of membrane-interaction QSAR, used to predict the interaction of compounds with biological membranes. In this method the phospholipid regions of the cell membrane are viewed as a receptor with which a solute will demonstrate both structure-specific and non-specific interaction. The results of this study indicated that absorption across Caco-2 cells was dependent on the solubility of the drug, the extent of interaction with the monolayer, and the flexibility of the molecule. Compounds that show reduced intramolecular hydrogen bonding within the membrane possess increased flexibility. It was proposed that this conformational flexibility allows a solute molecule to navigate through the membrane more readily, so that again increased hydrogen bonding results in less permeability. Additionally, strong intermolecular hydrogen bonding or electrostatic interaction with polar head groups of lipid membrane results in lower permeability.

## C.  Metabolism

Methods to predict the route and extent of metabolism include *in vitro* and *in silico* techniques. *In vitro* assays to determine metabolic stability or drug-drug interactions are typically carried out using hepatocytes or microsomes; details of these *in vitro* assay procedures are described by Li (2001). However, *in silico* prediction and optimization methods are more useful when dealing with large datasets.

The paper by Smith et al. (1996) provides information on which common functional groups act as substrates for specific enzymes of the $P_{450}$ superfamily. The 3 most common metabolizing enzymes in man, CYP2D6, CYP2C9, and CYP3A4, are discussed in terms of the preferred topography and presence of specific functional groups of their substrates. In general, metabolism by these enzymes is believed to be dependent on the topography of the active site, steric hindrance of access to this site, and the ease of abstracting electrons or hydrogen atoms from the substrate molecule. Examples are given for modifications to drug structures that act to improve drug stability. For instance, replacing a phenolic function with the bioisosteric indole-NH function has been shown to prevent glucuronidation and hence increase bioavailability of certain drugs.

The extent to which the liver successfully eliminates a xenobiotic from the blood is determined by the intrinsic clearance of the liver ($Cl_h$) and the rate at which the xenobiotic is presented to it (i.e., the hepatic blood flow [$Q_h$]). This gives the overall hepatic extraction ratio (E) for the compound, as shown in Equation 11.3:

$$E = \frac{Cl_h}{Q_h}$$

(11.3)

For highly metabolized compounds, the rate at which the drug is cleared from the liver approximately equals the hepatic blood flow; and poorly metabolized compounds, clearance approximates to zero. Poulin and Krishnan (1998; 1999) suggested that setting E equal to one for highly metabolized drugs and zero for poorly metabolized drugs could be used to generate plausible envelopes for venous drug concentrations. These values were combined with tissue:blood and blood:air PCs, predicted from molecular structure. For certain volatile organic chemicals, this novel approach enabled blood concentration-time profiles to be reasonably estimated. The approach only uses molecular structure to generate input parameters that could be used in mechanistically based toxicokinetic models, which may be useful for hazard assessment of this type of chemical.

Thompson (2001) provides a useful review article describing three decades experience of optimizing metabolic stability to aid drug design. *In vitro*, *in vivo*, and *in silico* methods to predict metabolism are all presented with their relative merits. Preventing breakdown of the drug at specific sites by rational modifications to the drug structure is also discussed. Similarly, van de Waterbeemd et al. (2001a) discuss methods of predicting metabolism and strategies to reduce rapid drug degradation in addition to the optimization of other key pharmacokinetic parameters.

Drugs that are susceptible to metabolism by the same enzyme or drugs that promote or inhibit the action of specific enzymes can affect the blood concentration of a co-administered compound; this is referred to as a drug-drug interaction. Ekins and Wrighton (2001) review *in silico* methods that have been used to predict such interactions.

## D. Affinity for Efflux Transporters

In many studies of absorption across the gut wall or uptake into the brain, any compound that has affinity for efflux transporters tends to be highlighted as an outlier from these data sets. Transporters are also expressed in other tissues and can affect overall drug distribution. *In vitro* systems of cell lines expressing P-glycoprotein (P-gp) can be used to detect substrates for this transporter, but these test systems are relatively slow and require synthesis of the drug candidate. Predicting *a priori* which drugs would have affinity for these transporters would help to screen out, at an earlier stage, drugs that are likely to be ineffective.

P-gp contains two homologous, transmembrane regions responsible for efflux and a cytosolic domain for binding and hydrolyzing adenosine tri-phosphate (ATP). Drugs that inhibit P-gp ATPase are unlikely to be transported out of the cell. In a study of 22 drug-like compounds Osterberg and Norinder (2000) found that ATPase activity increased as surface area (as defined by MolSurf); and the number and strength of hydrogen bonds also increased, making the compounds more susceptible to efflux. The review article by Ekins et al. (2000) provides numerous examples of studies undertaken on inhibitors and substrates of P-gp. Generally, the presence of carbonyl groups is associated with P-gp inhibition. Hydrogen bond acceptors and basic nitrogen groups are important for inhibitors, whereas hydrogen bond donors have been noted in substrates for this transporter.

## E.  Bioavailability

The four pharmacokinetic properties given above all have a significant effect on overall oral bioavailability. In terms of modeling the parameters it is easier to consider each of these as separate entities and optimize the individual properties. For example, it may become apparent that although the dissolution and absorption properties are good, bioavailability is low because of extensive metabolism. Efforts to improve bioavailability should concentrate on tactics to reduce metabolism. As the global problem has been distilled, it is clear where the emphasis lies for drug optimization. There have been several papers published that attempt to model overall oral bioavailability as a composite parameter. Because prediction of finite bioavailability values would be extremely diffi-cult, the models tend to rely on predicting classes of activity (i.e., high/moderate/low or accept-able/unacceptable bioavailability). Although less specific information is gained from these models, they provide an excellent tool for the preliminary screening of large databases.

Hirono et al. (1994a) studied 188 organic compounds separated into classes of aromatic, nonaromatic and heterocyclic. Using the method of fuzzy adaptive least squares, equations were generated using molecular weight, log $K_{ow}$, parameters based on the presence or absence of structural fragments or moieties, and the number of specific atoms or functional groups present. The equations enabled the compounds to be assigned to one of three classes for oral bioavailability (i.e., < 50%, 50 to 89%, and >89%), although a large number of descriptors were used for each correlation.

Andrews et al. (2000) also used the approach of correlating bioavailability with substructural fragments. A fingerprint of 608 substructure counts was generated for each of the 591 compounds studied. Recursive partitioning was initially used to classify the data after two splits into groups with different bioavailabilities. The resultant pair-wise descriptor interactions were also used as parameters in a subsequent regression model. This study is unusual as the actual bioavailability values were predicted by the regression equation, rather than classifying the data into groups. A model using 85 descriptors was able to predict the oral bioavailability of 591 compounds with an $R^2$ value of 0.71. The cross-validated $R^2$ values were 0.63 using leave-one-out (LOO) and 0.58 when one fifth of the data was omited in each iteration. The data were converted to classes of good (>80%) or poor (<20%) bioavailability. Classification by this method was compared to the results for classi-fication using the rule of five to predict good oral absorption. The model of Andrews et al. (2000) reduced the number of false negatives from 5 to 3% and false positives from 78 to 53% when compared to the Lipinski method. However, this does come at the cost of a much more complex model.

Yoshida and Topliss (2000) used the ORMUCS (ORdered MUlticategorical Classification method using the Simplex technique) to assign a series of 232 diverse drugs to 1 of 4 bioavailability ratings. Lipophilicity (log $D_{6.5}$) was found to be an important parameter. As they observed a trend of increasing bioavailability going from basic, to neutral, to acidic compounds, the parameter $\Delta$log D (i.e., log $D_{6.5}$ – log $D_{7.4}$) was also used as a descriptor, along with 15 descriptors for specific structural features. For 232 compounds in the training set, 71% were correctly classified and 97% were either correct or within 1 class. For the external test set of 40 compounds, 60% were correctly classified with 95% correct or within 1 class. This method has the advantage of using calculable parameters, known to be important for absorption and metabolism, that allow screening of com-pounds yet to be synthesised.

Veber et al. (2002) also proposed a method for rapid preliminary screening of large data sets. For 1100 drug candidates studied in the rat, good bioavailability was predicted for those having ≤10 rotatable bonds and a polar surface area of ≤140$A^2$ (or ≤12 hydrogen bond donors or acceptors). When this rule was applied to human oral bioavailability data, 86% of 277 drugs were correctly selected as having good oral bioavailability, compared with 93% correctly selected by the rule of 5. Interestingly, the authors note that the molecular weight cutoff of 500 is not itself a determinant; it is the number of rotatable bonds that is important. They suggest that compounds of higher molecular weight may prove to be useful, providing the number of rotatable bonds is below 10. The reason the higher-molecular-weight compounds are poorly absorbed is related more to the higher number of rotatable bonds present in larger molecules. Further investigation may result in fewer potentially useful candidates being screened out in the early stages of development.

The most accurate method of determining oral bioavailability is to make *in vivo* measurements. This method requires both labor- and animal-intensive experimental methods. Prediction using *in vitro* methods does provide some advantages, but normally only considers a single biological endpoint for each assay. In the work of Mandagere et al. (2002), the authors use a combination of two *in vitro* methods to study the composite endpoint of oral bioavailability. The work combines Caco-2 permeation data with metabolic stability data and correctly assigns the 21 drugs studied to the classes of high, medium, or low bioavailability. Although a logical and promising method for predicting oral bioavailability, it still requires experimental results and therefore would not be amenable to very high throughput, preliminary screening.

## F.  Alternative Routes of Administration

Although the oral route is the most common route of drug delivery, investigation into the prediction of uptake by other routes is also important. Dermal penetration and inhalation are not only relevant as methods of drug delivery, but in terms of toxicology they represent important routes of exposure.

Abraham et al. (1999) generated a QSAR equation for the permeation of 47 aqueous solutes across human skin. This demonstrated that dermal permeability increased with increasing molar refraction and McGowan volume and decreased as hydrogen bond acidity, basicity, and dipolarity/polarizability increased.

Potts and Guy (1992) studying permeation of 93 compounds through excised human skin found that a simple 2-parameter model explained 67% of the variation in the data. Log $K_{ow}$ had a positive influence and molecular mass a negative influence on skin penetration. Cronin et al. (1999) studied an expanded set of 114 compounds with similar results. Although the number of lone-pair, hydrogen bond acceptors was found to correlate negatively with skin permeability, in this data set it was noted that this was caused by the influence of only a small number of compounds. Log $K_{ow}$ and molecular mass were more significant in determining dermal penetration overall.

Gute et al. (1999) discuss the importance of being able to predict dermal penetration directly from structure, using readily calculable parameters, for use in hazard assessment. Five types of parameters were used: topostructural, topochemical, geometric, quantum chemical, and physicochemical. A hierarchical approach was used starting with the simplest indices and adding those of increasing complexity (i.e., the best topostructural indices were selected then, retaining the best of these indices, the topochemical indices were added and the best of these were selected to include in the model). Physicochemical parameters were used at each stage to compare their results with the theoretical descriptors. Models were also generated using the individual sets of parameters. The results indicated that topostructural, topochemical, geometric, and physicochemical indices were all equally effective in predicting dermal penetration, typically accounting for 65 to 70% of the variation in the data. The authors also found that molecular weight was a better predictor of permeability than lipophilicity, postulating that this was because of the compounds' having to cross through aqueous pores rather than through the dermal membranes.

The findings of du Plessis et al. (2002) in part supported those of Abraham et al. (1999). For a small number of phenols they demonstrated that increasing dipolar and hydrogen bonding capacity reduced dermal permeation. They also reported that symmetrical phenols showed higher permeation than asymmetric phenols, attributed to the fact that the symmetry reduces the dipolar and hydrogen bonding properties of the molecules. This indicates that the number and specific position of substituent groups may be significant in determining dermal penetration, although these results were generated from only a small data set.

The importance of exposure via the inhalational route for volatile organic compounds, particularly in industrial settings, is discussed by Meulenberg and Vijverberg (2000). In this paper the authors investigate the relative partitioning of volatile organic compounds between various tissues and air. The assumption is made that affinity for different tissues is based on relative affinity for lipophilic and hydrophilic components of the tissue. Tissue:air PCs are derived from a knowledge of saline:air and oil:air PCs of the chemicals. This type of analysis may be particularly useful in toxicological studies and partitioning into specific tissues is discussed in more detail in a later section.

## G.  Binding to Plasma Proteins

Irrespective of the route of administration, the drug or toxicant will be carried to its site of action or elimination via the blood; consequently, interaction with blood proteins is significant for drug design and toxicology studies, as binding will affect the overall distribution of the compound. Saiakhov et al. (2000) provide a detailed discussion of plasma-protein binding sites for drugs and the affinity of specific functional groups for these sites. Hydrophobic forces and the ability to form hydrogen bonds are key factors in determining the extent of binding. The authors also discuss the use of Multiple Computer-Automated Structure Evaluation (M-CASE) to generate a model for drug-protein binding. Although overall correlation with log $K_{ow}$ was poor, for the total 154 drugs studied, good correlations were found with small subsets of the data. This supports previous findings that log $K_{ow}$ may correlate positively with binding within limited data sets. Increased polarity of drugs, as measured by their hardness-softness index, tended to decrease binding.

Rowley et al. (1997) also showed protein binding increased with log $K_{ow}$ for a series of glycine *N*-methyl-D-aspartate (NMDA) receptor antagonists. Their study was undertaken to investigate why compounds of this series, that appeared promising *in vitro*, were not as effective as anticipated *in vivo*. The results of the study suggested that increasing log $K_{ow}$ increased drug binding to plasma proteins. This limited the amount of drug entering the brain, as a bound drug cannot cross the BBB, reducing the central effects of the drugs. This study is pertinent because it demonstrates directly that optimizing pharmacodynamic effects *in vitro*, without consideration of pharmacokinetics, can lead to disappointing results *in vivo*. The authors note that prediction of BBB penetration may be improved by incorporation of a protein binding term.

Davis et al. (2000) discuss the problem of correcting certain pharmacokinetic values for the fraction unbound in plasma, advising caution in the approach as it may lead to spurious correlations. Values for clearance or volume of distribution of compounds have been reported in previous studies to correlate with log $K_{ow}$. In some of these studies the correlations use values for clearance or volume of distribution that have been corrected for the fraction unbound in plasma. However, the fraction unbound in plasma is itself dependent on log $K_{ow}$. Therefore, it is important to check that these reported correlations are genuine and not merely a reflection of the relationship between plasma protein binding and log $K_{ow}$.

## H.  Tissue:Blood Partition Coefficients

Tissue:blood PCs indicate the relative affinity of compounds for the various tissues of the body compared to blood. The values are determined by the relative lipophilic/hydrophilic nature of the compound and relative affinity for the macromolecules found in tissue and blood. Each individual tissue will make up a specific balance of water, neutral lipid, phospholipid, and protein. Partitioning therefore is determined by the relative affinity of the compound for the specific tissue constituents.

De Jongh et al. (1997) derived tissue:blood PCs for 24 volatile organic compounds in liver, muscle, fat, kidney, and brain of humans and for 42 compounds in the fat, liver, and muscle of the rat, using the octanol-water PC. Deviations between tissue-blood and octanol-water models are accounted for using calibration coefficients for each tissue. Consequently, the model takes into account relative proportions of lipid and water in the individual tissues and whether or not the lipid fractions in tissues behave similarly to octanol. A second calibration coefficient is also included to take account of other interactions with tissue components, such as protein binding; this is described as an empirical approximation as a single coefficient cannot reasonably account for all possible interactions. Tissue:blood PCs tend to be higher in the rat than human, and it was proposed that this may be due to increased binding to haemoglobin in rat blood. Such interspecies differences can cause problems when scaling animal to human data. The same caveat was given by Meulenberg and Vijverberg (2000).

The study of Vaes et al. (1998) demonstrated the importance of considering both partitioning between an aqueous and an organic phase and hydrogen bonding parameters when predicting membrane/water interactions. Although octanol-water partitioning is frequently used to predict membrane-water partitioning, there are significant differences between the two. Octanol is a bulk solvent, whereas phospholipids are a highly organized bilayer, each possessing different lengths of alkyl chain and different polar head groups, which will affect hydrogen bonding properties. The authors conclude that high membrane/water PCs will be exhibited by hydrophobic chemicals acting as good hydrogen bond donors, as these will interact favorably with negatively charged phospholipid head groups. Compounds acting as hydrogen bond acceptors were assumed to interact unfavorably with the membrane, and hence have lower membrane/water PCs.

Balaz and Lukacova (1999) attempted to model the partitioning of 36 non-ionizable compounds in 7 tissues. Amphiphilic compounds, or those possessing extreme log $K_{ow}$ values, tended to show complex distribution kinetics because of their slow membrane transport. However for the non-amphiphilic, non-ionizable compounds with non-extreme log $K_{ow}$ values studied it should be possible to characterize their distribution characteristics based on tissue:blood PCs. Distribution is dependent on membrane accumulation, protein binding, and distribution in the aqueous phase. As these features are global rather than dependent on specific 3D structure, distribution is not expected to be structure-specific. In this study, tissue compositions in terms of their protein, lipid, and water content were taken from published data. This information was used to generate models indicating that partitioning was a non-linear function of the compound's lipophilicity and the specific tissue composition.

Meulenberg and Vijverberg (2000) derived relationships between tissue:air PCs and oil:air plus saline:air PCs. For 137 volatile organic compounds PCs in 7 tissues could be described by linear combination of oil:air and saline:air PCs with tissue-specific regression coefficients, the authors suggest the method may be useful in risk assessment. The assumption is made that tissue partitioning is determined by lipophilic and hydrophilic interactions with individual tissue constituents. Although tissue-specific coefficients are used, these are derived empirically and are not based on prior knowledge of tissue composition. This makes a simpler model, but does not allow for mechanistic interpretation.

Poulin and Krishnan (1995) developed a method to predict tissue:blood PCs for incorporation into physiologically based pharmacokinetic (PBPK) models. Tissue:blood partitioning was calculated as an additive function of partitioning into the water, neutral lipids and phospholipids constituent of individual tissues. These were calculated using published values for lipid and water content of tissues and the octanol-water PC of the compounds. Poulin and Krishnan (1998; 1999) used this method to predict tissue:blood PCs that were subsequently incorporated into a quantitative structure-toxicokinetic model. The prediction of tissue:plasma PCs to describe distribution processes and as input parameters for PBPK models has been extensively researched by Poulin and coworkers; a great deal of further information can be obtained from their references (Poulin and Theil, 2000; Poulin et al., 2001; Poulin and Theil, 2002a; Poulin and Theil, 2002b).

Another useful reference source is the review article by Payne and Kenny (2002), who compare the methods used to predict blood:air, tissue:air, or tissue:blood PCs of volatile organic chemicals. The relative advantages and disadvantages of each method are discussed, as well as the accuracy of each method in predictions made for individual organs in humans and rats.

## I.  Blood-Brain Barrier Partition Coefficient

Of all the tissue:blood PCs, none has been more intensively studied than the partitioning between the blood and the brain. This is a reflection of the importance of the brain as a site of action of drugs and as a site of potentially serious toxic side-effects of drugs, designed to act elsewhere, or for environmental pollutants.

Several data sets of BBB PCs exist, and each has generally been studied by a number of authors, leading to a large number of publications on this subject. One factor that is consistently expressed as being particularly significant is the hydrogen bonding capacity of the molecules. It has been demonstrated that BBB partitioning tends to decrease as the hydrogen bonding capacity of the molecules increases (Young et al., 1988; Abraham et al., 1994; Norinder et al., 1998; Abraham et al., 1999; Luco 1999; Platts et al., 2001).

Various calculations of the polar surface area (PSA) have also been used in prediction of BBB partitioning; these have been shown by several authors to have a negative correlation, as reviewed by Norinder and Haberlein (2002). One disadvantage to the use of polar surface area as a parameter is the time taken to generate a reasonable geometry-optimized 3D molecular structure and then generate the surface area for that specific conformation. In 2000, Ertl et al. reported the use of a surface area that could be rapidly calculated from the sum of fragment-based contributions to the overall surface. This was referred to as the TPSA. There were 34,810 drug-like molecules from the World Drug Index used to determine the contributions to the polar surface made by specific polar fragments. Forty-three polar atom types were identified and their contribution to the PSA tabulated. The sum of the fragmental values gives the total calculated PSA for a molecule. These values have been shown to be highly correlated with values determined from the surface of the optimized 3D structures. BBB partitioning when correlated with TPSA produced very similar results to those obtained using traditional 3D PSA calculations.

In addition to the key role of hydrogen bonding, many authors have reported other features that appear to be significant in determining BBB partitioning. Abraham et al. (1994) demonstrated that BBB partitioning of 57 compounds increased with excess molar refraction and McGowan volume and decreased with dipolarity/polarizability. Similar findings were reported in a study of 157 compounds by Platts et al. (2001), who also demonstrated a significant negative correlation with the presence of carboxylic acid groups. Lombardo et al. (1996) demonstrated a positive correlation between BBB partitioning and the computed solvation free energy in water, using the AMSOL program. Keseru and Molnar (2001) also found a similar correlation, but used a more rapid method of calculating the free energy. Kaliszan and Markuszewski (1996) found BBB partitioning in general increased with increasing log $K_{ow}$ and decreased as molecular weight or water-accessible volume increased. Salminen et al. (1997) showed a negative correlation with molecular volume and a positive correlation with log $K_{ow}$. The presence of an amino-nitrogen was favorable, whereas the presence of a carboxyl group was unfavorable to brain penetration.

Norinder et al. (1998), using descriptors generated from MolSurf, showed that BBB partitioning was also influenced positively by increased lipophilicity and the presence of polarizable surface electrons. Luco (1999) showed that parameters describing polarity of the molecules and to a lesser extent, overall molecular shape were significant. Crivori et al. (2000) used VolSurf (a procedure to generate descriptors from the 3D molecular field) to estimate whether or not a compound could enter the brain, with 90% correct prediction. Rose et al. (2002) generated a QSAR model for BBB partitioning using two E-state indices and a molecular connectivity index. E-state indices integrate both electronic and topological information for molecules. The indices are derived from the valence state electronegativity of an atom in relation to its local topological character, describing electron accessibility. Hydrogen accessibility is represented by hydrogen E-state indices. The parameters incorporated in the QSAR model indicate a negative correlation with hydrogen bond donating ability, degree of skeletal branching, and electronegativity. A positive correlation with the hydrogen

E-state descriptor for aromatic CH groups was also indicated, suggesting non-polar aromatic regions may have an influence on partitioning.

The review paper by Norinder and Haeberlein (2002) discusses several published data sets along with present state of knowledge of factors that affect BBB partitioning. They review the findings of many of the papers listed above in more detail. After considering the available data the authors report two very simple rules for prediction of BBB partitioning. Firstly, if the number of nitrogen plus oxygen atoms in a molecule (N + O) is ≤5, then it has a high chance of entering the brain. Secondly, if log $K_{ow}$ – (N + O) is positive, then log BBB partitioning is positive. Such simple and easily calculable parameters may prove to be very useful in the preliminary screening of very large data sets or corporate libraries.

## J.   Volume of Distribution

The approach of Hirono et al. (1994a) used to predict oral bioavailability, as discussed above, was also used to develop a model to predict volume of distribution (Hirono et al., 1994b). Fuzzy adaptive least squares analysis using log $K_{ow}$, molecular weight, and the presence or absence of specific structural fragments was used to allocate 373 miscellaneous drugs to one of three classes of volume of distribution (i.e., <0.5 Lkg$^{-1}$, 0.5 – 1.9 Lkg$^{-1}$ and ≥2 Lkg$^{-1}$). Large numbers of descriptors were used in generating the correlations.

Lombardo et al. (2002) present a method for the prediction of volume of distribution for neutral and basic compounds. The model uses an experimental value for log $D_{7.4}$, the fraction ionized at pH 7.4 and an experimentally determined value for the fraction of drug unbound in human plasma. From these values the fraction unbound in tissues is predicted and used to predict the overall volume of distribution in humans. A training set of 64 compounds was used; for the test set of 14 compounds, the predicted values were, in general, within an approximately twofold error of the actual values. The method has the advantage of not requiring results from animal experiments with the associated time constraints, ethical issues, and problems of scaling results from animal to man. However, it does require results from plasma protein binding studies, and therefore cannot be carried out purely *in silico*. Attempts to predict volume of distribution from computed parameters including surface area, charge, volume, and hydrogen bonding parameters were not successful in this case.

According to Poulin and Theil (2002a), the volume of distribution of a xenobiotic depends on the sum of products of all tissue:plasma PCs and the respective tissue and plasma volumes. Values for tissue and plasma volumes are available in the literature and much work has been done on predicting tissue:plasma PCs (see above). The authors' prediction of tissue:plasma PCs uses knowledge of the drug's lipophilicity, plasma protein binding affinity, and specific information on tissue composition. The paper provides specific details of how PCs are calculated for adipose and non-adipose tissues. It also provides a useful summary of information on the composition of the major organs in terms of overall weight and fractional content of water, neutral lipids, phospholipids, and interstitial space. Application of their method to 123 structurally diverse acidic, basic, and neutral compounds gave predicted volumes of distribution within a factor of 2 of the experimental value for 80% of the compounds. These compounds were believed to distribute to tissues homogenously by passive diffusion or remain in extracellular space. For the remaining compounds additional distribution processes, particularly those relevant for cationic-amphiphilic bases, were not adequately described. As a result, volume of distribution values were not accurately predicted for those compounds.

## K.   Clearance

For many drugs that do not undergo metabolism, renal clearance is the primary route of elimination from the body. At a simplistic level drugs may be categorized into those that are extensively metabolized or those that are primarily excreted unchanged in the urine (i.e., nonextensively metabolized). Predicting the fraction excreted unchanged in urine provides valuable information as to the likely route of elimination from the body and the overall susceptibility of the drug to metabolism. In the work of Manga et al. (2003) a hierarchical model was developed that allowed for the classification of drugs into extensively or nonextensively metabolized compounds using physicochemical properties alone. Classifying the biological data into groups is a commonly used approach that circumvents the problems of high biological and interlaboratory variation that is typical of this type of heterogeneous data set.

The final hierarchical model generated in this study was a three-level decision tree using different classification techniques. In the first step compounds possessing $\log D_{6.5} > 0.3$ were predicted to be extensively metabolized (i.e., <25% is excreted unchanged in urine). (It was noted that compounds possessing $\leq 1$ hydrogen bond donor were also extensively metabolized, and this could be used equally well as the first step in the decision tree.) In the second and third steps discriminant functions were generated that used the counts of acid groups, OH moieties, hydrogen bond donors, sum of net positive and negative charges on the compound, COSMIC total energy, electronic energy, and ionization potential. The data set was further split according to the results generated from these two functions. The model correctly assigned over 90% of the 160 drugs of the training set and 85% of the 40 drugs in the test set to the classes of extensively or nonextensively metabolized compounds.

## L. Half-Life

As half-life is a composite parameter dependent on clearance and volume of distribution, it is easier to model and optimize each constituent parameter individually, but there have been some attempts to model half-life overall. Historically half-life values in man have been predicted by scaling the results obtained from animal experiments. Generally this method has been reasonably satisfactory (Bachman et al., 1996). Attempts have been made to improve these predictions by the incorporation of additional physicochemical parameters. As the percentage of adipose in rat is only 7%, compared to 23% in man, the incorporation of $\log K_{ow}$ has been investigated. Sarver et al. (1997) showed that incorporation of $\log K_{ow}$ as a parameter only improved the correlation between human and rat data for compounds with extremely high $\log K_{ow}$ values, and that generally rat data were a good predictor of human data. The use of animals is extremely costly and time consuming, and alternative methods are required.

Quinones et al. (2000) reported the successful use of neural networks to predict the half-life of a series of 30 antihistamines. The input for the network was derived from the output of CODES, a routine that generates descriptors for a structure based on atom nature, bonding, and connectivity. Attempts to correlate the half-life with the physicochemical parameters $\log K_{ow}$, $pK_a$, molecular weight, molar refractivity, molar volume, parachor, and polarity were unsuccessful. In a subsequent study by Quinones-Torrelo et al. (2001), the authors correlated the half-life of 18 antihistamines with their retention in a biopartitioning micellar chromatography system with a resultant correlation coefficient ($R^2_{adj}$) value of 0.89. The correlation is explained in that the retention in this system is dependent on hydrophobic, electronic, and steric properties, which are also important in determining half-life.

## V. USE OF IMMOBILIZED ARTIFICIAL MEMBRANES AND LIPOSOMES TO PREDICT TRANSPORT PROPERTIES OF DRUGS OR TOXICANTS

Several studies have been undertaken to measure the retention times of compounds on chromatography columns comprising immobilized liposomes or artificial membranes. Liposomes comprise aqueous compartments surrounded by lipid bilayers, whereas the artificial membranes comprise monolayers of phospholipids (Lundahl and Beigi, 1997). Interaction between the compound and the phospholipids present mimics the interaction of compounds with cell membranes. Retention on these columns has been shown to correlate with certain transport properties of xenobiotics (e.g., permeation through Caco-2 cells or absorption following oral administration [Ong et al., 1996]). The models are representative only of passive diffusion of xenobiotics across lipid bilayers. As a result, retention will not correlate with absorption for compounds that use the paracellular route or active transport processes (Lundahl and Beigi, 1997).

Norinder and Osterberg (2000) comment that log $K_{ow}$ itself is a relatively poor predictor of drug transport properties. They discuss the applicability of liposome partitioning, immobilized artificial membrane (IAM), and liposome chromatography as predictors of drug transport. Liposomes were considered to provide a better model, as they possess an ordered structure, capable of electrostatic interaction, showing similarities with cell membrane structure. Schaper et al. (2001) similarly state that although octanol can interact with solutes via hydrophobic and hydrogen bonding interactions, it does not account for the interfacial interaction between membrane phospholipids and solutes. The use of IAM or liposome partitioning is a better predictor of drug absorption.

The degree of ionization of a compound will have a marked effect on its partitioning. The distribution profiles for acids and bases obtained using liposome partitioning were not the same as would be predicted from octanol-water partitioning. Amphiphilic drug molecules may interact with the polar head groups of the IAM in a similar manner to the interaction with the phospholipid bilayer of cells. The degree of ionization of a drug will affect its amphiphilicity and its interaction with these head groups. Amphiphilicity has a greater effect on drug transport rates than lipophilicity itself. The amphiphilic and lipophilic nature of a drug can be studied using these IAMs, enabling better prediction of drug transport (Balaz, 2000) and drug distribution (Barton et al., 1997).

Quinones-Torrelo et al. (1999; 2001) have demonstrated a correlation of pharmacokinetic properties with results from micellar liquid chromatography. In this method micellar solutions of nonionic surfactants are used as the mobile phase in reverse-phase liquid chromatography. Interactions between the mobile and stationary phases are purported to correspond to the membrane/water interface of biological barriers as hydrophobic, steric, and electronic interactions are important for both. For a series of 18 antihistamines Quinones-Torrelo et al. (2001) showed that both volume of distribution and half-life values were better correlated with retention on these columns than with the classical log $K_{ow}$ descriptor.

Results from studies using IAMs or liposomes may prove a useful screening tool for the prediction of transport properties of drug candidates or toxicants. Interaction between the compounds and phospholipids present in these systems is more representative of the *in vivo* situation than computer-generated parameters or results from simple oil/water partitioning systems. These methods do require prior synthesis of compound and take longer to perform than *in silico* screens. Although they do provide useful information, this has to be weighed against the cost of performing the experiments.

## VI. A GLOBAL PERSPECTIVE ON THE PREDICTION OF PHARMACOKINETIC PROPERTIES

Looking at Figure 11.1, it is apparent that the pharmacokinetic optimization of drug candidates, or the assessment of risk posed by a chemical entity, is a complex, multifaceted problem. Generally speaking, there are three levels at which information is obtained and utilized.

1. Low throughput — This is slow, labor, intensive and costly, providing very detailed information on few compounds per iteration. Techniques used here tend to include information obtained from *in vitro* and *in vivo* biological assays with extrapolation to human subjects. Although much information is obtained for individual compounds, the data sets are generally too small to be useful in external prediction for new compounds.
2. Medium throughput — This level may also use data from biological assays, but using techniques that allow many compounds to be analyzed simultaneously. Medium-sized data sets are amenable

to computer modeling, and QSAR techniques may identify trends within data that may allow for some prediction outside of the test series.

3. High throughput — Generally these involve *in silico* screening methods, capable of categorizing thousands to hundreds of thousands of compounds as to those possessing desirable or undesirable characteristics.

The choice of using low-, medium- or high-throughput techniques depends on the particular question being dealt with and the required outcome of the study. Each of these three levels is considered individually in this section, although the boundaries among the three levels cannot be simply defined.

## A.  Low Throughput

Low rates of throughput provide the most detailed and informative data, but at a high cost in terms of time, expense, and, frequently, animal usage. At the fundamental level, individual pharmacokinetic parameters, such as clearance, volume of distribution, or half-life, are measured in an alternative animal species and the results extrapolated to man. A scaling algorithm is used that typically takes account of inter-species differences in body weight and surface area. Bachman et al. (1996) discuss the successful application of this method to predict toxicokinetic parameters of xenobiotics. Although a commonly used approach, there are many problems associated with direct extrapolation of pharmacokinetic parameters. For example, following an oral dose, the extent of paracellular absorption is affected by the size of the hydrophilic pores in the gut through which molecules may pass. The size of these pores varies between species, and cutoff values for the maximum size of molecule that may permeate will not be the same across all species. The expression of the many metabolizing enzymes involved in drug or toxicant degradation is highly variable between species. The rate and extent of metabolism, as well as the specific metabolites formed, therefore differ. The extent of binding to plasma proteins has also been shown to vary among species, affecting the overall distribution of xenobiotics. Interspecies differences in adipose and lean tissue content will also affect disposition and storage properties.

More intensive in terms of animal usage are the experiments used to generate parameters for inclusion in full PBPK models. These models follow the time course of a drug or toxicant in all major organs of the body. Typically, following dosing, groups of animals are killed at a series of time points, and the amount of compound present in the individual tissues is measured directly. The key tissues studied are generally blood, liver, lung, heart, spleen, pancreas, stomach, testes, brain, kidney, skin, muscle, bone, and adipose. Much detailed pharmacokinetic information can be gained from direct measurement of compound levels in tissues as time progresses. The method requires the use of many animals and complex assay procedures to extract and quantify the amounts of individual compounds present.

The throughput of *in vivo* methods can be increased by the use of cassette dosing techniques, where rather than administering a single compound to an animal, several compounds are co-administered. The concentrations of all compounds in the tissues can then be measured simultaneously. This method uses low doses of each compound and assumes that the presence of one compound will have no effect on the pharmacokinetics of another or on the physiology of the animal. In these studies a balance must be obtained between administering sufficient compounds to enable accurate measurements of concentrations and giving a low enough dose such that the compounds do not interfere with each other or have an adverse effect on the animal. This can be problematic, particularly on administration of a group of compounds that exert the same biological effect, bind to the same macromolecules, or are metabolized by the same enzyme system.

Killing groups of animals at given time points obviously requires large numbers of animals to be used in these studies. One promising development to reduce the number of animals necessary is the use of microdialysis and ultrafiltration as membrane sampling techniques, discussed by Garrison et al. (2002). In these procedures only one group of animals is used for the entire time-course study. Probes are inserted directly into the tissue of interest and a small sample is withdrawn, from which the concentration of compound in the tissue can then be determined. This allows measurements for all time points to be made in a single animal and helps to reduce inter-animal variability in data.

From these techniques, very detailed information is obtained for a few compounds per iteration. This level of detail is generally unnecessary early in the drug development process and, as the number of compounds is limited, the information gained is not widely applicable to prediction of properties of diverse compounds. The information is useful when, for example, considering how a specific disease state would affect the pharmacokinetics of a compound. PBPK models use information relating to the functioning of the tissues and the rate at which blood presents compounds to that tissue. For example, predictions could be made as to how the pharmacokinetics of a compound would be affected when administered to a patient suffering from renal or hepatic failure etc.

Because physical measurement of parameters for input into full PBPK models is so labor intensive, prediction of the relevant parameters would be beneficial. The time course of a drug in any organ will be influenced by the blood flow to that organ, perfusion, and specific composition in terms of water content, constituent lipids, proteins and fat content of the organ. These parameters are already available in the literature, for several species. The tissue:blood PC and the ability of certain organs to metabolize or eliminate the compound will also be influential. These parameters have been measured for many compounds and several studies are currently being undertaken to predict these types of parameters (Poulin and Krishnan, 1995, 1998, 1999; Poulin and Theil, 2000, 2002a, 2002b; Poulin et al., 2001). The ability to predict parameters for incorporation into PBPK models will enable more rapid screening of compounds, as well as allow for mechanistic interpretation of pharmacokinetic profiles. It is hoped these predictive models will become more accurate and applicable to a wider range of compounds. Ultimately, it may become possible to recognize the influence of specific functional groups on distribution, uptake, storage, biotransformation, etc. within individual organs. Integrating data from quantitative structure-pharmacokinetic relationship (QSPkR) analyses within PBPK models may help to make this possible Fouchecourt et al. (2001).

## B. Medium Throughput

Moving away from direct *in vivo* measurements toward predictive models and the incorporation of QSPkR information increases the rate at which compounds can be screened for pharmacokinetic properties. Many of the models available in the literature, as discussed above, use medium-throughput techniques (i.e., QSPkR analyses on data sets comprising hundreds of compounds). The classical approach of investigating the specific structure and properties of a molecule and correlating these parameters with a pharmacokinetic endpoint has proven successful for a range of pharmacokinetic properties. In particular, the wide application of these techniques to BBB partitioning and gastrointestinal absorption has been able to elucidate which properties of molecules make compounds more or less susceptible to uptake in these systems. At the medium throughput level, as compound numbers tend to be in the hundreds, it is still possible to perform detailed, computationally expensive analysis of molecular structure. Detailed 3D information can be incorporated into the models and may aid prediction and understanding of the molecular properties important in determining pharmacokinetic profiles. 3D QSAR techniques used in predicting absorption, distribution, metabolism, and excretion (ADME) properties has been reviewed by Ekins et al. (2000). A range of programs available for the prediction of various pharmacokinetic parameters is listed by Livingstone (2003).

Rapid *in vitro* screens, such as using isolated hepatocytes or microsomes to ascertain the rate of production and nature of metabolites, are also possible in medium throughput. Any form of *in vitro* or *in vivo* testing requires the compound to be synthesized *a priori*. Although much useful information can be obtained by taking direct measurements, this prohibits the screening of compounds yet to be synthesized — a desirable endpoint for more rapid screening.

## C. High Throughput

With the advance of combinatorial chemistry and high throughput screening *in vitro* assays, basic pharmacokinetic information can be gained on thousands of compounds simultaneously. These techniques still require the time and expense of synthesizing the compounds *a priori*. A great advance in drug discovery and hazard assessment is the development of *in silico* screens that allow predictions to be made for hundreds of thousands of compounds within a relatively short period of time. As the predictions are made using information based on molecular structure only, prior synthesis of the compound is not necessary; this allows large corporate libraries of theoretical compounds to be screened.

When performing high throughput *in silico* screens, the physicochemical parameters incorporated in the models must be amenable to rapid generation. Although parameters obtained from detailed 3D molecular analysis can be generated for moderately sized data sets, once the data sets become larger, the time taken to generate and optimize 3D structures becomes a limiting factor. Osterberg and Norinder (2001) discuss the use of parameters generated from ACD/logP and ACD/Chemsketch (Advanced Chemistry Development, Canada) in the prediction of drug transport properties. These programs generate physicochemical parameters from 2D graphs, assuming additive contributions from atoms, groups, or fragments. The parameters generated from these programs were found to produce good statistical models with data sets relating to solubility, IAM chromatography, Caco-2 permeability, human intestinal perfusion, BBB partitioning, and P-gp ATPase activity. The authors compare the findings to those produced when using the MolSurf methodology. The MolSurf program is a computationally expensive routine taking account of dynamic behavior and electronic properties of molecules. Molecular descriptors related to properties such as lipophilicity, polarity, polarizability, and hydrogen bonding ability can be calculated and subsequently incorporated into models to predict transport properties. The authors have successfully used this approach to predict BBB partitioning and P-gp ATPase activity (Norinder et al., 1998; Osterberg and Norinder, 2000). Although MolSurf does produce useful descriptors, its use is better suited to smaller data sets, whereas programs used to generate 1D or 2D descriptors are more useful for larger data sets.

The work of Ertl et al. (2000) also demonstrates that it is not always necessary to use parameters from 3D analysis; these can be substituted for more rapidly calculable parameters. In this paper the authors showed that the PSA of a molecule, calculated using the sum of fragment contributions available from tables, produced practically identical results when compared to calculating the surface from a geometry-optimised 3D structure. This means that sufficiently accurate PSA values can be incorporated into screens without requiring excessive computational time.

In low- and medium-throughput studies, the data sets are frequently limited, so that prediction of finite pharmacokinetic parameters is reasonable. As the data sets become larger, the variation in the data, which is frequently accumulated from a number of sources, becomes more problematic. As the data sets become increasingly chemically diverse, prediction of finite values becomes less realistic. When screening large numbers of compounds, it is often more useful to be able to readily classify the data into specific groups (i.e., high/medium/low or acceptable/unacceptable pharmacokinetic values). In preliminary screening of very large data sets, this is often as much information as is necessary before proceeding to the next step of the development process. Rank orders can be established for groups of compounds for those that are likely or unlikely to show the desired characteristics.

Simplistic models with few rules have proven to be highly beneficial in predicting basic pharmacokinetic properties without extensive computational analysis. The Lipinkski "rule of 5" for predicting which compounds are likely to show good or poor absorption properties is one example that has found wide application in industry and there are others. Egan et al. (2000) reported that compounds possessing a PSA of $<148.1 A^2$ and log $K_{ow}$ above 5.88 would be poorly absorbed. Veber et al. (2002) proposed a simple classification system for oral bioavailability, where compounds

with ≤10 rotatable bonds and a PSA of ≤140A$^2$ (or ≤12 hydrogen bond donors or acceptors) would
s'        ' bioavailability. Similarly, two very simple rules were proposed by Norinder and Hae-
                    ~essment of blood-brain barrier partitioning (i.e., if [N + O] ≤5, then the
                    :r the brain, and if log $K_{ow}$ – (N + O) is positive, then log BBB is positive
                    ).

rules can be easily incorporated into rapid screening programs. Although
            certainly be misclassified, they still represent a very useful preliminary
            information is obtained the rules can be improved and refined to increase
            ctions. Absorption and BBB partitioning are two of the most extensively
            :tic properties. In future it is hoped that other simple rule systems could be
        pharmacokinetic properties. Output from a combination of these simple screens
        synthetic efforts or be used to rank predicted toxicity levels of other chemicals.
        being able to screen compounds more rapidly, another advantage of predicting
        .inetic properties *in silico*, is that the models are generated with human data. In
        h as prediction of metabolism, this may offer an advantage over scaling animal
        ivents the problems of interspecies differences (Li, 2001).
        iy other areas of structure-activity relationship analysis, neural networks have been
        iroblem of predicting pharmacokinetic parameters with varying degrees of success.
        al networks may detect patterns in data, mechanistic interpretation of the results can
        7hen dealing with high throughput screening more transparent approaches, such as
        ie systems discussed above, tend to be more useful. In drug design, for example, it
        sier to modify candidate structures to comply with simple rules such as the "rule of 5"
        ld be to modify the structure according to the outcome of a neural network.
        ; prediction of metabolism, commercial programs have been developed that allow the
        s of enzymatic attack to be identified for a molecule, indicating potential routes of
        ,m and likely metabolites. The use of such programs in the prediction of metabolism has
        iewed by Langowski and Long (2002), and this topic is discussed in more detail elsewhere
        )ook (see Chapter 10).
        larger data sets conventional methods such as multilinear regression may not be appropriate
        ,e there are nonlinearities in the data, threshold values are required for certain features, or
        are interactions between features. For large, complex data sets, recursive partitioning can be
        re useful technique because it can account for these types of problems (Hawkins et al., 1997).
        irsive partitioning procedures, such as Formal Inference-based Recursive Modeling (FIRM),
        de complete data sets into subgroups according to the values of parameters used as predictors.
        : output of these hierarchical models is in the form of a decision tree. In terms of large-scale
        prediction of pharmacokinetic properties such an approach can be used to divide compounds, for
example, into groups of high or low bioavailability, long or short half-life, and likelihood of crossing
the BBB. This approach has already been used successfully by Manga et al. (2003) to predict the
extent to which a compound would undergo biotransformation. For high-throughput screening, the
parameters used to split the data need to be readily calculable to keep computational time to a
minimum. However, these techniques can also be applied to smaller data sets, allowing inclusion
of more complex parameters. Iterative splitting of the data allows for mechanistic interpretation of
the features that affect specific pharmacokinetic properties, making recursive partitioning a very
useful approach to this type of problem.


## VII. SELECTION OF APPROACH

The approach taken to the prediction of pharmacokinetic parameters depends on several factors,
such as the size of the data set under consideration and the type of problem being investigated,
whether it is library design, lead optimization, or derivation of specific individual parameters for

incorporation into full PBPK models. For small data sets of novel drugs late in the development process, it is feasible to obtain precise pharmacokinetic information directly from *in vivo* studies. Although labor intensive and time consuming, this can provide essential information on a small number of compounds. This level of detail is often unnecessary, particularly early in the development process. A primary consideration in determining the most appropriate approach for any given data set is, not only how many compounds are present, but also what level of detail is essential for the study. Working with medium-sized data sets, rapid *in vitro* screens, or QSPkR analysis may provide sufficient information and permit investigation of a wide range of compounds per iteration. Once lead candidates have been identified, these techniques may be used to investigate the mechanism of interaction in more detail, allowing for further optimization of structure.

For many scenarios in preliminary drug development and hazard assessment, high-throughput, *in silico* screening can be the most practical solution. The techniques, as discussed above, allow for ready prioritization of compounds from very large data sets. Rank orders may be established for those compounds most likely to possess desirable or undesirable properties, and this can be input into the next stage of investigation. Discriminant functions or decision trees may be useful for categorizing compounds of interest, acting as filters to select the most promising drug candidates or highlighting potential toxicants, even among compounds yet to be synthesized. Drug development has been described as moving from an era of few, highly validated targets to a greater number of targets with a lower degree of validation (Manly et al., 2001). It is increasingly important to be able to rapidly screen larger numbers of compounds, with potential for interaction with these targets.

Once compounds have been synthesized and pharmacokinetic information is available, modeling of the pharmacokinetic parameters from structural information can provide further information for subsequent iterations of the process. As has been discussed previously, modeling pharmacokinetic parameters is a multifactorial process, and it is difficult to produce global QSPkR models for large data sets. However, if the global approach is unsuccessful, it is possible that subdivision of the data set into smaller groups may still allow trends within the data to be identified. Distillation of the global problem into fragments that can be studied more easily in isolation allows more information to be accumulated and assimilated into later models. As data of this type proliferate, integration of results from more studies will help to elucidate the mechanisms involved. Increasing crossover of results from pharmacokinetic and toxicokinetic analysis would also be beneficial. For example, information on structural features that are found to increase bioaccumulation in toxicological studies may be usefully incorporated into drug development in terms of optimizing half-life of drugs. When approaching the prediction of pharmacokinetic parameters, it is important to have a clear and structured view of all the information and various modeling tools available.

## VIII. FURTHER CONSIDERATIONS IN SELECTION OF APPROACH

Aside from considering the size of the data set and the level of detail required from the investigation, there are several other considerations when choosing an approach.

One key issue is the requirement for animal testing. The report and recommendations of the European Centre for the Validation of Alternative Methods (ECVAM) workshop (Leahy et al., 1997) clearly state that a major priority is to reduce the number of animals used in pharmacokinetic studies in drug development. As much information as possible should be obtained from alternative sources, such as computer modeling or using data already generated for similar compounds.

The end user of the results of the study is an additional factor to consider. In drug development programs recommendations for structural modification should be logical and easy to implement by medicinal chemists involved in designing the next sequence of compounds. Lipinski's "rule of 5" (Lipinski et al., 1997) has no doubt found such wide application in industry, because the rule can be so easily interpreted and applied to new compounds. The paper by Lipinski (2000) shows a simple graph distributed to chemists that can be used to determine whether the solubility of a

**Table 11.2    Recommendations for Selecting an Approach to Estimate Pharmacokinetic Parameters**

| Size of Data Set | Level of Detail Necessary/Intended Outcome of Study | Possible Approaches |
|---|---|---|
| Small | Detailed information on individual pharmacokinetic parameters<br>Full PBPK models<br>Lead optimization | Direct *in vivo* measurements with extrapolation to man as necessary<br>(single compound or cassette dosing)<br>Use of existing literature information or models for analogs<br>Incorporation of known or predictable parameters (e.g., tissue:blood PCs) |
| Medium | Reasonably accurate estimation of individual parameters<br>Quantitative prediction for related series of compounds<br>Identification of clear trends within data sets | Medium throughput *in vitro* assays<br>Generation or use of existing QSPkR models (may include complex 3-D or other computationally expensive parameters)<br>Use of commercial software to predict relevant parameters (i.e., aqueous solubility, likely metabolic fate, etc.) |
| Large | General predictions of pharmacokinetic parameters for rapid screens<br>Rank ordering<br>Classification into groups with acceptable/unacceptable properties<br>Selection of candidates for synthesis<br>Generation of in-house rules or refinement of existing screens for acceptable limits for pharmacokinetic parameters<br>Lead generation<br>Library design | Generation or use of existing rule-based screens<br>Refinement of known rule-based screens<br>Use of hierarchical models, discriminant functions or decision trees to classify data<br>Generation of QSPkR models (replacing complex or 3D parameters with more rapidly calculable 1D and 2D parameters wherever possible)<br>Database searching for analogous compounds |

**Further Recommendations:**
Use validated alternatives to animal testing wherever possible.
Consider all alternatives for estimation in terms of reliability, accuracy, time required, and cost efficiency.
Develop predictive models that allow for *in silico* screening, rather than necessitating prior synthesis of compound.
Analyze literature for both pharmacokinetic and toxicokinetic parameter estimation, to identify models that already exist or ones that could be suitably modified for the parameter of interest
Ensure the results of the study will be readily interpretable by the end user, allowing flexibility for intuitive input into subsequent stages of the project

compound is acceptable or unacceptable in terms of absorption of a therapeutically effective oral dose of the drug. This provides readily accessible information to the people involved. Egan and Lauri (2002) comment that, "clear and concise presentation in a manner appropriate to the audience...is crucial." It is also important that proposed rules or models presented to the development team are sufficiently flexible to accommodate the team's intuition and personal experience.

Carr and Hann (2002) suggest that instead of screening for compounds that exactly comply with a set of rules, the limits are set lower, to allow for subsequent optimization. For example, rather than screening for compounds with a molecular weight below 500, a screen could be run at a cut-off of 450. This would then allow for modifications to be made to the structure, such as the addition of a particular functional group, without violating the rule stated for maximum molecular weight. This allows greater flexibility for input based on the experience of more team members.

A summary of the factors to consider when selecting the most appropriate approach to use when estimating pharmacokinetic parameters is given in Table 11.2.

## IX. TIMING OF PHARMACOKINETIC SCREENS

Traditionally in drug development, compounds were optimized in terms of activity prior to analysing pharmacokinetic properties. Currently there is a logical move toward trying to optimize both pharmacokinetics and biological activity simultaneously.

In earlier drug development, compounds were frequently designed as analogs of known agonists or antagonists at biological receptors. Consequently, as the drugs were similar to compounds with acceptable pharmacokinetic properties, these compounds were also likely to demonstrate reasonable pharmacokinetic profiles. Also, *in vivo* testing was generally performed at an earlier stage of development, so that any drug with unacceptable pharmacokinetics would be screened out at an early stage. Drug development has now moved away from this approach, to the study of much more diverse chemical libraries. If novel compounds are less likely to possess good pharmacokinetics inherently, there is a need to screen for pharmacokinetics and activity, in parallel, wherever practicable. The problem of trying to optimize a structure to demonstrate acceptable pharmacokinetics, after optimizing for activity, has frequently been underestimated (Manly et al., 2001). The advantages of simultaneously optimizing both pharmacodynamic and pharmacokinetic properties by the integration of parallel screens have been discussed by several authors (Smith et al., 1996; Darvas and Dorman, 1999; Thompson, 2000; Li, 2001; Grass and Sinko, 2002). Using this approach, the drugs developed should show good average properties for pharmacodynamics and pharmacokinetics. This avoids the problem of designing a compound that shows high affinity for its receptor site, but is later discovered to have insurmountable pharmacokinetic problems. A reported 50.4% of drug candidates fail during the latter stages of development for this reason (Egan et al., 2000). Screening out candidates with these problems earlier on would be highly beneficial from an economic viewpoint (Oprea, 2002).

Another feasible approach would be to screen all compounds for pharmacokinetic properties, and then only select drug candidates from compounds existing in pharmacokinetically viable chemistry space, (Lipinski, 2000).

Similarly, the investigation of potential toxicity of drug candidates is another consideration. Although side-effects are clearly more difficult to predict, screens for known toxicophores could also be incorporated earlier in the process and the information integrated with that from pharmacokinetic and pharmacodynamic screens.

The difficulties of optimizing a compound for pharmacodynamic effect, reasonable pharmacokinetic profile, and lack of toxicity are in no way underestimated. Minor modifications to a compound's structure to improve pharmacokinetic properties could result in the drug being ineffective at its site of action, or could even result in an alternative pharmacokinetic problem arising. In theory, for a given series of compounds, it may be recognized that optimization of certain parameters may not be a restricting factor. If all compounds within a series under consideration possess similar log $K_{ow}$ values and suitable absorption characteristics, it may only be necessary to look to optimization of the metabolic profile to improve bioavailability. This alone can be onerous if one considers, for example, the problem of metabolic switching as discussed by Thompson (2001). A drug may be susceptible to metabolism by one specific pathway; adding a group to hinder access to this site may simply result in it being susceptible to attack by a different enzyme. Alternatively, modifying the drug to increase log $K_{ow}$ and so avoid renal clearance, may merely render it more susceptible to metabolic degradation overall (Smith et al., 1996; Waterbeemd et al., 2001b). A minor modification to the structure may result in a different pharmacokinetic problem arising, not withstanding the effect on pharmacodynamics or toxicity.

## X. SUMMARY

To date there has only been limited success in modeling global pharmacokinetic and toxicokinetic parameters from physicochemical properties for large data sets. More recently there has been some success with predicting input parameters that may be incorporated into PBPK models. The possibility of being able to predict the full time course of a drug or toxicant in the body, is certainly a very interesting proposition for the future. The ultimate aim is to produce accurate, reliable, and fully validated models that can be applied to rapid prediction of pharmacokinetic parameters of drug candidates, or primary hazard assessment of toxicants. An advantage of developing full PBPK models is that additional information can then be incorporated, such as the effect of reduced hepatic metabolism in patients with liver disease or the effects of prolonged exposure to toxicants in industrial or environmental settings.

Logical organization of information already available is essential. The use of flowcharts or diagrams, such as Figure 11.1, can be used to identify where there are current gaps in the knowledge for series of compounds under investigation. This can be used to direct further study, (i.e., to ascertain whether or not a particular parameter can be reasonably predicted from information already known) or to identify which compounds would yield the most informative results if experimental measurement were unavoidable. It is also important to keep the global issue in mind. Deconstructing the problem to more easily resolvable fragments may be a useful approach, but should be used to gradually build up the full picture of all the interrelated factors that contribute to the pharmacokinetic profile.

Another approach of great potential is the improvement of the rule-based screens, such as those by proposed by Lipinski et al. (1997), Norinder and Haeberlein (2002), and Veber et al. (2002). Although useful screening tools in themselves, undoubtedly these models will be amenable to refinement and improvement as more data become available.

As QSAR and molecular modeling techniques developed, great advances were made in the prediction of pharmacodynamic and toxic effects of compounds. As improvements were made in technology, the identification and structural elucidation of more biological targets became possible. Unfortunately, the application of this technology to pharmacokinetic and toxicokinetic issues has tended to lag behind. As interest in these applications increase, it is likely that greater improvements will be made in the ability to predict accurately the time course of xenobiotics in the human body. The improvements in technology for high-throughput screening and data mining will witness a proliferation in available kinetic data, and a coordinated approach will be required to harness and utilize all of this information rationally. A higher degree of crossover between groups interested in kinetics from the pharmacological viewpoint and those interested from the toxicological perspective would also be beneficial. In drug discovery, more widely available information on failed drug candidates would also be useful, as this would aid identification of unfavorable characteristics in drug candidates. High quality, accurate experimental data are mandatory for the production of any useful model.

Optimization of the array of pharmacokinetic parameters necessary for the production of an effective new drug is undoubtedly an onerous task. For every drug commercially available an optimum balance has already been achieved in the physicochemical properties of the drug. What is needed is an approach to unravel the information held within these structures and to assimilate the information into subsequent studies, without stifling the creativity of development scientists.

For compounds of environmental concern, as the mechanisms responsible for uptake, distribution, storage, and elimination from the body are elucidated, more accurate and reliable models will be developed for predicting the toxicokinetic fate of chemicals.

## REFERENCES

Abraham, M.H., Chadha, H.S., and Mitchell, R.C., Hydrogen bonding. 33. Factors that influence the distribution of solutes between blood and brain, *J. Pharm. Sci.*, 83, 1257–1268, 1994.

Abraham, M.H., Chadha, H.S., Martins, F., Mitchell, R.C., Bradbury, M.W., and Gratton, J., Hydrogen bonding part 46: A review of the correlation and prediction of transport properties by an LFER method: physicochemical properties, brain penetration and skin permeability, *Pest. Sci.*, 55, 78–88, 1999.

Abraham, M.H., Zhao, Y.Y., Le, J., Hersey, A., Luscombe, C.N., Reynolds, D.P., Beck, G., Sherbourne, B., and Cooper, I., On the mechanism of human intestinal absorption, *Eur. J. Med. Chem.*, 37, 595–605, 2002.

Andrews, C.W., Bennett, L., and Yu, L.X., Predicting human oral bioavailability of a compound: development of a novel quantitative structure-bioavailability relationship, *Pharm. Res.*, 17, 639–644, 2000.

Bachman, K., Pardoe, D., and White, D., Scaling basic toxicokinetic parameters from rat to man, *Environ. Health Perspect.*, 104, 400–407, 1996.

Balaz, S., Lipophilicity in transbilayer transport and subcellular pharmacokinetics, *Perspect. Drug Discovery Design*, 19, 157–177, 2000.

Balaz, S. and Lukacova, V., A model-based dependence of the human tissue/blood partition coefficients of chemicals on lipophilicity and tissue composition, *Quant. Struct.-Act. Relat.*, 18, 361–368, 1999.

Barton, P., Davis, A.M., McCarthy, D.J. and Webborn, P.J.H., Drug-phospholipid interactions. 2. Predicting the sites of drug distribution using n-octanol/water and membrane/water distribution coefficients, *J. Pharm. Sci.*, 86, 1034–1039, 1997.

Benet, L.Z., Pharmacokinetics. I. Absorption, distribution and excretion, in *Basic and Clinical Pharmacology*, 3rd ed., Katzung, B.G., Ed., Appleton and Lange, Norwalk, CT, 1987.

Carr, R. and Hann, M., The right road to drug discovery, *Modern Drug Discovery*, 45–48, April 2002.

Crivori, P., Cruciani, G., Carrupt P.A., and Testa B., Predicting blood-brain permeation from the three-dimensional molecular structure, *J. Med. Chem.*, 43, 2204-2216, 2000.

Cronin, M.T.D., Dearden, J.C., Moss, G., and Murray-Dickson, G., Investigation of the mechanism of flux across human skin *in vitro* by quantitative structure-permeability relationships, *Eur. J. Pharm. Sci.*, 7, 325–330, 1999.

Darvas, F. and Dorman, G., Early integration of ADME/Tox parameters into the design process of combinatorial libraries, *Chimica Oggi/Chemistry Today*, 17, 10–13, 1999.

Davis, A.M., Salt, D.W., and Webborn, P.J.H., Induced correlations in the use of unbound/intrinsic pharmacokinetic parameters in quantitative structure pharmacokinetic relationships with lipophilicity, *Quant. Struct.-Act. Relat.*, 19, 574–580, 2000.

DeJongh, J., Verhaar, H.J.M., and Hermens, J.L.M., A quantitative property-property (QPPR) approach to estimate *in vitro* tissue-blood partition coefficients of organic chemicals in rats and humans, *Arch. Toxicol.*, 72, 17–25, 1997.

du Plessis, J., Pugh, W.J., Judefeind, A., and Hadgraft, J., Physico-chemical determinants of dermal drug delivery: effects of the number and substitution pattern of polar groups, *Eur. J. Pharm. Sci.*, 16, 107–112, 2002.

Egan, W.J. and Lauri, G., Prediction of intestinal permeability, *Adv. Drug Delivery Rev.*, 54, 273–289, 2002.

Egan, W.J., Merz Jr., K.M., and Baldwin, J.J., Prediction of drug absorption using multivariate statistics, *J. Med. Chem.*, 43, 3867–3877, 2000.

Ekins, S., Waller, C.L., Swaan, P.W., Cruciani, G., Wrighton, S. A., and Wikel, J. H., Progress in predicting human ADME parameters in silico, *J. Pharmacol. Toxicol. Meth.*, 44, 251–272, 2000.

Ekins, S. and Wrighton, S.A., Application of in silico approaches to predicting drug-drug interactions, *J. Pharmacol. Toxicol. Meth.*, 45, 65–69, 2001.

Ertl P., Rohde, B., and Selzer, P., Fast calculation of molecular polar surface area as a sum of fragment-based contributions and its application to the prediction of drug transport properties, *J. Med. Chem.*, 43: 3714-3717.

Fouchecourt, M.O., Beliveau, M., and Krishnan, K., Quantitative structure-pharmacokinetic relationship modelling, *Sci. Total Environ.*, 274, 125–135, 2001.

Fujiwara, S., Yamashita, F., and Hashida, M., Prediction of Caco-2 cell permeability using a combination of MO-calculation and neural network, *Int. J. Pharm.*, 237, 95–105, 2002.

Garrison, K.E., Pasas, S.A., Cooper, J.D., and Davies, M.I., A review of membrane sampling from biological tissues with applications in pharmacokinetics, metabolism and pharmacodynamics, *Eur. J. Pharm. Sci.*, 17, 1–12, 2002.

Grass, G.M. and Sinko, P.J., Physiologically-based pharmacokinetic simulation modelling, *Adv. Drug Delivery Rev.*, 54, 433–451, 2002.

Gute, B.D., Grunwald, G.D., and Basak, S.C., Prediction of the dermal penetration of polycyclic aromatic hydrocarbons (PAHs): a hierarchical QSAR approach, *SAR QSAR Environ. Res.*, 10, 1–15, 1999.

Hawkins, D.M., Young, S.S., and Rusinko, A., III, Analysis of a large structure-activity data set using recursive partitioning, *Quant. Struct.-Act. Relat.*, 16, 296–302, 1997.

Hirono, S., Nakagome, I., Hirano, H., Matsushita, Y., Yoshi, F., and Moriguchi, I., Non-congeneric structure pharmacokinetic property correlation studies using fuzzy adaptive least-squares: oral bioavailability, *Biol. Pharm. Bull*, 17, 306–309, 1994a.

Hirono, S., Nakagome, I., Hirano, H., Yoshi, F., and Moriguchi, I., Non-congeneric structure pharmacokinetic property correlation studies using fuzzy adaptive least-squares: volume of distribution, *Biol. Pharm. Bull.*, 17, 686–690, 1994b.

Hosoya, K., Ohtsuki, S., and Terasaki, T., Recent advances in the brain-to-blood efflux transport across the blood-brain barrier, *Int. J. Pharm.*, 248, 15–29, 2002.

Huuskonen, J., Rantanen, J., and Livingstone, D., Prediction of aqueous solubility for a diverse set of organic compounds based on atom-type electrotopological state indices, *Eur. J. Med. Chem.*, 35, 1081-1088, 2000.

Kaliszan, R. and Markuszewski, M., Brain/blood distribution described by a combination of partition coefficient and molecular mass, *Int. J. Pharm.*, 145, 9–16, 1996.

Keseru, G.M. and Molnar, L., High-throughput prediction of blood-brain partitioning: a thermodynamic approach, *J. Chem. Inf. Comput. Sci.*, 41, 120–128, 2001.

Kobayashi, M., Sada, N., Sugawara, M., Iseki, K., and Miyazaki, K., Development of a new system for prediction of drug absorption that takes into account drug dissolution and pH change in the gastro-intestinal tract, *Int. J. Pharm.*, 221, 87–94, 2001.

Kulkarni, A., Han, Y., and Hopfinger, A.J., Predicting Caco-2 cell permeation coefficients of organic molecules using membrane-interaction QSAR analysis, *J. Chem. Inf. Comput. Sci.*, 42, 331–342, 2002.

Langowski, J. and Long, A., Computer systems for the prediction of xenobiotic metabolism, *Adv. Drug Delivery Rev.*, 54, 407–415, 2002.

Leahy, D.E., Duncan, R., Ahr, H.J., Bayliss, A., de Boer, G., Darvas, F., Fentem, J.H., Fry, J.R., Hopkins, R., Houston, J.B., Karlsson, J., Kedderis, G.L., Pratten, M.K., Prierto, P., Smith, D., and Straughan, D.W., Pharmacokinetics in early drug research, *Alt. Lab. Anim. (ATLA)*, 25, 17–31, 1997.

Li, A.P., Screening for human ADME/Tox drug properties in drug discovery, *Drug Discovery Today*, 6, 357–366, 2001.

Lipinski, C.A., Lombardo, F., Dominy, B.W., and Feeney, P.J., Experimental and computational approaches to estimate solubility and permeability in drug discovery and development settings, *Adv. Drug Delivery Rev.*, 23, 3-25, 1997.

Lipinski, C.A., Drug-like properties and the causes of poor solubility and poor permeability, *J. Pharm. Toxicol. Meth.*, 44, 235–249, 2000.

Liu, R. and So, S.S., Development of quantitative structure-property relationship models for early ADME evaluation in drug discovery. 1. Aqueous solubility, *J. Chem. Inf. Comput. Sci.*, 41, 1633–1639, 2001.

Livingstone, D.J., Theoretical property predictions, *Curr. Top. Med. Chem.*, 3, 1171–1192, 2003.

Lombardo, F., Blake, J.F., and Curatolo, W.J., Computation of brain-blood partitioning of organic solutes via free energy calculations, *J. Med. Chem.*, 39, 4750–4755, 1996.

Lombardo, F., Obach, R.S., Shalaeva, M.Y., and Gao, F., Prediction of volume of distribution values in humans for neutral and basic drugs using physicochemical measurements and plasma protein binding data, *J. Med. Chem.*, 45, 2867–2876, 2002.

Luco, J.M., Prediction of the brain-blood distribution of a large set of drugs from structurally derived descriptors using partial least squares (PLS) modeling, *J. Chem. Inf. Comput. Sci.*, 39, 396–404, 1999.

Lundahl, P. and Beigi, F., Immobilised liposome chromatography of drugs for model analysis of drug-membrane interactions, *Adv. Drug Delivery Rev.*, 23, 221–227, 1997.

Mandagere, A.K., Thompson, T.N., and Hwang, K.K., Graphical model for estimating oral bioavailability of drugs in humans and other species from their Caco-2 permeability and *in vitro* liver enzyme metabolic stability rates, *J. Med. Chem.*, 45, 304–311, 2002.

Manga, N., Duffy, J.C., Rowe, P.H. and Cronin, M.T.D., A hierarchical QSAR model for urinary excretion of drugs in humans as a predictive tool for biotransformation, *QSAR Combinatorial Sci.*, 22, 263–273, 2003.

Manly, C.J., Louise-May, S., and Hammer, J.D., The impact of informatics and computational chemistry on synthesis and screening, *Drug Discovery Today*, 6, 1101–1110, 2001.

Meulenberg, C.J.W. and Vijverberg, H.P.M., Empirical relations predicting human and rat tissue:air partition coefficients of volatile organic compounds, *Toxicol. Appl. Pharm.*, 165, 205–216, 2000.

Norinder, U. and Haeberlein, M., Computational approaches to the prediction of the blood-brain distribution, *Adv. Drug Delivery Rev.*, 54, 291–313, 2002.

Norinder, U. and Osterberg, T., The applicability of computational chemistry in the evaluation and prediction of drug transport properties, *Perspect. Drug Discovery Design*, 19, 1–18, 2000.

Norinder, U., Sjoberg, P., and Osterberg, T., Theoretical calculation and prediction of brain-blood partitioning of organic solutes using MolSurf parameterization and PLS statistics, *J. Pharm. Sci.*, 87, 952–959, 1998.

Ong, S., Liu, H., and Pidgeon, C., Immobilised-artificial-membrane chromatography: measurements of membrane partition coefficients and predicting drug membrane permeability, *J. Chromatogr. A*, 728, 113–128, 1996.

Oprea, T.I., Virtual screening in lead discovery: a viewpoint, *Molecules*, 7, 51–62, 2002.

Osterberg, T. and Norinder, U., Theoretical calculation and prediction of P-glycoprotein-interacting drugs using MolSurf parameterization and PLS statistics, *Eur. J. Pharm. Sci.*, 10, 295–303, 2000.

Osterberg, T. and Norinder, U., Prediction of drug transport processes using simple parameters and PLS statistics: the use of ACD/LogP and ACD/ChemSketch descriptors, *Eur. J. Pharm. Sci.*, 12, 327–337, 2001.

Pagliara, A., Reist, M., Geinoz, S., Carrupt, P-A., and Testa, B., Evaluation and prediction of drug permeation, *J. Pharm. Pharmacol.*, 51, 1339–1357, 1999.

Payne, M.P. and Kenny, L.C., Comparison of models for the estimation of biological partition coefficients, *J. Toxicol. Environ. Health, Part A*, 65, 897–931, 2002.

Platts, J.A., Abraham, M.H., Zhao, Y.H., Hersey, A., Ijaz, L., and Butina, D., Correlation and prediction of a large blood-brain distribution data set: an LFER study, *Eur. J. Med. Chem.*, 36, 719–730, 2001.

Potts, R.O. and Guy, R.H., A predictive algorithm for skin permeability: the effects of molecular size and hydrogen bond activity, *Pharm. Res.*, 12, 1628–1633, 1992.

Poulin, P. and Krishnan, K., An algorithm for predicting tissue:blood partition coefficients of organic chemicals from n-octanol:water partition coefficient data, *J. Toxicol. Environ. Health*, 46, 117–129, 1995.

Poulin, P. and Krishnan, K., A quantitative structure-toxicokinetic relationship model for highly metabolised chemicals, *Alt. Lab. Anim. (ATLA)*, 26, 45–59, 1998.

Poulin, P. and Krishnan, K., Molecular structure-based prediction of the toxicokinetics of inhaled vapors in humans, *Int. J. Toxicol.*, 18, 7–18, 1999.

Poulin, P., Schoenlein, K., and Theil, F.P., Prediction of adipose tissue:plasma partition coefficients for structurally unrelated drugs, *J. Pharm. Sci.*, 90, 436–447, 2001.

Poulin, P. and Theil, F.P., A priori prediction of tissue:plasma partition coefficients of drugs to facilitate the use of physiologically-based pharmacokinetic models in drug discovery, *J. Pharm. Sci.*, 89, 16–35, 2000.

Poulin, P. and Theil, F.P., Prediction of pharmacokinetics prior to *in vivo* studies. 1. Mechanism-based prediction of volume of distribution, *J. Pharm. Sci.*, 91, 129–156, 2002a.

Poulin, P. and Theil, F.P., Prediction of pharmacokinetics prior to *in vivo* studies. 2. Generic physiologically based pharmacokinetic models of drug disposition, *J. Pharm. Sci.*, 91, 1358–1370, 2002b.

Quinones, C., Caceres, J., Stud, M., and Martinez, A., Prediction of drug half-life values of anti-histamines based on the CODES/neural network model, *Quant. Struct.-Act. Relat.*, 19, 448–454, 2000.

Quinones-Torrelo, C., Sagrado, S., Villanueva-Camanas, R.M., and Medina-Hernandez, M.J., Development of predictive retention-activity relationship models of tricyclic antidepressants by micellar liquid chromatography, *J. Med. Chem.*, 42, 3154–3162, 1999.

Quinones-Torrelo, C., Sagrado, S., Villanueva-Camanas, R.M., and Medina-Hernandez, M.J., Retention pharmacokinetic and pharmacodynamic parameter relationships of antihistamine drugs using biopartitioning micellar chromatography, *J. Chromatogr. B*, 761, 13–26, 2001.

Raevsky, O., Fetisov, V., Trepalina, E.P., McFarland, J.W., and Schaper, K.J., Quantitative estimation of drug absorption in humans for passively transported compounds on the basis of their physico-chemical parameters, *Quant. Struct.-Act. Relat.*, 19, 366–374, 2000.

Rose, K., Hall, L.H. and Kier, L.B., Modeling blood-brain barrier partitioning using the electrotopological state, *J. Chem. Inf. Comput. Sci.*, 42, 651–666, 2002.

Rowley, M., Kulagowski, J.J., Watt, A.P., Rathbone, D., Stevenson, G.I., Carling, R.W., Baker, R., Marshall, G.R., Kemp, J.A., Foster, A.C., Grimwood, S., Hargreaves, R., Hurley, C., Saywell, K.L. Tricklebank, M.D. and Leeson, P.D., Effect of plasma protein binding on *in vivo* activity and brain penetration of glycine/NMDA receptor antagonists, *J. Med. Chem.*, 40, 4053–4068, 1997.

Saiakhov, R.D., Stefan, L.R., and Klopman, G., Multiple computer-automated structure evaluation model of the plasma protein binding affinity of diverse drugs, *Perspect. Drug Discovery Design* 19, 133–155, 2000.

Salminen, T., Pulli, A., and Taskinen, J., Relationship between immobilised artificial membrane chromatographic retention and the brain penetration of structurally diverse drugs, *J. Pharm. Biomed. Anal.*, 15, 469–477, 1997.

Sarver, J.G., White, D., Erhardt, P., and Bachman, K., Estimating xenobiotic half-lives in humans from rat data: influence of log P, *Environ. Health Perspect.*, 105, 1204–1209, 1997.

Schaper, K.J., Zhang, Z., and Raevsky, O., pH dependent partitioning of acidic and basic drugs into liposomes: a QSAR analysis, *Quant. Struct.-Act. Relat.*, 20, 46–54, 2001.

Smith, D.A., Jones, B.C. and Walker, D.K., Design of drugs involving the concepts and theories of drug metabolism and pharmacokinetics, *Med. Res. Rev.*, 16, 243-266, 1996.

Testa, B., Crivori, P., Reist, M., and Carrupt, P.A., The influence of lipophilicity on the pharmacokinetic behaviour of drugs: concepts and examples, *Perspect. Drug Discovery Design*, 19, 179–211, 2000.

Thompson, T.N., Early ADME in support of drug discovery: the role of metabolic stability studies, *Curr. Drug Metab.*, 1, 215–241, 2000.

Thompson, T.N., Optimization of metabolic stability as a goal of modern drug design, *Med. Res. Rev.*, 21, 412–449, 2001.

Vaes, W.H.J., Ramos, E.U., Verhaar, H.J.M., Cramer, C.J., and Hermens, J.L.M., Understanding and estimating membrane/water partition coefficients: approaches to derive quantitative structure property relationships, *Chem. Res. Toxicol.*, 11, 847–854, 1998.

van de Waterbeemd, H., Smith, D.A., Beaumont, K., and Walker, D.K., Property-based design: optimisation of drug absorption and pharmacokinetics, *J. Med. Chem.*, 44, 1313–1333, 2001a.

van de Waterbeemd, H., Smith, D., and Jones, B.C., Lipophilicity in PK design: methyl, ethyl, futile, *J. Comput.-Aided Mol. Design*, 15, 273–286, 2001b.

Veber, D.F., Johnson, S.R., Cheng, H-Y., Smith, B.R., Ward, K.W., and Kopple, K.D., Molecular properties that influence the oral bioavailability of drug candidates, *J. Med. Chem.*, 45, 2615–2623, 2002.

Wessel, M.D., Jurs, P.C., Tolan, J.W., and Muskal, S.M., Prediction of human intestinal absorption of drug compounds from molecular structure, *J. Chem. Inf. Comput. Sci.*, 38, 726–735, 1998.

Yoshida, F. and Topliss, J.G., QSAR model for drug human oral bioavailability, *J. Med. Chem.*, 43, 2575–2585, 2000.

Young, R.C., Mitchell, R.C., Brown, T.H., Ganellin, R., Griffiths, R., Jones, M., Rana, K.K., Saunders, D., Smith, I.R., Sore, N.E., and Wilks, T.J., Development of a new physicochemical model for brain penetration and its application to the design of centrally acting $H_2$ receptor histamine antagonists, *J. Med. Chem.*, 31, 656–671, 1988.

Zhao, Y.H., Le, J., Abraham, M.H., Hersey, A., Eddershaw, P.J., Luscombe, C.N., Boutina, D., Beck, G., Sherborne, B., Cooper, I., and Platts, J.A., Evaluation of human intestinal absorption data and subsequent derivation of a quantitative structure-activity relationship (QSAR) with the Abraham descriptors, *J. Pharm. Sci.*, 90, 749–784, 2001.

# QSARs for Environmental Toxicity and Fate

CHAPTER 12

# Development and Evaluation of QSARs for Ecotoxic Endpoints: The Benzene Response-Surface Model for *Tetrahymena* Toxicity

T. Wayne Schultz and Tatiana I. Netzeva

## CONTENTS

## I. INTRODUCTION

As the uses of toxicological-based quantitative structure-activity relationships (QSARs) move into the arenas of priority setting, risk assessment, and chemical classification and labeling the demands for a better understanding of the foundations of these QSARs are increasing. Specifically, issues of quality, transparency, domain identification, and validation have been recognized as topics of particular interest (Schultz and Cronin, 2003).

Quality QSAR can only be constructed and validated with quality data, but quality in a QSAR is more than a high coefficient of determination. Transparency has several different meanings as it applies to QSARs. First, transparency means that the data, both biological and chemical, that are used in QSAR development and validation are available for examination. Second, models, which are developed with descriptors that quantify the pivotal aspects of toxic expression, are considered to be mechanistic-based, fundamental, and more easily interpreted, and thus transparent. Transparency can also mean the amount of process information obtainable from the statistical methodology; it goes from the black boxes of neural networks to interpretable multiple linear regression. Since the use of a particular QSAR is only valid within its domain (Schultz and Cronin, 2003), identification of that domain is critical to QSAR acceptability.

In this present analysis concerns about quality, transparency, and domain identification are addressed in the validation of a previous developed QSAR. This QSAR examines the prediction of ectotoxic potency for population growth impairment to the aquatic ciliate *Tetrahymena pyriformis* by substituted benzenes.

## II. BACKGROUND

The basic concept of QSAR as applied to toxicology has been reviewed several times; the most recent efforts include that of Walker and Schultz (2002). There are three elements to a QSAR: the toxicological data, the descriptor data, and the statistical method of linking the two data sets (Schultz et al., 2002). The function of a toxicological QSAR is to predict toxicity accurately. To meet this goal knowledge of the toxicological and chemical information on which the model is based is essential. A number of computer-assisted statistical methods are available for the development of QSAR models. Each method has advantages, disadvantages, and practical constraints.

Issues of quality, transparency, and domain may apply to each of the three components of a QSAR and may be multifactoral because of interactions among components. The development of a toxicity-based QSAR is an integrated process requiring a working knowledge in chemistry, toxicology, and statistics. Determining the quality of a QSAR is frequently a difficult task. In part, this is because structure-toxicity relationships are simple approximations of complex processes that are not comprehended well (Nendza and Russom, 1991). Transparency is a critical issue for regulatory acceptance and wider use of QSARs (Blaauboer et al., 1999). It is worth noting that the use of mechanism-based descriptors, while transparent, differs from the QSAR being based on a mechanism of toxic action. The latter is biochemical based, while the former, at least as it related to aquatic toxicity, is physicochemical and quantum chemical based.

One approach to developing QSARs has been the use of congeneric series of chemicals. While it is easy in the case of a congeneric series to identify the chemical domain, the congeneric series-derived QSAR is of little predictive value precisely because of the narrow structural domain on which they are based (Kaiser et al., 1999). Even within homologous series, efforts such as selecting derivatives with markedly different substitutents can be made to optimize molecular diversity and thus the domain.

### A. Toxicity Data

Central to the issues of quality, transparency, and domain identification as they relate to toxicological QSAR is biological data. High quality toxicity data on a structurally diverse set of molecules are required to formulate and validate high quality QSARs. Quality toxicity data typically come from standardized assays measured in a consistent manner, with a clear and unambiguous endpoint, and low experimental error. In such cases, quality is associated with values, which are accurate, consistent with other data within the same set, and consistent with data for other similar endpoints. In the case of comparisons between endpoints, it is as important for data to be consistent between endpoints as for the inconsistencies to be consistent.

The inhibition of growth of the ciliated protozoan *T. pyriformis* database (Schultz, 1997) is considered to be a high quality data set (Bradbury et al., 2003). It has been developed in a single laboratory over more than two decades. While numerous workers using slight variations in the static protocol and nominal concentrations have generated the data, the data set still remains an excellent primary source of information; it is also unique in terms of its size, molecular diversity, and quality. Moreover, these data have been compiled for the express purpose of QSAR development and validation.

All toxicity measurements are subject to experimental error. The reality of toxicity testing is that however standardized the protocol, it is not possible to obtain precise potency data. Therefore, toxicity values are often reported as the mean from a series of replicates. However, different toxicological measurements have different amounts of error associated with them. Toxicity assessments made in a single laboratory by a single protocol tend to be the most precise. Even within such testing, there is varying reproducibility between toxicants. In a study of *T. pyriformis* toxicity data, it was observed that the variability in measured values was greater for chemicals considered to be reactive, than for those thought to act through a narcosis mode of action (Seward et al., 2001).

## B.  Chemical Descriptor Data

The primary supposition of any toxicological QSAR is that the potency of a compound is dependent upon its molecular structure, which is typically quantified by chemical properties (Schultz et al., 2002). Chemical descriptors include a variety of types, including atom, substituent, and molecular parameters. The most transparent of these are the molecular-based empirical and quantum chemical descriptors. Empirical descriptors are measured descriptors and include physicochemical properties such as hydrophobicity (Dearden, 1990). Quantum chemical properties are theoretical descriptors and include charge and energy values (Karelson et al., 1996). Physicochemical and quantum chemical descriptors are for the most part easily interpretable with regard to how that property may be related to toxicity. The classic example of this, the partitioning of a toxicant between aqueous and lipid phases, has been used as a measure of hydrophobicity for over a century (Livingstone, 2000).

From the perspective of *T. pyriformis* population growth inhibition, there are limited controlling events (e.g., bio-uptake). One is able to develop probabilistic models where the analysis of single aspects of the system is replaced by the study of time-ensemble averaging of a range of procedures. Such an approach allows one to development a QSAR without regular knowledge of the living system under investigation. Historically in the modeling of *T. pyriformis* toxicity, this ploy has worked well because it has been possible to identify global actions (e.g., bio-uptake) that appear to be autonomous from specific molecular events.

Toxicity is a multivariate process based on events that are not well understood. For the purposes of modeling aquatic toxicity such as fish acute toxicity or *Tetrahymena* population growth impairment, the limited number of controlling aspects means that not every toxicological process must be evaluated, or even understood, in order to get useful QSARs. Experiences (Veith et al., 1983) have pointed toward the use of descriptors, which quantify information on a key process (i.e., bio-uptake). These experiences have shown that combinations of select descriptors can provide information on an integrated group of toxicological processes (Mekenyan and Veith, 1993). These might include macroscale measurements, such as measures of hydrophobicity and electrophilic reactivity, and microscale measurements of key processes such as steric hindrance (Karabunarliev et al., 1996a). The latter turn out to be especially useful for explaining observed variability in reactive-based ecotoxicity.

Like toxicity assessments, descriptor values used in QSARs are also subject to variability. This fact is sometimes unnoticed, especially when values for descriptors are produced by software packages (Benfenati et al., 2001). In a study of the molecular orbital properties of pyridines, Seward et al. (2001) demonstrated that a mean of nine values was required to obtain consistent values for the energies of the highest occupied molecular orbital and lowest unoccupied molecular orbital. Moreover, Benfenati et al. (2001) demonstrated variability of up to 23% in conformationally dependent descriptors.

## C. Statistical Methods

Some type of statistical technique is required to link the toxic potencies of the series of chemicals to their molecular descriptors. These techniques range from linear least squares regression analyses, to multivariate techniques including the use of principal component analysis and partial least squares, and to neural networks and genetic algorithms (see Chapter 7 and Livingstone [1995]). These statistical techniques vary in their transparency (i.e., the amount of process information obtainable from the statistical methodology). The automatic self-adapting methodologies of genetic algorithms and neural networks are largely black boxes, whereas multiple linear regression equations are, at least from physicochemical and quantum chemical viewpoints, unambiguous.

The best models to predict aquatic toxicity are ones that are simple and interpretable. A regression-based QSAR established with fundamental descriptors maximizes the interpretability of the model, while at the same time maintaining simplicity. Such QSARs are easily updated, capable of mechanistic-based interpretation, portable from one user to another, and allow the user to observe and comprehend how the prediction of toxic potency is made (Schultz and Cronin, 2003).

## III. MATERIALS AND METHODS

### A. Test Chemicals

More than 400 substituted benzenes representing several mechanisms of toxic action were evaluated. The molecules were obtained commercially (Aldrich Chemical Co., Milwaukee, WI; MTM Research Chemicals or Lancaster Synthesis Inc., Windham, NH). In the large majority of cases purity was greater than 95%.

### B. Biological Data

Population growth impairment testing with the common ciliate, *T. pyriformis* (strain GL-C), was conducted following the protocol described by Schultz (1997). This 40-h assay is static in design and uses population density quantified spectrophotometrically at 540 nm as its endpoint. The test protocol allows for 8 to 9 cell cycles in controls. Following range finding, each chemical was tested in three replicate tests (or assays). Two controls were used to provide a measure of the acceptability of the test by indicating the suitability of the medium and test conditions as well as a basis for interpreting data from other treatments. The first control had no test substance, but was inoculated with *T. pyriformis*. The other, a blank, had neither test substance nor inoculum. Each test replicate consisted of six to ten different concentrations of each test material with duplicate flasks of each concentration. Only replicates with control-absorbency values greater than 0.60 but less than 0.90 were used in the analyses.

### C. Molecular Descriptors

Hydrophobicity was quantified by the logarithm of the 1-octanol-water partition coefficient (log $K_{ow}$) values. The hydrophobicity values were measured or estimated by the ClogP (ver 3.55) software (BIOBYTE Corp., Claremont, CA, USA). The acceptor superdelocalizabilities were determined as a sum of the ratios between the squared eigenvectors (coefficients) of the *i*-th atomic orbital in the $j$-th unoccupied molecular orbital and the eigenvalue (energy) of the $j$-th unoccupied molecular orbital, multiplied by two. The calculations were performed using the Austin Model 1 (AM1) method implemented in MOPAC 93 (Fujitsu Ltd., Windows 95/98/NT/2k adaptation and MO indices by J. Kaneti [1988–1994] MO-QC). The maximum acceptor superdelocalizabilities ($A_{max}$) were extracted by in-house macros in Microsoft Word and Excel.

## D. Statistical Analyses

The 50% growth inhibitory concentration ($IGC_{50}$) was determined for each compound tested by Probit Analysis using the Statistical Analysis System (SAS) software (SAS Institute, 1989). The y-values were absorbencies normalized as percentage of control. The x-values were the toxicant concentrations in mg/L. QSARs were developed using the regression procedures of MINITAB version 13.0 (MINITAB Inc., State College, PA) and Statistical Package for Social Sciences (SPSS version 10.0.5) software (SPSS Inc., Chicago IL, USA). Log $(IGC_{50})^{-1}$ values reported as mM were used as the dependent variable. Log $K_{ow}$ and electrophilicity ($A_{max}$) acted as the independent variables. Resulting models were measured for fit by the coefficient of determination adjusted to the degrees of freedom ($R^2$adj). The uncertainty in the model was noted as the square root of the mean square for errors, while the predictivity of the model was noted as the $R^2$ pred. determined by the leave-one-out method (see Chapter 7). Outliers were identified as compounds with a standardized residual greater than three (Lipnick, 1991).

## E. Data Selection

For structure-toxicity models data were confined to selected domains. Specifically, substructures not included in these evaluations were carboxylic acids, catechols, hydroquinones, and benzoquinones. The training set, the response-plane model, consisted of the 215 substituted benzenes for which measured toxic response data (i.e., $IGC_{50}$) prior to saturation were reported by Schultz (1999). The distribution of the training set chemicals based on their electrophilicity measured as $A_{max}$ is shown in Figure 12.1(a). The validation set was selected from an initial group of 450 candidates limited to commercially available substituted benzenes within the descriptor domain of the training set. Final selection of the validation set was based on attaining a data set that mimicked the $A_{max}$ distribution training set. The distribution of the 177 validation set chemicals based on their $A_{max}$ values is shown in Figure 12.1(b), which compares very favorably with Figure 12.1(a).

# IV. RESULTS

## A. Initial Benzene Response-Surface Model

Earlier work by Schultz (1999) examined the toxicity (log $(IGC_{50})^{-1}$) of a heterogeneous series of 218 substituted benzenes (200 benzenes for training and 18 for external validation). Because of the use of a different algorithm for the determination of $A_{max}$ values, previously reported data on benzene toxicity were re-evaluated. The data for toxicity along with hydrophobicity and newly calculated electrophilicity are reported in Table 12.1. Toxicity values varied uniformly over four orders of magnitude (from –1.13 to 2.82 on a log scale). Hydrophobicity varied over about six orders of magnitude (from –0.55 to 5.76 on a log scale). Reactivity measured by $A_{max}$ varied on a linear scale from 0.280 to 0.385.

To investigate the influence of the change of the algorithm for $A_{max}$ calculation on the coefficients in the model, only the compounds, considered in Schultz (1999) for training ($n = 200$) were used in the analysis (see Table 12.1). The compounds, being not toxic at saturation as well as those detected as outliers in Schultz (1999), were excluded prior to the modeling. The resulting equation:

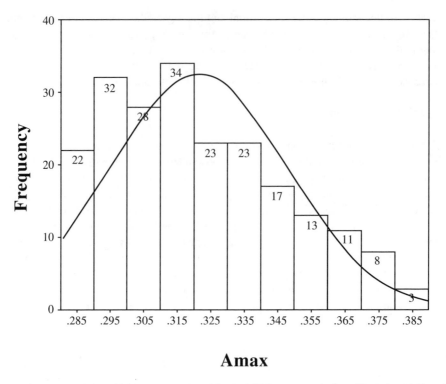

**Figure 12.1**   Histogram charts of (a) compounds used for the initial response-surface (Equation 12.3) and (b) for external validation (Equation 12.4).

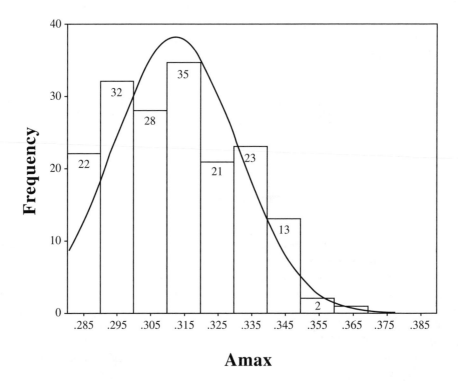

**Figure 12.1**   (continued).

**Table 12.1** Toxicity to *T. pyriformis* (Log $[IGC_{50}]^{-1}$), Octanol-Water Partition Coefficient (Log $K_{ow}$), and Maximum Acceptor Superdelocalizability ($A_{max}$) Values for the Compounds Published by Schultz (1999)

| No. | CAS | Name | Log $(IGC_{50})^{-1}$ | Log $K_{ow}$ | $A_{max}$ |
|---|---|---|---|---|---|
| 1 | 71-43-2 | Benzene | −0.12 | 2.13 | 0.280 |
| 2 | 106-42-3 | 4-xylene | 0.25 | 3.15 | 0.283 |
| 3 | 120055-09-6 | 1-Phenyl-2-butanol | −0.16 | 2.02 | 0.284 |
| 4 | 108-88-3 | Toluene | 0.25 | 2.73 | 0.284 |
| 5 | 104-51-8 | *n*-Butylbenzene | 1.25 | 4.26 | 0.284 |
| 6 | 538-68-1 | *n*-Amylbenzene | 1.79 | 4.90 | 0.284 |
| 7 | 100-46-9 | Benzylamine | −0.24 | 1.09 | 0.284 |
| 8 | 98-82-8 | Isopropylbenzene | 0.69 | 3.66 | 0.285 |
| 9 | 2430-16-2 | 6-Phenyl-1-hexanol | 0.87 | 3.30 | 0.285 |
| 10 | 10521-91-2 | 5-Phenyl-1-pentanol | 0.42 | 2.77 | 0.285 |
| 11 | 103-05-9 | $\alpha,\alpha$-Dimethylbenzenepropanol | −0.07 | 2.42 | 0.285 |
| 12 | 3360-41-6 | 4-Phenyl-1-butanol | 0.12 | 2.35 | 0.285 |
| 13 | 122-97-4 | 3-Phenyl-1-propanol | −0.21 | 1.88 | 0.285 |
| 14 | 100-51-6 | Benzyl alcohol | −0.83 | 1.05 | 0.285 |
| 15 | 98-85-1 | (*Sec*)phenethyl alcohol | −0.66 | 1.42 | 0.285 |
| 16 | 768-59-2 | 4-Ethylbenzyl alcohol | 0.07 | 2.13 | 0.285 |
| 17 | 2722-36-3 | 3-Phenyl-1-butanol | 0.01 | 2.11 | 0.286 |
| 18 | 22144-60-1 | (R+-)-1-Phenyl-1-butanol | −0.01 | 2.47 | 0.286 |
| 19 | 3597-91-9 | 4-Biphenylmethanol | 0.92 | 2.99 | 0.287 |
| 20 | 5707-44-8 | 4-Ethylbiphenyl | 1.97 | 5.06 | 0.288 |
| 21 | 92-52-4 | Biphenyl | 1.05 | 3.98 | 0.288 |
| 22 | 1565-75-9 | (±)-2-Phenyl-2-butanol | 0.06 | 2.34 | 0.288 |
| 23 | 5342-87-0 | (±)-1,2-Diphenyl-2-propanol | 0.80 | 3.23 | 0.290 |
| 24 | 29338-49-6 | 1,1-Diphenyl-2-propanol | 0.75 | 2.93 | 0.290 |
| 25 | 95-64-7 | 3,4-Dimethylaniline [b] | −0.16 | 1.86 | 0.293 |
| 26 | 1877-77-6 | 3-Aminobenzyl alcohol [b] | −1.13 | −0.55 | 0.293 |
| 27 | 4344-55-2 | 4-Butoxyaniline | 0.61 | 2.59 | 0.293 |
| 28 | 39905-50-5 | 4-Pentyloxyaniline | 0.97 | 3.12 | 0.293 |
| 29 | 39905-57-2 | 4-Hexyloxyaniline | 1.38 | 3.65 | 0.293 |
| 30 | 106-49-0 | 4-Methylaniline | −0.05 | 1.39 | 0.293 |
| 31 | 99-88-7 | 4-Isopropylaniline [b] | 0.22 | 2.47 | 0.293 |
| 32 | 587-02-0 | 3-Ethylaniline | −0.03 | 1.94 | 0.294 |
| 33 | 589-16-2 | 4-Ethylaniline | 0.03 | 1.96 | 0.294 |
| 34 | 108-44-1 | 3-Methylaniline | −0.28 | 1.40 | 0.294 |
| 35 | 104-13-2 | 4-Butylaniline | 1.07 | 3.18 | 0.294 |
| 36 | 103-63-9 | (2-Bromoethyl)benzene | 0.42 | 3.09 | 0.294 |
| 37 | 95-53-4 | 2-Methylaniline | −0.16 | 1.43 | 0.294 |
| 38 | 24544-04-5 | 2,6-Diisopropylaniline | 0.76 | 3.18 | 0.294 |
| 39 | 62-53-3 | Aniline | −0.23 | 0.90 | 0.295 |
| 40 | 578-54-1 | 2-Ethylaniline | −0.22 | 1.74 | 0.295 |
| 41 | 579-66-8 | 2,6-Diethylaniline | 0.31 | 2.87 | 0.295 |
| 42 | 100-68-5 | Thioanisole | 0.18 | 2.74 | 0.296 |
| 43 | 150-76-5 | 4-Methoxyphenol | −0.14 | 1.34 | 0.298 |
| 44 | 527-54-8 | 3,4,5-Trimethylphenol | 0.93 | 2.87 | 0.298 |
| 45 | 100-44-7 | Benzyl chloride | 0.06 | 2.30 | 0.298 |
| 46 | 104-93-8 | 4-Methylanisole | 0.25 | 2.81 | 0.299 |
| 47 | 697-82-5 | 2,3,5-Trimethylphenol | 0.36 | 2.92 | 0.299 |
| 48 | 527-60-6 | 2,4,6-Trimethylphenol | 0.42 | 2.73 | 0.299 |
| 49 | 98-54-4 | 4-(*Tert*)butylphenol | 0.91 | 3.31 | 0.300 |
| 50 | 80-46-6 | 4-(*Tert*)-pentylphenol | 1.23 | 3.83 | 0.300 |
| 51 | 2416-94-6 | 2,3,6-Trimethylphenol | 0.28 | 2.67 | 0.300 |
| 52 | 103-73-1 | Phenetole | −0.14 | 2.51 | 0.300 |
| 53 | 100-66-3 | Anisole | −0.10 | 2.11 | 0.300 |
| 54 | 105-67-9 | 2,4-Dimethylphenol | 0.14 | 2.35 | 0.300 |
| 55 | 127-66-2 | 2-Phenyl-3-butyn-2-ol | −0.18 | 1.88 | 0.300 |
| 56 | 106-44-5 | *p*-Cresol (4-methylphenol) | −0.16 | 1.97 | 0.300 |

**Table 12.1 (continued)** Toxicity to *T. pyriformis* (Log $[IGC_{50}]^{-1}$), Octanol-Water Partition Coefficient (Log $K_{ow}$), and Maximum Acceptor Superdelocalizability ($A_{max}$) Values for the Compounds Published by Schultz (1999)

| No. | CAS | Name | Log $(IGC_{50})^{-1}$ | Log $K_{ow}$ | $A_{max}$ |
|-----|-----|------|------|------|------|
| 57 | 123-07-9 | 4-Ethylphenol | 0.21 | 2.50 | 0.300 |
| 58 | 645-56-7 | 4-Propylphenol | 0.64 | 3.20 | 0.300 |
| 59 | 620-17-7 | 3-Ethylphenol | 0.29 | 2.50 | 0.300 |
| 60 | 104-40-5 | Nonylphenol | 2.47 | 5.76 | 0.300 |
| 61 | 108-39-4 | *m*-Cresol (3-methylphenol) | −0.08 | 1.98 | 0.300 |
| 62 | 95-48-7 | *o*-Cresol (2-methylphenol) | −0.29 | 1.98 | 0.301 |
| 63 | 90-00-6 | 2-Ethylphenol | 0.16 | 2.47 | 0.301 |
| 64 | 108-95-2 | Phenol | −0.35 | 1.50 | 0.301 |
| 65 | 1745-81-9 | 2-Allylphenol | 0.33 | 2.55 | 0.301 |
| 66 | 591-50-4 | Iodobenzene | 0.36 | 3.25 | 0.301 |
| 67 | 106-47-8 | 4-Chloroaniline | 0.05 | 1.83 | 0.302 |
| 68 | 529-19-1 | 2-Tolunitrile | −0.24 | 2.21 | 0.302 |
| 69 | 501-94-0 | 4-Hydroxyphenethyl alcohol [b] | −0.83 | 0.52 | 0.303 |
| 70 | 615-65-6 | 2-Chloro-4-methylaniline | 0.18 | 2.41 | 0.303 |
| 71 | 95-51-2 | 2-Chloroaniline | −0.17 | 1.88 | 0.304 |
| 72 | 500-66-3 | 5-Pentylresorcinol | 1.31 | 3.42 | 0.305 |
| 73 | 150-19-6 | 3-Methoxyphenol | −0.33 | 1.58 | 0.305 |
| 74 | 136-77-6 | 4-Hexylresorcinol[a] | 1.80 | 3.45 | 0.306 |
| 75 | 88-04-0 | 4-Chloro-3,5-dimethylphenol | 1.20 | 3.48 | 0.306 |
| 76 | 106-38-7 | 4-Bromotoluene | 0.47 | 3.50 | 0.306 |
| 77 | 1585-07-5 | 1-Bromo-4-ethylbenzene | 0.67 | 4.03 | 0.306 |
| 78 | 623-12-1 | 4-Chloroanisole | 0.60 | 2.79 | 0.307 |
| 79 | 59-50-7 | 4-Chloro-3-methylphenol | 0.80 | 3.10 | 0.307 |
| 80 | 108-46-3 | 1,3-Dihydroxybenzene | −0.65 | 0.80 | 0.307 |
| 81 | 108-86-1 | Bromobenzene | 0.08 | 2.99 | 0.308 |
| 82 | 106-48-9 | 4-Chlorophenol | 0.54 | 2.39 | 0.308 |
| 83 | 540-38-5 | 4-Iodophenol [b] | 0.85 | 2.90 | 0.311 |
| 84 | 156-41-2 | 2(4-Chlorophenyl)ethylamine [b] | 0.14 | 2.00 | 0.311 |
| 85 | 104-86-9 | 4-Chlorobenzylamine | 0.16 | 1.81 | 0.311 |
| 86 | 554-00-7 | 2,4-Dichloroaniline | 0.56 | 2.78 | 0.311 |
| 87 | 108-90-7 | Chlorobenzene[a] | −0.13 | 2.84 | 0.311 |
| 88 | 108-42-9 | 3-Chloroaniline | 0.22 | 1.88 | 0.312 |
| 89 | 99-51-4 | 1,2-Dimethyl-4-nitrobenzene | 0.59 | 2.91 | 0.314 |
| 90 | 5736-91-4 | 4-(Pentyloxy)benzaldehyde | 1.18 | 3.89 | 0.315 |
| 91 | 99-99-0 | 4-Nitrotoluene | 0.65 | 2.37 | 0.315 |
| 92 | 122-03-2 | 4-Isopropylbenzaldehyde | 0.67 | 2.92 | 0.316 |
| 93 | 83-41-0 | 1,2-Dimethyl-3-nitrobenzene | 0.56 | 2.83 | 0.316 |
| 94 | 108-43-0 | 3-Chlorophenol | 0.87 | 2.50 | 0.317 |
| 95 | 99-08-1 | 3-Nitrotoluene | 0.42 | 2.45 | 0.317 |
| 96 | 88-72-2 | 2-Nitrotoluene | 0.26 | 2.30 | 0.317 |
| 97 | 106-37-6 | 1,4-Dibromobenzene[a] | 0.68 | 3.79 | 0.317 |
| 98 | 100-52-7 | Benzaldehyde | −0.20 | 1.48 | 0.317 |
| 99 | 121-32-4 | 3-Ethoxy-4-hydroxybenzaldehyde | 0.02 | 1.58 | 0.317 |
| 100 | 121-33-5 | 3-Methoxy-4-hydroxybenzaldehyde [b] | −0.03 | 1.21 | 0.318 |
| 101 | 70-70-2 | 4′-Hydroxypropiophenone [b] | 0.12 | 2.03 | 0.318 |
| 102 | 120-83-2 | 2,4-Dichlorophenol | 1.04 | 3.17 | 0.318 |
| 103 | 1009-14-9 | Valerophenone | 0.56 | 3.17 | 0.318 |
| 104 | 93-55-0 | Propiophenone | −0.07 | 2.19 | 0.318 |
| 105 | 495-40-9 | Butyrophenone | 0.21 | 2.77 | 0.318 |
| 106 | 90-02-8 | 2-Hydroxybenzaldehyde | 0.42 | 1.81 | 0.318 |
| 107 | 1671-75-6 | Heptanophenone | 1.56 | 4.23 | 0.318 |
| 108 | 98-86-2 | Acetophenone | −0.46 | 1.63 | 0.318 |
| 109 | 98-95-3 | Nitrobenzene | 0.14 | 1.85 | 0.318 |
| 110 | 1674-37-9 | Octanophenone | 1.89 | 4.75 | 0.318 |
| 111 | 95-82-9 | 2,5-Dichloroaniline | 0.58 | 2.75 | 0.318 |

**Table 12.1 (continued)** Toxicity to *T. pyriformis* (Log [IGC$_{50}$]$^{-1}$), Octanol-Water Partition Coefficient (Log K$_{ow}$), and Maximum Acceptor Superdelocalizability (A$_{max}$) Values for the Compounds Published by Schultz (1999)

| No. | CAS | Name | Log (IGC$_{50}$)$^{-1}$ | Log K$_{ow}$ | A$_{max}$ |
|---|---|---|---|---|---|
| 112 | 95-75-0 | 3,4-Dichlorotoluene | 1.07 | 3.95 | 0.318 |
| 113 | 99-09-2 | 3-Nitroaniline | 0.03 | 1.43 | 0.319 |
| 114 | 626-43-7 | 3,5-Dichloroaniline | 0.71 | 2.90 | 0.319 |
| 115 | 7530-27-0 | 4-Bromo-6-chloro-2-cresol [b] | 1.28 | 3.61 | 0.319 |
| 116 | 95-50-1 | 1,2-Dichlorobenzene | 0.53 | 3.38 | 0.319 |
| 117 | 555-03-3 | 3-Nitroanisole | 0.72 | 2.17 | 0.321 |
| 118 | 119-61-9 | Benzophenone | 0.87 | 3.18 | 0.321 |
| 119 | 65262-96-6 | 3-Chloro-5-methoxyphenol | 0.76 | 2.50 | 0.322 |
| 120 | 100-14-1 | 4-Nitrobenzyl chloride [b] | 1.18 | 2.45 | 0.323 |
| 121 | 615-58-7 | 2,4-Dibromophenol [b] | 1.40 | 3.25 | 0.323 |
| 122 | 5922-60-1 | 2-Amino-5-chlorobenzonitrile | 0.44 | 1.79 | 0.323 |
| 123 | 552-41-0 | 2-Hydroxy-4-methoxyacetophenone | 0.55 | 1.98 | 0.324 |
| 124 | 591-35-5 | 3,5-Dichlorophenol | 1.56 | 3.61 | 0.325 |
| 125 | 104-88-1 | 4-Chlorobenzaldehyde | 0.40 | 2.13 | 0.325 |
| 126 | 134-85-0 | 4-Chlorobenzophenone | 1.50 | 3.97 | 0.325 |
| 127 | 108-70-3 | 1,3,5-Trichlorobenzene[a] | 0.87 | 4.19 | 0.325 |
| 128 | 636-30-6 | 2,4,5-Trichloroaniline | 1.30 | 3.69 | 0.325 |
| 129 | 90-90-4 | 4-Bromobenzophenone [b] | 1.26 | 4.12 | 0.326 |
| 130 | 120-82-1 | 1,2,4-Trichlorobenzene | 1.08 | 4.02 | 0.326 |
| 131 | 88-06-2 | 2,4,6-Trichlorophenol | 1.41 | 3.69 | 0.326 |
| 132 | 616-86-4 | 4-Ethoxy-2-nitroaniline | 0.76 | 2.39 | 0.326 |
| 133 | 2973-76-4 | 5-Bromovanillin | 0.62 | 1.92 | 0.326 |
| 134 | 100-29-8 | 4-Nitrophenetole | 0.83 | 2.53 | 0.328 |
| 135 | 89-59-8 | 4-Chloro-2-nitrotoluene | 0.82 | 3.05 | 0.328 |
| 136 | 585-79-5 | 1-Bromo-3-nitrobenzene | 1.03 | 2.64 | 0.328 |
| 137 | 3217-15-0 | 4-Bromo-2,6-dichlorophenol [b] | 1.78 | 3.52 | 0.329 |
| 138 | 83-42-1 | 2-Chloro-6-nitrotoluene | 0.68 | 3.09 | 0.329 |
| 139 | 3481-20-7 | 2,3,5,6-Tetrachloroaniline | 1.76 | 4.10 | 0.330 |
| 140 | 619-24-9 | 3-Nitrobenzonitrile | 0.45 | 1.17 | 0.330 |
| 141 | 95-95-4 | 2,4,5-Trichlorophenol | 2.10 | 3.72 | 0.330 |
| 142 | 95-94-3 | 1,2,4,5-Tetrachlorobenzene | 2.00 | 4.63 | 0.331 |
| 143 | 89-62-3 | 4-Methyl-2-nitroaniline [b] | 0.37 | 1.82 | 0.331 |
| 144 | 121-73-3 | 1-Chloro-3-nitrobenzene | 0.73 | 2.47 | 0.332 |
| 145 | 88-74-4 | 2-Nitroaniline | 0.08 | 1.85 | 0.332 |
| 146 | 634-83-3 | 2,3,4,5-Tetrachloroaniline | 1.96 | 4.27 | 0.333 |
| 147 | 118-79-6 | 2,4,6-Bromophenol | 1.91 | 4.08 | 0.334 |
| 148 | 7149-70-4 | 2-Bromo-5-nitrotoluene | 1.16 | 3.25 | 0.334 |
| 149 | 3819-88-3 | 1-Fluoro-3-iodo-5-nitrobenzene | 1.09 | 3.15 | 0.335 |
| 150 | 88-75-5 | 2-Nitrophenol | 0.67 | 1.77 | 0.335 |
| 151 | 121-87-9 | 2-Chloro-4-nitroaniline | 0.75 | 2.05 | 0.336 |
| 152 | 42454-06-8 | 5-Hydroxy-2-nitrobenzaldehyde | 0.33 | 1.75 | 0.336 |
| 153 | 576-55-6 | 3,4,5,6-Tetrabromo-2-cresol | 2.57 | 4.97 | 0.336 |
| 154 | 58-90-2 | 2,3,4,6-Tetrachlorophenol | 2.18 | 3.88 | 0.337 |
| 155 | 350-46-9 | 1-Fluoro-4-nitrobenzene | 0.10 | 1.89 | 0.338 |
| 156 | 771-60-8 | Pentafluoroaniline | 0.26 | 1.87 | 0.338 |
| 157 | 577-19-5 | 1-Bromo-2-nitrobenzene | 0.75 | 2.51 | 0.338 |
| 158 | 90-59-5 | 3,5-Dibromosalicylaldehyde | 1.65 | 3.42 | 0.338 |
| 159 | 618-62-2 | 3,5-Dichloronitrobenzene | 1.13 | 3.09 | 0.339 |
| 160 | 610-78-6 | 4-Chloro-3-nitrophenol | 1.27 | 2.46 | 0.339 |
| 161 | 4901-51-3 | 2,3,4,5-Tetrachlorophenol[a] | 2.72 | 4.21 | 0.339 |
| 162 | 2227-79-4 | Thiobenzamide | 0.09 | 1.50 | 0.339 |
| 163 | 100-00-5 | 1-Chloro-4-nitrobenzene | 0.43 | 2.39 | 0.340 |
| 164 | 2357-47-3 | α,α,α-4-Tetrafluoro-3-toluidine | 0.77 | 2.51 | 0.341 |
| 165 | 88-73-3 | 1-Chloro-2-nitrobenzene | 0.68 | 2.52 | 0.343 |
| 166 | 7147-89-9 | 4-Chloro-6-nitro-3-cresol [b] | 1.63 | 2.93 | 0.343 |

Table 12.1 (continued)   Toxicity to *T. pyriformis* (Log $[IGC_{50}]^{-1}$), Octanol-Water Partition Coefficient (Log $K_{ow}$), and Maximum Acceptor Superdelocalizability ($A_{max}$) Values for the Compounds Published by Schultz (1999)

| No. | CAS | Name | Log $(IGC_{50})^{-1}$ | Log $K_{ow}$ | $A_{max}$ |
|---|---|---|---|---|---|
| 167 | 87-86-5 | Pentachlorophenol | 2.07 | 5.18 | 0.343 |
| 168 | 99-65-0 | 1,3-Dinitrobenzene | 0.76 | 1.49 | 0.345 |
| 169 | 121-14-2 | 2,4-Dinitrotoluene | 0.87 | 1.98 | 0.345 |
| 170 | 6641-64-1 | 4,5-Dichloro-2-nitroaniline | 1.66 | 3.21 | 0.345 |
| 171 | 771-61-9 | Pentafluorophenol [b] | 1.63 | 3.23 | 0.345 |
| 172 | 608-71-9 | Pentabromophenol | 2.66 | 4.85 | 0.346 |
| 173 | 350-30-1 | 3-Chloro-4-fluoronitrobenzene | 0.80 | 2.74 | 0.347 |
| 174 | 100-25-4 | 1,4-Dinitrobenzene | 1.30 | 1.47 | 0.347 |
| 175 | 99-54-7 | 3,4-Dichloronitrobenzene | 1.16 | 3.12 | 0.348 |
| 176 | 89-61-2 | 2.5-Dichloronitrobenzene | 1.13 | 3.03 | 0.349 |
| 177 | 2683-43-4 | 2,4-Dichloro-6-nitroaniline | 1.26 | 3.33 | 0.349 |
| 178 | 79544-31-3 | 3,4-Dinitrobenzyl alcohol | 1.09 | 0.59 | 0.349 |
| 179 | 611-06-3 | 2,4-Dichloronitrobenzene | 0.99 | 3.09 | 0.350 |
| 180 | 3209-22-1 | 2,3-Dichloronitrobenzene | 1.07 | 3.05 | 0.350 |
| 181 | 528-29-0 | 1,2-Dinitrobenzene | 1.25 | 1.69 | 0.351 |
| 182 | 103-72-0 | Phenyl isothiocyanate | 1.41 | 3.28 | 0.352 |
| 183 | 88-30-2 | 3-Trifluoromethyl-4-nitrophenol | 1.65 | 2.77 | 0.352 |
| 184 | 305-85-1 | 2,6-Iodo-4-nitrophenol [b] | 1.81 | 3.52 | 0.353 |
| 185 | 609-89-2 | 2,4-Chloro-6-nitrophenol [b] | 1.75 | 3.07 | 0.354 |
| 186 | 18708-70-8 | 1,3,5-Trichloro-2-nitrobenzene | 1.43 | 3.69 | 0.354 |
| 187 | 89-69-0 | 1,2,4-Trichloro-5-nitrobenzene | 1.53 | 3.47 | 0.354 |
| 188 | 17700-09-3 | 1,2,3-Trichloro-4-nitrobenzene | 1.51 | 3.61 | 0.357 |
| 189 | 6361-21-3 | 2-Chloro-5-nitrobenzaldehyde | 0.53 | 2.25 | 0.357 |
| 190 | 653-37-2 | Pentafluorobenzaldehyde | 0.82 | 2.39 | 0.357 |
| 191 | 709-49-9 | 2,4-Dinitro-1-iodobenzene | 2.12 | 2.50 | 0.359 |
| 192 | 117-18-0 | 2,3,5,6-Tetrachloronitrobenzene [a] | 1.82 | 4.38 | 0.360 |
| 193 | 329-71-5 | 2,5-Dinitrophenol | 1.04 | 1.86 | 0.361 |
| 194 | 97-02-9 | 2,4-Dinitroaniline | 0.72 | 1.72 | 0.361 |
| 195 | 879-39-0 | 2,3,4,5-Tetrachloronitrobenzene | 1.78 | 3.93 | 0.361 |
| 196 | 771-69-7 | 1,2,3-Trifluoro-4-nitrobenzene | 1.89 | 2.01 | 0.362 |
| 197 | 6306-39-4 | 1,2-Dichloro-4,5-dinitrobenzene | 2.21 | 2.93 | 0.365 |
| 198 | 606-22-4 | 2,6-Dinitroaniline | 0.84 | 1.79 | 0.366 |
| 199 | 534-52-1 | 4,6-Dinitro-2-methylphenol | 1.73 | 2.12 | 0.366 |
| 200 | 4097-49-8 | 4-(*Tert*)butyl-2,6-dinitrophenol | 1.80 | 3.61 | 0.367 |
| 201 | 584-48-5 | 1-Bromo-2,4-dinitrobenzene | 2.31 | 2.29 | 0.368 |
| 202 | 51-28-5 | 2,4-Dinitrophenol (nonneutralized) | 1.06 | 1.54 | 0.368 |
| 203 | 28689-08-9 | 1,5-Dichloro-2,3-dinitrobenzene | 2.42 | 2.85 | 0.369 |
| 204 | 3531-19-9 | 6-Chloro-2,4-dinitroaniline | 1.12 | 2.46 | 0.370 |
| 205 | 1817-73-8 | 2-Bromo-4,6-dinitroaniline | 1.24 | 2.61 | 0.372 |
| 206 | 314-41-0 | 2,3,4,6-Tetrafluoronitrobenzene | 1.87 | 1.86 | 0.372 |
| 207 | 573-56-8 | 2,6-Dinitrophenol | 0.83 | 1.33 | 0.372 |
| 208 | 97-00-7 | 1-Chloro-2,4-dinitrobenzene | 2.16 | 2.14 | 0.374 |
| 209 | 70-34-8 | 2,4-Dinitro-1-fluorobenzene | 1.71 | 1.47 | 0.375 |
| 210 | 880-78-4 | Pentafluoronitrobenzene [a] | 2.43 | 2.00 | 0.378 |
| 211 | 20098-38-8 | 1,4-Dinitrotetrachlorobenzene | 2.82 | 3.44 | 0.380 |
| 212 | 327-92-4 | 1,5-Difluoro-2,4-dinitrobenzene [a] | 2.08 | 1.31 | 0.384 |
| 213 | 2678-21-9 | 1,3-Dinitro-2,4,5-trichlorobenzene | 2.60 | 3.05 | 0.385 |
| 214 | 6284-83-9 | 1,3,5-Trichloro-2,4-dinitrobenzene hemihydrate | 2.19 | 2.97 | 0.385 |
| 215 | 1930-72-9 | 4-Chloro-3,5-dinitrobenzonitrile [a] | 2.66 | 1.37 | 0.393 |
| 216 | 82-68-8 | Pentachloronitrobenzene | NTAS[c] | 4.64 | 0.364 |
| 217 | 608-93-5 | Pentachlorobenzene | NTAS | 5.17 | 0.339 |
| 218 | 1825-21-4 | Pentachloroanisole | NTAS | 5.45 | 0.339 |

[a] Outliers to Equation 12.2 reported by Schultz (1999).
[b] Compounds used for external validation in Schultz (1999).
[c] Not toxic at saturation.

**Table 12.2** Toxicity to *T. pyriformis* (Log [IGC$_{50}$]$^{-1}$), Octanol-Water Partition Coefficient (Log K$_{ow}$) and maximum Acceptor Superdelocalizability (A$_{max}$) Values for the Compounds Used for External Validation

| No. | CAS | Name | Log (IGC$_{50}$)$^{-1}$ | Log K$_{ow}$ | A$_{max}$ |
|---|---|---|---|---|---|
| 1 | 14898-87-4 | 1-Phenyl-2-propanol | −0.62 | 1.97 | 0.283 |
| 2 | 589-18-4 | 4-Methylbenzyl alcohol | −0.49 | 1.58 | 0.284 |
| 3 | 705-73-7 | (±)-1-Phenyl-2-pentanol | 0.16 | 2.55 | 0.284 |
| 4 | 536-60-7 | 4-Isopropylbenzyl alcohol | 0.18 | 2.53 | 0.284 |
| 5 | 3261-62-9 | 2-(4-tolyl)ethylamine | −0.04 | 1.78 | 0.284 |
| 6 | 104-84-7 | 4-Methyl benzylamine | −0.01 | 2.81 | 0.284 |
| 7 | 587-03-1 | 3-Methylbenzyl alcohol | −0.24 | 1.60 | 0.285 |
| 8 | 104-54-1 | 3-Phenyl-2-propen-1-ol | −0.08 | 1.95 | 0.285 |
| 9 | 877-65-6 | 4-(*Tert*)butylbenzyl alcohol | 0.48 | 2.93 | 0.285 |
| 10 | 699-02-5 | 4-Methylphenethyl alcohol | −0.26 | 1.68 | 0.285 |
| 11 | 618-36-0 | 1-Phenylethylamine | −0.18 | 1.40 | 0.285 |
| 12 | 89-95-2 | 2-Methylbenzyl alcohol | −0.43 | 1.55 | 0.285 |
| 13 | 100-86-7 | 2-Methyl-1-phenyl-2-propanol | −0.41 | 1.86 | 0.285 |
| 14 | 589-08-2 | *n*-Methylphenethylamine | −0.41 | 1.43 | 0.285 |
| 15 | 582-22-9 | β-Methylphenethylamine | −0.28 | 1.68 | 0.285 |
| 16 | 22135-49-5 | (S±)-1-Phenyl-1-butanol | −0.09 | 2.47 | 0.286 |
| 17 | 93-54-9 | (±)1-Phenyl-1-propanol | −0.43 | 1.94 | 0.286 |
| 18 | 60-12-8 | Phenethyl alcohol | −0.59 | 1.36 | 0.286 |
| 19 | 1123-85-9 | 2-Phenyl-1-propanol | −0.40 | 1.58 | 0.286 |
| 20 | 617-94-7 | 2-Phenyl-2-propanol | −0.57 | 1.81 | 0.288 |
| 21 | 89104-46-1 | 2-Phenyl-1-butanol | −0.11 | 2.11 | 0.288 |
| 22 | 91-01-0 | Benzhydrol | 0.50 | 2.67 | 0.289 |
| 23 | 622-32-2 | Benzaldoxime | −0.11 | 1.75 | 0.291 |
| 24 | 108-69-0 | 3,5-Dimethylaniline | −0.36 | 1.91 | 0.293 |
| 25 | 769-92-6 | 4-(*Tert*)butylaniline | 0.36 | 2.70 | 0.293 |
| 26 | 95-68-1 | 2,4-Dimethylaniline | −0.29 | 1.68 | 0.293 |
| 27 | 2046-18-6 | 4-Phenylbutyronitrile | 0.15 | 2.21 | 0.293 |
| 28 | 88-05-1 | 2,4,6-Trimethylaniline | −0.05 | 2.31 | 0.293 |
| 29 | 645-59-0 | 3-Phenylpropionitrile | −0.16 | 1.72 | 0.294 |
| 30 | 30273-11-1 | 4-(*Sec*)-butylaniline | 0.61 | 2.87 | 0.294 |
| 31 | 87-59-2 | 2,3-Dimethylaniline | −0.43 | 1.81 | 0.294 |
| 32 | 140-29-4 | Benzyl cyanide | −0.36 | 1.56 | 0.294 |
| 33 | 95-78-3 | 2,5-Dimethylaniline | −0.33 | 1.83 | 0.294 |
| 34 | 1823-91-2 | α-Methylbenzyl cyanide | 0.01 | 1.87 | 0.294 |
| 35 | 643-28-7 | 2-Isopropylaniline | 0.12 | 2.12 | 0.294 |
| 36 | 87-62-7 | 2,6-Dimethylaniline | −0.43 | 1.84 | 0.294 |
| 37 | 103-69-5 | *n*-Ethylaniline | 0.07 | 2.16 | 0.295 |
| 38 | 1821-39-2 | 2-Propylaniline | 0.08 | 2.42 | 0.295 |
| 39 | 100-61-8 | *n*-Methylaniline | 0.06 | 1.66 | 0.295 |
| 40 | 1199-46-8 | 2-Amino-4-(*tert*)butylphenol | 0.37 | 2.44 | 0.295 |
| 41 | 90-04-0 | 2-Methoxyaniline | −0.69 | 1.18 | 0.295 |
| 42 | 1008-88-4 | 3-Phenylpyridine | 0.47 | 2.53 | 0.296 |
| 43 | 5344-90-1 | 2-Aminobenzyl alcohol | −1.07 | −0.17 | 0.296 |
| 44 | 101-82-6 | 2-Benzylpyridine | 0.38 | 2.71 | 0.296 |
| 45 | 1138-52-9 | 3,5-Di-*tert*-butylphenol | 1.64 | 5.13 | 0.298 |
| 46 | 5651-88-7 | Phenyl propargyl sulfide | 0.54 | 3.30 | 0.298 |
| 47 | 622-62-8 | 4-Ethoxyphenol | 0.01 | 1.81 | 0.298 |
| 48 | 122-94-1 | 4-Butoxyphenol | 0.70 | 2.90 | 0.298 |
| 49 | 2116-65-6 | 4-Benzylpyridine | 0.63 | 2.62 | 0.298 |
| 50 | 1008-89-5 | 2-Phenylpyridine | 0.27 | 2.63 | 0.299 |
| 51 | 95-65-8 | 3,4-Dimethylphenol | 0.12 | 2.23 | 0.299 |
| 52 | 585-34-2 | 3-(*Tert*)butylphenol | 0.74 | 3.30 | 0.300 |
| 53 | 108-68-9 | 3,5-Dimethylphenol | 0.11 | 2.35 | 0.300 |
| 54 | 1879-09-0 | 6-(*Tert*)butyl-2,4-dimethylphenol | 1.16 | 4.30 | 0.300 |
| 55 | 99-89-8 | 4-Isopropylphenol | 0.47 | 2.90 | 0.300 |
| 56 | 618-45-1 | 3-Isopropylphenol | 0.61 | 2.90 | 0.300 |

Table 12.2 (continued)    Toxicity to *T. pyriformis* (Log $[IGC_{50}]^{-1}$), Octanol-Water Partition
Coefficient (Log $K_{ow}$) and Maximum Acceptor Superdelocalizability ($A_{max}$) Values
for the Compounds Used for External Validation

| No. | CAS | Name | Log $(IGC_{50})^{-1}$ | Log $K_{ow}$ | $A_{max}$ |
|---|---|---|---|---|---|
| 57 | 526-75-0 | 2,3-Dimethylphenol | 0.12 | 2.42 | 0.300 |
| 58 | 95-87-4 | 2,5-Dimethylphenol | 0.14 | 2.34 | 0.300 |
| 59 | 498-00-0 | 4-Hydroxy-3-methoxybenzyl alcohol | −0.70 | 0.29 | 0.300 |
| 60 | 88-69-7 | 2-Isopropylphenol | 0.61 | 2.88 | 0.301 |
| 61 | 53222-92-7 | 3-Amino-2-cresol | −0.55 | 0.70 | 0.301 |
| 62 | 95-69-2 | 4-Chloro-2-methylaniline | 0.35 | 2.36 | 0.301 |
| 63 | 97-54-1 | 2-Methoxy-4-propenylphenol | 0.75 | 3.31 | 0.302 |
| 64 | 90-72-2 | 2,4,6-Tris(dimethylaminomethyl)phenol | −0.52 | 0.92 | 0.302 |
| 65 | 348-54-9 | 2-Fluoroaniline | −0.37 | 1.26 | 0.302 |
| 66 | 3544-25-0 | 4-Aminobenzyl cyanide | −0.76 | 0.34 | 0.302 |
| 67 | 626-01-7 | 3-Iodoaniline | 0.65 | 2.90 | 0.302 |
| 68 | 4360-47-8 | Cinnamonitrile | 0.16 | 1.95 | 0.303 |
| 69 | 456-47-3 | 3-Fluorobenzyl alcohol | −0.39 | 1.25 | 0.305 |
| 70 | 2237-30-1 | 3-Cyanoaniline | −0.47 | 1.07 | 0.306 |
| 71 | 371-41-5 | 4-Fluorophenol | 0.02 | 1.77 | 0.307 |
| 72 | 615-43-0 | 2-Iodoaniline | 0.35 | 2.32 | 0.307 |
| 73 | 372-19-0 | 3-Fluoroaniline | −0.10 | 1.30 | 0.307 |
| 74 | 1570-64-5 | 4-Chloro-2-methylphenol | 0.70 | 2.78 | 0.307 |
| 75 | 14143-32-9 | 4-Chloro-3-ethylphenol | 1.08 | 3.51 | 0.308 |
| 76 | 1124-04-5 | 2-Chloro-4,5-dimethylphenol | 0.69 | 3.10 | 0.309 |
| 77 | 500-99-2 | 3,5-Dimethoxyphenol | −0.09 | 1.64 | 0.309 |
| 78 | 14191-95-8 | 4-Hydroxybenzyl cyanide | −0.38 | 0.90 | 0.309 |
| 79 | 2374-05-2 | 4-Bromo-2,6-dimethylphenol | 1.16 | 3.63 | 0.309 |
| 80 | 18982-54-2 | 2-Bromobenzyl alcohol | 0.10 | 1.97 | 0.309 |
| 81 | 615-74-7 | 2-Chloro-5-methylphenol | 0.54 | 2.65 | 0.309 |
| 82 | 367-12-4 | 2-Fluorophenol | 0.19 | 1.67 | 0.309 |
| 83 | 100-10-7 | 4-(Dimethylamino)benzaldehyde | 0.23 | 1.81 | 0.310 |
| 84 | 106-41-2 | 4-Bromophenol | 0.68 | 2.59 | 0.311 |
| 85 | 95-79-4 | 5-Chloro-2-methylaniline | 0.50 | 2.36 | 0.311 |
| 86 | 95-74-9 | 3-Chloro-4-methylaniline | 0.39 | 2.41 | 0.311 |
| 87 | 87-60-5 | 3-Chloro-2-methylaniline | 0.38 | 2.36 | 0.312 |
| 88 | 1875-88-3 | 4-Chlorophenethyl alcohol | 0.32 | 1.90 | 0.312 |
| 89 | 873-76-7 | 4-Chlorobenzyl alcohol | 0.25 | 1.96 | 0.312 |
| 90 | 6627-55-0 | 2-Bromo-4-methylphenol | 0.60 | 2.85 | 0.312 |
| 91 | 603-71-4 | 1,3,5-Trimethyl-2-nitrobenzene | 0.86 | 3.22 | 0.313 |
| 92 | 873-63-2 | 3-Chlorobenzyl alcohol | 0.15 | 1.94 | 0.314 |
| 93 | 95-56-7 | 2-Bromophenol | 0.33 | 2.33 | 0.314 |
| 94 | 4421-08-3 | 4-Hydroxy-3-methoxybenzonitrile | −0.03 | 1.42 | 0.315 |
| 95 | 619-25-0 | 3-Nitrobenzyl alcohol | −0.22 | 1.21 | 0.315 |
| 96 | 16532-79-9 | 4-Bromophenyl acetonitrile | 0.60 | 2.43 | 0.315 |
| 97 | 874-90-8 | 4-Methoxybenzonitrile | 0.10 | 1.70 | 0.315 |
| 98 | 36436-65-4 | 2-Hydroxy-4,5-dimethylacetophenone | 0.71 | 2.86 | 0.316 |
| 99 | 135-02-4 | 2-Anisaldehyde | 0.15 | 1.72 | 0.316 |
| 100 | 95-88-5 | 4-Chlororesorcinol | 0.13 | 1.80 | 0.316 |
| 101 | 18358-63-9 | Methyl-4-methylaminobenzoate | 0.31 | 2.16 | 0.316 |
| 102 | 67-36-7 | 4-Phenoxybenzaldehyde | 1.26 | 3.96 | 0.317 |
| 103 | 621-59-0 | 3-Hydroxy-4-methoxybenzaldehyde | −0.14 | 0.97 | 0.317 |
| 104 | 3218-36-8 | 4-Biphenylcarboxaldehyde | 1.12 | 3.38 | 0.317 |
| 105 | 4460-86-0 | 2,4,5-Trimethoxybenzaldehyde | −0.10 | 1.19 | 0.317 |
| 106 | 1137-41-3 | 4-Benzoylaniline | 0.68 | 2.46 | 0.317 |
| 107 | 5991-31-1 | 3-Anisaldehyde | 0.23 | 1.71 | 0.317 |
| 108 | 7778-83-8 | *n*-Propyl cinnamate | 1.23 | 3.52 | 0.318 |
| 109 | 103-36-6 | (*Trans*)ethyl cinnamate | 0.99 | 2.99 | 0.318 |
| 110 | 942-92-7 | Hexanophenone | 1.19 | 3.70 | 0.318 |
| 111 | 538-65-8 | *n*-Butyl cinnamate | 1.53 | 4.05 | 0.318 |
| 112 | 140-53-4 | 4-Chlorobenzyl cyanide | 0.66 | 2.47 | 0.319 |

**Table 12.2 (continued)  Toxicity to *T. pyriformis* (Log [IGC$_{50}$]$^{-1}$), Octanol-Water Partition Coefficient (Log K$_{ow}$) and Maximum Acceptor Superdelocalizability (A$_{max}$) Values for the Compounds Used for External Validation**

| No. | CAS | Name | Log (IGC$_{50}$)$^{-1}$ | Log K$_{ow}$ | A$_{max}$ |
|---|---|---|---|---|---|
| 113 | 103-26-4 | (*Trans*)methyl cinnamate | 0.58 | 2.62 | 0.319 |
| 114 | 94-30-4 | Ethyl-4-methoxybenzoate | 0.77 | 2.81 | 0.319 |
| 115 | 937-39-3 | Phenylacetic acid hydrazide | −0.48 | 0.14 | 0.319 |
| 116 | 100-83-4 | 3-Hydroxybenzaldehyde | 0.08 | 1.38 | 0.320 |
| 117 | 87-65-0 | 2,6-Dichlorophenol | 0.73 | 2.64 | 0.320 |
| 118 | 2495-37-6 | Benzyl methacrylate | 0.65 | 2.53 | 0.320 |
| 119 | 6521-30-8 | Isoamyl-4-hydroxybenzoate | 1.48 | 3.97 | 0.320 |
| 120 | 2491-32-9 | Benzyl-4-hydroxyphenyl ketone | 1.07 | 3.22 | 0.321 |
| 121 | 120-51-4 | Benzyl benzoate | 1.45 | 3.97 | 0.321 |
| 122 | 1137-42-4 | 4-Benzoylphenol | 1.02 | 3.07 | 0.321 |
| 123 | 5428-54-6 | 2-Methyl-5-nitrophenol | 0.66 | 2.35 | 0.321 |
| 124 | 621-42-1 | 3-Acetamidophenol | −0.16 | 0.73 | 0.322 |
| 125 | 3034-34-2 | 4-Cyanobenzamide | −0.38 | 0.48 | 0.322 |
| 126 | 86-00-0 | 2-Nitrobiphenyl | 1.30 | 3.77 | 0.322 |
| 127 | 7120-43-6 | 5-Chloro-2-hydroxybenzamide | 0.59 | 2.13 | 0.323 |
| 128 | 554-84-7 | 3-Nitrophenol | 0.51 | 2.00 | 0.324 |
| 129 | 626-19-7 | Phenyl-1,3-dialdehyde | 0.18 | 1.36 | 0.324 |
| 130 | 5798-75-4 | Ethyl-4-bromobenzoate | 1.33 | 3.50 | 0.325 |
| 131 | 89-84-9 | 2,4-Dihydroxyacetophenone | 0.25 | 1.41 | 0.325 |
| 132 | 1016-78-0 | 3-Chlorobenzophenone | 1.55 | 3.97 | 0.325 |
| 133 | 17696-62-7 | Phenyl-4-hydroxybenzoate | 1.37 | 3.49 | 0.327 |
| 134 | 93-99-2 | Phenyl benzoate | 1.35 | 3.59 | 0.327 |
| 135 | 131-57-7 | 2-Hydroxy-4-methoxybenzophenone | 1.42 | 3.58 | 0.327 |
| 136 | 2700-22-3 | Benzylidene malononitrile | 0.64 | 2.15 | 0.328 |
| 137 | 620-88-2 | 4-Nitrophenyl phenyl ether | 1.58 | 3.83 | 0.330 |
| 138 | 136-36-7 | Resorcinol monobenzoate | 1.11 | 3.13 | 0.330 |
| 139 | 14548-45-9 | 4-Bromophenyl-3-pyridyl ketone | 0.82 | 2.96 | 0.331 |
| 140 | 121-89-1 | 3′-Nitroacetophenone | 0.32 | 1.42 | 0.332 |
| 141 | 99-61-6 | 3-Nitrobenzaldehyde | 0.11 | 1.47 | 0.332 |
| 142 | 4553-07-5 | Ethyl phenylcyanoacetate | −0.02 | 1.63 | 0.332 |
| 143 | 91-23-6 | 2-Nitroanisole | −0.07 | 1.73 | 0.332 |
| 144 | 4920-77-8 | 3-Methyl-2-nitrophenol | 0.61 | 2.29 | 0.332 |
| 145 | 844-51-9 | 2,5-Diphenyl-1,4-benzoquinone | 1.48 | 3.16 | 0.332 |
| 146 | 610-15-1 | 2-Nitrobenzamide | −0.72 | −0.12 | 0.332 |
| 147 | 2905-69-3 | Methyl-2,5-dichlorobenzoate | 0.81 | 3.16 | 0.332 |
| 148 | 552-89-6 | 2-Nitrobenzaldehyde | 0.17 | 1.74 | 0.333 |
| 149 | 119-33-5 | 4-Methyl-2-nitrophenol | 0.57 | 2.15 | 0.333 |
| 150 | 131-55-5 | 2,2′,4,4′-Tetrahydroxybenzophenone | 0.96 | 2.92 | 0.333 |
| 151 | 555-16-8 | 4-Nitrobenzaldehyde | 0.20 | 1.56 | 0.333 |
| 152 | 700-38-9 | 5-Methyl-2-nitrophenol | 0.59 | 2.31 | 0.333 |
| 153 | 90-60-8 | 3,5-Dichlorosalicylaldehyde | 1.55 | 3.07 | 0.334 |
| 154 | 69212-31-3 | 2-(Benzylthio)-3-nitropyridine | 1.72 | 3.42 | 0.335 |
| 155 | 99-77-4 | Ethyl-4-nitrobenzoate | 0.71 | 2.33 | 0.335 |
| 156 | 874-42-0 | 2,4-Dichlorobenzaldehyde | 1.04 | 3.08 | 0.335 |
| 157 | 13608-87-2 | 2′,3′,4′-Trichloroacetophenone | 1.34 | 3.21 | 0.336 |
| 158 | 835-11-0 | 2,2′-Dihydroxybenzophenone | 1.16 | 3.47 | 0.336 |
| 159 | 619-50-1 | Methyl-4-nitrobenzoate | 0.39 | 1.94 | 0.336 |
| 160 | 2973-19-5 | 2-Chloromethyl-4-nitrophenol | 0.75 | 2.42 | 0.338 |
| 161 | 402-45-9 | α,α,α-Trifluoro-4-cresol | 0.62 | 2.82 | 0.340 |
| 162 | 5292-45-5 | Dimethylnitroterephthalate | 0.43 | 1.66 | 0.340 |
| 163 | 637-53-6 | Thioacetanilide | −0.01 | 1.71 | 0.341 |
| 164 | 601-89-8 | 2-Nitroresorcinol | 0.66 | 1.56 | 0.341 |
| 165 | 1689-84-5 | 3,5-Dibromo-4-hydroxybenzonitrile | 1.16 | 2.88 | 0.341 |
| 166 | 440-60-8 | Pentafluorobenzyl alcohol | −0.20 | 1.82 | 0.342 |
| 167 | 42087-80-9 | Methyl-4-chloro-2-nitrobenzoate | 0.82 | 2.41 | 0.342 |
| 168 | 1493-27-2 | 1-Fluoro-2-nitrobenzene | 0.23 | 1.69 | 0.343 |

**Table 12.2 (continued)   Toxicity to *T. pyriformis* (Log [IGC$_{50}$]$^{-1}$), Octanol-Water Partition Coefficient (Log K$_{ow}$) and Maximum Acceptor Superdelocalizability (A$_{max}$) Values for the Compounds Used for External Validation**

| No. | CAS | Name | Log (IGC$_{50}$)$^{-1}$ | Log K$_{ow}$ | A$_{max}$ |
|-----|-----|------|-------------------------|--------------|-----------|
| 169 | 393-39-5 | α,α,α-4-Tetrafluoro-o-toluidine | −0.02 | 2.51 | 0.343 |
| 170 | 704-13-2 | 3-Hydroxy-4-nitrobenzaldehyde | 0.27 | 1.47 | 0.345 |
| 171 | 3460-18-2 | 2,5-Dibromonitrobenzene | 1.37 | 3.41 | 0.346 |
| 172 | 613-90-1 | Benzoyl cyanide | 0.31 | 1.91 | 0.347 |
| 173 | 78056-39-0 | 4,5-Difluoro-2-nitroaniline | 0.75 | 2.19 | 0.348 |
| 174 | 364-74-9 | 2,5-Difluoronitrobenzene | 0.33 | 1.86 | 0.349 |
| 175 | 827-23-6 | 2,4-Dibromo-6-nitroaniline | 1.62 | 3.63 | 0.352 |
| 176 | 3011-34-5 | 4-Hydroxy-3-nitrobenzaldehyde | 0.61 | 1.48 | 0.352 |
| 177 | 532-55-8 | Benzoyl isothiocyanate | 0.10 | 1.91 | 0.364 |

$$\log (IGC_{50})^{-1} = 0.529 \ (0.027) \log K_{ow} + 17.6 \ (0.93) \ A_{max} - 6.30 \ (0.31) \tag{12.1}$$

$$n = 188, \ R^2(adj) = 0.800, \ R^2(pred) = 0.795, \ s = 0.338, \ F = 376, \ Pr > F = 0.0001$$

was not significantly different in terms of the coefficient of log K$_{ow}$ and the constant in the equation than the result, reported by Schultz (1999):

$$\log (IGC_{50})^{-1} = 0.50 \log K_{ow} + 9.85 \ A_{max} - 3.47 \tag{12.2}$$

$$n = 197; \ R^2 = 0.816, \ s = 0.34; \ F = 429$$

However, the value of the regression coefficient of A$_{max}$, in the new model (Equation 12.1) is almost eight units greater than in the earlier study (Equation 12.2).

Evaluation of all 215 compounds with measurable toxicity, reported by Schultz (1999), with the newly calculated A$_{max}$ showed that only one compound (4-chloro-3,5-dinitrobenzonitrile, Res. = 1.16, St. res. = 3.24) is a statistically significant outlier and this was excluded from the data set. The resulting model is shown below:

$$\log (IGC_{50})^{-1} = 0.513 \ (0.026) \ (\log K_{ow}) + 18.40 \ (0.94) \ (A_{max}) - 6.50 \ (0.31) \tag{12.3}$$

$$n = 214, \ R^2(adj) = 0.793, \ R^2(pred) = 0.787, \ s = 0.359, \ F = 408, \ Pr > F = 0.0001$$

A plot of observed vs. predicted toxicity is presented in Figure 12.2.

## B.  Evaluation of the Benzene Response-Surface Model

Toxicity, along with hydrophobicity and electrophilicity data, for an additional set of 177 substituted benzenes is provided in Table 12.2. Toxicity values varied uniformly over about three orders of magnitude (from −1.07 to 1.72 on a log scale). Hydrophobicity varied over five orders of magnitude (from −0.17 to 5.13 on a log scale). Reactivity measured by A$_{max}$ varied on a linear scale from 0.283 to 0.364.

Least-squares regression analysis of these data yields the equation:

$$\log (IGC_{50})^{-1} = 0.550 \ (0.014) \log K_{ow} + 13.5 \ (0.68) \ A_{max} - 5.10 \ (0.21) \tag{12.4}$$

$$n = 174, \ R^2(adj) = 0.930, \ R^2(pred) = 0.928, \ s = 0.159, \ F = 1150, \ Pr > F = 0.0001.$$

A plot of observed vs. predicted toxicity is presented in Figure 12.3.

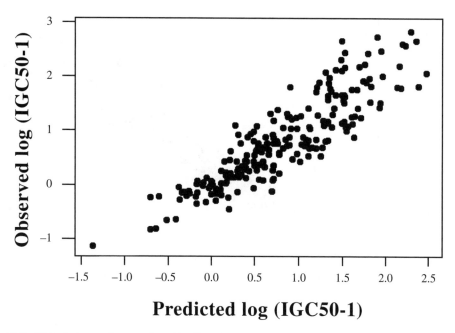

**Figure 12.2** Plot of observed vs. predicted by Equation 12.3 log $(IGC_{50})^{-1}$ values.

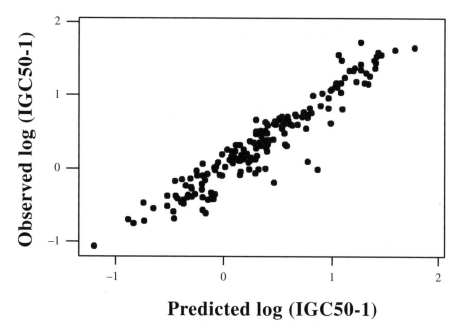

**Figure 12.3** Plot of observed vs. predicted by Equation 12.4 log $(IGC_{50})^{-1}$ values.

Three of the 177 derivatives (pentafluorobenzyl alcohol, Res = –0.67, St. res = –3.61; $\alpha,\alpha,\alpha$-4-tetrafluoro-2-toluidine, Res = –0.88, St. res = –4.74; and benzoyl isothiocyanate, Res = –0.68, St. res. = –3.73) were observed to be statistical outliers to Equation 12.4 and were not included in the model above.

The comparison of the coefficients and statistics of Equation 12.3 and Equation 12.4 reveals them to be very similar with the exception of the regression coefficient of $A_{max}$, which differs with almost 5 units between the models.

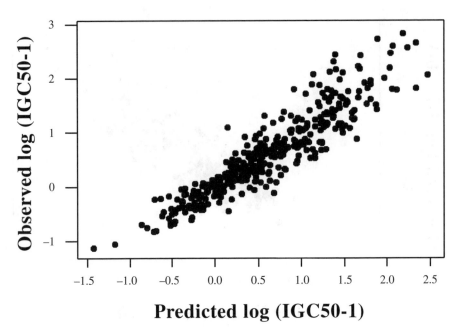

**Figure 12.4**   Plot of observed vs. predicted by Equation 12.5 log $(IGC_{50})^{-1}$ values.

## C.  Combined Benzene Response-Surface Model

Least-squares regression analysis of the combined data from Equation 12.3 and Equation 12.4 yields the equation:

$$\log (IGC_{50})^{-1} = 0.545 \ (0.015) \log K_{ow} + 16.2 \ (0.62) \ A_{max} - 5.91 \ (0.20) \tag{12.5}$$

$$n = 384, \ R^2(adj) = 0.859, \ R^2(pred) = 0.856, \ s = 0.275, \ F = 1163, \ Pr > F = 0.0001.$$

A plot of observed vs. predicted toxicity is presented in Figure 12.4.

Four of the 388 derivatives (3,4-dinitrobenzyl alcohol, Res = 0.95, St. res. = 3.28; 1,5-difluoro-2,4-dinitrobenzene, Res = 0.93, St. res. = 3.26; 1-bromo-2,4-dinitrobenzene, Res. = 0.93, St. res. = 3.23; and pentafluoronitrobenzene, Res. = 1.04, St. res. = 3.60) were observed to be statistical outliers to Equation 12.5 and were not included in the analysis.

## V. DISCUSSION

All toxicity-related QSARs require validation to ensure they are capable of making accurate predictions of toxicity for compounds not included in the training set. The best means of validation is by way of an external data set. This is the most demanding method because it requires additional testing and attention to the selection of compounds for validation. Efforts should be made to have chemical diversity within the training set and the chemicals in the validation set similar to those in the training set (Golbraikh and Tropsha, 2002). The training chemicals should represent the depth and breadth of all existing chemicals within the domain. The validating chemicals should also represent the distribution of existing chemicals within the training domain. In this exercise reactivity quantified by $A_{max}$ was used to assess both diversity for training and representation for validation.

It is important to note that validation should not be confused with the statistical fit of a QSAR. The latter can be assessed by many easily available statistical terms (e.g., $R^2$(adj), $s$, F, etc.), which reflect the ability of the QSAR to mimic the data. While a poor statistical fit to a QSAR results in a model with little or no predictive value, a significantly good statistical fit does not necessarily imply that the QSAR will predict toxic potency accurately for an untested compound.

There are two general types of validation of QSARs: horizontal and vertical. Horizontal validation is performed within a data set by using a distinct training set for model development and a separate validation, or test set, for assessing the predictive capability of the QSAR. Examples of such validation can be found in the literature. Schultz et al. (1987) limited the number of compounds tested to 29 *para*-substituted phenols, but was at the same time able to maximize the physiochemical space of the data set (i.e., the domain). The result was a high statistical fitted QSAR ($R^2 = 0.911$), which revealed hydrophobicity, electrophilicity, and hydrogen bonding to be the important properties in *T. pyriformis* population growth impairment by phenols (Schultz et al., 1987). Cronin et al. (2002) validated the model and domain when experimental toxicity values for the majority of more than 200 phenols could be predicted accurately by a multiple regression-based QSAR developed using descriptors of molecular hydrophobicity, electrophilicity, and ionization.

Also important to the validation process of QSARs is vertical validation. In this instance, quantitatively similar QSARs are developed with similar descriptors but using data for a different toxic endpoint. For example, the investigation of Karabunarliev et al. (1996b) modeled acute aquatic toxicity data for the fathead minnow *Pimephales promelas*. The compounds considered in the analysis were confined to substituted benzenes, and descriptors limited to log $K_{ow}$ and $A_{max}$. The fish toxicity QSAR (log $[LC_{50}]^{-1} = 0.62$ log $K_{ow} + 9.17 A_{max} - 3.21$; $n = 122$; $R^2 = 0.83$; $s = 0.16$; $F = 292$) of Karabunarliev et al. (1996b) was very similar in terms of slope, intercept, and statistical fit to the QSAR presented in Equation 12.2. The fact that different endpoints provide very similar QSARs indicates that the QSAR is valid across protocols. This shows the universality of the model.

The first goal of the horizontal validation is to prove the robustness of the model. This means to determine where can others use the model (i.e., within what boundaries) to yield reliable results. Bearing in mind that the external validation is the only way to establish the real predictivity of the model (Golbraikh and Tropsha, 2002), an external set of 177 commercially available chemicals was collected. The compounds, included in Equation 12.4, were used for validation of the initial response-surface model presented with Equation 12.3. Ideally, the correlation between the observed activity and that predicted by the model should have a high correlation coefficient, an intercept of zero and a slope of one. To be able to compare the predictivity of different models, regression fit through the origin was performed. For 174 compounds this yielded a model with $R^2 = 0.936$, $s = 0.179$, and a slope of 0.938 ($\pm 0.019$). For regression through the origin (the no-intercept model), $R^2$ measures the proportion of the variability in the dependent variable about the origin explained by regression and cannot be compared to $R^2$ for models that include an intercept.

To validate the model presented by Equation 12.4, the 214 compounds included in equation (12.3) were used. For this second validation the model between observed and predicted toxicity showed $r^2 = 0.886$, $s = 0.379$, and a slope of 1.091 ($\pm 0.027$). Because both the intercepts were very close to one, it may be assumed that both the models are capable of predicting correctly the toxicity of new compounds with descriptors within the defined ranges. It can be noted that Equation 12.3 performs a little better than Equation 12.4 because of the higher $R^2$, lower $s$, and slope that is closer to 1 in the statistical fit between the observed and predicted toxicity. To take advantage of the availability of data for larger number of compounds, the model presented by Equation 12.5 was developed. The slopes of the regression line between the observed toxicity values and those calculated by Equation 12.5 for both the subsets ($n = 174$ and $n = 214$) were 0.959 ($\pm 0.017$) and 1.036 ($\pm 0.024$), respectively. Because this is not a real validation with external data sets, the validity of the latter model needs to be additionally demonstrated.

The horizontal validation aims not only to assess the robustness of the model, but also to identify the crucial aspects that are susceptible to change and can affect the result. They include all the three components of a QSAR, but only the influence of the descriptor variability, and particularly the changes in the electronic term in the response-surface analysis was the focus of the validation in this study. As was shown from the comparison between Equations 12.1 and 12.2, the change of the algorithm for calculation of the electronic descriptors hides potential risk for miss-prediction of toxicity using quantum mechanical descriptors from different sources. In this study the observed difference in the coefficient of $A_{max}$ is produced by the fact that in the 1999 paper the algorithm for the calculation of the acceptor superdelocalizabilities (Karabunarliev et al., 1996b) includes a shift in the summed energies of the unoccupied molecular orbitals in order to avoid a problem with energy levels with opposite sign. Comparison of the coefficients and statistical values of Equation 12.3 with those of Equation 12.4 revealed stronger similarities for all values. These similarities were without doubt due in large part to the great care taken to mimic the descriptor domain of the training set, especially $A_{max}$. The difference of almost five in the coefficient of $A_{max}$ can be explained by the fact that there are fewer compounds with high $A_{max}$ ($A_{max} > 0.34$) values in the external validation set, as well as the presence of a large number of narcotics, for which the quantum mechanical term is less significant than for the more reactive chemicals.

Toxicity, and hence the modeling of toxic potency, is intrinsically a non-linear phenomenon. To expect to be able to model all compounds or even all compounds within a chemical class (e.g., benzenes), even for a single endpoint like *Tetrahymena* population growth impairment, with a single relationship is naive. In narcosis, a clear boundary is observed between whether a compound exerts aquatic toxicity and its hydrophobicity. Highly hydrophobic, liquids or ones that are solid with high melting points have no noticeable toxicity mainly because the compounds exhibit insufficient water soluble to elicit toxicity. Other boundaries, while less easily explained are often noted. Indications of a potential boundary are typically first observed in the form of outliers. Outliers are often explained by physicochemical or protocol limitation (e.g., volatility in a static system) or a change in mechanism of toxic action (e.g., narcotic vs. electrophiles) (Schultz et al., 1998).

Outliers are useful in QSAR development as they assist in establishing the chemical domain of the model. As noted by Egan and Morgan, (1998) outliers from a statistical relationship are data that do not fit the model, or are poorly predicted by it. There are several potential reasons for a chemical being an outlier from a QSAR. Customarily, such compounds have been recognized as acting by a different mechanism of action from the other chemicals, which are well modeled by the QSAR. For instance, the domain of the phenol model was established by Cronin et al. (2002) who noted clear outliers to the model, including phenols capable of redox cycling or being metabolized. Moreover, Cronin and co-workers (2002) also noted that carboxyl-derivatives whose toxicity was unduly impacted by ionization (Muccini et al., 1999) were outliers to the general phenol model. These two groups of benzenes carboxylic acids and tautomerizing quinones and semiquinones are also outside the domain of the present QSARs.

Outliers also may be the result of variability in the experimental measurement of toxicity. Seward et al. (2001) noted that there was less likelihood of reproducing the experimental toxicity of reactive electrophilic chemicals as compared to compounds acting via a narcosis mechanism of toxicity. Even within a chemical domain there may be different levels of confidence in the predicted activity. For example, the predictivity can be sensitive to the distribution of $A_{max}$ values. In examining the distribution within the 214 derivatives used in the training set, it is observed that there are 54 derivatives with $A_{max}$ values between 0.28 and 0.30, 62 chemicals with $A_{max}$ values between 0.30 and 0.32, and 46 compounds with $A_{max}$ values between 0.32 and 0.34. However, only 30 derivatives were included with $A_{max}$ values between 0.34 and 0.36 and even less, 22 compounds, were included with $A_{max}$ values greater than 0.36. Even fewer benzenes with $A_{max}$ values in these higher ranges were available for inclusion in the validation set. It may be proposed that the predicted toxic potency of benzene derivatives with $A_{max}$ values between 0.28 and 0.34 can be taken with a greater level of confidence than compounds with $A_{max}$ values greater than 0.34. The latter is supported by the fact that the 4 statistical outliers to Equation 12.5 (i.e., 3,4-dinitrobenzyl alcohol, 1,5-difluoro-2,4-dinitrobenzene, 1-bromo-2,4-dinitrobenzene, and pentafluoronitrobenzene) all have $A_{max}$ values greater than 0.34.

In summary, external validation of a transparent and fully interpretable two-descriptor (one for bio-uptake and one for electro[nucleo]philic reactivity) regression-based model for benzene potency, derived from quality data has been presented. Both the training and the test set were carefully selected to span a large array of reactivity. Both exhibited similar variability in toxicity, hydrophobicity, and electro(nucleo)philicity. The robustness of the models was demonstrated by the similarity of the coefficients and the statistical criteria between models derived on same mechanistic descriptors with different sets of compounds. The closeness to unity of the slopes in the regression through the origin between observed and calculated toxicity for an external validation set provided additional evidence for the good predictivity of the models. The impact of the algorithm for calculation of quantum mechanical descriptors, and particularly acceptor superdelocalizabilities for the weighting of this descriptor was demonstrated. The boundaries and the level of confidence within a certain domain were also discussed.

## REFERENCES

Benfenati, E., Piclin, N., Roncaglioni, A., and Vari, M.R., Factors influencing predictive models for toxicology, *SAR QSAR Environ. Res.*, 12, 593-603, 2001.

Blaauboer, B.J., Barratt, M.D., and Houston, J.B., The integrated use of alternative methods in toxicological risk evaluation — ECVAM Integrated Testing Strategies task force report 1, *Alt. Lab. Anim. (ATLA)*, 27, 229–237, 1999.

Bradbury, S.P., Russom, C.L., Ankley, G.T., Schultz, T.W., and Walker, J.D., Overview of data and conceptual approaches for derivation of quantitative structure-activity relationships for ecotoxicological effects of organic chemicals, *Environ. Toxicol. Chem.*, 22, 1789–1798, 2003.

Cronin, M.T.D., Aptula, A.O., Duffy, J.C., Netzeva, T.I., Rowe, P.H., Valkova, I.V., and Schultz, T.W., Comparative assessment of methods to develop QSARs for the prediction of the toxicity of phenols to *Tetrahymena pyriformis*, *Chemosphere*, 49, 1201–1221, 2002.

Dearden, J.C., Physico-chemical descriptors, in *Practical Applications of Quantitative Structure-Activity Relationships (QSAR) in Environmental Chemistry and Toxicology*, Karcher, W. and Devillers, J., Eds., Kluwer Academic, Dordrecht, Netherlands, 1990, pp. 25–59.

Egan, W.J. and Morgan, S.L., Outlier detection in multivariate analytical chemical data, *Anal. Chem.*, 70, 2372–2379, 1998.

Golbraikh, A. and Tropsha, A., Beware of q²!, *J. Mol. Graphics Modeling*, 20, 269–276, 2002.

Kaiser, K.L.E., Dearden, J.C., Klein, W., and Schultz, T.W., A note of caution to users of ECOSAR, *Water Qual. Res. J. Can.*, 34, 179–182, 1999.

Karabunarliev, S., Mekenyan, O.G., Karcher, W., Russom, C.L., and Bradbury, S.P., Quantum-chemical descriptors for estimating the acute toxicity of electrophiles to the fathead minnow (*Pimephales promelas*): an analysis based on molecular mechanisms, *Quant. Struct.-Act. Relat.*, 15, 302–310, 1996a.

Karabunarliev, S., Mekenyan, O.G., Karcher, W., Russom, C.L., and Bradbury, S.P., Quantum-chemical descriptors for estimating acute toxicity of substituted benzenes to the guppy (*Poecilia reticulata*) and fathead minnow (*Pimephales promelas*), *Quant. Struct.-Act. Relat.*, 15, 311–320, 1996b.

Karelson, M., Lobanov, V.S., and Katritzky, A.R., Quantum-chemical descriptors in QSAR/QSPR studies, *Chem. Rev.*, 96, 1027–1043, 1996.

Lipnick, R.L., Outliers, their origin and use in the classification of molecular mechanisms of toxicity, *Sci. Total Environ.*, 109/110, 131–154, 1991.

Livingstone, D.J., *Data Analysis for Chemists: Applications to QSAR and Chemical Product Design*, Oxford University Press, Oxford, 1995.

Livingstone, D.J., The characterization of chemical structures using molecular properties: a survey, *J. Chem. Inf. Comput. Sci.*, 40, 195–209, 2000.

Mekenyan, O.G. and Veith, G.D., Relationships between descriptors for hydrophobicity and soft electrophilicity in predicting toxicity, *SAR QSAR Environ. Res.*, 2, 335–344, 1993.

Muccini, M., Layton, A.C., Sayler, G.S., and Schultz, T.W., Aquatic toxicities of halogenated benzoic acids to *Tetrahymena pyriformis*, *Bull. Environ. Contamination Toxicol.*, 62, 616–622, 1999.

Nendza, M. and Russom, C.L., QSAR modeling of the ERL-D fathead minnow acute toxicity database, *Xenobiotica*, 21, 147–170, 1991.

SAS Institute Inc., *SAS/STAT User's Guide*, Version 6, 4th ed., Vol. 2, North Carolina, 1989.

Schultz, T.W., TETRATOX: *Tetrahymena pyriformis* population growth impairment endpoint — A surrogate for fish lethality, *Toxicol. Methods*, 7, 289–309, 1997.

Schultz, T.W., Structure-toxicity relationships for benzenes evaluated with *Tetrahymena pyriformis, Chem. Res. Toxicol.*, 12, 1262–1267, 1999.

Schultz, T.W. and Cronin, M.T.D., Essential and desirable characteristics of ecotoxicity QSARs, *Environ. Toxicol. Chem.*, 22, 599–607, 2003.

Schultz, T.W., Cronin, M.T.D., Walker, J.D., and Aptula, A.O., Quantitative structure-activity relationships in (QSARs) in toxicology: a historical perspective, *J. Mol. Struct. (Theochem)*, 622, 1–22, 2002.

Schultz, T.W., Riggin, G.W., and Wesley, S.K., Structure-activity relationships for para-substituted phenols, in *QSAR in Environmental Toxicology-II,* Kaiser, K.L.E., Ed., D. Reidel Publishing, Dordrecht, 1987, pp. 333–345.

Schultz, T.W., Sinks, G.D., and Bearden, A.P., QSARs in aquatic toxicology: a mechanism of action approach comparing toxic potency to *Pimephales promelas, Tetrahymena pyriformis,* and *Vibrio fischeri,* in *Comparative QSAR,* Devillers, J., Ed., Taylor and Francis, London, 1998, pp. 52–109.

Seward, J.R., Sinks, G.D., and Schultz, T.W., Reproducibility of toxicity across mode of toxic action in the *Tetrahymena* population growth impairment assay, *Aquatic Toxicol.*, 53, 33–47, 2001.

Veith, G.D., Call, D.J., and Brooke, L.T., Structure-toxicity relationships for the fathead minnow, *Pimephales promelas*: Narcotic industrial chemicals, *Can. J. Fishery Aquatic Sci.*, 40: 473-748, 1983.

Walker, J.D. and Schultz, T.W., Structure activity relationships for predicting ecological effects of chemicals, in *Handbook of Ecotoxicology,* 2nd ed., Hoffman, D.J., Rattner, B.A., Burton, G.A., Jr., and Cairns, J., Jr., Eds., CRC Press, Boca Raton, FL, 2002, pp. 893–910.

CHAPTER **13**

# Receptor-Mediated Toxicity: QSARs for Estrogen Receptor Binding and Priority Setting of Potential Estrogenic Endocrine Disruptors

Weida Tong, Hong Fang, Huixiao Hong, Qian Xie, Roger Perkins, and Daniel M. Sheehan

## CONTENTS

## I. INTRODUCTION

Nuclear receptors (NRs) are a superfamily of ligand-dependent transcription factors that mediate the effects of hormones and other endogenous ligands to regulate the expression of specific genes. Members of the NR superfamily, which may number in the hundreds, include receptors for various steroid hormones (estrogen, androgen, progesterone, and several corticosteroids), retinoic acid (the retinoic acid receptor $\alpha$, $\beta$, and $\gamma$ isoforms, and the retinoid X receptor $\alpha, \beta$, and $\gamma$ isoforms), thyroid hormones, vitamin D, and dietary lipids (the peroxisome proliferator activated receptor [PPAR] $\alpha$, $\beta$, and $\gamma$ isoforms). A large number of orphan NRs have also been identified whose cognate ligands are still unknown (Giguere, 1999). Diminished or excessive production of a particular hormone or target-cell insensitivity to a hormone are among the major problems related to human endocrine dysfunction diseases (Zubay et al., 1995).

Receptor-mediated effects are stimulated and inhibited not only by endogenous cognate ligands for each NR, but also by exogenous substances including natural products and synthetic chemicals. There are a large number of ligands, diverse in both structure and source, which act through the NRs to produce receptor-mediated effects. The NRs and their ligands have thus attracted broad scientific interest, particularly in the pharmaceutical industry for drug discovery and in toxicology and environmental science for risk assessment as, for example, pertaining to endocrine disrupting chemicals (EDCs).

Among numerous NRs, the estrogen receptor (normally abbreviated to ER) and its ligands are probably most studied. This is because the ER plays a vital role in a wide variety of essential physiological processes (Duax et al., 1996). Estrogens elicit many cellular responses in target tissues and can exert both positive and negative effects on health and reproductive function. For example, estrogens are used beneficially for fertility control (oral contraception) and for relief of menopausal symptoms (estrogen replacement therapy). The adverse developmental effects of diethylstilbestrol (DES) are demonstrated by human fetal sensitivity to estrogenic chemicals.

Estrogens regulate the expression of specific genes and the secretion of certain hormones. They coordinate diverse and complex processes such as cell proliferation, cell differentiation, and tissue organization through pleiotropic actions. Once estrogens reach the bloodstream, they may remain free or bind to serum estrogen-binding proteins such as $\alpha$-fetoprotein (AFP) in rodents (Baker et al., 1998; Sheehan and Young, 1979) or sex hormone binding globulin (SHBG) in humans (Sheehan and Young 1979). Only the free (unbound) hormone is able to diffuse into the target cells, where it binds to the ER to form a hormone-receptor complex. The prevailing model suggests that this complex then interacts with an estrogen response element (ERE) of target genes and activates the transcriptional machinery (Gillesby and Zacharewski, 1998; Norris et al., 1997).

Quantitative structure-activity relationship (QSAR) models have proven their utility, from both the pharmaceutical and toxicological perspectives, for the identification of chemicals that might interact with ER. While their primary function in the pharmaceutical enterprise is lead discovery and optimization for high-affinity ER ligands, QSAR models can play an essential role in toxicology as a priority-setting tool for risk assessment.

QSAR modeling employs statistical approaches to correlate or rationalize variations in the biological activity of a series of chemicals with variations in their molecular structures. The first step in developing a traditional QSAR model is the acquisition of a training set of chemicals that have known activities. Second, descriptors representing the molecular structure of individual chemicals (i.e., hydrophobicity, structural fragments, charged surface area, the number of hydrogen bonds, solubility, etc.) are calculated. Then, a correlation between descriptors and activity for the training set is evaluated by employing various statistical approaches to determine the most statistically significant relationship (the QSAR model). A proper validation is required to ensure the model's predictivity for the chemicals not used in the training set. With adequately validated performance, such models can be used to predict activities of untested chemicals.

Obtaining a good quality QSAR model depends heavily on many factors in the approach, particularly on the quality of biological data, descriptor selection, and statistical methods (see Chapter 19 for more details). Given the fact that any QSAR approach has strengths and weaknesses, the careful selection of a specific model, or a combination of models, also needs to be emphasized, and is often specific to the particular application in question.

In this chapter, we first summarize our motives and effort to develop a robust training set (the National Center for Toxicological Research [NCTR] data set) for QSARs, which covers a broad range of ER binding affinity and structural diversity. We will then propose a systematic procedure to pre-process molecular structure that is particularly important for QSAR studies utilizing toxicological data. Next, we will present several QSAR approaches currently used in our labs. The strengths and weaknesses of these methods will be also discussed. We will then focus on a strategy for the validation of the QSAR models, a topic that has received sparse attention in computational toxicology despite its critical importance. The review concludes by integrating these models into an integrated Four-Phase approach that could be useful for priority setting of large number of chemicals according to their potential estrogenic endocrine disruption.

For the sake of clarification, the term QSAR is used broadly in this review to include methodologies that predict activity on an ordinal or categorical scale rather than on only a quantitative scale (Perkins et al., 2003; Tong et al., 2003).

## II. NCTR ER DATA SET: A ROBUST TRAINING SET FOR QSARS

Although a predictive QSAR model is dependent on a number of factors, a training set with high-quality biological data is the first step in developing a useful QSAR model. It is desirable that the biological data come from the same assay protocol. Data error adds noise to the correlation of structure with activity. The rules of thumb for a good biological data for the training set are (1) a smooth dose-response relationship, (2) a reproducible potency (or affinity), (3) an activity range that spans two or more orders of magnitude from the least active to the most active chemical in the series, and (4) data values that are evenly distributed across the range of activity. It is important to note that most toxicity data do not meet all these criteria because of the nature of toxicological research, in which case care should be taken in interpreting QSAR results.

A robust QSAR model to predict the activity of a wide variety of chemical structures must start with a training set that contains a sufficiently large number of chemicals with diverse structure that reflects, to some degree, the data set to be evaluated. Despite decades of studies of estrogens, we found that the existing data are inadequate to construct robust QSAR models. For example, in the past few years, a number of QSAR models have been developed for ligand binding to the ER (Bradbury et al., 1996; Waller et al., 1996; Wiese et al., 1997; Sadler et al., 1998; Zheng and Tropsha 2000), including some of our early work (Tong et al., 1997a;1997b; 1998; Xing et al., 1999). Unfortunately, most of these QSAR models were developed based on data sets available in the literature; these data sets were both too small and lacked structural diversity (Sadler et al., 1998; Wiese et al., 1997; Tong et al., 1997a). Although these models yield good statistical results in the training and cross-validation steps and explain some structural determinants for ER binding, they have limited applicability in predicting the ER-ligand binding affinity of chemicals that cover a wide range of structural diversity.

In order to obtain an adequate training set to develop a more robust QSAR model, we developed a rat ER binding assay (Blair et al., 2000; Branham et al., 2002). For many years the ER competitive binding assay was considered the gold standard. Many variants have since been developed, leading to some significant differences in results. Our ER binding assay was rigorously validated, and provides high-quality data for model development. Each experimental value is replicated at least twice. We assayed 232 chemicals to obtain a training set for model development (Table 13.1). The

Table 13.1    The NCTR Data Set, Containing ER Binding Data (RBA)
for 232 Diverse Chemicals

| Name | CAS | Log RBA |
|---|---|---|
| Diethylstilbestrol (DES) | 56-53-1 | 2.60 |
| Hexestrol | 84-16-2 | 2.48 |
| Ethynylestradiol | 57-63-6 | 2.28 |
| 4-OH-Tamoxifen | 68047-06-3 | 2.24 |
| 17β-Estradiol ($E_2$) | 50-28-2 | 2.00 |
| 4-OH-Estradiol | 5976-61-4 | 1.82 |
| Zearalenol | 71030-11-0 | 1.63 |
| ICI 182780 | 129453-61-8 | 1.57 |
| Dienestrol | 84-17-3 | 1.57 |
| α-Zearalanol | 55331-29-8 | 1.48 |
| 2-OH-Estradiol | 362-05-0 | 1.47 |
| Monomethyl ether diethylstilbestrol | | 1.31 |
| 3,3′-Dihydroxyhexestrol | 79199-51-2 | 1.19 |
| Droloxifene | 82413-20-5 | 1.18 |
| ICI 164384 | | 1.16 |
| Dimethylstilbestrol | 552-80-7 | 1.16 |
| Moxestrol | 34816-55-2 | 1.14 |
| 17-Deoxyestradiol | 53-63-4 | 1.14 |
| Estriol | 50-27-1 | 0.99 |
| Monomethyl ether hexestrol | 13026-26-1 | 0.97 |
| 2,6-Dimethyl hexestrol | | 0.95 |
| Estrone | 53-16-7 | 0.86 |
| 3-(p-Phenol)-4-(p-tolyl)-hexane | | 0.60 |
| 17α-Estradiol | 57-91-0 | 0.49 |
| Dihydroxymethoxychlor olefin | 14868-03-2 | 0.42 |
| Mestranol | 72-33-3 | 0.35 |
| Zearalanone | 5975-78-0 | 0.32 |
| Tamoxifen | 10540-29-1 | 0.21 |
| Toremifene | 89778-26-7 | 0.14 |
| α,α-Dimethyl-β-ethyl allenolic acid | 65118-81-2 | −0.02 |
| Coumestrol | 479-13-0 | −0.05 |
| 4-Ethyl-7-OH-3-(4-methoxyphenyl)coumarin | 5219-17-0 | −0.05 |
| Nafoxidine | 1845-11-0 | −0.14 |
| Clomiphene | 911-45-5 | −0.14 |
| 6α-OH-Estradiol | 1229-24-9 | −0.15 |
| β-Zearalanol | 42422-68-4 | −0.19 |
| 3-Hydroxy-estra-1,3,5(10)-trien-16-one | 3601-97-6 | −0.29 |
| 3-Deoxyestradiol | 2529-64-8 | −0.30 |
| 3,6,4′-Trihydroxyflavone | | −0.35 |
| Genistein | 446-72-0 | −0.36 |
| 4,4′-Dihydroxystilbene | 659-22-3 | −0.55 |
| HPTE | 2971-36-0 | −0.60 |
| Monohydroxymethoxychlor olefin | 75938-34-0 | −0.63 |
| 2,3,4,5-Tetrachloro-4′-biphenylol | | −0.64 |
| Norethynodrel | 68-23-5 | −0.67 |
| 2,2′,4,4′-Tetrahydroxybenzil | 5394-98-9 | −0.68 |
| β-Zearalenol | | −0.69 |
| Equol | 531-95-3 | −0.82 |
| 4′,6-Dihydroxyflavone | 63046-09-3 | −0.82 |
| Monohydroxymethoxychlor | 28463-03-8 | −0.89 |
| 3-β-Androstanediol | 571-20-0 | −0.92 |
| Bisphenol B | 77-40-7 | −1.07 |
| Phloretin | 60-82-2 | −1.16 |
| Diethylstilbestrol dimethyl ether | 7773-34-4 | −1.25 |
| 2′,4,4′-Trihydroxychalcone | 961-29-5 | −1.26 |
| 4,4′-(1,2-Ethanediyl)bisphenol | 6052-84-2 | −1.44 |
| 2,5-Dichloro-4′-biphenylol | 53905-28-5 | −1.44 |

Table 13.1 (continued)   The NCTR Data Set, Containing ER Binding
Data (RBA) for 232 Diverse Chemicals

| Name | CAS | Log RBA |
|---|---|---|
| 16β-OH-16-Methyl-3-methyl-estradiol | 5108-94-1 | −1.48 |
| Aurin | 603-45-2 | −1.50 |
| Nordihydroguaiaretic acid | 500-38-9 | −1.51 |
| Nonylphenol | 25154-52-3 | −1.53 |
| Apigenin | 520-36-5 | −1.55 |
| Kaempferol | 520-18-3 | −1.61 |
| Daidzein | 486-66-8 | −1.65 |
| 3-Methyl-estriol | 3434-79-5 | −1.65 |
| 4-Dodecylphenol | 104-43-8 | −1.73 |
| 2-Ethylhexyl-4-hydroxybenzoate | 5153-25-3 | −1.74 |
| 4-tert-Octylphenol | 140-66-9 | −1.82 |
| Phenolphthalein | 77-09-8 | −1.87 |
| Kepone | 143-50-0 | −1.89 |
| Heptyl 4-hydroxybenzoate | 1085-12-7 | −2.09 |
| Bisphenol A | 80-05-7 | −2.11 |
| Naringenin | 480-41-1 | −2.13 |
| 4-Chloro-4′-biphenylol | 28034-99-3 | −2.18 |
| 3-Deoxy-estrone | 53-45-2 | −2.20 |
| 4-Cumyl phenol | 599-64-4 | −2.30 |
| 4-n-Octylphenol | 1806-26-4 | −2.31 |
| Fisetin | 528-48-3 | −2.35 |
| 3′,4′,7-Trihydroxy isoflavone | 485-63-2 | −2.35 |
| Biochanin A | 491-80-5 | −2.37 |
| 4′-Hydroxychalcone | 2657-25-2 | −2.43 |
| 2,2′-Methylenebis(4-chlorophenol) | 97-23-4 | −2.45 |
| 4,4′-Dihydoxy-benzophenone | 611-99-4 | −2.46 |
| Benzyl 4-hydroxybenzoate | 94-18-8 | −2.54 |
| 4-Hydroxychalcone | 20426-12-4 | −2.55 |
| 2,4-Hydroxybenzophenone | 131-56-6 | −2.61 |
| 4′-Hydroxyflavanone | 6515-37-3 | −2.65 |
| 3α-Androstanediol | 1852-53-5 | −2.67 |
| 4-Phenethylphenol | 6335-83-7 | −2.69 |
| Prunetin | 552-59-0 | −2.74 |
| Doisynoestrol | 15372-34-6 | −2.74 |
| Myricetin | 529-44-2 | −2.75 |
| 2-Chloro-4-biphenylol | 92-04-6 | −2.77 |
| Triphenylethylene | 58-72-0 | −2.78 |
| 3′-Hydroxyflavanone | | −2.78 |
| Chalcone | 94-41-7 | −2.82 |
| 2,4′-DDT | 789-02-6 | −2.85 |
| 4-Heptyloxyphenol | 13037-86-0 | −2.88 |
| Dihydrotestosterone | 521-18-6 | −2.89 |
| Formononetin | 485-72-3 | −2.98 |
| Bis(4-hydroxyphenyl)methane | 620-92-8 | −3.02 |
| 4-Hydroxybiphenyl | 92-69-3 | −3.04 |
| Baicalein | 491-67-8 | −3.05 |
| 6-Hydroxyflavanone | 4250-77-5 | −3.05 |
| n-Butyl 4-hydroxybenzoate | 94-26-8 | −3.07 |
| 4,4′-Sulfonyldiphenol | 80-09-1 | −3.07 |
| Morin | 480-16-0 | −3.09 |
| Diphenolic acid | 126-00-1 | −3.13 |
| 1,3-Diphenyltetramethyldisiloxane | 56-33-7 | −3.16 |
| n-Propyl 4-hydroxybenzoate | 94-13-3 | −3.22 |
| Ethyl 4-hydroxybenzoate | 120-47-8 | −3.22 |
| Phenol red | 143-74-8 | −3.25 |
| 3,3′,5,5′-Tetrachloro-4,4′-biphenyldiol | 13049-13-3 | −3.25 |
| 4-Tert-Amylphenol | 80-46-6 | −3.26 |

**Table 13.1 (continued)   The NCTR Data Set, Containing ER Binding Data (RBA) for 232 Diverse Chemicals**

| Name | CAS | Log RBA |
|---|---|---|
| 4-*Sec*-butylphenol | 99-71-8 | −3.37 |
| 4-Chloro-3-methylphenol | 59-50-7 | −3.38 |
| 6-Hydroxyflavone | 6665-83-4 | −3.41 |
| Methyl 4-hydroxybenzoate | 99-76-3 | −3.44 |
| 4-(Benzyloxyl)phenol | 103-16-2 | −3.44 |
| 3-Phenylphenol | 580-51-8 | −3.44 |
| 2-*Sec*-butylphenol | 89-72-5 | −3.54 |
| 4-*Tert*-butylphenol | 98-54-4 | −3.61 |
| 2,4′-Dichlorobiphenyl | 34883-43-7 | −3.61 |
| 2-Cholor-4-methyl phenol | 6640-27-3 | −3.66 |
| Phenolphthalin | 81-90-3 | −3.67 |
| 4-Chloro-2-methyl phenol | 1570-64-5 | −3.67 |
| 7-Hydroxyflavanone | 6515-36-2 | −3.73 |
| 2-Ethylphenol | 620-17-7 | −3.87 |
| Rutin | 153-18-4 | −4.09 |
| 4-Ethylphenol | 123-07-9 | −4.17 |
| 4-Cresol | 106-44-5 | −4.50 |
| Vinclozolin | 50471-44-8 | Inactive |
| Vanillin | 121-33-5 | Inactive |
| Triphenyl phosphate | 115-86-6 | Inactive |
| Trans-4-hydroxystilbene | 6554-98-9 | Inactive |
| Trans,trans-1,4-diphenyl-1,3-butadiene | 886-65-7 | Inactive |
| Thalidomide | 50-35-1 | Inactive |
| Testosterone | 58-22-0 | Inactive |
| Taxifolin | 480-18-2 | Inactive |
| Suberic acid | 505-48-6 | Inactive |
| Sitosterol | 83-46-5 | Inactive |
| Simazine | 122-34-9 | Inactive |
| Sec-butylbenzene | 135-98-8 | Inactive |
| Quercetin | 117-39-5 | Inactive |
| Prometon | 1610-18-0 | Inactive |
| Progesterone | 57-83-0 | Inactive |
| Phenol | 108-95-2 | Inactive |
| 4,4′-Methoxychlor olefin | 2132-70-9 | Inactive |
| 4,4′-Methoxychlor | 72-43-5 | Inactive |
| 4,4′-DDT | 50-29-3 | Inactive |
| 4,4′-DDE | 72-55-9 | Inactive |
| 4,4′-DDD | 72-54-8 | Inactive |
| 2,4′-DDE | 3424-82-6 | Inactive |
| 2,4′-DDD | 53-19-0 | Inactive |
| Nerolidol | 7212-44-4 | Inactive |
| *n*-Butylbenzene | 104-51-8 | Inactive |
| Naringin | 10236-47-2 | Inactive |
| Mirex | 2385-85-5 | Inactive |
| Metolachlor | 51218-45-2 | Inactive |
| Melatonin | 73-31-4 | Inactive |
| Lindane (gamma-HCH) | 58-89-9 | Inactive |
| Isoeugenol | 97-54-1 | Inactive |
| Hexyl alcohol | 111-27-3 | Inactive |
| Hexachlorobenzene | 118-74-1 | Inactive |
| Hesperetin | 520-33-2 | Inactive |
| Heptaldehyde | 111-71-7 | Inactive |
| Heptachlor | 76-44-8 | Inactive |
| Genistin | 529-59-9 | Inactive |
| Folic acid | 59-30-3 | Inactive |
| Flavone | 525-82-6 | Inactive |
| Flavanone | 487-26-3 | Inactive |

Table 13.1  (continued)    The NCTR Data Set, Containing ER Binding
Data (RBA) for 232 Diverse Chemicals

| Name | CAS | Log RBA |
|------|-----|---------|
| Eugenol | 97-53-0 | Inactive |
| 17β-Hydroxyetiocholan-3-one | 571-22-2 | Inactive |
| Ethyl cinnamate | 103-36-6 | Inactive |
| Epitestesterone | 481-30-1 | Inactive |
| Endosulfan, technical grade | 115-29-7 | Inactive |
| Dopamine | 51-61-6 | Inactive |
| di-n-Butyl phthalate (DBuP) | 84-74-2 | Inactive |
| Dimethyl phthalate | 131-11-3 | Inactive |
| di-iso-Butyl phthalate (DIBP) | 84-69-5 | Inactive |
| Diisononylphthalate | 28553-12-0 | Inactive |
| Diethyl phthalate | 84-66-2 | Inactive |
| Dieldrin | 60-57-1 | Inactive |
| Dibenzo-18-crown-6 | 14187-32-7 | Inactive |
| Di(2-Ethylhexyl) adipate | 103-23-1 | Inactive |
| Dexamethasone | 50-02-2 | Inactive |
| Corticosterone | 50-22-6 | Inactive |
| Cinnamic acid | 140-10-3 | Inactive |
| Cineole | 470-82-6 | Inactive |
| Chrysin | 480-40-0 | Inactive |
| Chrysene | 218-01-9 | Inactive |
| Cholesterol | 57-88-5 | Inactive |
| Chlordane | 57-74-9 | Inactive |
| Catechin | 154-23-4 | Inactive |
| Carbofuran | 1563-66-2 | Inactive |
| Carbaryl | 63-25-2 | Inactive |
| Caffeine | 58-08-2 | Inactive |
| Butylbenzylphthalate | 85-68-7 | Inactive |
| Bis(n-octyl) phthalate | 117-84-0 | Inactive |
| Bis(2-hydroxyphenyl)methane | 2467-02-9 | Inactive |
| Bis(2-ethylhexyl)phthalate | 117-81-7 | Inactive |
| Benzylalcohol | 100-51-6 | Inactive |
| Atrazine | 1912-24-9 | Inactive |
| Amaranth | 915-67-3 | Inactive |
| Aldrin | 309-00-2 | Inactive |
| Aldosterone | 52-39-1 | Inactive |
| Alachlor | 15972-60-8 | Inactive |
| 7-Hydroxyflavone | 6665-86-7 | Inactive |
| 6-Hydroxy-2′-methoxyflavone | | Inactive |
| 4-Aminophenyl ether | 101-80-4 | Inactive |
| 4-Amino butylbenzoate | 94-25-7 | Inactive |
| 4′,6,7-Trihydroxy isoflavone | 17817-31-1 | Inactive |
| 4,4′-Methylenedianiline | 101-77-9 | Inactive |
| 4,4′-Methylenebis(N,N-dimethylaniline) | 101-61-1 | Inactive |
| 4,4′-Methylenebis(2,6-di-tert-butylphenol) | 118-82-1 | Inactive |
| 4,4′-Dichlorobiphenyl | 2050-68-2 | Inactive |
| 4,4′-Diaminostilbene | 54760-75-7 | Inactive |
| 3,3′,4,4′-Tetrachlorobiphenyl | 32598-13-3 | Inactive |
| 2-OH-4-Methoxy-benzophenone | 131-57-7 | Inactive |
| 2-Hydroxy biphenyl | 90-43-7 | Inactive |
| 2-Furaldehyde | 98-01-1 | Inactive |
| 2-Ethylphenol | 90-00-6 | Inactive |
| 2-Chlorophenol | 95-57-8 | Inactive |
| 2,4-D (2,4-dichlorophenoxyacetic acid) | 94-75-7 | Inactive |
| 2,4,5-T | 93-76-5 | Inactive |
| 2,3-Benzofluorene | 243-17-4 | Inactive |
| 2,2′-Dihydroxy-benzophenone | 835-11-0 | Inactive |
| 2,2′-Dihydroxy-4-methoxy-benzophenone | 131-53-3 | Inactive |

Table 13.1 (continued)   The NCTR Data Set, Containing ER Binding
Data (RBA) for 232 Diverse Chemicals

| Name | CAS | Log RBA |
|------|-----|---------|
| 2,2′,4,4′-Tetrachlorobiphenyl | 2437-79-8 | Inactive |
| 1,8-Octanediol | 629-41-4 | Inactive |
| 1,6-Dimethylnaphthalene | 575-43-9 | Inactive |
| 1,3-Dibenzyltetramethyldisiloxane | 1833-27-8 | Inactive |

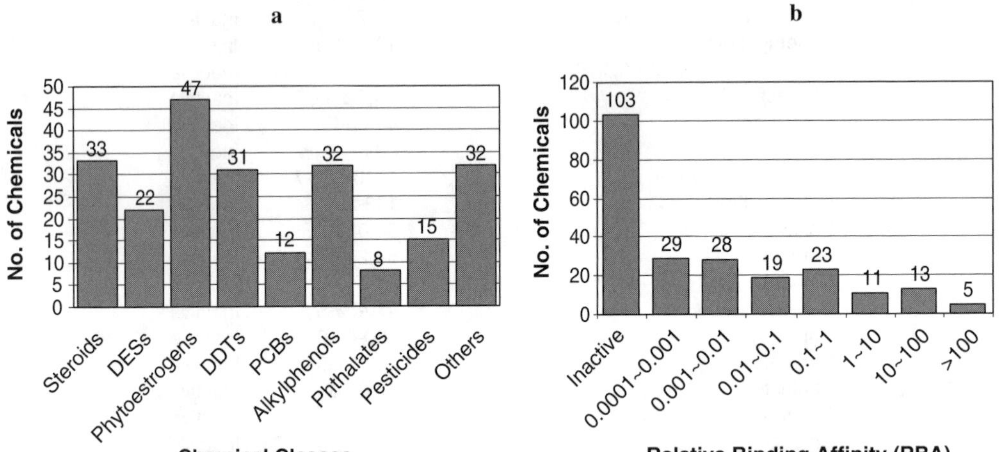

Figure 13.1   The distribution of (a) chemical classes and (b) binding activity for the NCTR dataset. The oestrogen receptor binding activity is represented as RBA. The RBA for the endogenous ligand, 17β-estradiol, was set to 100.

ER binding activity is represented by the Relative Binding Affinity (RBA), where the RBA value for the endogenous ER ligand, 17β-estradiol ($E_2$), was set to 100. This NCTR data set contains chemicals that were selected to cover the structural diversity of chemicals (Figure 13.1a) that bind to ER with an activity distribution ranging over six orders of magnitude (Figure 13.1b), both of which are important for the prediction of structurally diverse estrogens. This NCTR data set has been used extensively to build and validate a series of QSAR models for ER binding.

## III. SYSTEMATIC PROCEDURE FOR PREPROCESSING MOLECULAR STRUCTURES

It is common to represent a chemical with a single structure formula in QSARs as mixtures are difficult to model. Moreover, most computational chemistry programs accept only discrete organic chemicals for descriptor calculation. Unfortunately, most toxicological databases contain chemicals that are not necessarily discrete organic chemicals, and sometimes are mixtures. It is important to process molecular structures of a toxicological data set by correcting chemicals with separate entities (e.g., mixtures, organic salts, presence of $H_2O$ and HCl, etc.), and eliminate such chemicals for which descriptors cannot be calculated. It is important to note that such preprocessing might lead to a prediction that may not reflect the real activity of a chemical.

A general hierarchical process was developed for preprocessing molecular structures in our lab:

Ensure the validity of a structure by using structural information without obvious errors in the structural description. Some structural errors are identified by chance, and others appear in the process of 2D to 3D structure conversion, geometry optimisation, and descriptor calculation. We pay specific attention to structures with chiral centers, where problems may occur because most 2D structures in chemical databases do not have an indicator for the chiral center.

We removed the following chemicals from a QSAR study:

*Inorganic chemicals* (i.e., $NaCO_3$, $K_2SO_4$) and *pure chemical elements* (i.e., Ni, P) are difficult to model and parameterize; meaningful (or useful) descriptors for these chemicals may not be calculated from most computational chemistry programs.

*Polymers* have a wide distribution of molecular weights. The single relevant structure for polymers is often difficult to determine.

*Chelates* and *organometallic chemicals* contain a metal, for which most software cannot generate meaningful (or useful) descriptors. Moreover, it is difficult to determine the toxicologically relevant entities for these chemicals. For example, we cannot remove metals from a chelate leaving only the organic part for modeling.

Mixtures were processed as follows:

For structures with solvent molecules, the solvent molecules are removed to obtain a single structure.

Technical-grade chemicals, dyes, and DL-isomers are not a good first choice for a training set because they contain more than one molecule that cannot be represented by a single structure. We avoided using these types of chemicals for the training set. However, these chemicals were predicted in our project, where a single structure representing the main composition of the chemical is used.

Organic salts were processed to remove counter ions (e.g., $K^+$, $Na^+$, $SO4^{2-}$, $NH4^+$, $SO_3^{2-}$, $NO_3^-$, $Cl^-$) to obtain a single structure. The structure was then neutralized by adding or deleting protons.

Charged molecules were neutralized with acidic and basic groups by adding or removing protons. This prevents the generation of structural differences caused by different protonation states, which might lead to differences in calculated molecular descriptors.

## IV. QSAR MODELS FOR ER BINDING

There are a number of QSAR approaches useful for predicting receptor binding affinity. These range from simple rejection filters for drug-like chemical identification to more sophisticated QSAR models used in lead optimization. We have constructed various types of QSAR models for ER binding.

## A. Rejection Filters

Rejection filters are useful to exclude chemicals rapidly from further evaluation. For instance, Lipinski's "rule of five" is the most commonly used set of rejection filters to eliminate nondrug-like chemicals in the early stage of drug discovery (Lipinski et al., 1997). The name does not imply there are five rules; it derives from the fact that the criteria in each rule are a numerical multiple of five. The rule states that a nondrug-like chemical is more likely to have one of the following characteristics in its structure: (1) more than 5 hydrogen bond donors, (2) more than 10 hydrogen bond acceptors, (3) molecular weight > 500, or (4) log $K_{ow}$ > 5. Currently, the rigidity of a structure as measured by the number of rotatable bonds is found important for oral bioavailability chemicals (Veber et al., 2002). This may be a useful complement to the "rule of 5" for the identification of potential drug-like compounds. However, the application of these rules as rejection filters to the

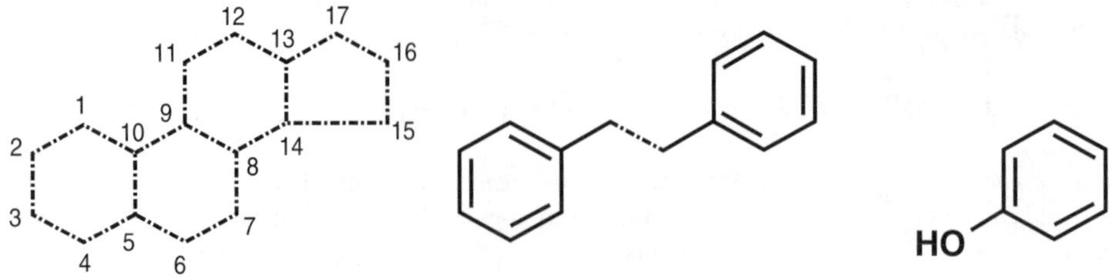

**Steroid Skeleton**
C1-C17, any
bond-type

**DES Skeleton**
two phenyl rings linked by two
carbons with any bond-type

**Phenolic ring**

**Figure 13.2** Three structural alerts. Chemicals with any of these 2D substructural features should receive proper attention in priority setting because they are commonly observed in most ligands to ER.

NCTR data set, resulted in high rates of false positives and false negatives (Hong et al., 2002). Hence, the rule of 5 has not been found useful for the prediction of ER binding.

Various physicochemical parameters were investigated as potential rejection filters. Considering that filters are normally used as one of the earliest stages of modeling, the ideal filters should (1) not generate any false negatives and (2) be able to reduce significantly the number of chemicals for further evaluation. Two rejection filters, molecular weight range and lack of a ring motif, were found to satisfy the two criteria. Chemicals matching any one of these two filters were excluded from subsequent models. The first rejection filter is a molecular weight range, set to <94 or >1000. The molecular weight of phenol, 94, was considered as the lowest limit for a chemical to bind to ER, whereas a molecular weight of 1000 was considered as the upper limit. The second rejection filter requires that an ER ligand contain at least one ring structure of any size. This structural rejection filter was developed following a large literature survey and based on the fact that there are no known estrogens lacking a ring (Fang et al., 2001).

## B.  Structural Alerts

Structural alerts are key 2D structural fragments of the molecule associated with ER binding. Figure 13.2 depicts the three structural alerts — the steroid, DES and phenolic skeletons — that were used to identify potential ER binders. Each alert independently characterized a unique structural feature important for ER binding. Chemicals containing any of these structural alerts are considered to be potential ER binders.

These three substructures were selected as structural alerts because the length and width of both the steroid and DES skeletons conform well to the dimensions of the ER binding pocket. In addition, while most endogenous hormones contain the steroid skeleton, most strong estrogens have two benzene rings separated by two carbon atoms (Fang et al., 2001). It has long been understood that the phenolic ring is often associated with estrogenic activity (Anstead et al., 1997). The contribution of the phenolic ring in ER binding is much more significant than any other structural feature (Fang et al., 2001). By overlaying the crystal structures of four ligand-ER complexes (E2-ER, 4-hydroxytamoxifen-ER, raloxifene-ER, and DES-ER complexes) based on their common protein residues at the binding site, it was found that the phenolic rings of all four ligands are closely positioned at the same location, allowing hydrogen bond interactions with Glu 353, and Arg 394 of the receptor and a water molecule (Figure 13.3) (Shi et al., 2001).

When these structural alerts were applied to the NCTR data set, most of the chemicals known to be active contained at least one of the structural alerts. For example, of 131 active chemicals, 110 (84%) of the chemicals contained the phenolic ring, 30 contained (23%) the DES skeleton,

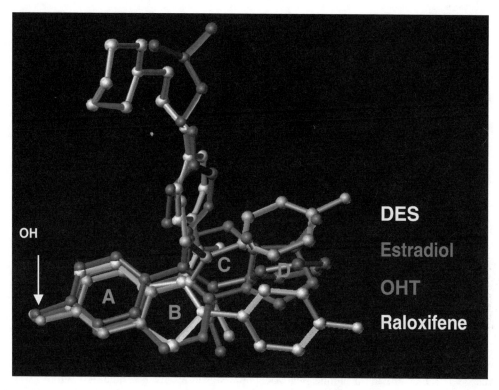

**Figure 13.3**   Relative positions of E$_2$, DES, raloxifene, and 4-OH-tamoxifen in the ER.

and 22 contained (17%) the steroid skeleton. A total of 95% (124/131) of the active chemicals matched one or more of these structural alerts.

## C.  Pharmacophore Queries

In the drug design industry, 3D pharmacophores have proven valuable as queries for lead discovery, whether applied alone or in conjunction with 2D structural alerts. A pharmacophore is a combination of a few molecular features (e.g., H-donor, H-acceptor, hydrophobic centers, and associated geometry) in 3D space that are required for a molecule to exhibit a certain type of biological activity (Hong et al., 1997). A query-matched chemical is considered positive and segregated for further evaluation. One of the advantages of pharmacophore searching is that it can identify chemicals whose 3D structures are similar to the template structure (normally, a highly active chemical) that may not be discernable by chemists in two dimensions.

The bound ligand-ER crystal structures (Brzozowski et al., 1997; Shiau et al., 1998) guided the design of pharmacophore queries for our project. We started with four 3D structures as the templates — E2, raloxifene, 4-OH-tamoxifen, and DES — in the ER-bound conformations. Then, all possible molecular features, as well as molecular shape, were delineated from those templates and combined to form pharmacophore queries using any combination of three to six features. These queries were sorted according to their discriminatory power to separate active from inactive chemicals in the NCTR datatset, where a chemical with any of multiple 3D conformations (up to 100) (Smellie et al., 1995a, 1995b, 1995c) matching the query was considered to be active. Queries with high discriminatory power were evaluated further for their biological relevance based on the knowledge of our careful structure-activity relationships examination of the binding affinities to ER of a large number of chemicals (Fang et al., 2001) in conjunction with analysis of the ligand-ER crystal structures (Brzozowski et al., 1997; Shiau et al., 1998). Finally, a Tanimoto similarity

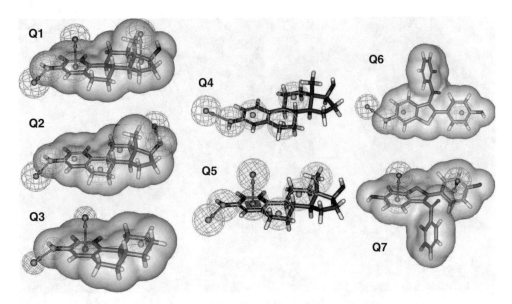

**Figure 13.4**  Seven pharmacophore queries. The mesh balls represent H-bond acceptor sites, hydrophobic centers and aromatic centers. The solid white surfaces represent shape constrain.

score was used to determine the uniqueness of each query. Through this process, we identified seven queries that provide a unique pharmacophore signatures for ER binding (Figure 13.4). A chemical could match none, a few, or many of the seven separate queries. We found that the number of matches increased in direct proportion to measured activity recorded in the training set. Thus, the number of pharmacophore matches could be used to rank chemicals in accordance with potential activity.

## D.  Classification Models

Classification is a supervised learning technique that provides categorical prediction (e.g., active/inactive). A number of classification methods were evaluated to categorize chemicals as ER binders or non-binders in our project. Although the methods differ in a number of ways, they generally produce similar results (Shi et al., 2002). We found that the nature of the descriptors used, and more specifically the effectiveness of descriptors to encode the structural features of the molecules related to the ER binding activity, is far more important than the specific method employed. The selection of biologically relevant descriptors is the critical step to develop a robust model. We also found a Genetic Algorithm to be the preferred method to segregate the most biological relevant descriptors from a large set of descriptors.

As an example, using the best 10 descriptors selected by the Genetic Function Approximation approach (Clark and Westhead, 1996; Forrest, 1993) from 153 descriptors, we were able to construct a Decision Tree model consisting of 5 meaningful descriptors:

The Phenolic Ring Index indicates the presence or absence of the phenolic ring in a chemical. This is considered to be the most important structural feature for ER binding (Anstead et al., 1997).

The Shadow-XY fraction is a geometric descriptor related to the breadth of a molecule (Rohrbaugh and Jurs 1987). This is consistent with our observation that substitution of a bulky group at $7\alpha$ and $11\beta$ position of $E_2$ increased the breadth of a chemical and enhanced ER binding (Fang et al., 2001).

The log $K_{ow}$, Jurs-PNSA-2 (Jurs-partial negative surface area-2) and Jurs-RPCS (Relative Positive Change Surface) reflect the hydrophobicity (log $K_{ow}$) (Leffler and Grunwald 1963) and

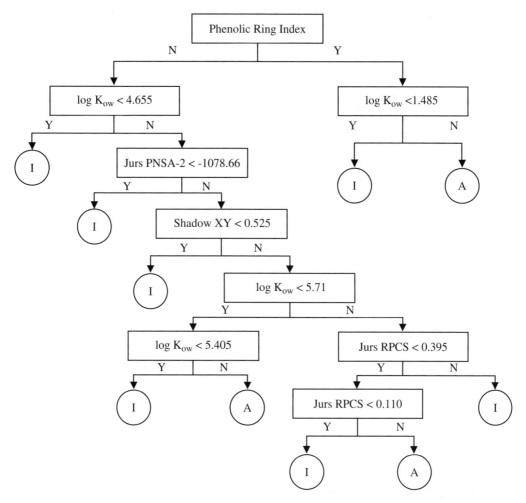

**Figure 13.5**  Decision Tree model. The model displays a series of yes/no (Y/N) rules to classify chemicals into active (A) and inactive (I) categories based on five descriptors: Phenolic Ring Index, log $K_{ow}$, Jurs PNSA-2, Shadow XY, and Jurs RPCS. The squares represent the rules, while the circles represent the categorical results.

charged surface area (Jurs-PNSA-2, Jurs-RPCS) (Stanton and Jurs 1990), which are, in principle, critical for all receptor mediated activity.

The Decision Tree model using five optimal descriptors is summarized in Figure 13.5. The model identified the Phenolic Ring Index as the most important descriptor for ER binding. If chemicals contained a phenolic moiety but also had log $K_{ow}$ values larger than 1.49, they were more likely to be ER binders. In contrast, chemicals without a phenolic moiety were less likely to be ER binders unless they were relatively large and hydrophobic and were broad with a significant charged surface area.

## E.  3D-QSAR/CoMFA Model

We have evaluated three different techniques to generate QSAR models, namely Comparative Molecular Field Analysis (CoMFA), COmprehensive Descriptors for Structural and Statistical Analysis (CODESSA), and Hologram QSAR (HQSAR). More specifically they were evaluated for their utility (predictivity, speed, accuracy, and reproducibility) to predict ER binding activity quantitatively (Tong et al., 1998; Shi et al., 2001). Common to the three QSAR methods is the

**Table 13.2    Predictions from the CoMFA Model for the Data from Kuiper et al. (1997; 1998)**

| Name | Exp. log RBA | CoMFA Predicted | CoMFA Residual |
|------|------|------|------|
| 2,2′,3,3′,4′,5,5′-Heptachloro-4-biphenylol | −1.498 | −1.792 | 0.294 |
| 2,2′,3,3′,4′,5-Hexachloro-4-biphenylol | −1.650 | −1.920 | 0.270 |
| 2,2′,3′,4,4′,5,5′-Heptachloro-3-biphenylol | −1.549 | −1.710 | 0.161 |
| 2,2′,3,4′,5,5′6-Heptachloro-4-biphenylol | −1.498 | −2.184 | 0.686 |
| 2,2′,3,4′,5,5′-Hexachloro-4-biphenylol | −2.023 | −2.219 | 0.196 |
| 2,2′,3′,4′,5′-Pentachloro-4-biphenylol | −1.498 | −0.820 | 0.678 |
| 2,2′,3′,4′,6′-Pentachloro-4-biphenylol | −1.014 | −1.293 | 0.279 |
| 2,2′,3′,5′,6′-Pentachloro-4-biphenylol | −1.549 | −1.187 | 0.362 |
| 2,2′,4′,6′-Tetrachloro-4-biphenylol | −1.014 | −1.376 | 0.362 |
| 2′,3,3′,4′,5′-Pentachloro-4-biphenylol | −1.458 | −1.076 | 0.382 |
| 2,3,3′,4′,5-Pentachloro-4-biphenylol | −2.023 | −2.536 | 0.513 |
| 2′,3,3′,4′,5-Pentachloro-4-biphenylol | −2.507 | −1.894 | 0.613 |
| 2′,3,3′,4′,6-Pentachloro-4-biphenylol | −1.387 | −1.125 | 0.262 |
| 2′,3,3′,5′,6′-Pentachloro-4-biphenylol | −1.720 | −1.291 | 0.429 |
| 2′,3,4′,6′-Tetrachloro-4-biphenylol | −1.236 | −1.615 | 0.379 |
| 2,4,6-Trichloro-4′-biphenylol | −0.106 | −1.496 | 1.390 |
| 5-Androstenediol | −0.489 | −0.790 | 0.301 |
| 16α-Bromo-estradiol | 1.408 | 0.868 | 0.540 |
| 16-Keto-estradiol | −0.378 | 1.104 | 1.482 |
| 17-Epi-estriol | 0.984 | −0.112 | 1.096 |
| 2-OH-estrone | −0.187 | 0.305 | 0.492 |
| Raloxifene | 1.367 | −0.521 | 1.888 |
| Zearalenone | 0.368 | 0.210 | 0.158 |
| 4,4′-Biphenol | <−2.51 | −2.622 | — |
| Ipriflavone | <−2.51 | −5.358 | — |

Predictive $q_{pred}^2 = 0.62$

derivation of a regression model using the partial least squares (PLS) methods. The differences among these QSAR techniques are primarily in the type of the descriptors used to represent chemical structure. CoMFA employs steric and electrostatic field descriptors that encode detailed information regarding 3D intermolecular interaction. CODESSA calculates molecular descriptors on the basis of 2D and 3D structures and quantum-chemical properties. HQSAR uses molecular holograms constructed from counts of substructural molecular fragments. For three relatively small data sets under investigation, the QSAR models generated using the CoMFA and HQSAR techniques demonstrated comparable high quality for potential use to identify ER ligands (Tong et al., 1998). CoMFA and HQSAR were further investigated and compared, particularly for their predictivity, using the NCTR data set and two other test sets (Shi et al., 2001). We found that CoMFA performed better for the training set as well as for the prediction of the activity of compounds in two test sets (Table 13.2 and Table 13.3).

To develop a CoMFA model, the molecules of interest must first be aligned to maximize superposition of their steric and electrostatic fields. Although a statistically robust CoMFA model is dependent on a number of factors, proper alignment is essential to produce a valid QSAR model. For chemical congeners, the alignment rule is normally defined based on the maximum common substructure among the training set chemicals, which usually leads to a statistically robust CoMFA model. The drawback for such models is the unreliability in predicting the activities of chemicals whose structures are not similar to the training set, such as those in the models reported previously (Tong et al., 1997a, 1997b, 1998; Bradbury et al., 1996; Waller et al., 1996; Wiese et al., 1997; Xing et al., 1999; Sadler et al., 1998; Zheng and Tropsha, 2000).

In contrast, a CoMFA model based on a structurally diverse data set provides more robust predictions. However, the critical and most difficult part of developing a CoMFA model, such as

Table 13.3    Predictions from the CoMFA Models for the Data from Waller
et al. (1996)

| Name | Exp. log RBA | CoMFA Predicted | Residual |
|---|---|---|---|
| 2-Tert-butylphenol | −4.546 | −3.946 | 0.600 |
| 3-Tert-butylphenol | −4.819 | −3.181 | 1.638 |
| 2,4,6-Trichloro-4′-biphenylol | −0.158 | −1.496 | 1.338 |
| 2-Chloro-4,4′-biphenyldiol | −0.610 | −1.532 | 0.922 |
| 2,6-Dichloro-4′-biphenylol | −1.110 | −1.905 | 0.795 |
| 2,3,5,6-Tetrachloro-4,4′-biphenyldiol | −2.180 | −0.572 | 1.608 |
| 2,2′,3,3′,6,6′-Hexachloro-4-biphenylol | −2.739 | −1.917 | 0.822 |
| 2,2′3,4′,6,6′-Hexachloro-4-biphenylol | −2.596 | −1.985 | 0.611 |
| 2,2′,3,6,6′-Pentachloro-4-biphenylol | −1.966 | −2.280 | 0.314 |
| 2,2′,5,5′-Tetrachloro-biphenyl | −2.667 | −3.956 | 1.289 |
| 2,2′,4,4′,5,5′-Hexachloro-biphenyl | −2.834 | −3.282 | 0.448 |
| 2,2′,4,4′,6,6′-Hexachloro-biphenyl | −1.870 | −2.982 | 1.111 |
| 2,2′,3,3′,5,5′-Hexachloro-6′-biphenylol | −2.691 | −2.176 | 0.515 |
| 4′-Deoxyindenestrol | −1.371 | −0.010 | 1.361 |
| 4′-Deoxyindenestrol | −0.230 | 0.629 | 0.859 |
| 5′-Deoxyindenestrol | −0.587 | −0.382 | 0.204 |
| 5′-Deoxyindenestrol | 0.353 | 0.267 | 0.086 |
| Indenestrol A (R) | 1.078 | 0.473 | 0.605 |
| Indenestrol A (S) | 2.386 | 0.993 | 1.393 |
| R5020 | −1.811 | −1.413 | 0.398 |
| Zearalenone | 0.912 | 0.210 | 0.702 |
| DACT | NA[a] | −6.258 | — |
| Hydroxyflutamide | NA | −3.224 | — |
| M1 | NA | −3.526 | — |
| M2 | NA | −5.672 | — |

Predictive $q^2_{pred} = 0.71$

[a] NA = No Activity

the one in our study, is choosing the most appropriate set of alignment rules for the structurally diverse training set. Fortunately, crystal structures of four ligands binding to the ER have been published (Brzozowski et al., 1997; Shiau et al., 1998). These aided our derivation of rational CoMFA alignment rules. Our CoMFA model, based on the crystal structure-guided alignment, is statistically robust. The CoMFA-calculated vs. experimental RBAs (as logs) for the NCTR training set chemicals are plotted in Figure 13.6. The conventional $r^2$ and cross-validated $q^2_{LOO}$ were 0.91 and 0.66, respectively, indicating that the CoMFA model was both internally consistent and highly predictive.

## V. MODEL VALIDATION

The goodness-of-fit of a model can be assessed using various statistical measures. Concordance, specificity, and sensitivity (Hong et al., 2002) are commonly used to assess the quality of a classification model, while a quantitative regression (or nonlinear) model is measured using a $r^2$ value (a correlation coefficient) (Tong et al., 1997a). A qualitative or quantitative model is generally deemed statistically significant if concordance or $r^2 \geq 0.9$, respectively. However, the current challenge in QSAR is no longer to construct a statistically sound model using the training set, but to develop a model with the *capability* to predict accurately the activity of untested chemicals. Therefore, the issue becomes how we quantify and validate this capability of a model. Here, we summarize several common approaches that we have used routinely to assess the predictive capability of a QSAR model (further information on validation issues is provided in Chapters 19 and 20). Specifically, the issues associated with these approaches are discussed.

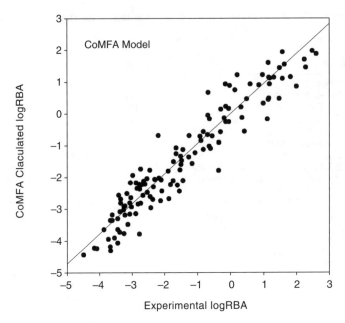

**Figure 13.6**  Plot of CoMFA-calculated log relative binding affinity (log RBA) vs. experimental log RBA for the NCTR dataset. $r^2 = 0.91$; $q^2 = 0.66$.

## A.  Cross-Validation

Most experts in the QSAR field, as well as the present authors, assert that a model's predictive capability must be demonstrated using some sort of cross-validation procedure. In such a procedure, a fraction of chemicals in the training set are excluded, and then their activity is predicted by a new model developed from the remaining chemicals. The process has to be repeated until all chemicals in the training set are left out once. When each chemical is left out one at a time, and the process repeated for each chemical, this is known as leave-one-out (LOO) cross-validation. A quantitative model with a value of cross-validated $q^2_{LOO} > 0.5$ is normally considered to possess significant predictive ability. If the training set is divided into $N$ groups with approximately equal numbers of chemicals, the process is called leave-$N$-out (LNO) cross-validation. The leave-10-out procedure is commonly used to assess the predictive capability of a classification model.

For example, our CoMFA model showed a $q^2_{LOO} = 0.66$, demonstrating a predictive capability (Shi et al., 2001). In addition to LOO, a more thorough LNO validation was also conducted for this CoMFA model. Unlike $q^2_{LOO}$ that provides a single value from the process, $q^2_{LNO}$ varies for a selected $N$ group in each run because of the random nature of selection of chemicals in the process. We ran LNO 100 times for each random $N$ groups of chemicals ($N = 2$–$10$, $13$, $20$, and $50$) to ensure a valid statistical analysis. Consequently, the standard deviation (SD) of $q^2_{LNO}$ value can be used to assess the model's stability for the prediction of the activity of diverse chemicals. As shown in Figure 13.7, the mean $q^2_{LNO}$ values for the CoMFA model decreased as the number of left-out chemicals were increased. However, even when 50% ($N = 2$) of the training chemicals were randomly omitted, the worst CoMFA model selected from 100 random repetitions still gave a highly robust $q^2_{LNO} > 0.5$. This indicates that the CoMFA model is robust for predicting structurally diverse chemicals.

For the classification models, instead of using a routine leave-10-out procedure, a more extensive validation procedure was applied in our project to assess the model's predictive capability. This is a variant of the LNO procedure, where the data set is divided randomly into 2 groups, 2/3 for training and 1/3 for testing. The classification models were constructed using the training set, and

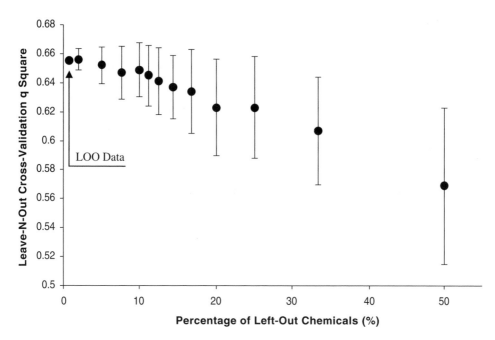

**Figure 13.7** Leave-N-out cross-validation results for CoMFA.

subsequently used to predict the test set. The process was repeated 2000 times. The validation results for the Decision Tree classification model are illustrated in Figure 13.8. In the training step, a narrowed distribution was observed in a concordance range of 89 to 99%, indicating that all 2000 models show comparable results in measured overall accuracy. In contrast, a much wider distribution of concordance was found in the prediction of the 2000 test sets. The difference between the best and worst prediction was as large as 30%, three times greater than for the training set. The results indicate that the differences in widths and peaks of the concordance distribution between training and testing may be another way to assess a model's predictivity.

It is important to note that both LOO and LNO methods test the stability of the model, through perturbation of the correlation coefficients, by consecutively omitting chemicals. The methods only assess the internal extrapolation in the training set, and have limited indication in predicting untested chemicals, specifically for those that are structurally different from the training chemicals.

## B. External Validation

When additional data are available, a QSAR model should be validated by predicting the activity of other chemicals not used in the training set, but whose activities are known (i.e., the test set). This is called external validation. The major difference between the cross-validation and external validation is that the chemicals selected in the latter case are in a sense random. This provides a more robust evaluation of the model's predictive capability for untested chemicals than cross-validation. We feel strongly that the confidence in a model's predictive capability can be tested and validated when robust prediction has been demonstrated with an external test set. Further details regarding a formal framework for the validation of QSARs are provided in Chapter 20.

In most cases, since a data set used for modeling barely contains enough chemicals to create a statistically robust model in the first place, one rarely enjoys the luxury of setting aside a sufficient number of test chemicals for use in external validation (10 to 20% of the data set is recommended). One approach for the selection of a test set is to identify data sets in the literature with the same type of activity that may or may not be generated using the same assay protocol for the training

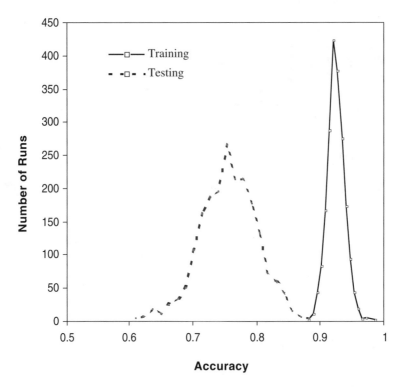

**Figure 13.8** An extensive cross-validation procedure to validate the Decision Tree classification model. In this method, the NCTR dataset was divided into 2 groups, 2/3 for training and 1/3 for testing. The process was repeated 2000 times. Concordance was calculated based on the misclassifications divided by the number of training chemicals for the training models and the misclassifications divided by the testing chemicals for prediction.

set. In such a case, care must be taken to avoid interlaboratory and assay variability among the different sources of data. It is desirable that a potential test set has activity data generated from an assay protocol that is close to the one for the training set to keep the variability at minimum.

For example, there are a number of data sets reported in the literature for estrogenic activity obtained experimentally from *in vitro* assays. These include the ER binding assay (Kuiper et al., 1997; 1998), the yeast-based reporter gene assay (Gaido et al., 1997) and the E-SCREEN assay (Soto et al., 1994; 1995; 1999). We found that considerable variability in activity exists among these assays (Fang et al., 2000). Data sets from other ER binding assays were selected as primary sources for the test sets in our project because they are consistent with the assay used for the training set: the data sets reported by Kuiper et al. (1998) and Waller et al. (1996). In Kuiper et al.'s study, the pure human ERα was used, whereas the mouse uterine cytosol that primarily contains ERα was used by Waller et al.. Despite this consistency, we found that the absolute activity value of these two test sets is different from the NCTR data set, particular for the Waller data set. Thus, the activity value of the selected test sets cannot be used as is for external validation. Their activity data need to be normalized to the training set (the NCTR data set), and then the normalized activity value of a chemical can then be compared with the value predicted from the model to assess the performance of that model. In the normalization process, the selected test set was first correlated with the NCTR data set on the basis of the shared chemicals in both data sets, and then the activity value for each chemical not in the NCTR data set was normalized to the NCTR data set based on the correlation and the equation derived from it. It is worthwhile to mention that, in such applications, the selected test set should include a sufficient number of chemicals shared with the training set to establish a statistically significant linear correlation to ensure a valid normalization.

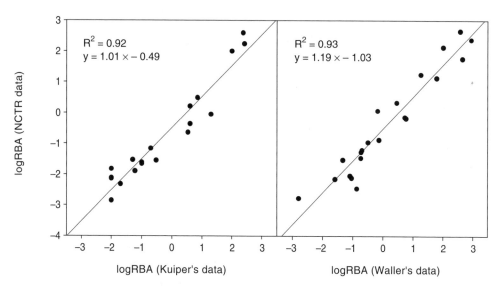

**Figure 13.9**  Correlation of the NCTR data with Kuiper et al.'s and Waller et al.'s datasets.

Good linear correlations were observed between the Kuiper and NCTR data sets ($r^2 = 0.92$) based on 19 shared chemicals, and between the Waller and NCTR datatsets ($r^2 = 0.93$) on 21 common active chemicals (Figure 13.9). Both Kuiper et al.'s and Waller et al.'s data sets contain 25 chemicals, respectively, that are not assayed in the NCTR data set. These were normalized using the correlation equations shown in Figure 13.9 and these RBA values were used to test the QSAR models. Tables 13.2 and 13.3 summarize the results for the ability of the CoMFA model to predict the RBAs of these two external test sets. The predictive $q^2_{pred}$ results for two test sets are quite similar to the $q^2_{LOO}$ of the training set. For about 75% of the chemicals (Tables 13.2 and 13.3), the residuals (the difference between predicted and observed activity) were less than 1.0 (i.e., less than a 10-fold difference in activity). For 6 chemicals that are reported to be inactive, or to have undetectable activity, in the original references, the CoMFA model predicted 5 of them with activity (log RBA) of less than –3.0 (100,000-fold below the endogenous estrogen, $E_2$). 4,4'-Biphenol shows undetectable activity from the maximum experimentally determined limit in Kuiper et al.'s data set, while it is active with log RBA = –1.7 in the ER binding assay using mouse uterine cytosol (Sadler et al., 1998). The CoMFA model predicted it to be active with log RBA = –2.62.

Another important consideration in the selection of a test set is to ensure that the chemicals in the data set relate to the real problem in question. It should be emphasized that the QSAR models developed in our project are used primarily to predict the activity of environmental chemicals, mostly pesticides and industrial chemicals. A data set reported by Nishihara et al. (Nishihara et al., 2000) was also selected as a test set. This data set contains 517 chemicals tested with the yeast two-hybrid assay, of which over 86% are pesticides and industrial chemicals. Only 463 chemicals were used for this validation study after structure processing. Only 62 chemicals were categorized as active on the basis of having on activity greater than 10% of $10^{-7}$M $E_2$, as defined in the original paper (Nishihara et al., 2000). The majority of the chemicals were inactive, which is similar to the real-world situation where inactive chemicals are expected from a large proportion of those in the environment.

It is important to note that the yeast two-hybrid assay used to develop the Nishihara et al. data set differs from the NCTR binding assay. The ER competitive binding assay measures the binding affinity of a chemical for ER, while the yeast two-hybrid assay measures ER binding-dependent transcriptional and translational activity. These two assays differ in their sensitivity to distinguish active from inactive chemicals, particularly for weak estrogens and antiestrogens (Fang et al., 2000).

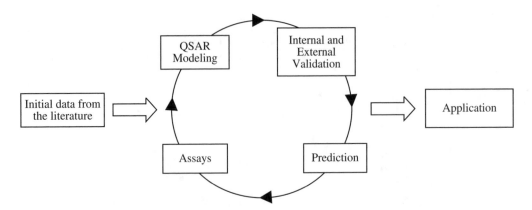

**Figure 13.10**    Depiction of the recursive process used by NCTR to develop QSAR models for predicting ER binding. The process starts with an initial set of chemicals from the literature for QSAR modeling. Next, the preliminary QSAR models are used prospectively to define a set of chemicals that will further improve the model's robustness and predictive capability. The new chemicals are assayed, and these data are then used to challenge and refine the QSAR models. Validation of the model is critical. The process emphasizes the living model concept.

We compared the assay results for 80 common chemicals from both the Nishihara et al. and NCTR data sets; inconsistent assay results were observed for 12 chemicals. Specifically, of 30 active chemicals in the Nishihara et al. data set, one chemical was found inactive in the NCTR data set; of 50 inactive chemicals in the Nishihara et al. data set, 11 chemicals were found active in the NCTR data set. These observations show that even using the experimental data from the ER binding assay (the NCTR data set) to predict the experimental results from the yeast two-hybrid assay (the Nishihara et al. data set), there may be about a 15% (12/80) discrepancy, or 3.3% (1/30) false negative rate and 22% (11/50) false positive rate. Care should be taken in interpreting the QSAR validation results using this data set (Hong et al., 2002).

## C.  Validation and Living Model

No matter how rigorous the validation procedure of a model, it is important to realize that the model may give incorrect predictions for some chemicals because it is impossible to determine the entire chemistry space. Any QSAR model has to be considered as a *living* model that is improved when new data are available.

In our project, a recursive process has been adopted by integrating assay and QSAR modeling to determine chemicals for the training set and the resulting model. A model validation step is specifically emphasized in this procedure (Tong et al., 2001; Perkins et al., 2001). The process is highly interdisciplinary, involving computational chemists, biologists, and experimental toxicologists. As depicted in Figure 13.10, the process starts with an initial set of chemicals from the literature for QSAR modeling (Tong et al., 1997a, 1997b, 1998). Next, the preliminary QSAR models are used prospectively to define and rationalize a set of chemicals that may further improve the model's robustness and predictive capability. These new chemicals are assayed, and the data are then used to challenge and refine the QSAR model.

Several benefits accrue from the integration of the experimental and modeling efforts. Immediate feedback can be given to the experimentalists so that suspected assay problems can be investigated rapidly. Also, as the models evolve, the modelers can select the chemicals for subsequent testing, based on considerations of structural diversity and activity range. Each new assay data point obtained directly from the lab becomes a challenge to the evolving model; the result is either further confirmation of its validity, identification of a limitation, or the prediction of an outlier. Failure of the model also provides important information, such as identification of the need for new data based on a rational understanding of the dependence of activity on structure. Regardless of the cause of model failure, a research hypothesis is spawned in each iteration; this should lead to new data or an improved training set and an improvement to the living model.

## VI. QSAR AS A PRIORITY-SETTING TOOL FOR REGULATORY APPLICATION

### A. Endocrine Disrupting Chemicals: Issues

A large number of environmental chemicals are suspected of disrupting endocrine function by mimicking or antagonizing natural hormones in experimental animals, wildlife, and humans. There is growing concern among the scientific community, government regulators and the public that endocrine-disrupting chemicals (EDCs) in the environment are adversely affecting human and wildlife health (Hileman, 1994; 1997). Adverse outcomes have been observed in experimental animals and wildlife; potential effects in humans include reproductive and developmental toxicity, carcinogenesis, immunotoxicity, and neurotoxicity (Kavlock et al., 1996). EDCs may exert adverse effects through a variety of mechanisms, such as ER-mediated mechanisms of toxicity.

The scientific debate surrounding EDCs has grown contentiously, in part to the fact that some suspected EDCs are produced in high volume, and most chemicals are economically important. These public and regulatory concerns led to government regulatory actions (U.S. Congress, 1996a; 1996b) and expanded research across Europe, Japan, and North America. The U.S. Congress in 1996 mandated that the Environmental Protection Agency (EPA) should develop a strategy for screening and testing a large number of chemicals found in drinking water (U.S. Congress, 1996b) and food additives (U.S. Congress, 1996a) for their endocrine disruption potential. In response to congressional action, the EPA established the Endocrine Disruptor Screening and Testing Advisory Committee (EDSTAC) which includes scientific expertise from government, academia, and industry. EDSTAC recommended a two-tier (Tier 1 screening and Tier 2 testing) strategy to screen and test estrogenic, androgenic, and thyroidal activities of a large number of chemicals. To accomplish this, chemicals will be screened (Tier 1) using a multiple-endpoint strategy that includes more than 20 different *in vitro* and *in vivo* assays recommended by EDSTAC (Gray, 1998). Although more than ~87,000 chemicals were initially selected for evaluation, many are polymers, leaving about ~58,000 chemicals for evaluation in Tier 1. The number that will progress to the testing step (Tier 2) (Patlak, 1996) is not known. Processing chemicals through both tiers will require many years and extensive resources. The EPA has adopted an approach requiring priority setting before Tier 1 (www.epa.gov/scipoly/oscpendo/).

Among the types of hormonal activities, estrogenic activities have been the most studied. Estrogenic endocrine disruption can result from a variety of biological mechanisms. In our early research, we found a strong linear correlation for ER binding affinities among a diverse group of chemicals assayed with ER from the rat uterine cytosol and human ERα (Fang et al., 2000). Furthermore, the rat ER binding data also correlate strongly with the results from assays measuring estrogenicity using a downstream event (i.e., a yeast-based reporter gene assay and MCF-7 cell proliferation assay). More importantly, chemicals positive in uterotrophic responses (*in vivo* estrogenic activity) are also positive in the ER binding assay, indicating that binding affinity is a good predictor of *in vivo* activity with few false negatives observed (Zacharewski, 1998). These findings demonstrate that ER binding is the major determinant for estrogenic EDCs, and therefore the prediction of the rat ER binding affinity provides one important piece of information for priority setting.

### B. QSAR and Priority Setting

The objective of priority setting is to rank order a large number of chemicals from the most to least important chemicals so they may be entered into more resource-intensive experimental evaluations. There are a number of criteria that can be used for this purpose, such as production volume, persistence and fate in the environment, and human exposure levels. Most of the 58,000 chemicals to be evaluated in Tier 1 have no biological data. QSARs were recommended by the EDSTAC as the primary source of biological effect information for priority setting.

How and which QSAR approaches should be selected for priority setting is heavily dependent on the application domain. Priority setting using QSAR is applied widely in the process of drug discovery. The objective of priority setting in the pharmaceutical industry is to increase the chance of finding active chemicals or hits that are more likely to be developed into leads. Hence, false positives are of great concern. In contrast, a good priority setting method for regulatory application should generate a small fraction of false negatives (chemicals predicted to fail to bind to their receptor, but which actually bind). False negatives constitute a crucial error, because they will receive a relatively lower priority for evaluation in Tier 1, and may remain in use for many years. Furthermore, the methods should provide reasonable quantitative accuracy for true positives, as those with higher affinities will generally be of higher priority.

## C. NCTR Four-Phase System

The QSAR approaches described in Section IV have strengths and weaknesses, and they all produce a degree of prediction error. We found that a model, particularly one that provides only a yes/no prediction, can be optimized to minimize either the overall prediction error, or the false negatives or positives individually. It appears that decreasing false negatives, by modifying the criteria, normally increases overall cost due to an increase in false positives. Using a single model for priority setting poses difficulties for regulatory application. We have adopted an approach to rationally combine different QSAR models into a sequential Four-Phase scheme according to the strength of each type of model. A progressive phase paradigm is used as a screen to reduce the number of chemicals to be considered in each subsequent phase. These four phases work in a hierarchical manner to reduce the size of a data set incrementally while, at the same time, increase precision of prediction. Within each phase, different models have been selected to work in a complementary fashion to represent key activity-determining structural features to minimize the rate of false negatives. The overall architecture of the NCTR Four-Phase system for identification of ER ligands is outlined below and illustrated in Figure 13.11.

### 1. Phase I: Filtering

Two filters, the molecular weight range and ring-structure indicator (Section IV.A), were selected in this phase to eliminate chemicals very unlikely to bind to ER. As shown in Table 13.4, these two rejection filters correctly eliminated 6 inactive chemicals from the NCTR data set and 98 from the Nishihara et al. data set, respectively. No false negatives were introduced using these two rejection filters. For the Nishihara et al. data set, the data size was reduced by about 21%. This demonstrated that, for real-world applications, these two rejection filters might significantly reduce the number of chemicals for further evaluation with a minimum risk of introducing false negatives.

### 2. Phase II: Active/Inactive Assignment

This phase classifies chemicals passing from the previous phase into active and inactive categories. Three structural alerts (Section IV.B), seven pharmacophore queries (Section IV.C), and the Decision Tree classification model (Section IV.D) were used in parallel to discriminate active from inactive chemicals. To ensure the lowest false negative rate in this phase, a chemical predicted to be active by any of these 11 models is subsequently evaluated in Phase III, whereas only those predicted to inactive by all these models are eliminated for further evaluation. Since structural alert, pharamacophore and Decision Tree methods incorporate and weight differently the various structural features that endow a chemical with the ability to bind the ER; the combined outputs derived

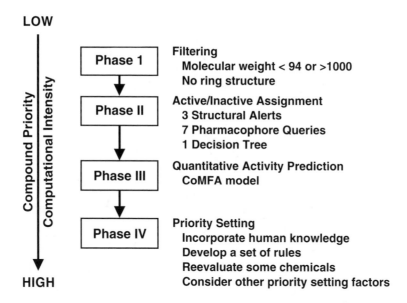

**Figure 13.11**    Overview diagram of the NCTR Four-Phase approach for priority setting. In Phase I, chemicals with molecular weight < 94 or > 1000 or containing no ring structure will be rejected. In Phase II, three approaches (structural alerts, pharmacophores, and classification methods) that include a total of 11 models are used to make a qualitative activity prediction. In Phase III, a 3D QSAR/CoMFA model is used to make a more accurate quantitative activity prediction. In Phase IV, an expert system is expected to make a decision on priority setting based on a set of rules. Different phases are hierarchical; different methods within each phase are complementary.

**Table 13.4    Results of Two Rejection Filters for the NCTR, Nishihara et al., and Walker et al. Data Sets**

| Data Sets | Data Size | Eliminated by Molecular Weight | | Eliminated by Ring | | Number (%) of Eliminated Chemicals |
|---|---|---|---|---|---|---|
| | | Active | Inactive | Active | Inactive | |
| NCTR | 232 | 0 | 0 | 0 | 6 | 6 (2.6%) |
| Nishihara et al. | 463 | 0 | 28 | 0 | 89 | 98 (21.2%) |
| Walker et al. | 58,230 | 16,048 | | 1495 | | 16,689 (28.7%) |

*Note:* Table lists the number of chemicals eliminated by molecular weight range or lack of ring criteria as well as their combination. No active chemicals were rejected by these two filters.

from the three approaches are complementary in minimising false negatives. All active chemicals in the NCTR, Waller et al. (1996), Kuiper et al. (1998) and Nishihara et al. (2000) data sets were identified by combining these 11 models.

## 3. Phase III: Quantitative Predictions

In this Phase, the CoMFA model (Section IV.E) was used to make a more accurate quantitative activity prediction for chemicals from Phase II. Chemicals with higher predicted binding affinity are given higher priority for further evaluation in Phase IV. The CoMFA model demonstrated good statistical reliability in both cross-validation and external validation (Shi et al., 2001).

## 4. Phase IV: Rule-Based Decision-Making System

In this final stage of the integrated system, we believe that a set of rules needs to be developed as a knowledge-based or expert system to make a priority setting decision. The system is useful

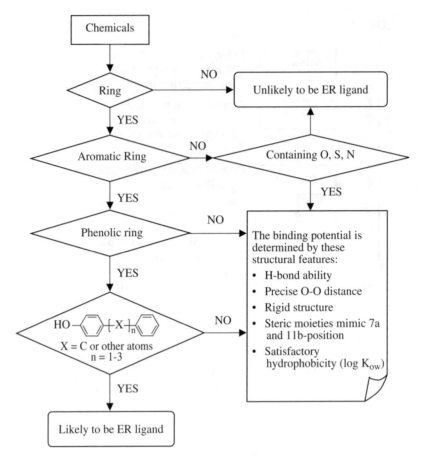

**Figure 13.12**    Flowchart for the identification of ER ligands using a set of "if-then" rules: (1) if a chemical
contains no ring structure, then it is unlikely to be an ER ligand; (2) if a chemical has a
nonaromatic ring structure, then it is unlikely to be an ER ligand if it does not contain an O, S,
N, or other heteroatoms for H-bonding. Otherwise, its binding potential is dependent on the
occurrence of the key structural features; (3) if a chemical has a non-OH aromatic structure,
then its binding potential is dependent on the occurrence of the key structural features; and (5)
if a chemical contains a phenolic ring, then it tends to be an ER ligand if it contains any additional
key structural features. For the chemicals containing a phenolic ring separated from another
benzene ring with the bridge atoms ranging from none to three, it will most likely be an ER ligand.

only after incorporating accumulated human knowledge and expertise (i.e., rules). This system can
make decisions on individual chemicals based on the rules in its knowledge base. Computational
chemists, toxicologists, and regulatory reviewers should jointly develop and define the rules. The
following are suggestions for the design of priority setting rules, which are subject to changes to
accommodate new information:

1. Special attention needs to be placed on the following chemicals, which may need to be re-evaluated
   by assaying or modeling according to the "if-then" scheme depicted in Figure 13.12 (Fang et al.,
   2001)
   a. The chemical is predicted to be inactive, but its structure has been modified during structural
      pre-processing.
   b. The chemical is structurally dissimilar to all those that have been used to train and test the
      models.
   c. The chemical is active in Phase II, but inactive in Phase III.
2. Information on the level of human exposure and production, environmental fate, and other public-
   health-related parameters can be used independently or jointly for priority setting.

Table 13.5    Size Reduction of Two
              Environmental Data Sets
              Processed by the NCTR
              Four-Phase System

|                         | HPV-Inerts | Walker |
|-------------------------|------------|--------|
| Original data size      | 457        | 58,391 |
| After Phase I and II    | 15.7%      | 12.0%  |
| After Phase III         | 9.8%       | —      |

The NCTR Four-Phase system has been validated by a number of existing data sets, including the E-SCREEN assay data (Soto et al., 1995), the yeast two-hybrid reporter gene assay data (Nishirara et al., 2000), and other data sets (Gaido et al., 1997; Coldham et al., 1997; Waller et al., 1996; Routledge and Sumpter 1997; Routledge et al., 1998; Harris et al., 1997). The system has produced no false negatives for these data sets. This integrated scheme is being extended at NCTR to include endpoints of other endocrine disrupting mechanisms, such as androgen receptor (AR) binding.

## D.  Regulatory Application

The NCTR Four-Phase system was recently applied to two environmental data sets recognized by EPA as representative subsets of potential EDCs:

1.  HPV-Inerts data set — It contains 623 high production volume inerts (human papilloma virus-[HPV] Inerts), which is a subset of the Toxic Substances Control Act (TSCA) Inventory. The EPA is including HPV-Inerts in version 2 of the Endocrine Disruption Priority Setting Database (EDPSD2), and there was a need to prioritize HPV-Inerts for further experimental evaluation. Of 623 chemicals, 166 chemicals were either mixtures or their structures were not available, excluding them from prediction. The activities (or otherwise) of 457 chemicals were predicted by this system.
2.  Walker data set — Walker et al., developed a database that contains a large and diverse collection of known pesticides and industrial chemicals, as well as some food additives and drugs (Walker et al., 2003). The database contains 92,964 Chemical Abstract Service (CAS) Registry numbers of chemicals that will probably have to be evaluated for their potential endocrine disruption. A final data set of 58,391 chemicals was processed by our system after eliminating those chemicals for which structures were not available (Walker et al., 2003) and/or 3D structures could not be generated (Hong et al., 2002).

Table 13.5 summarizes the priority setting results for the two data sets using the NCTR Four-Phase system. When only the Phase I and II protocols are used, the system dramatically reduced the number of potential estrogens by some 80 to 85%, demonstrating its effectiveness in eliminating these most unlikely ER binders from further expensive experimentation. The Phase III CoMFA model further reduces the data size by about 5 to 10%. More importantly, the quantitative binding affinity prediction from Phase III provides an important rank-order value for priority setting.

## VII. QSAR APPLICATION IN PERSPECTIVE

QSARs have been applied extensively in a wide range of scientific disciplines, including chemistry, biology, and toxicology (Hansch and Leo, 1995; Hansch et al., 1995b). In both drug discovery and environmental toxicology (Bradbury, 1995), QSAR models are now regarded as a scientifically credible tool for predicting and classifying the biological activities of untested chemicals. As we enter the new millennium, QSAR has become inexorably imbedded as an essential tool in the pharmaceutical industry, from lead discovery and optimization to lead development (Kubinyi et al., 1998). For example, there is a growing trend to use QSAR early in the drug discovery process as a screening and enrichment tool to eliminate from further development those chemicals lacking drug-like properties (Lipinski et al., 1997) or those chemicals predicted to elicit a toxic response. This developing scenario portends the spread of QSAR beyond the pharmaceutical industry to human and environmental regulatory authorities for use in toxicology (Bradbury 1994; Russom et al., 1995; Benigni and Richard 1998; Schultz and Seward, 2000; Hansch et al., 1995a; Shi et al., 2001; Tong et al., 2002).

Improvements in computer hardware and software have enabled technologies in QSAR development during the past decade. Within the pharmaceutical industry alone, the enormous financial incentives to accelerate the drug discovery process and to improve the odds of success by enriching the drug pipeline with more effective and less toxic candidates are powerful driving forces. These have led to improved QSAR approaches and associated software. The integration of QSAR modeling with recent advances in hardware and software for data storage and management has further stimulated its wider implementation (Bone et al., 1999). The algorithms used in QSAR software have also improved markedly, particularly with respect to the large and growing pool of descriptors used to characterize molecular structure and properties.

Although the primary objective of QSARs is to construct a predictive model, QSAR research offers numerous additional benefits beyond prediction (McKinney et al., 2000) such as: (1) leveraging existing structure-activity data, (2) providing insights into mechanisms of action (e.g., agonist vs. antiagonist) or identifying alternative mechanisms (e.g., metabolism), (3) identifying key structural features associated with high/low activity, (4) suggesting new design strategies and synthetic targets, (5) narrowing the dose range for a planned assay, (6) assisting in generation of new hypotheses to guide further research, and (7) revealing chemicals that deviate from the QSAR model and therefore from the presumed biological model.

It is important to point out that any QSAR model will produce some degree of error. This is partially due to the inherent limitation to predict a biological activity solely based on chemical structure. One can argue from the principles of chemistry that the molecular structure of a chemical is key to understanding its physicochemical properties and ultimately its biological activity and influence on organisms. Because both molecular structure and physicochemical properties are associated with the chemical itself, the relationship between structure and physicochemical properties should be apparent and more accessible using QSARs. In contrast, the biological activity of a chemical is an induced response that is influenced by numerous factors dictated by the level of biological complexity of the system under investigation. The relationship between structure and activity is more implicit and poses a more challenging problem in QSAR applications.

In principle, a chemical can be represented in three distinct, but also related structural representations: 2D substructures, 3D pharmacophores, and physicochemical properties. If a biological mechanism is mainly related to the chemical structure (which is probably the case for receptor binding such as estrogenicity), QSARs become meaningful using the aforementioned structural features. However, we often found that, even for a simple mechanism such as ER-binding, some features may well represent binding dependencies for one structural class, while other features will better represent binding dependencies for a different structural class (Fang et al., 2001). In such cases, caution must be taken in interpreting QSAR results for the chemical classes that are not well represented in the training set. This is critical when a QSAR model is used for regulatory application.

While the results presented in this chapter clearly show both the feasibility and utility of using QSARs for various applications, it is important to note that predictions from any model are intrinsically no better than the experimental data employed for modeling. Any limitations of the assay used to generate the training data apply equally to the model's predictions.

For the models providing yes/no predictions only, it is important to point out that false negatives and false positives depend on the defined cut-off value to distinguish active from inactive compounds. As the cut-off value is lowered it is likely that error will increase even for a well-designed and executed assay. The increased experimental error in close proximity to the cut-off value will be transferred to the QSAR model, which will increase false prediction for chemicals with activity in this region.

For example, as defined by EPA, the cut-off log RBA value to distinguish ER binders from non-ER binders is set to −3.0 for the NCTR Four-Phase system development. There are 31 chemicals exhibiting binding affinity in our assay that were characterized as inactive under this cut-off value, of which 14 chemicals fall below 0.3 log units of the cut-off value (an error of 0.5 log units is expected for a good assay). Since the real activity classification for these chemicals are unknown, assigning them as inactive in QSAR modeling will introduce errors in prediction for chemicals with similar structure. Caution must be taken in interpreting QSAR prediction results for these chemicals. Apparently, high confidence in prediction can be only placed on these chemicals with log RBA at least larger than −2.5 in our Four-Phase system.

## ACKNOWLEDGMENT

The authors gratefully acknowledge funding from the Food and Drug Administration's Office of Women's Health that enabled the NCTR Endocrine Disruptor Knowledge Base project to begin and make early rapid progress. The American Chemistry Council (ACS), through a Cooperative Research and Development Agreement (CRADA), supported the estrogen and androgen receptor competitive binding assays and 3D QSAR model development. Finally, through an Interagency Agreement (IAG), the EPA provides support for predictive model development and validation.

## REFERENCES

Anstead, G.M., Carlson, K.E., and Katzenellenbogen, J.A., The estradiol pharmacophore: ligand structure-estrogen receptor binding affinity relationships and a model for the receptor binding site, *Steroids,* 62, 268–303, 1997.

Baker, M.E., Medlock, K.L., and Sheehan, D.M., Flavonoids inhibit estrogen binding to rat α-fetoprotein, *Proc. Soc. Exp. Biol. Med.,* 217: 317-321.

Benigni, R. and Richard, A.M., Quantitative structure-based modeling applied to characterization and prediction of chemical toxicity, *Methods,* 14, 264–276, 1998.

Blair, R.M., Fang, H., Branham, W.S., Hass, B., Dial, S.L., Moland, C.L., Tong, W., Shi, L., Perkins, R., and Sheehan, D.M., The estrogen receptor relative binding affinities of 188 natural and xenochemicals: structural diversity of ligands, *Toxicol. Sci.,* 54, 138–153, 2000.

Bone, R.G.A., Firth, M.A., and Sykes, R.A., SMILES extensions for pattern matching and molecular transformations: applications in chemoinformatics, *J. Chem. Inf. Comput. Sci.,* 39, 846–860, 1999.

Bradbury, S.P., Predicting modes of toxic action from chemical structure: an overview, *SAR QSAR Environ. Res.,* 2, 89–104, 1994.

Bradbury, S.P., Quantitative structure-activity-relationships and ecological risk assessment: an overview of predictive aquatic toxicology research, *Toxicol. Lett.,* 79, 229–237, 1994.

Bradbury, S.P., Mekenyan, O.G., and Ankley, G.T, Quantitative structure-activity relationships for polychlorinated hydroxybiphenyl estrogen receptor binding affinity: an assessment of conformer flexibility, *Environ. Toxicol. Chem.,* 15, 1945–1954, 1996.

Branham, W.S., Dial, S.L., Moland, C.L., Hass, B., Blair, R., Fang, H., Shi, L., Tong, W., Perkins, R., and Sheehan, D.M., Phytoestrogens and mycoestrogens bind to the rat uterine estrogen receptor, *J. Nutr.,* 132, 658–664, 2002.

Brzozowski, A.M., Pike, A.C.W., Dauter, Z., Hubbard, R.E., Bonn, T., Engstrom, O., Ohman, L., Greene, G.L., Gustafsson, J.A. and Carlquist, M., Molecular basis of agonism and antagonism in the oestrogen receptor, *Nature,* 389, 753–758, 1997.

Clark, D.E. and Westhead, D.R., Evolutionary algorithms in computer-aided molecular design, *J. Comput. Aided Mol. Design,* 10, 337–358, 1996.

Coldham, N.G., Dave, M., Sivapathasundaram, S., McDonnell, D.P., Connor, C., and Sauer, M.J., Evaluation of a recombinant yeast cell estrogen screening assay, *Environ. Health Perspect.,* 105, 734–742, 1997.

Duax, W.L., Ghosh, D., Pletnev, V., and Griffin, J.F., Three-dimensional structures of steroids and their protein targets, *Pure Appl. Chem.,* 68, 1297–1302, 1996.

Fang, H., Tong, W.D., Perkins, R., Soto, A.M., Prechtl, N.V. and Sheehan, D.M., Quantitative comparisons of in vitro assays for estrogenic activities, *Environ. Health Perspect.,* 108, 723–729, 2000.

Fang, H., Tong, W.D., Shi, L., Blair, R., Perkins, R., Branham, W.S., Dial, S.L., Moland, C.L., and Sheehan, D.M., Structure-activity relationships for a large diverse set of natural, synthetic, and environmental estrogens, *Chem. Res. Toxicol.,* 14, 280–294, 2000.

Forrest, S., Genetic algorithms — principles of natural-selection applied to computation, *Science,* 261, 872–878, 1993.

Gaido, K.W., Leonard, L.S., Lovell, S., Gould, J.C., Babai, D., Portier, C.J., and McDonnell, D.P., Evaluation of chemicals with endocrine modulating activity in a yeast-based steroid hormone receptor gene transcription assay, *Toxicol. Appl. Pharmacol.,* 143, 205–212, 1997.

Giguere, V., Orphan nuclear receptors: From gene to function, *Endocrine Rev.,* 20, 689–725, 1999.

Gillesby, B.E. and Zacharewski, T.R., Exoestrogens: mechanisms of action and strategies for identification and assessment, *Environ. Toxicol. Chem.,* 17, 3–14, 1998.

Gray, L.E., Tiered screening and testing strategy for xenoestrogens and antiandrogens, *Toxicol. Lett.,* 103, 677–680, 1998.

Hansch, C., Hoekman, D., Leo, A., Zhang, L.T. and Li, P., The expanding role of quantitative structure-activity-relationships (QSAR) in toxicology, *Toxicol. Lett.,* 79, 45–53, 1995a.

Hansch, C., Telzer, B.R. and Zhang, L., Comparative QSAR in toxicology: examples from teratology and cancer-chemotherapy of aniline mustards, *Crit. Rev. Toxicol.,* 25, 67–89, 1995b.

Hansch, C. and Leo, A., *Exploring QSAR: Fundamentals and Applications in Chemistry and Biology,* The American Chemical Society, Washington, D.C., 1995.

Harris, C.A., Henttu, P., Parker, M.G., and Sumpter, J.P., The estrogenic activity of phthalate esters *in vitro,* *Environ. Health Perspect.,* 105, 802–811, 1997.

Hileman, B., Environmental estrogens linked to reproductive abnormalities, *Chem. Eng. News,* 72, 19–23, 1994.

Hileman, B., Hormone disrupter research expands, *Chem. Eng. News,* 75, 24–25, 1997.

Hong, H.X., Neamati, N., Wang, S.M., Nicklaus, M.C., Mazumder, A., Zhao, H., Burke, T.R., Pommier, Y., and Milne, G.W.A., Discovery of HIV-1 integrase inhibitors by pharmacophore searching, *J. Med. Chem.,* 40, 930–936, 1997.

Hong, H.X., Tong, W.D., Fang, H., Shi, L. M., Xie, Q., Wu, J., Perkins, R., Walker, J.D., Branham, W., and Sheehan, D.M., Prediction of estrogen receptor binding for 58,000 chemicals using an integrated system of a tree-based model with structural alerts, *Environ. Health Perspect.,* 110, 29–36, 2002.

Kavlock, R.J., Daston, G.P., DeRosa, C., Fenner-Crisp, P., Gray, L.E., Kaattari, S., Lucier, G., Luster, M., Mac, M.J., Maczka, C., Miller, R., Moore, J., Rolland, R., Scott, G., Sheehan, D.M., Sinks, T. and Tilson, H.A., Research needs for the risk assessment of health and environmental effects of endocrine disruptors: a report of the US EPA-sponsored workshop, *Environ. Health Perspect.,* 104 (Suppl. 4), 715–740, 1996.

Kubinyi, H., Folkers, G. and Martin, Y.C., 3D QSAR in drug design: recent advances. Preface, *Perspect. Drug Discovery Design,* 12, V–VII, 1998.

Kuiper, G.G.J.M., Carlsson, B., Grandien, K., Enmark, E., Haggblad, J., Nilsson, S. and Gustafsson, J.A., Comparison of the ligand binding specificity and transcript tissue distribution of estrogen receptors alpha and beta, *Endocrinology,* 138, 863–870, 1997.

Kuiper, G.G.J.M., Lemmen, J.G., Carlsson, B., Corton, J.C., Safe, S.H., van der Saag, P.T., van der Burg, B., and Gustafsson, J.A., Interaction of estrogenic chemicals and phytoestrogens with estrogen receptor beta, *Endocrinology,* 139, 4252–4263, 1997.

Leffler, J.E. and Grunwald, E., *Rates and Equilibrium Constants of Organic Reaction,* John Wiley and Sons, New York, 1963.

Lipinski, C.A., Lombardo, F., Dominy, B.W., and Feeney, P.J., Experimental and computational approaches to estimate solubility and permeability in drug discovery and development settings, *Adv. Drug Delivery Rev.,* 23, 3–25, 1997.

McKinney, J.D., Richard, A., Waller, C., Newman, M.C., and Gerberick, F., The practice of structure activity relationships (SAR) in toxicology, *Toxicol. Sci.*, 56, 8–17, 2000.

Nishihara, T., Nishikawa, J., Kanayama, T., Dakeyama, F., Saito, K., Imagawa, M., Takatori, S., Kitagawa, Y., Hori, S., and Utsumi, H., Estrogenic activities of 517 chemicals by yeast two-hybrid assay, *J. Health Sci.*, 46, 282–298, 2000.

Norris, J. D., Fan, D.J., Kerner, S.A., and McDonnell, D.P., Identification of a third autonomous activation domain within the human estrogen receptor, *Mol. Endocrinol.*, 11, 747–754, 1997.

Patlak, M., A testing deadline for endocrine disrupters, *Environ. Sci. Tech.*, 30, 540A–544A, 1996.

Perkins, R., Anson, J., Blair, R., Branham, W.S., Dial, S., Fang, H., Hass, B.S., Moland, C., Shi, L., Tong, W., Welsh, W., Walker, J.D., and Sheehan, D.M., The endocrine disruptor knowledge base (EDKB), a prototype toxicological knowledge base for endocrine disrupting chemicals, in *Handbook on Quantitative Structure Activity Relationships (QSARs) for Predicting Chemical Endocrine Disruption Potentials,* Walker, J.D., Ed., SETAC Press, Pensacola, FL, 2003 (in press).

Perkins, R., Fang, H., Tong, W., and Welsh, W.J., Quantitative structure-activity relationship methods: perspectives on drug discovery and toxicology, *Environ. Toxicol. Chem.*, 22, 1666–1679, 2001.

Rohrbaugh, R.H. and Jurs, P.C., Descriptions of molecular shape applied in studies of structure activity and structure/property relationships, *Analytica Chimica Acta,* 199, 99–109, 1987.

Routledge, E.J., Parker, J., Odum, J., Ashby, J., and Sumpter, J.P., Some alkyl hydroxy benzoate preservatives (parabens) are estrogenic, *Toxicol. Appl. Pharmacol.*, 153, 12–19, 1998.

Routledge, E.J. and Sumpter, J.P., Structural features of alkylphenolic chemicals associated with estrogenic activity, *J. Biol. Chem.*, 272, 3280–3288, 1997.

Russom, C.L., Bradbury, S.P., and Carlson, A.R., Use of knowledge bases and QSARs to estimate the relative ecological risk of agrichemicals: a problem formulation exercise, *SAR QSAR Environ Res,* 4, 83–95, 1995.

Sadler, B.R., Cho, S.J., Ishaq, K.S., Chae, K., and Korach, K.S., Three-dimensional quantitative structure-activity relationship study of nonsteroidal estrogen receptor ligands using the comparative molecular field analysis cross-validated $r^2$-guided region selection approach, *J. Med. Chem.*, 41, 2261–2267, 1998.

Schultz, T.W., and Seward, J.R., Health-effects related structure-toxicity relationships: a paradigm for the first decade of the new millennium, *Sci. Total Environ.*, 249, 73–84, 2000.

Sheehan, D.M. and Young, M., Diethylstilbestrol and estradiol binding to serum albumin and pregnancy plasma of rat and human, *Endocrinology,* 104, 1442–1446, 1979.

Shi, L., Tong, W., Fang, H., Perkins, R., Wu, J., Tu, M., Blair, R., Branham, W., Walker, J., Waller, C.L., and Sheehan, D.M., An integrated "4-phase" approach for setting endocrine disruption screening priorities: Phase I and II predictions of estrogen receptor binding affinity, *SAR QSAR Environ. Res.*, 13, 69–88, 2002.

Shi, L.M., Fang, H., Tong, W.D., Wu, J., Perkins, R., Blair, R.M., Branham, W.S., and Sheehan, D.M., QSAR models using a large diverse set of estrogens, *J. Chem. Inf. Comput. Sci.,* 41, 186–195, 2001.

Shiau, A.K., Barstad, D., Loria, P.M., Cheng, L., Kushner, P.J., Agard, D.A., and Greene, G.L., The structural basis of estrogen receptor/coactivator recognition and the antagonism of this interaction by tamoxifen, *Cell,* 95, 927–937, 1998.

Smellie, A., Kahn, S.D. and Teig, S.L., Analysis of conformational coverage. 1. Validation and estimation of coverage, *J. Chem. Inf. Comput. Sci.,* 35, 285–294, 1995a.

Smellie, A., Kahn, S.D. and Teig, S., Analysis of conformational coverage. 2. Application of conformational models, *J. Chem. Inf. Comput. Sci.,* 35, 295–304, 1995b.

Smellie, A., Teig, S.L., and Towbin, P., POLING: promoting conformational variation, *J. Computational Chem.,* 16, 171–187, 1995c.

Soto, A.M., Chung, K.L., and Sonnenschein, C., The pesticides endosulfan, toxaphene, and dieldrin have estrogenic effects on human estrogen-sensitive cells, *Environ. Health Perspect.*, 102, 380–383, 1994.

Soto, A.M., Michaelson, C., Prechtl, N., Weill, B., and Sonnenschein, C., *In vitro* endocrine disruptor screening, in *Environmental Toxicology and Assessment*, Henshel, D., Ed., American Society for Testing and Materials, West Conshohocken, PA, 1999.

Soto, A.M., Sonnenschein, C., Chung, K.L., Fernandez, M.F., Olea, N., and Serrano, F.O., The e-screen assay as a tool to identify estrogens: an update on estrogenic environmental pollutants, *Environ. Health Perspect.*, 103 (Suppl. 7), 113–122, 1995.

Stanton, D.T. and Jurs, P.C. (1990) Development and use of charged partial surface-area structural descriptors in computer-assisted quantitative structure property relationship studies, *Analytical Chemistry*, 62: 2323-2329.

Tong, W., Lowis, D.R., Perkins, R., Chen, Y., Welsh, W.J., Goddette, D.W., Heritage, T.W., and Sheehan, D.M., Evaluation of quantitative structure-activity relationship methods for large-scale prediction of chemicals binding to the estrogen receptor, *J. Chem. Inf. Comput. Sci.*, 38, 669–677, 1998.

Tong, W.D., Perkins, R., Fang, H., Hong, H., Xie, Q., Branham, W., Sheehan, D., and Anson, J., Development of quantitative structure-activity relationships (QSARs) and their use for priority setting in testing strategy of endocrine disruptors, *Regul. Res. Perspect.*, 1, 1–16, 2002.

Tong, W.D., Perkins, R., Strelitz, R., Collantes, E.R., Keenan, S., Welsh, W.J., Branham, W.S., and Sheehan, D.M., Quantitative structure-activity relationships (QSARs) for estrogen binding to the estrogen receptor: predictions across species, *Environ. Health Perspect.*, 105, 1116–1124, 1997a.

Tong, W.D., Perkins, R., Xing, L., Welsh, W.J., and Sheehan, D.M., QSAR models for binding of estrogenic compounds to estrogen receptor alpha and beta subtypes, *Endocrinology*, 138, 4022–4025, 1997b.

Tong, W., Welsh, W.J., Shi, L., Fang, H., and Perkins, R., Structure-activity relationship approaches and applications, *Environ. Toxicol. Chem.*, 22, 1680–1695, 2003.

United States Congress, *U.S.C.*, Vol. 21, p. 346a(p), 1996a.

United States Congress, *U.S.C.*, Vol. 42, p. 300j-17, 1996b.

Veber, D.F., Johnson, S.R., Cheng, H.Y., Smith, B.R., Ward, K.W., and Kopple, K.D., Molecular properties that influence the oral bioavailability of drug candidates, *J. Med. Chem.*, 45, 2615–2623, 2002.

Walker, J.D., Waller, C.W., and Kane, S., The endocrine disruption priority setting database (EDPSD): a tool to rapidly sort and prioritize chemicals for endocrine disruption screening and testing, in *Handbook on Quantitative Structure Activity Relationships (QSARs) for Predicting Chemical Endocrine Disruption Potentials*, Walker, J.D., Ed., SETAC Press, Pensacola, FL, 2003 (in press).

Waller, C.L., Oprea, T.I., Chae, K., Park, H.K., Korach, K.S., Laws, S.C., Wiese, T. E., Kelce, W.R., and Gray, L.E., Jr., Ligand-based identification of environmental estrogens, *Chem. Res. Toxicol.*, 9, 1240–1248, 1996.

Wiese, T.E., Polin, L.A., Palomino, E., and Brooks, S.C., Induction of the estrogen specific mitogenic response of MCF-7 cells by selected analogues of estradiol-17 beta: A 3D QSAR study, *J. Med. Chem.*, 40, 3659–3669, 1997.

Xing, L., Welsh, W.J., Tong, W., Perkins, R., and Sheehan, D.M., Comparison of estrogen receptor alpha and beta subtypes based on comparative molecular field analysis (CoMFA), *SAR QSAR Environ. Res.*, 10, 215–237, 1999.

Zacharewski, T., Identification and assessment of endocrine disruptors: Limitations of *in vivo* and *in vitro* assays, *Environ. Health Perspect.*, 106 (Suppl. 2), 577–582, 1998.

Zheng, W. and Tropsha, A., Novel variable selection quantitative structure-property relationship approach based on the k-nearest-neighbor principle, *J. Chem. Inf. Comput. Sci.*, 40, 185–194, 2000.

Zubay, G.L., Parson, W.W., and Vance, D.E., *Principles of Biochemistry*, Wm. C. Brown Publishers, Dubuque, IA, 1995.

**CHAPTER 14**

# Prediction of Persistence

Monika Nendza

## CONTENTS

## I. INTRODUCTION

A knowledge and understanding of persistence is critical to evaluate hazard and risk from chemicals released into the environment. The established assessment schemes generally assume higher risk with increasing persistence of compounds. Persistence is not a simple singular endpoint, but a complex multifactorial phenomenon comprising (bio)degradation/transformation, accumulation, distribution, and transport processes. The major factors influencing the persistence of a chemical are:

1. Its chemical structure
2. The environmental conditions into which it is released
3. The bioavailability of the compound

Besides the parent compounds, the fate and effects of metabolites and degradation products also need to be considered. Because of its complex composite nature, and the general lack of representative measured environmental half-lives, persistence is evaluated mostly from data on the individual contributing processes. This chapter concentrates on predictive tools for environmental degradation parameters (i.e., abiotic transformations and biodegradation). More detail on the use of catabolic biosensors to derive quantitative structure-activity relationships (QSARs) is given in Chapter 17, whereas accumulation and distribution are described in Chapters 15 and 16, respectively. The degradation parameters are treated in this chapter in three sections:

1. A section on methodology provides some background on relevant degradation/transformation processes, influential factors, data characteristics, and availability.
2. A second section on QSAR models describes the principal QSAR techniques and approaches to process-related modeling of environmental degradation, with an emphasis on application ranges/limits and validity assessments.
3. A third section on algorithms and software attempts to overview the currently available, preferably computerized, predictive degradation/transformation models, as well as their applicability and reliability. Any recommendations for particular programs are dependent on the suitability of the program for specific questions (e.g., specific compounds and transformation), and also continuous improvements and new developments.

## A. Definitions

*Persistence* relates to the length of time that (a predefined fraction of) a substance remains in a particular environment before it is physically transported to another compartment and/or chemically or biologically transformed.

*Primary degradation* produces organic derivatives of a contaminant. The resulting one or more products exhibit their own properties, reactivities, fates, and effects. The metabolites may be either less toxic (detoxification) or even more hazardous than the parent compound (toxification).

*Mineralization* yields the complete (ultimate) degradation of an organic chemical to stable inorganic forms of C, H, N, P, etc.

*Abiotic degradation* transforms organic compounds by chemical reactions such as oxidation, reduction, hydrolysis, and photodegradation. Abiotic degradation processes do not usually achieve a complete breakdown of the chemical (mineralization).

*Biodegradation* is the transformation by microorganisms of organic compounds by enzymatic reactions such as oxidation, reduction, and hydrolysis. Depending on the ambient conditions, different modes and rates of biodegradation may predominate and may render a chemical readily biodegradable at one site, but not at another because of different degradative capacities. Note that microbial transformation reactions are usually the only processes by which a xenobiotic organic compound may be mineralized in the environment, whereas abiotic reactions commonly yield other organic degradation products.

*Bioavailability* of a substance depends on its chemical or physical reactivity with various environmental components and its ability to be absorbed through the gastrointestinal tract, respiratory tract and/or skin of susceptible species. The bioavailability determines the fraction of emitted compounds that is able to interact with the biosystem of organisms per unit time.

## II. METHODOLOGY

In environmental exposure assessments, level II and III multimedia models require compartmental half-lives ($t_{1/2}$) for air, water, soil, and sediment. The most relevant compartments vary with the chemical being evaluated, but experimental data for every compartment are generally not available. To evaluate the environmental residence time of compounds adequately, the various degradation pathways have to be considered in combination. The extent, rates, and byproducts of the individual processes should preferably be integrated. The degradation reactions may take a considerable time with half-lives of the parent compounds ranging from seconds to years. Pseudo-first-order kinetics are usually applied to describe the time course of the transformations (i.e., the reaction rates depend solely on the concentration of the contaminants), whereas the concentrations of the environmental reaction partners (e.g., water, radicals in the atmosphere) are assumed to be sufficiently large to be taken as constant. Besides the concentration and (bio)availability of the compounds, the ambient temperature and redox conditions may influence the transformation rates. Microbial biodegradations (enzymatic conversions) generally increase exponentially with increasing temperatures; within the environmentally relevant temperature range, the effect of temperature on most abiotic transformations is limited.

Depending on the different established test protocols, the type and amount of data differ greatly for the endpoints discussed here. Abiotic degradation data are typically determined within small, more or less closely related, series of chemicals and quantified as reaction rate constants ([pseudo]first order: [s$^{-1}$], second order: [mol$^{-1}$s$^{-1}$]). Such continuous quantitative activity scales allow for Linear Free Energy Relationship (LFER) modeling according to the Hansch approach (Hansch, 1968a, 1968b) and recognition of changes in structural features related to the rate-limiting steps of the transformations. In contrast, most biodegradation data are from screening tests with yes/no results that relate to passing or failing a predefined threshold. Although kinetic data from laboratory and field biodegradation tests become more and more available, they are still too scarce and heterogeneous for modeling purposes. The various testing protocols for the experimental determination of biodegradability differ considerably in duration of the adaptation and incubation period, kind of inoculum used (e.g., pure cultures, surface water, sewage water, soil), inoculum size and density, treatment and concentration of the test compounds, parameters measured (BOD, COD, $CO_2$ production, DOC, etc.), and the different path-levels evaluated (pathlevels from 15 to 90% are used to classify degradable and nondegradable substances). The categorical experimental classification limits the adequate ranking particularly of compounds of intermediate biodegradability. The associated modeling techniques, such as cluster and discriminant analyses, usually employ indicator variables for the absence/presence of predefined features of the structures. In this way, the models may indicate substructures or properties contributing to the biodegradability of compounds, but they do not (intend to) relate to underlying, rate-limiting processes. Because of the inherent diversity of the endpoints and parameters concerned, biodegradability is not a uniform principal property of the chemical contaminants. Corresponding to the severely empirical nature of experimental biodegradability assessments, it is necessary to realise that biodegradability is not a well-defined parameter. The significance of biodegradation estimates cannot be expected to exceed that of the underlying data, and they should be regarded rather as indicators of probabilities toward higher or lower biodegradability.

The degradation rates of chemicals, either determined in the laboratory under ideal conditions or estimated from structural information, can only indicate which of several possible transformation pathways is most likely to occur. Because degradability assessed this manner may not correspond to the degradation occurring in the environment, the extrapolation of laboratory data to the field is extremely difficult. Under given local conditions, different degradative reactions, abiotic and also biotic, at varying rates may predominate, yielding different metabolites with differing fate. The assessment of the transformation of chemicals can be valid only for a stringently defined environmental scenario. An extrapolation to more general conditions is feasible only with regard to comparing different compounds on a relative scale (i.e., to identify the substances with the least likelihood of persistence). These inherent uncertainties have to be accounted for when evaluating the degradability of chemicals, no matter the provenance of the data.

## A. Abiotic Degradation in the Air Compartment (Atmosphere)

The (indirect) photodegradation of chemicals (i.e., reactions with reactive species formed by photochemical processes) has been recognized as the major transformation pathway for chemicals in the troposphere. The electrophilic addition of tropospheric radicals constitutes the principal degradation pathway. The relevant reactive species are hydroxyl radicals (OH•) and ozone ($O_3$) during daytime and $NO_3$• radicals at night. Hydroxyl radicals react with almost any kind of organic compounds by abstraction of hydrogen atoms from C–H and O–H bonds, by addition to double bonds and by addition to aromatic rings. This degradation pathway is the dominant one controlling the atmospheric persistence of chemicals: up to 90% of organic compounds are transformed more rapidly by reactions with hydroxyl radicals than by any other process. Only few compound classes are inert under tropospheric conditions; the perhalogenated alkanes are the most important.

## B.  Abiotic Degradation in the Water Phase

Oxidation-reduction (redox) reactions, along with hydrolysis and acid-base reactions, account for the vast majority of chemical reactions that occur in aquatic environmental systems. Factors that affect redox kinetics include environmental redox conditions, ionic strength, pH-value, temperature, speciation, and sorption (Tratnyek and Macalady, 2000). Sediment and particulate matter in water bodies may influence greatly the efficacy of abiotic transformations by altering the truly dissolved (i.e., non-sorbed) fraction of the compounds — the only fraction available for reactions (Weber and Wolfe, 1987). Among the possible abiotic transformation pathways, hydrolysis has received the most attention, though only some compound classes are potentially hydrolyzable (e.g., alkyl halides, amides, amines, carbamates, esters, epoxides, and nitriles [Harris, 1990; Peijnenburg, 1991]). Current efforts to incorporate reaction kinetics and pathways for reductive transformations into environmental exposure models are due to the fact that many of them result in reaction products that may be of more concern than the parent compounds (Tratnyek et al., 2003).

## C.  Biodegradation

Biodegradation can be an effective mechanism to transform organic compounds in water, soil, and sediment. It is a complex multistep process involving uptake, intracellular transport and enzymatic reactions in microorganisms. Depending on the ambient conditions, different modes and rates of biodegradation may predominate. These are influenced by factors related to the chemical (substrate), the microorganisms, and the environment (Scow, 1990; Schwarzenbach et al., 1993, Pedersen et al., 1994). In different environmental conditions, a given chemical may be biodegraded by different pathways, resulting in different degrees of persistence. Bacteria have a variety of enzyme systems, but specific enzymes for transforming xenobiotics are generally absent. The required enzymes are either always present at certain concentrations and activities (constitutive enzymes) or have to be induced, expressed, or transferred by plasmids during an adaptive lag phase (inducible enzymes). Primary metabolic reactions are mediated mostly by non-specific enzyme systems, which catalyze oxidation, reduction, and hydrolysis, with the different transformations occurring either consecutively or simultaneously (competitive) at different sites of the substrate. For large compounds with a molecular weight more than 500, which cannot react with the intracellular bacterial enzymes because of hindered membrane transfer, biodegradation is generally negligible; abiotic degradation (e.g., hydrolysis) may be the only degradation pathway.

## III. QSAR MODELS

The fundamentals and background of QSAR modeling and predictions have been detailed previously (e.g., Nendza, 1998). Regarding the different physicochemical bases of transformation processes, different approaches to their estimation have been taken. The intention should always be to use process-related models. Thermodynamic principles underlie the relationships between abiotic one-step reaction rates and physicochemical descriptors of the structures. Mechanistic modeling is much more intricate for multistep biodegradations, where the (varying) rate-limiting interactions may not be identified totally. Besides some LFER-type correlations, most degradation models use group-contribution methods. Factors and descriptors (e.g., fragment indicators that may be applicable to a variety of chemical classes) are used to quantify the contributions of certain substructures to the degradability of the compounds. Depending particularly on the quality of the experimental input data and also the descriptor selection, the models are of differing validity and application range. Numerous QSARs for estimating degradation have been derived, with reviews given by, for example, Scow (1990), Organization for Economic Co-operation and Development (1993a; 1993b), Peijnenburg (1994), Howard and Meylan (1997), Damborsky et al. (1998), Nendza (1998), Atkinson (2000), Raymond et al. (2001), Canonica and Tratnyek (2003), Cronin et al. (2003), Jaworska et al. (2003), and Tratnyek et al. (2003).

## A. Abiotic Degradation in the Air Compartment (Atmosphere)

The measurement of rate constants in the gas phase is restricted due to the difficult and time- and cost-consuming experiments. These experiments are generally limited to compounds with at least some water solubility ($S_{aq} > 1$ μmol/L) and volatility ($P_V > 1$ Pa); QSARs may be used to obtain estimates of rate constants. Estimation of the reaction rate constants for the OH• or NO$_3$• radicals from the corresponding vertical ionization energies (Güsten et al., 1984; Sabljić and Güsten, 1990) still requires experimental input data. Some models use calculated quantum-chemical descriptors (Klamt, 1993; Medven et al., 1996). The widely recommended method to estimate reaction rate constants for the reaction of organic compounds with hydroxyl radicals has been derived by Atkinson (1987; 1988). The underlying principle is to account for molecular fragments that are likely to be attacked by reactive species. The total hydroxyl reaction rate constants ($k_{tot.}$) are calculated from rate constants of four important types of reactions: H atom abstraction from C–H and O–H bonds ($k_{H-abst.}$), addition of hydroxyl radicals to C–C multiple bonds ($k_{add.(C=C)}$), addition to aromatic rings ($k_{add.(aromat.)}$), and reactions with N, S, or P ($k_{N,S,P}$):

$$k_{tot.} = k_{H-abst.} + k_{add.(C=C)} + k_{add.(aromat.)} + k_{N,S,P} \tag{14.1}$$

The rates of the four types of reactions are estimated using additivity fragment constants. By considering exclusively substructures that are liable to transformations, ignoring persistent moieties, this approach is principally assumed to correspond to the worst case. If none of the substructures considered in the model is present in the molecules, zero degradation is assumed. The factors (group contributions) relate to the respective reaction rates of the degradable fragments with reactive species, and degradation as a result of further functional groups is neglected. In a validation study, the method (Atmospheric Oxidation Program [AOP], 1990) was used to calculate rate constants of 369 compounds and to compare those with experimental data (Müller and Klein, 1991). Considering that almost all of the available experimental data were included in the development of the model, the experimental and calculated rate constants differed by more than a factor of 2 for only 34 compounds (9.2%). Major discrepancies occurred for N–O compounds, P- and S-containing compounds, and some aromatic compounds. The good overall agreement of experimental and calculated data justifies the established application of this program for the estimation of rate constants (Organization for Economic Co-operation and Development, 1993a).

## B. Abiotic Degradation in the Water Phase

Estimation methods for the hydrolysis rates of several types of carboxylic acid esters, carbamates, aromatic nitriles, and phosphoric acid esters have been reported. Hydrolysis rates are subject to substituent effects, and consequently LFERs, as represented by Hammett or Taft correlations, have been applied to their estimations. Reviews (e.g., Harris, 1990; Peijnenburg, 1991; Nendza, 1998) reveal that QSARs are available only for a few compound classes and are based mostly on limited sets of experimental data. The alkaline hydrolysis rates depend on the electronic and steric features of the leaving group (alcohol moiety), and generally decrease with the size and branching of the alcohol and increase with electron-withdrawing groups on the alcohol. The respective QSARs are strictly limited to homologous compounds because of the class-specific sensitivity of the chemical reaction centre, as indicated by the greatly varying slopes and intercepts of the functions. The available hydrolysis QSARs are limited in their application range, but are of acceptable validity when applied in accordance with their restrictions.

Most aquatic oxidation reactions are attributable to well-defined chemical oxidants. As a result, model systems can be designed where second-order rate constants can be determined precisely for families of organic congeners. The comparatively high quality of these data allows mechanistic models of electron transfer to describe aquatic oxidations of environmental interest. Kinetic studies of these processes have produced many QSARs, mostly simple empirical correlations with common convenient descriptors such as the Hammett constant ($\sigma$), half-wave oxidation potential ($E_{1/2}$), energies of the highest occupied molecular orbital ($E_{HOMO}$), or rate constants for other oxidation reactions as descriptors (Canonica and Tratnyek, 2003). Their predictive power has lead to engineering applications in water treatment and remediation.

Estimation methods for reductive transformations (e.g., dehalogenation or nitro reduction reactions) are limited because it is not yet possible to predict the rates of reductive transformations quantitatively. The choice of appropriate descriptors is complicated by the variability in rate-limiting steps with contaminant structure and environmental conditions. Most QSARs for reduction reactions have been developed as diagnostic tools to determine reduction mechanisms and pathways. So far, only a few of these QSARs provide sufficiently precise predictions and are sufficiently general in scope that they might be useful to predict environmental fate (Tratnyek et al. 2003). They mostly use LFER-type correlations or quantum-chemically derived parameters (e.g., Peijnenburg et al., 1991; Rorije et al., 1995; Scherer et al., 1998; Tratnyek and Macalady, 2000) and many of them are compiled in a recent review by Tratnyek et al. (2003).

## C. Biodegradation

Despite the considerable uncertainties regarding the relevance of the different activity data sets for modeling, numerous QSARs for biodegradation have been reported, with reviews given by, for example, Parsons and Govers (1990), Scow (1990), Organization for Economic Co-operation and Development (1993a; 1993b), Peijnenburg (1994), Peijnenburg and Karcher (1995), Damborsky et al. (1998), Nendza (1998), and Jaworska et al. (2003). The available models are all based on the assumption that structural features of a compound may indicate its biodegradability; these include chain length, degree of branching, saturation state of the carbon chain and oxidation state of the terminal groups for the determination of biodegradability of acyclic compounds. The type, number, and position of substituents, and also the number of rings, are relevant for microbial transformations of aromatic compounds. Additionally, physicochemical properties may affect the degradability of the chemicals, relating to processes such as transport into the microbial cell. Electronic parameters can be used to explain different transformation mechanisms caused by differing polarity of chemicals. The electron density on the aromatic ring, which is dependent on the functional groups, determines the ring cleavage of an aromatic system. Because of the multitude of processes involved in biodegradation, no single-descriptor model can accurately predict the biodegradability of a broad range of chemicals. If the reaction mechanism (i.e., the explicit reaction equation and stochiometry) of the degradation process is known, thermodynamic descriptors may be used; these have the advantage of allowing for variable environmental conditions such as the presence of water, concentrations of reactants and products, temperature, and redox conditions (Govers et al., 1995).

To test the applicability of published QSARs for biodegradation, several validation exercises have been conducted. The most comprehensive validation exercise (Degner, 1991; Organization for

Economic Co-operation and Development, 1993b) does not cover very recent models, but highlights many aspects that are still valid. To test their predictive power more than 70 models were compiled and the respective estimates compared with experimental data for more than 600 diverse chemicals, uniformly obtained by the MITI-I (Japanese Ministry of International Trade and Industry) procedure (Organization for Economic Co-operation and Development, 1984; 1989). In spite of the shortcomings of this test, the available data pool proved useful. The MITI-I degradation test takes an intermediate position in terms of degradative capacity and can be regarded as representative of degradation tests for category 1 (ready biodegradability). These data do not comply with all QSARs because of the different endpoints modeled. However, this validation study provides a realistic approximation of the predictive power of the various models to discriminate readily and non-readily degradable compounds. Even if a sufficiently large uniform data set on other biodegradation parameters became available, the principal results of this study would probably not change drastically.

Only a few of the available QSAR models to estimate biodegradability provide an adequate level of agreement between calculated and experimental data. Classifications were considered adequate if more than 75% of the MITI-I data could be discriminated into readily and non-readily degradable substances for any arbitrary path-level (for details of this validation study for the individual QSARs see Organization for Economic Co-operation and Development [1993b]). The reasons for misclassifications by many models can be ascribed partly to the inconsistent original data, the lack of homogeneity in the endpoint, and the selection of limited sets of homologous test compounds to derive the respective QSARs. Many of the models could not be validated with the MITI-I data as they apply to chemical classes whose compounds are either all readily degradable or all nonreadily degradable in the MITI-I degradation test, or the models were developed for chemical classes that were not included in the MITI-I data set.

From the literature, 64 regression models for specific compound classes were retrieved, of which 35 could be tested with the MITI-I data, but only 7 QSARs were successfully validated. These models were derived with four to eight homologous substances and, because of their specificity, they are suitable for application to corresponding substances only. The number of validated QSAR models for specific compound classes is too low to make predictions solely on this basis; in the MITI-I data set they were applicable for estimating the biodegradability of only 3% of the chemicals.

More general group-contribution models for the classification of larger data sets were also subjected to validation with the MITI-I data. They have been derived mostly on the basis of substructure indicators: increasing biodegradability has been demonstrated, for example, in the presence of functional groups such as carboxyl-, hydroxyl-, and methyl groups and decreasing degradability in the presence of nitro-, amino-, cyano-, and halogen groups. The respective criteria may be combined by rules of logic to support expert judgment of the biodegradability of environmental contaminants. Non-linear substructure models are also supposed to account for the contribution of group interactions to the chemicals' degradability. A principal limitation is that substructure models are not applicable to compounds with structural elements absent from the original data set. Among the 9 models tested, most yielded less than 70% correct classifications of the MITI-I data, except the substructure model described by Niemi et al. (1987), which provided 76% coincident predictions (Organization for Economic Co-operation and Development, 1993b). A feature of all these models, which gives cause for concern, is a marked difference in their recognition of degradable and non-degradable compounds. Although they classify 50 to 83% (mean: 73%) of the degradable substances correctly, the predictions are poor for most of the non-degradable chemicals; the success rate is as low as 10% for some models (range 10 to 70%, mean: 47%). This imbalance indicates a major problem with the application of biodegradation models: predictions that compounds are readily degradable have a reliability of less than 50% of being correct and are therefore useless and should be discarded. Only if a compound is predicted non-degradable by such models is there a good probability that it really is non-degradable, and the predicted result may be used with some confidence. The rejection of QSAR-predicted ready degradability as being untrustworthy is based on the rationale that incorrect predictions for readily degradable compounds as persistent may be filed as over-protective assessments. However, the opposite classification of persistent chemicals as readily degradable may result in substantial hazard for man and the environment.

The most recently developed estimation methods for biodegradation consist of expert systems that make use of artificial intelligence approaches. These so-called knowledge-based expert systems tend to act as a repository of expert knowledge regarding phenomena or processes that, like biodegradation, can be described by a set of rules. The library of rules or transformations is organized in a hierarchy ordered by their likelihood of being executed. These mechanistic models that (attempt to) predict the biodegradation pathways are qualitative in nature but they can be linked to other models to provide a quantitative assessment. The available expert systems to predict biodegradation have been reviewed by Jaworska et al. (2003).

## IV. ALGORITHMS AND SOFTWARE

A number of algorithms and estimation software are available for most environmental degradation parameters. These models vary in their degree of user friendliness. A large database of several thousand QSARs for a total of more than 15,000 datasets in physical organic and biomedicinal chemistry and toxicology (www.clogP.pomona.edu/medchem/chem/qsar-db/) has been collected by Hansch and coworkers at Pomona College for the purpose of lateral validation and comparative analyses of models (Hansch, 1993). A comprehensive collection of environmental parameter estimation programs is provided in the U.S. Environmental Protection Agency's (EPA) EPISUITE (www.epa.gov/opptintr/exposure/docs/episuite.htm) (see Chapter 19 for more details on EPISUITE). The Estimations Programs Interface for Windows (EPIWIN) is an interface programme that transfers a single Simplified Molecular Line Entry System (SMILES) notation (Anderson et al., 1987) to ten separate structure estimation programs and captures their output. Structures in scientific file formats other than SMILES notations can be imported directly into EPIWIN. Input by Chemcial Abstract Service (CAS) number via the SMILECAS database (a database of 103,000 CAS numbers with associated SMILES notations) is also possible. The ten stand-alone programmes that are part of the EPISUITE of programs include:

1. AOPWIN — Estimates atmospheric oxidation rates
2. HYDROWIN — Estimates aqueous hydrolysis rates (acid- and base-catalyzed)
3. BIOWIN— Estimates biodegradation probability

Experimental data, if available from the physicochemical properties database, are provided with the estimates.

## A. Abiotic Degradation in the Air Compartment (Atmosphere)

According to current results and recommendations from validation exercises (Müller and Klein, 1991; Güsten et al., 1995, European Commission, 1996; Howard and Meylan, 1997) models based on the method of Atkinson are suitable for estimating overall atmospheric oxidation rate constants and associated half-lives. The latest computerised version of the AOPWIN program in EPA's EPISUITE can be downloaded for free (www.epa.gov/opptintr/exposure/docs/episuite.htm). The program is very easy to use and when run separately (without EPIWIN) AOPWIN can show the calculation equations. The accuracy of the estimates must be evaluated by the user on a case-by-case basis, considering that the validation study by Müller and Klein (1991) revealed generally satisfactory agreement (deviations < factor 2) between experimental and calculated rate constants for:

1.  C H compounds (not aromatic)
2.  C H O compounds
3.  C H N compounds (not NOx)
4.  C halogen compounds (< 3 halogens per C atom)
5.  C H S compounds
6.  C H O S compounds

whereas major discrepancies occurred for:

1.  –NO compounds
2.  –NO$_2$ compounds
3.  –NO$_3$ compounds
4.  C H P compounds
5.  C H O P S compounds
6.  Aromatic compounds

Further indications on the predictive power of the program for certain chemical classes can be obtained from the tables of measured and calculated rate constants within the accuracy documentation of AOPWIN.

## B. Abiotic Degradation in the Water Phase

Hydrolysis is relevant only for a limited number of chemical classes (e.g., alkyl halides, amides, amines, carbamates, esters, epoxides, and nitriles). If the chemical of interest does not fall within these groups, zero degradation by this pathway has to be assumed. Only if the compound is potentially hydrolyzable QSARs may be used if experimental data are not available, or for the evaluation of experimental data. The Technical Guidance Document of the European Commission (1996) recommends five class-specific hydrolysis models for brominated alkanes, esters, carbamates, and benzonitriles. The HYDROWIN program of U.S. EPA's EPISUITE (www.epa.gov/ opptintr/exposure/docs/episuite.htm) calculates second-order acid- or base-catalysed hydrolysis rate constants at 25°C for esters, carbamates, halomethanes, alkyl halides, and epoxides. Acid- and base-catalyzed half-lives are calculated for pH 7 or pH 8. The prediction methodology was developed by EPA's Office of Pollution Prevention and Toxics based on data from Mill et al. (1987). The class-specific equations are given in the HYDROWIN help file, but no statistics are provided. Based on a different methodology, the SPARC online calculator (ibmlc2.chem.uga.edu/sparc/) uses models describing intra- and inter-molecular interactions. These are linked by the appropriate thermodynamic relationships to provide estimates of reactivity parameters under the desired conditions. The hydrolysis options include acid- and base-catalyzed transformation rates and also neutral catalysis for variable conditions (temperature, solvent or mixture, catalyst). The full output delivers the mechanistic information from the calculations. SPARC has been considered the most accurate predictive model for hydrolysis (Tratnyek, 2002). Examples of some QSAR models for estimating hydrolysis rates are listed in Table 14.1. They are limited in their application range, but generally assumed acceptable when operated in accordance with their restrictions. It should be noted, however, that none of the available models has been cross validated or externally validated.

QSARs for aquatic oxidation reactions are specific for chemical classes and the respective oxidant. Such models have been reviewed thoroughly by Canonica and Tratnyek (2003). Their application to the estimation of environmental persistence — aside from engineered remediation — is limited because of insufficient knowledge regarding the fractional contributions of these processes to the total degradation.

Over 50 equations for nitro reductions and reductive dehalogenations have been compiled and reviewed by Tratnyek et al. (2003). They reflect the various mechanisms and model systems, assuming the exact chemical agents that are responsible for reduction in the environment are rarely quantifiable.

**Table 14.1    Examples of QSAR Models for the Estimation of Hydrolysis Rates for Specific Chemical Classes ($K_{hyd.}$ in mol$^{-1}$s$^{-1}$)**

| Chemical Class | | r | n | Reference |
|---|---|---|---|---|
| Benzonitriles (p-subst.) | log $k$ = 1.64 (±0.42) σ(para) − 1.37 (±0.17) | 0.926 | 14 | Masunaga et al. (1993) |
| Benzoic esters | log $K_{hyd.}$= 1.17 σ + 2.26 | 0.996 | 18 | Harris (1990) |
| Phosphoric acid esters | log $K_{hyd.}$ = 1.4 Σ σ − 0.47 | 0.995 | 4 | Harris (1990) |
| Phosphoric acid esters | log $K_{hyd.}$ = −9.65 ρ + 2.85 $E_{S(alcohol)}$ + 4.89 | 0.95 | 19(?) | Johnson et al. (1985) |
| Phthalate esters | log $K_{hyd.}$ = 4.59 σ* + 1.52 $E_S$ − 1.02 | 0.986 | 5 | Wolfe et al. (1980) |
| N-Methyl-N-phenylcarbamates | log $K_{hyd.}$ = −0.26 p$K_{a(alcohol)}$ − 1.3 | 1.0 | 3(!) | Wolfe et al. (1978) |
| N-Phenylcarbamates | log $K_{hyd.}$ = −1.3 p$K_{a(alcohol)}$ + 13.6 | 0.99 | 20 | Wolfe et al. (1978) |
| N-Methylcarbamates | log $K_{hyd.}$ = −0.91 p$K_{a(alcohol)}$ + 9.3 | 0.99 | 6 | Wolfe et al. (1978) |
| N,N-Dimethylcarbamates | log $K_{hyd.}$ = −0.17 p$K_{a(alcohol)}$ − 2.6 | 0.89 | 7 | Wolfe et al. (1978) |
| Carbamates (R2 = H) | log $k$ = 2.39 σ*(R1, R2) + 0.96 σ(X1) + 7.97 σ* (R3) + 2.81 σ(X2) + 0.275 | 0.986 | 62 | Drossman et al. (1988) |
| Alkyl/Phenylcarbamates | log $k$ = 7.99 σ*(R3) + 0.31 σ(X2) + 3.14 $E_S$(R1, R2) + 0.442 | 0.950 | 18 | Drossman et al. (1988) |
| Esters | log $k$ = 0.98 $E_S$(R) + 0.25 $E_S$(R′) + 2.24 σ*(R) + 2.24 σ*(R′) + 2.09 σ(x) + 1.21 σ(x′) + 2.69 | 0.987 | 103 | Drossman et al. (1988) |
| Aromatic nitriles (3,4-subst.) | log $K_{corr.}$ = 0.54 log $K_{ow}$ + 0.57 σ − 5.28 | 0.925 | 17 | Peijnenburg et al. (1993) |
| Aromatic nitriles (2-subst.) | log $K_{corr.}$ = −0.46 log $K_{OW}$ + 1.26 σ − 4.56 | 0.981 | 7 | Peijnenburg et al. (1993) |
| Brominated alkanes | log $k$(i)/$k$(o) = −11.9 (±3.5) σ(I) | 0.877 | 16 | Vogel and Reinhard (1986) |

*Note:*   r = correlation coefficient; n = number of compounds analyzed, ρ = net charge on P, $K_{corr.}$ = hydrolysis rate corrected for fraction of compound sorbed to sediment. $K_{corr.}$ represents a sediment-catalyzed transformation and rates of hydrolysis in the overlying water phase will be much lower than predicted using these QSARs.

## C.  Biodegradation

Current generally applicable biodegradation models focus on the estimation of readily and nonreadily biodegradability in screening tests. This is because most experimental data are from such tests (e.g., MITI-I). There are far fewer data that are both quantitative and environmentally relevant (i.e., measured half-lives or rate constants). However, individual transformations and pathways are well documented in the literature. This allows for development of explicitly mechanistic models, making use of established group-contribution approaches, hierarchic rule-based expert systems, and probabilistic evaluation of possible transformation pathways.

The following description of several algorithms and software is partly based on the comparative evaluation of model performance by Rorije et al. (1999) and the recent review of broadly applicable methods for predicting biodegradation by Jaworska et al. (2003). Table 14.2 summarizes the main features of some models.

Based on the information gained during the validation exercise, Degner (1991 [Organization for Economic Co-operation and Development, 1993b]) suggested use of a set of QSARs supplemented with rules (decision trees) to guide the selection of appropriate models. The principle of the hierarchic model approach is that a set of discriminant criteria can be used to identify the most appropriate QSARs for a given compound. First, the compounds have to be assigned to a chemical class based on structural characteristics. This categorization of substances, according to their parent structure as well as substructures, is intended to group compounds with a similar degradation pattern. Within the classes defined in this manner, the biodegradability may then be related to the same structural descriptors. For acyclic compounds and mono-aromatic compounds several substructure-based QSARs were developed. They revealed sufficient predictive power and were successfully externally validated with 488 of the MITI-I data, especially for the prediction of readily

**Table 14.2 Comparison of the Main Features of Group-Contribution Approaches, MultiCASE/META, BESS and CATABOL for Predicting Biodegradation**

| | Group-Contribution Methods | MultiCASE/ META | BESS | CATABOL |
|---|---|---|---|---|
| Predefined fragments | + | + | − | − |
| Predefined transformations | − | + | + | + |
| Prediction of metabolic pathways | − | + | + | + |
| Metabolic transformation weighed | − | −(+)[a] | − | + |
| Identification of metabolites | − | − | − | + |
| Hierarchy of transformations | N/A | − | − | + |
| Quantitative endpoint assessed | + | − | − | + |

[a] Fragments are weighted.
*Note:* N/A = not available.

Reprinted with permission from Jaworska et al. (2003). Copyright SETAC, Pensacola, FL, U.S.A.

biodegradation (Rorije et al., 1997, 1999). For acyclic compounds, 93.7% of the predictions for readily and 80.9% for non-readily biodegradable were correct. For mono-aromatic compounds 75.0% of the predictions for 'readily' and 90.9% for non-readily biodegradable were correct. Unfortunately, respective schemes for further compound classes are not available.

Based on 894 substances with biodegradation assessed according to MITI-I test protocol (388 readily, 506 non-readily), Loonen et al. (1999) developed a multivariate partial least squares (PLS) model for the prediction of compounds that are readily biodegradable. The chemicals were characterized by a set of 127 predefined structural fragments (Eakin et al., 1974). The most important fragments as identified by the PLS analysis were found to conform with the generally known mechanisms of biodegradation. They include the presence of a long alkyl chain, hydroxy, ester and acid groups (enhancing biodegradation), and the presence of one or more aromatic rings and halogen substituents (retarding biodegradation). The model was evaluated by means of internal cross validation and repeated external validation. More than 85% of the model predictions were correct for the complete data set. The non-readily predictions were slightly better than the readily biodegradable predictions (86 vs. 84%). The average percentage of correct predictions from four external validation studies was 83%. Model optimization by the additional inclusion of fragment-fragment interactions (n = 706) on the basis of their PLS regression coefficients, improved the model predicting capabilities to 89%. A comparative evaluation of the model (Rorije et al., 1999) revealed the PLS model to show the best performance in external validation (82.7%) combined with the broadest applicability of the model.

The BIOWIN group-contribution method of the EPA's EPISUITE (www.epa.gov/opptintr/ exposure/docs/episuite.htm) features six linear and non-linear regression models based on weight-of-evidence evaluations of 264 chemicals in the BIODEG database (Howard et al., 1987; 1992), 200 semiquantitative estimates of rates of primary and ultimate biodegradation [Boethling et al., 1994]) and MITI-I test data for 884 discrete organic substances (Tunkel et al., 2000). The calculated probabilities cannot themselves be used to predict biodegradation pathways, but it is often possible to gain insight into pathways from analysis of the positive and negative fragments in the parent molecule (which are identified and listed by the software). Validation of the linear probability model for the prediction of the outcome of the MITI-I test for ready biodegradability (Langenberg et al., 1996, Rorije et al., 1999) yielded only 56.2% correct overall predictions (n = 488), resulting from acceptable predictions of non-readily biodegradable compounds (93.5% correct predictions, n = 169), but major deficiencies in reliably predicting readily biodegradable compounds (36.4% correct predictions, n = 319).

A very recent validation exercise with 305 premanufacture notice (PMN) structures, for which Organization for Economic Co-operation and Development readily biodegradable data had been submitted (Boethling et al., 2003), revealed that the overall accuracy of four of the six BIOWIN models (MITI Linear, MITI Nonlinear, Survey Ultimate, Survey Primary) was in the 80+% range

(62 to 70% correct readily predictions, 91 to 95% correct non-readily predictions). The original linear and non-linear probability models did not perform as well, similar to the previous study (Rorije et al. 1999), with an overall accuracy of less than 70% (44 to 63% correct readily predictions and 78 to 85% correct nonreadily predictions). Poorly predicted substances included $N$-hetero-cyclics, cycloaliphatics, and polyethoxylates in particular. Some improvements in performance of all of the models could be achieved by optimizing the classification criteria (i.e., numerical model output values discriminating readily and non-readily biodegradable substances). For the time being, it is recommended to use BIOWIN model estimates only when the four MITI and Survey models agree (accuracy greater than 90%). Recently the BIOWIN survey models have also been proposed for applications requiring screening-level estimation of biodegradation half-lives, including EPA's Waste Minimization Prioritization Tool (WMPT; www.epa.gov/epaoswer/hazwaste/minimize/), as well as multimedia modeling to identify substances with potential persistent, bioaccumulative, and toxic (PBT) properties. The main example of the latter is the PBT Profiler (www.epa.gov/ pbt/tool-box.htm), which is undergoing formal review at the U.S. EPA and, if the review is successful, will become a freely accessible web-based program for pollution prevention.

The Multiple Computer Automated Structure Evaluation/ (MultiCASE/META) system combines a group-contribution model and an expert system to simulate aerobic biodegradation pathways (Klopman and Tu, 1997). The META expert system features 70 transformations that match 13 biophores. The hierarchy is based on weights of biophores in the parent substance as calculated by MultiCASE, and ascribing the same weights to associated transformations. Because of the neglect of substructures that inhibit biodegradation (biophobes), the model tends to overpredict biodegradability.

The MultiCASE approach has also been used to model anaerobic aquatic biodegradation rates (Rorije et al., 1998). Based on 79 chemicals, MultiCASE identified aromatic and aliphatic thiol, methoxy, alcohol, and carboxylic ester groups as anaerobic biophores, but no significant biophobes were found; this was again most likely because of the limited number of negatives for biodegradation in the dataset. In the MultiCASE methodology, substances with one or more biophores are predicted biodegradable unless the molecule contains one or more biophobes. The difficulties in finding biophobes (i.e., structures inhibiting biodegradation) negatively influences model performance because the model may fail for chemicals containing both easy and difficult to degrade substructures by predicting these chemicals as biodegradable. As with the new MITI-BIOWIN model, MultiCASE makes better predictions for biodegradable (89% correct) than non-biodegradable (64% correct) compounds. The fact that the selection of structural descriptors as performed by the MultiCASE model does not lead to a better performance is probably because of the multiple linear regression (MLR) implementation of the selected fragments. MLR combined with the high number of structural descriptors gives a fair chance of overfitting the data, leading to reduced performance in external validation (Rorije et al., 1999).

The Biodegradability Evaluation and Simulation System (BESS) features rules relating to biodegradation pathways documented in the literature (Punch et al., 1996). The hierarchy of the approximately 150 general rules and 2000 specific rules, based on expert knowledge, is organized in a tree structure, with major types of transformation at the top level. Each of these transformations represents a group of transformation subtypes. Each group can be a part of another group and the same rule can be in multiple groups. In this way, it is possible to match chemical structures with transformation reactions efficiently and to consider pathways other than the most probable one. BESS provides qualitative assessments; a chemical is assumed to be biodegradable if any transformation pathway is indicated by the system. The BESS pathways, mostly derived from biodegradation of surfactants, are not validated yet and the system still needs work to be operational for routine applications.

CATABOL is a knowledge-based expert system for the prediction of biotransformation pathways. It works in tandem with a probabilistic model that calculates the probabilities of the individual transformations and overall biochemical oxygen demand (BOD) and extent of $CO_2$ production (Jaworska et al., 2002). The model assesses biodegradation based on the entire pathway and not,

**Table 14.3    Performance Statistics of MITI-BIOWIN (Tunkel et al., 2000), the PLS Biodegradability Model (Loonen et al., 1999), and MultiCASE/META (Klopman et al., 1997) as Reported in Rorije et al. (1999) and CATABOL (Dimitrov et al., 2002) Models**

| | MITI-BIOWIN Nonlinear Method | PLS | MultiCASE/META | CATABOL |
|---|---|---|---|---|
| **Training** | | | | |
| NRB: Correct predicted/observed | 85% | N/A | N/A | 86% |
| RB: Correct predicted/observed | 80% | | | 91% |
| Total: Correct predicted/observed | 83% | | | 87% |
| **Validation[a]** | | | | |
| NRB: Correct predicted/observed | 82% | 85% | 80% | 82% |
| RB: Correct predicted/observed | 77% | 80% | 73% | 91% |
| Total: Correct predicted/observed | 81% | 83% | 77% | 83% |

[a] For MITI-BIOWIN, external validation with 295 MITI chemicals not used in training. For the remaining models, results are averages from 4 cross validations leaving out 25% of the data used in training.

*Note:*   NRB = non-readily biodegradable; RB = readily biodegradable; N/A = not available.

Reprinted with permission from Jaworska et al. (2003). Copyright SETAC, Pensacola, FL, U.S.A.

as with other models, the parent structure alone. CATABOL features more than 550 principle transformations (Dimitrov et al., 2002) that often encompass more than one real biodegradation step in order to improve the speed of predictions. The hierarchy of transformations, based on expert knowledge, is organized in sequential and branched pathways. Similar principal catabolic reactions (those yielding similar TOD and having similar targets) are grouped and assumed to have the same probability. Before executing the transformation of a target fragment, adjacent fragments are evaluated for possible effects on transformation probabilities (i.e., inhibiting fragments). Considering spontaneous biotic and abiotic as well as catabolic transformations, CATABOL generates one, the most probable, pathway. Through the analysis of the pathway and its critical steps based on individual transformation probabilities, CATABOL can identify potentially persistent catabolic intermediates and estimates their molar amounts.

Currently, CATABOL's transformation probabilities are parameterized for the MITI-I test. The experimental values correlate with the calculated BOD values ($r^2 = 0.69$) over the entire range (i.e., acceptable fit was observed for readily degradable, intermediate, and difficult-to-degrade substances). Using 60% TOD (theoretical oxygen deman) as a cutoff value, the model correctly predicts 86% of the readily biodegradable structures and 91% of the non-readily biodegradable structures in the training set. Cross-validation by leaving out 25% of the data 4 times resulted in $Q^2 = 0.86$ and 82% and 91% readily/nonreadily correct classifications, respectively. Development of CATABOL is ongoing and new transformations are being added.

For a comparison of performance of the nonlinear MITI-BIOWIN model (Tunkel et al., 2000), MultiCASE/META (Klopman et al., 1997), the PLS model (Loonen et al., 1999), and CATABOL (Jaworska et al., 2002; Dimitrov et al., 2002) training and validation statistics are listed in Table 14.3. The heterogeneity of validation methods and different sizes of training sets limit somewhat the comparability of the performance of the models (Jaworska et al., 2003). The MITI-BIOWIN model was developed using 589 chemicals and externally tested with the remaining 295 chemicals from the MITI-I database. The MultiCASE/META model was developed and cross validated with 759 chemicals, the PLS model was developed and cross validated with 894 chemicals, and the CATABOL model was developed and cross validated with 743 chemicals. In all three cases the cross validations (mean values reported) were performed by leaving out 25% chemicals four times. Overall, the models perform better in predicting non-readily biodegradable compounds. This can partly be explained by the fact that the presence of a biodegradation retarding fragment will prevent mineralization, while a biodegradation enhancing fragment can indicate a possible metabolic step, but does not necessarily lead to complete mineralization. The reason for non-readily biodegradability may be a structural fragment that is not present in the parent compound, but in (one of the) metabolites from the multitude of transformation processes that lead to ultimate degradation. As a consequence, it still remains that only if a compound is predicted non-degradable by the models, is there a good probability that it really is non-degradable and the predicted result may be used with some confidence.

## ACKNOWLEDGMENTS

Special thanks are due to Bob Boethling, Jiri Damborsky, Bjørn Hansen, Joanna Jaworska, Martin Müller, Willie Peijnenburg, Paul Tratnyek, and Kirsten Vormann for generously sharing information and recent materials on their work on environmental degradation.

## REFERENCES

Anderson, E., Veith, G.D., and Weininger, D., SMILES: a line notation and computerized interpreter for chemical structures, *Environ. Res. Brief*, EPA 600/M-87-021, 1987.

Atmospheric Oxidation Program (AOP) Version 1.31, Syracuse Research Corp., Syracuse, NY, 1990.

Atkinson, R., Structure-activity relationship for the estimation of rate constants for the gas-phase reactions of hydroxyl radicals with organic compounds, *Int. J. Chem. Kinetics*, 19, 799–828, 1987.

Atkinson, R., Estimation of gas-phase hydroxyl radical rate constants for organic chemicals, *Environ. Toxicol. Chem.*, 7, 435–442, 1988.

Atkinson, R., Atmospheric oxidation, in *Handbook of Property Estimation Methods for Chemicals*, Boethling, R.S. and Mackay, D., Eds., CRC Press, Boca Raton, FL, 2000, pp. 335–354.

Boethling, R.S., Howard, P.H., Meylan, W., Stiteler, W., Beauman, J., and Tirado, N., Group contribution method for predicting probability and rate of aerobic biodegradation, *Environ. Sci. Tech.*, 28, 459–465, 1994.

Boethling, R.S., Lynch, D.G., and Thom, G.C., Predicting ready biodegradability of premanufacture notice chemicals, *Environ. Toxicol. Chem.*, 22, 837–844, 2003.

Canonica, S. and Tratnyek, P.G., Quantitative structure-activity relationships for oxidation reactions of organic chemicals in water, *Environ. Toxicol. Chem.*, 22, 1743–1754, 2003.

Cronin, M.T.D., Walker, J.D., Jaworska, J.S., Comber, M.H.I., Watts, C.D., and Worth, A.P., Use of QSARs in international decision-making frameworks to predict ecologic effects and environmental fate of chemical substances, *Environ. Health Perspect.*, 111, 1376–1390, 2003.

Damborsky, J., Lynam, M.M., and Kuty, M., Structure-biodegradability relationships for chlorinated dibenzo-p-dioxins and dibenzofurans, in *Biodegradation of Dioxins and Furans*, Wittich, R.M., Ed., R.G. Landes, Austin, TX, 1998, pp. 163–226.

Degner, P., *Abschätzung der biologischen Abbaubarkeit mittels SARs*, Dissertation, Universität Duisburg, Germany, 1991.

Dimitrov, S., Breton, R., MacDonald, D., Walker, J.D., and Mekenyan, O., Quantitative prediction of biodegradability, metabolite distribution and toxicity of stable metabolites, *SAR QSAR Environ. Res.*, 13, 445–455, 2002.

Drossman, H., Johnson, H., and Mill, T., Structure activity relationships for environmental processes 1: hydrolysis of esters and carbamates, *Chemosphere*, 17, 1509–1530, 1988.

Eakin, D.R., Hyde, E., and Palmer, G., The use of computers with chemical structural information: the ICI CROSSBOW system, *Pest. Sci.*, 5, 319–326, 1974.

European Commission, Technical Guidance Document in Support of Commission Directive 93/67/EEC on Risk Assessment for New Notified Substances and Commission Regulation (EC) No 1488/94 on Risk Assessment for Existing Substances: Part III, Brussels, Belgium, 1996.

Govers, H.A.J., Parsons, J.R., Krop, H.B., and Cheung, C.L., Thermodynamic descriptors for (bio-)degradation, *Proc. Workshop Quant. Struct. Act. Relat. Biodegradation* (September 1994, Belgirate, Italy), Report No. 719101021, National Institute of Public Health and Environmental Protection, Bilthoven, The Netherlands, 1995.

Güsten, H., Klasinc, L., and Maric, D., Prediction of the abiotic degradability of organic compounds in the troposphere, *J. Atmos. Chem.*, 2, 83–93, 1984.

Güsten, H., Medven, Z., Sekusak, S., and Sabljić, A., Predicting tropospheric degradation of chemicals: from estimation to computation, *SAR QSAR Environ. Res.*, 4, 197–209, 1995.

Hansch, C., Physico-chemical parameters in drug design, *Ann. Rep. Med. Chem.*, 348–357, 1968a.

Hansch, C., The use of substituent constants in drug modification, *Il Farmaco Edition Science*, 23, 293–320, 1968b.

Hansch, C., Quantitative structure-activity relationships and the unnamed science, *Acc. Chem. Res.*, 26, 147–153, 1993.

Harris, J.C., Rate of hydrolysis, in *Handbook of Chemical Property Estimation Methods*, Lyman, W.J., Reehl, W.F., and Rosenblatt, D.H., Eds., American Chemical Society, Washington, D.C., 1990, pp. 7-1-7-48.

Howard, P.H., Boethling, R.S., Stiteler, W.M., Meylan, W.M., Hueber, A.E., Beauman, J.A., and Larosche, M.E., Predictive model for aerobic biodegradability from a file of evaluated biodegradation data, *Environ. Toxicol. Chem.*, 11, 593–603, 1992.

Howard, P.H., Hueber, A.E., and Boethling, R.S., Biodegradation data evaluation for structure/biodegradability relations, *Environ. Toxicol. Chem.*, 6, 1–10, 1987.

Howard, P.H. and Meylan, W.M., Prediction of physical properties, transport, and degradation for environmental fate and exposure assessments, in *Quantitative Structure-Activity Relationships in Environmental Sciences – VII*, Chen, F. and Schüürmann, G., Eds., SETAC Press, Pensacola, FL:, 1997, pp. 185–205.

Jaworska, J.S., Boethling, R.S., and Howard, P.H., Recent developments in broadly applicable structure-biodegradability relationships, *Environ. Toxicol. Chem.*, 22, 1710–1723, 2003.

Jaworska, J.S., Dimitrov, S., Nikolova, N., and Mekenyan, O., Probabilistic assessment of biodegradability based on metabolic pathways: CATABOL system, *SAR QSAR Environ. Res.*, 13, 307–323, 2002.

Johnson, H., Kenley, R.A., Rynard, C., and Golub, M.A,. QSAR for cholinesterase inhibition by organophosphorus esters and CNDO/2 calculations for organophosphorus ester hydrolysis, *Quant. Struct.-Act. Relat.*, 4, 172–180, 1985.

Klamt, A., Estimation of gas-phase hydroxyl radical rate constants of organic compounds from molecular orbital calculations, *Chemosphere*, 26, 1273–1289, 1993.

Klopman, G., and Tu, M., Structure-biodegradability study and computer-automated prediction of aerobic biodegradation of chemicals, *Environ. Toxicol. Chem.*, 16, 1829–1835, 1997.

Klopman, G., Tu, M., and Talafous, J., META 3: A genetic algorithm for metabolic transform priorities optimization, *J. Chem. Inf. Comput. Sci.*, 37, 329–334, 1997.

Langenberg, J.H., Peijnenburg, W.J.G.M. and Rorije, E. (1996) On the usefulness and reliability of existing QSBRs for risk assessment and priority setting, *SAR QSAR Environ.Res.*, 5, 1–16, 1996.

Loonen, H., Lindgren, F., Hansen, B., Karcher, W., Niemela, J., Hiromatsu, K., Takatsuki, M., Peijnenburg, W., Rorije, E., and Struijs, J., Prediction of biodegradability from chemical structure: modeling of ready biodegradation test data, *Environ. Toxicol. Chem.*, 18, 1763–1768, 1999.

Masunaga, S., Wolfe, N.L., and Carreira, K., Transformation of para-substituted benzonitriles in sediment and in sediment extract, *Water Sci. Tech.*, 28, 123–132, 1993.

Medven, Z., Güsten, H. and Sabljić, A., Comparative QSAR study on hydroxyl radical reactivity with unsaturated hydrocarbons: PLS vs. MLR, *J. Chemometrics*, 10, 135–147, 1996.

Mill, T., Haag, W., Penwell, P., Pettit, T., and Johnson, H., Environmental fate and exposure studies development of a PC-SAR for hydrolysis: esters, alkyl halides and epoxides, EPA Contract No. 68-02-4254, SRI International, Menlo Park, CA, 1987.

Müller, M. and Klein, W., Estimating atmospheric degradation processes by SARs, *Sci. Total Environ.*, 109/110, 261–273, 1991.

Nendza, M., *Structure-Activity Relationships in Environmental Sciences*, Chapman and Hall, London, 1998.

Niemi, G.J., Veith, G.D., Regal, R.R., and Vaishnav, D.D., Structural features associated with degradable and persistent chemicals, *Environ. Toxicol. Chem.*, 6, 515–527, 1987.

Organization for Economic Co-operation and Development, Guideline for Testing of Chemicals: Degradation, Paris, France, 1984.

Organization for Economic Co-operation and Development, Guideline for Testing of Chemicals: Degradation, Paris, France, 1989.

Organization for Economic Co-operation and Development, Application of Structure-Activity Relationships to the Estimation of Properties Important in Exposure Assessment, Environment Monograph No. 67, Paris, France, 1993a.

Organization for Economic Co-operation and Development, Structure-Activity Relationships for Biodegrada-
    tion, Environment Monograph No. 68, Paris, France, Organization for Economic Cooperation and
    Development.
Parsons, J.R. and Govers, H.A.J., Quantitative-structure activity relationships (QSARs) for biodegradation,
    *Ecotoxicology Environ. Saf.*, 19, 212–227, 1990.
Pedersen, F., Kristensen, P., Damborg, A., and Christensen, H.W., *Ecotoxicological Evaluation of Industrial
    Wastewater*, Ministry of the Environment, Copenhagen, Denmark, 1994.
Peijnenburg, W., Structure-activity relationships for biodegradation: a critical review, *Pure Appl. Chem.*, 66,
    1931–1941, 1994.
Peijnenburg, W.J.G.M., The use of quantitative structure-activity relationships for predicting rates of environ-
    mental hydrolysis processes, *Pure Appl. Chem.*, 63, 1667–1676, 1991.
Peijnenburg, W.J.G.M., de Beer, K.G.M. den Hollander, H.A., Stegeman, M.H.L., and Verboom, H., Kinetics,
    products, mechanisms and QSARs for the hydrolytic transformation of aromatic nitriles in anaerobic
    sediment slurries, *Environ. Toxicol. Chem.*, 12, 1149–1161, 1993.
Peijnenburg, W.J.G.M. and Karcher, W., *Proc. Workshop Quant. Struct. Act. Relat. Biodegradation* (September
    1994, Belgirate, Italy), Report No. 719101021, National Institute of Public Health and Environmental
    Protection, Bilthoven, The Netherlands, 1995.
Peijnenburg, W.J.G.M., 't Hart, M.J., den Hollander, H.A., van de Meent, D., Verboom, H., and Wolfe, N.L.,
    QSARs for predicting biotic and abiotic reductive transformation rate constants of halogenated
    hydrocarbons in anoxic sediment systems, *Sci. Total Environ.*, 109/110, 283–300, 1991.
Punch, B., Patton, A., Wight, K., Larson, R.J., Masscheleyn, P., and Forney, L., A biodegradability evaluation
    and simulation system (BESS) based on knowledge of biodegradation pathways, in *Biodegradability
    Prediction*, Peijnenburg, W.J.G.M. and J. Damborsky, J., Eds., Kluwer, Dordrecht, The Netherlands,
    1996, pp 65–73.
Raymond, J.W., Rogers, T.N., Shonnard, D.R., and Kline, A.A., A review of structure-based biodegradation
    estimation methods, *J. Hazardous Mater.*, B84, 189–215, 2001.
Rorije, E., Langenberg, J.H., Richter, J., and Peijnenburg, W.J.G.M., Modeling reductive dehalogenation with
    quantum-chemically derived descriptors, *SAR QSAR Environ. Res.*, 4, 237–252, 1995.
Rorije, E., Loonen, H., Müller, M., Klopman, G., and Peijnenburg, W.J.G.M., Evaluation and application of
    models for the prediction of ready biodegradability in the MITI-I test, *Chemosphere*, 38, 1409–1417,
    1999.
Rorije, E., Müller, M., and Peijnenburg, W.J.G.M., *Prediction of Environmental Degradation Rates for High
    Production Volume Chemicals (HPVC) using Quantitative Structure-Activity Relationships*, Report
    No. 719101030, National Institute of Public Health and Environmental Protection, Bilthoven, The
    Netherlands, 1997.
Rorije, E., Peijnenburg, W.J.G.M., and Klopman, G., Structural requirements for anaerobic biodegradation of
    organic chemicals: a fragment model analysis, *Environ. Toxicol. Chem.*, 17, 1943–1950, 1998.
Sabljić, A. and Güsten, H., Predicting the night-time $NO_3$ radical reactivity in the troposphere, *Atmos. Environ.*,
    24A, 73–78, 1990.
Scherer, M.M., Balko, B.A., Gallagher, D.A., and Tratnyek, P.G., Correlation analysis of rate constants for
    dechlorination by zero-valent iron, *Environ. Sci. Tech.*, 32, 3026–3033, 1998.
Schwarzenbach, R.P., Gschwend P.M., and Imboden, D.M., *Environmental Organic Chemistry*, John Wiley
    and Sons Inc, New York, 1993.
Scow, K.M., Rate of biodegradation, in *Handbook of Chemical Property Estimation Methods*, Lyman, W.J.,
    Reehl, W.F., and Rosenblatt, D.H., Eds., American Chemical Society, Washington, D.C., 1990,
    pp. 9-1–9-85.
Tratnyek, P.G., Personal communication, 2002.
Tratnyek, P.G. and Macalady, D.L. Oxidation-reduction reactions in the aquatic environment, in *Handbook of
    Property Estimation Methods for Chemicals*, Boethling, R.S. and Mackay, D., Eds., CRC Press, Boca
    Raton, FL, 2000, pp. 383–415.
Tratnyek, P.G., Weber, E.J., and Schwarzenbach, R.P., Quantitative structure-activity relationships for chemical
    reductions of organic contaminants, *Environ. Toxicol. Chem.*, 22, 1733–1742, 2003.
Tunkel, J., Howard, P.H., Boethling, R.S., Stiteler, W., and Loonen, H., Predicting ready biodegradability in
    the Japanese Ministry of International Trade and Industry test, *Environ. Toxicol. Chem.*, 19, 2478–2485,
    2000.

Vogel, T.M. and Reinhard, M., Reaction products and rates of disappearance of simple bromoalkanes, 1,2-dibromopropane, and 1,2-dibromoethane in water, *Environ. Sci. Tech.*, 20, 992–997, 1986.

Weber, E.J. and Wolfe, N.L., Kinetic studies of the reduction of aromatic azo compounds in anaerobic sediment/water systems, *Environ. Toxicol. Chem.*, 6, 911–919, 1987.

Wolfe, N.L., Steen, W.C., and Burns, L.A., Phthalate ester hydrolysis: linear free energy relationships, *Chemosphere*, 9, 403–408, 1980.

Wolfe, N.L., Zepp, R.G., and Paris, D.F., Use of structure-reactivity relationships to estimate hydrolytic persistence of carbamate pesticides, *Water Res.*, 12, 561–563, 1978.

# CHAPTER 15

# QSAR Modeling of Bioaccumulation

John C. Dearden

## CONTENTS

## I. INTRODUCTION

If we are to quantify the risks that environmental pollution poses, we need an understanding of the many factors that affect the distribution of a chemical in the environment. The fugacity model of Mackay et al. (1992) incorporates these factors, one of which is bioaccumulation. Connell (1990) has written a comprehensive review of the bioaccumulation of xenobiotic compounds.

Nendza (1998) has defined bioaccumulation as uptake by an organism of a chemical from the environment via any possible pathway, and this can be subdivided into biomagnification (uptake via the food chain) and bioconcentration (uptake from the surrounding milieu). As we shall see, it is the latter that has been the subject of by far the greater number of quantitative structure-activity relationship (QSAR) studies of bioaccumulation. The bioconcentration factor (BCF) is defined as:

$$BCF = \frac{\text{concentration of chemical in organism at steady state}}{\text{concentration of chemical in surrounding milieu}} \qquad (15.1)$$

In similar fashion, the biomagnification factor (BMF) is defined as:

$$\text{BMF} = \frac{\text{concentration of chemical in organism at steady state}}{\text{concentration of chemical in organism's diet}} \tag{15.2}$$

In both cases, the concentration in the organism (or diet) is expressed in units of mass per kilogram of organism (or diet), and the concentration in water is expressed in units of mass per unit volume. The weight of the organism (or diet) can be expressed on a wet weight, dry weight, or lipid weight basis.

Bioaccumulation is a partition phenomenon, as the above equations indicate. A common surrogate for partitioning of a chemical between an organism and water is its octanol-water partition coefficient, $K_{ow}$ (sometimes written simply as P), and it will be seen later that this property (in the form of its common logarithm, log $K_{ow}$) in many cases correlates well with bioaccumulation.

As will be shown, the large amount of QSAR modeling that has been carried out for bioaccumulation has almost exclusively involved non-ionic organic compounds. For strongly ionizing and inorganic chemicals, no QSARs are available (Macdonald et al., 2002), although Meylan et al. (1999) have proposed QSARs for some ionic compounds, namely carboxylic and sulfonic acids and salts, and some nitrogen compounds; Saarikoski and Viluksela (1982) have developed a bioconcentration QSAR incorporating a correction for ionization.

## II. BIOCONCENTRATION

### A. Methodology

For regulatory purposes, bioconcentration factors are often used to assess the uptake of both new and existing chemicals by organisms; Japan is probably the only country to require experimental determination of bioconcentration by the basic trophic level. Nevertheless, it is sometimes necessary to measure BCFs, and so a brief resumé is given here.

It is important to recognize that the concentration in the aqueous phase must be that in free solution (i.e., not including that sorbed on to organic matter in the water or on to the surface of the test vessel). In general, for chemicals that are not highly hydrophobic (log $K_{ow} < 5$), the total aqueous concentration can be taken as equal to the freely dissolved concentration. However, for very hydrophobic chemicals this may not be the case (Voutsas et al., 2002).

In Europe, the U.S., and Canada, a flowthrough method (European Centre for Ecotoxicology and Toxicology of Chemicals, 1996; U.S. Environmental Protection Agency, 1996) is used in which two groups of organisms of the species under investigation are exposed to water and a constant concentration of the test chemical, respectively, until steady state is achieved or for at least 28 to 30 days; this is followed by an elimination phase in which they are exposed to water only for a period of about twice the uptake period. During the tests, organisms and water are removed in geometric time series and analyzed. From these data the uptake and elimination rate constants are calculated, and the ratio of the two gives the BCF.

An Organization for Economic Co-operation and Development (OECD) monograph (OECD, 1981) describes static, semistatic, and flow-through methods; Gobas and Zhang (1995) have developed a method suitable for very hydrophobic chemicals.

It is important to note that there can be great variability in measured BCF values for a given chemical. For example, Nendza (1998) reported that published BCF values for pentachlorobenzene ranged from 900 to 250,000. She cited a number of factors that can contribute to such variability, including the test species; the size, age, and sex of the test species; purity of the test chemical; whether or not steady state is reached during the test; analytical method used; stability of the test chemical in water; presence of surfactants; pH and buffer capacity; water chemistry (e.g., hardness), co-solute effects; and the presence of suspended organic matter.

In general, experimental measurement of bioconcentration factors is time-consuming and expensive. To measure BCF values for the very large number of chemicals that are released into the environment, and that are of potential regulatory concern, is impracticable. Attention is turning more and more to the prediction of BCF values, and the technique of QSAR lends itself admirably to this.

## B. QSAR Models

To a first approximation, bioconcentration can be considered as the partitioning of a chemical between an organism and the surrounding aqueous milieu. It is therefore not surprising that many studies have been made of the relationship between BCF and the partition coefficient. Some such studies have involved specific chemical classes such as chlorinated polycyclic hydrocarbons (Schüürmann and Klein, 1988) and anilines (Zok et al., 1991), but a good number have involved diverse chemicals. There is considerable divergence in the correlations reported, which probably reflects differences in test conditions as mentioned (*vide ultra*) by Nendza (1998). Nendza (1991) cited six QSARs of the form:

$$\log \text{BCF} = a \log \text{K}_{ow} + b \tag{15.3}$$

in which the gradients ranged from 0.54 to 1.02 and the intercepts ranged from −1.95 to +0.61. Another factor that could be responsible for this wide variation is the range of log $\text{K}_{ow}$ values of the chemicals used in the studies. There is general agreement (Könemann and van Leeuwen, 1980; Connell and Hawker, 1988; Bintein et al., 1993; Nendza, 1991) that the rectilinear correlation between log BCF and log $\text{K}_{ow}$ fails for very hydrophobic chemicals (log $\text{K}_{ow} > 6$–7), and that bioconcentration is lower than expected. There are several possible reasons for this (Nendza 1998):

1. The aqueous solubility of such highly hydrophobic chemicals is too low.
2. The test period is too short for a steady state to be attained.
3. There is hindered membrane passage of large molecules.
4. Metabolism occurs (Hermens, 1995; Sabljić, 2001).
5. Degradation occurs.
6. There is inaccuracy in the determination of $\text{K}_{ow}$ and BCF for very hydrophobic chemicals (due in part to the total solute in the aqueous phase being greater than the truly dissolved concentration [Gobas et al., 1989]).
7. The octanol-water solvent pair fails to act as a good surrogate for lipid-water for very hydrophobic chemicals.
8. Specific substructural effects can occur, such as have been reported for 2,4-dinitrophenols (Deneer et al., 1987).

The effect of molecular size has recently been examined by Dimitrov et al. (2002, 2003), who showed that the maximal cross-sectional diameter of highly hydrophobic molecules had a significant negative effect on bioconcentration.

In connection with the test period being too short for steady state to be achieved, Hawker and Connell (1985) devised a QSAR that incorporated the test period ($t$, in hours):

$$\log \text{BCF} = 0.337 \log \text{K}_{ow} + \log (0.424\ t) \tag{15.4}$$

Because of these deviations from rectilinearity, Meylan et al. (1999) have proposed, from an analysis of 694 compounds, different rules for different log $\text{K}_{ow}$ ranges. For non-ionic compounds, they recommend the following relationships (compounds with very low and very high log $\text{K}_{ow}$ are assumed to have constant BCF):

$$\log K_{ow} < 1, \log BCF = 0.50 \tag{15.5}$$

$$\log K_{ow} \ 1\text{--}7, \log BCF = 0.77 \log K_{ow} - 0.70 + \Sigma F_i \tag{15.6}$$

$$\log K_{ow} > 7, \log BCF = -1.37 \log K_{ow} + 14.4 + \Sigma F_i \tag{15.7}$$

$$\log K_{ow} > 10.5, \log BCF = 0.50 \tag{15.8}$$

where $\Sigma F_i$ = a series of 12 correction factors relating to structural features. The proposed rules for ionic compounds (carboxylic and sulfonic acids and salts, and some nitrogen compounds) are different:

$$\log K_{ow} < 5, \log BCF = 0.5 \tag{15.9}$$

$$\log K_{ow} \ 5\text{--}6, \log BCF = 0.75 \tag{15.10}$$

$$\log K_{ow} \ 6\text{--}7, \log BCF = 1.75 \tag{15.11}$$

$$\log K_{ow} \ 7\text{--}9, \log BCF = 1.00 \tag{15.12}$$

$$\log K_{ow} > 9, \log BCF = 0.5 \tag{15.13}$$

The model is reported to give good predictions, but it should be noted that uncertainty about a $K_{ow}$ value (e.g., 6.9 or 7.1) can lead to different predictions.

Some other published log BCF-log $K_{ow}$ QSARs are listed in Table 15.1. For compounds with log $K_{ow}$ values below about 7, rectilinear log BCF-log $K_{ow}$ correlations are satisfactory (Connell, 1995; Devillers et al., 1996; Meylan et al., 1999), even for diverse compounds, despite the assertion of Streit (1992) that "thermodynamic considerations indicate that a unique log BCF/log $K_{ow}$ relationship cannot exist, since enthalpy changes vary for different classes of chemical substances."

For chemicals that are weakly ionic, Saarikoski and Viluksela (1982) showed that a correction for ionization gave a greatly improved correlation between log BCF and log $K_{ow}$.

Those QSARs shown in Table 15.1 with a $(\log K_{ow})^2$ term and that of the bilinear form (Kubinyi, 1976) attempt to allow for the observed levelling off or decrease of BCF values above log $K_{ow} \sim 7$, as do the series of QSARs developed by Meylan et al. (1999) (*vide ultra*). The QSAR developed by Dimitrov et al. (2002) gives a Gaussian-type correlation to account for log BCF approximating to 0.5 at low and high log $K_{ow}$ values.

It is clear from the above that while $K_{ow}$ is in general a reasonable surrogate for BCF, it is not ideal. This is further illustrated by a 100-chemical test-set (Table 15.2) of log BCF values, selected by taking approximately every third compound from the compilation of Lu et al. (2000); Figure 15.1 shows the correlation with log $K_{ow}$. The corresponding QSAR is:

$$\log BCF = 0.526 \log K_{ow} + 0.532 \tag{15.14}$$

$$n = 100, \ r^2 = 0.645, \ s = 0.820, \ F = 178.4$$

Many of the chemicals listed in Table 15.2 are chlorine-containing; this simply reflects the fact that many, perhaps the majority, of chemicals of environmental concern are chlorinated compounds.

Figure 15.1 shows the expected fall-off in log BCF values of many (but not all) chemicals at log $K_{ow}$ values > *ca.* 6, and several possible reasons for this have been discussed above. More unusually, three hydrophilic compounds (acrolein, acrylonitrile, and hydroquinone) have BCF

**Table 15.1    Some Bioconcentration QSARs Based on Log $K_{ow}$**

| QSAR | Organism | Chemical Class | n | $R^2$ | Reference |
|---|---|---|---|---|---|
| log BCF = 0.85 log $K_{ow}$ − 0.70 | Fish | Diverse | 59 | 0.90 | Lassiter (1975) |
| log BCF = 0.79 log $K_{ow}$ − 0.40 | Fish | Diverse | 122 | 0.865 | Veith and Kosian (1983) |
| log BCF = 1.00 log $K_{ow}$ − 1.32 | Fish | Diverse | 44 | 0.941 | Mackay (1982) |
| log BCF = 0.76 log $K_{ow}$ − 0.31 | Fish | Diverse | 38 | 0.593 | Schüürmann and Klein (1988) |
| log BCF = 0.90 log $K_{ow}$ − 0.80 | Fish | Diverse non-polar | 80 | 0.891 | Lu et al. (1999) |
| log BCF = 0.74 log $K_{ow}$ + 0.80 | Fish | Diverse | 66 | 0.840 | Escuder-Gilabert et al. (2001) |
| log BCF = 0.54 log $K_{ow}$ + 0.12 | Fish | Halogenated aromatics | 8 | 0.90 | Neely et al. (1974) |
| log BCF = 0.67 log $K_{ow}$ − 0.18 | Fish | Anilines | 9 | 0.872 | Zok et al. (1991) |
| log BCF = 0.91 log $K_{ow}$ − 1.975 log(6.8 x $10^{-7}$ $K_{ow}$ + 1) − 0.77 | Fish | Diverse | 154 | 0.90 | Bintein et al. (1993) |
| log BCF = 0.42 + 3.321 exp{− (log $K_{ow}$ − log $K_{ow0}$)$^2$/10.15}[a] | Fish | Diverse narcotic | 443 | 0.730 | Dimitrov et al. (2002) |
| log BCF = 3.41 log $K_{ow}$ − 0.26 (log $K_{ow}$)$^2$ − 5.51 | Fish | Chlorobenzenes | 6 | NG[b] | Könemann and van Leeuwen (1980) |
| log BCF = 0.90 log $K_{ow}$ − 1.32 | Daphnid | Diverse | 22 | 0.922 | Hawker and Connell (1986) |
| log BCF = 0.85 log $K_{ow}$ − 1.10 | Daphnid | Diverse | 52 | 0.913 | Geyer et al. (1991) |
| log BCF = 0.681 log $K_{ow}$ + 0.164 | Alga | Diverse | 41 | 0.814 | Geyer et al. (1984) |
| log BCF = 0.70 log $K_{ow}$ − 0.26 | Alga | Pesticides | 8 | 0.93 | Ellgehausen et al. (1980) |
| log BCF = 0.86 log $K_{ow}$ − 0.81 | Mussel | Diverse | 16 | 0.912 | Geyer et al. (1982) |

[a] log $K_{ow0}$ = optimal log $K_{ow}$ value (in this case 6.35).
[b] Not given.

values much greater than expected. The first two possess α,β-unsaturation and hydroquinone readily generates a free radical, so all three would be expected to be highly reactive. On the assumption that covalently bound forms of these chemicals would be determined along with the free chemical in the organism, the apparent BCF values would be higher than expected. In addition, the instability of two of these compounds (acrylonitrile and hydroquinone) in aqueous solution could lead to their aqueous concentrations being low, which would also increase their apparent BCF values.

If chemicals with log $K_{ow}$ > 6 are removed from the test set, together with 4-nonylphenol (log $K_{ow}$ = 5.76, log BCF = 2.45) and the 3 reactive chemicals mentioned above, there is a good correlation of log BCF with log $K_{ow}$, as shown in Figure 15.2:

$$\log BCF = 0.807 \log K_{ow} - 0.477 \tag{15.15}$$

$$n = 68, \ r^2 = 0.823, \ s = 0.510, \ F = 307.5$$

This correlation is very similar to that of Veith and Kosian (1983) given in Table 15.1.

It is interesting to speculate as to the reasons that some chemicals with log $K_{ow}$ > 6 (and also including 4-nonylphenol) have low measured BCF values, while others do not (see Figure 15.1). An inspection of Table 15.2 and Figure 15.1 shows that, by and large, the chemicals with low BCF values possess hydrogen bonding ability (10 out of 14 chemicals), whereas those whose BCF values appear from Figure 15.1 not to be low do not possess hydrogen bonding ability (13 out of 15 chemicals). It is known (Abraham et al., 1995) that hydrogen bonding reduces the rate of penetration of chemicals through the membranes of organisms. It is postulated that the apparently low BCF values of some highly hydrophobic chemicals is due to steady state's not being reached during the time period of the test, because of low penetration rates caused by hydrogen bonding. Another reason may be that compounds containing polar groups are more susceptible to metabolic attack.

Table 15.2 Measured Log BCF Values and Log BCF Values Predicted by Various Methods for 100-Compound Test Set

| Compound Name | CAS No. | Measured log BCF | Measured log $K_{ow}$ | BCFWIN ver. 2.14 | Veith and Kosian (1983) | Bintein et al. (1993) | Park and Lee (1993) | Dimitrov et al. (2002) | Dearden (this work) | Sabljić (2001) | Koch (1983) |
|---|---|---|---|---|---|---|---|---|---|---|---|
| Acrolein | 107028 | 2.54 | -0.01 | 0.50 | -0.41 | -0.80 | -1.18 | 0.48 | 0.53 | -1.54 | 0.92 |
| Acrylonitrile | 107131 | 1.68 | 0.25 | 0.50 | -0.20 | -0.56 | -0.28 | 0.50 | 0.66 | -1.62 | 0.87 |
| Aniline | 62533 | 0.41 | 0.90 | 0.50 | 0.31 | 0.03 | 0.24 | 0.60 | 1.01 | 0.52 | 1.88 |
| Anthracene | 120127 | 2.83 | 4.54 | 2.73 | 3.19 | 3.33 | 2.72 | 2.82 | 2.92 | 3.59 | 3.94 |
| Benzene | 71432 | 0.64 | 2.13 | 0.94 | 1.28 | 1.15 | 0.87 | 0.99 | 1.65 | 0.05 | 1.73 |
| Benzo[a]anthracene | 56553 | 4.00 | 5.76 | 3.74 | 4.15 | 4.17 | 3.69 | 3.63 | 3.56 | 4.61 | 5.05 |
| Benzo[a]pyrene | 50328 | 3.42 | 6.13 | 4.02 | 4.44 | 4.23 | 3.81 | 3.73 | 3.76 | 5.02 | 5.65 |
| α-BHC | 319846 | 2.95 | 3.72 | 2.49 | 2.54 | 2.60 | 3.80 | 2.10 | 2.49 | 5.11 | 4.82 |
| Biphenyl | 92524 | 2.64 | 3.98 | 2.36 | 2.74 | 2.83 | 2.55 | 2.33 | 2.63 | 2.60 | 3.36 |
| BPMC | 3766812 | 1.41 | 2.78 | 1.44 | 1.80 | 1.74 | 1.99 | 1.37 | 1.99 | 3.52 | 4.23 |
| Bromobenzene | 108861 | 1.70 | 2.99 | 1.60 | 1.96 | 1.93 | 1.48 | 1.51 | 2.10 | 1.85 | 2.43 |
| 4-Bromophenol | 106412 | 1.56 | 2.59 | 1.29 | 1.65 | 1.57 | 1.49 | 1.24 | 1.89 | 2.12 | 2.54 |
| Bromoxynil | 1689845 | 1.67 | 3.25[a] | 1.51 | 2.17 | 2.17 | 2.89 | 1.71 | 2.24 | 3.65 | 3.55 |
| 2-tert-Butoxyethanol | 7580850 | -0.22 | 0.39 | 0.50 | -0.09 | -0.43 | -0.15 | 0.52 | 0.74 | 2.54 | 2.36 |
| tert-Butyl isopropyl ether | 17348593 | 0.76 | 2.14 | 0.95 | 1.29 | 1.16 | -2.03 | 1.00 | 1.66 | 3.29 | 2.59 |
| 4-tert-Butylphenol | 98544 | 1.86 | 3.31 | 1.85 | 2.21 | 2.22 | 2.38 | 1.76 | 2.27 | 3.85 | 3.14 |
| Chlordane | 57749 | 4.58 | 6.16 | 4.09 | 4.47 | 4.23 | 5.49 | 3.73 | 3.77 | 4.15 | 6.74 |
| 3-Chloroaniline | 108429 | 0.34 | 1.88 | 0.75 | 1.09 | 0.92 | 0.87 | 0.88 | 1.52 | 1.51 | 2.26 |
| Chlorobenzene | 108907 | 1.85 | 2.84 | 1.49 | 1.84 | 1.80 | 1.36 | 1.41 | 2.03 | 1.08 | 2.10 |
| 4-Chlorobiphenyl | 2051629 | 2.69 | 4.61 | 2.85 | 3.24 | 3.39 | 3.07 | 2.88 | 2.96 | 3.32 | 3.74 |
| Chloroform | 67663 | 0.78 | 1.97 | 0.82 | 1.16 | 1.01 | 1.22 | 0.92 | 1.57 | 1.87 | 1.70 |
| 2-Chloronitrobenzene | 88733 | 2.10 | 2.24 | 1.02 | 1.37 | 1.25 | 1.35 | 1.05 | 1.71 | 1.70 | 2.50 |
| 2-Chlorophenol | 95578 | 2.33 | 2.15 | 0.96 | 1.30 | 1.17 | 1.18 | 1.00 | 1.66 | 1.29 | 2.21 |
| Chloropyrifos | 2921882 | 3.18 | 4.96 | 3.12 | 3.52 | 3.68 | 1.93 | 3.17 | 3.14 | 5.27 | 7.02 |
| p,p'-DDE | 72559 | 4.71 | 6.51 | 4.31 | 4.74 | 4.14 | 5.15 | 3.73 | 3.96 | 5.03 | 5.40 |
| p,p'-DDT | 50293 | 4.84 | 6.91 | 4.62 | 5.06 | 3.89 | 5.81 | 3.64 | 4.17 | 5.22 | 5.94 |
| Diazinon | 333415 | 1.80 | 3.81 | 2.23 | 2.61 | 2.68 | 0.92 | 2.18 | 2.54 | 5.25 | 7.17 |
| 4,4'-Dibromobiphenyl | 92864 | 4.19 | 5.72 | 4.32 | 4.12 | 4.16 | 3.80 | 3.61 | 3.54 | 4.70 | 4.77 |
| 3,4-Dichloroaniline | 95761 | 1.48 | 2.69 | 1.37 | 1.73 | 1.66 | 1.42 | 1.31 | 1.95 | 2.26 | 2.64 |
| 2,5-Dichlorobiphenyl | 34883391 | 4.20 | 5.10 | 3.85 | 3.63 | 3.78 | 3.56 | 3.27 | 3.21 | 3.88 | 4.12 |
| 2,3-Dichloronaphthalene | 2050751 | 4.04 | 4.42 | 3.32 | 3.09 | 3.22 | 2.72 | 2.72 | 2.86 | 3.46 | 3.59 |
| 2,3-Dichloronitrobenzene | 3209221 | 2.16 | 3.05 | 1.65 | 2.01 | 1.99 | 1.96 | 1.56 | 2.14 | 2.44 | 2.88 |
| 3,4-Dichloronitrobenzene | 99547 | 2.07 | 3.12 | 1.70 | 2.06 | 2.05 | 1.96 | 1.61 | 2.17 | 2.52 | 2.88 |
| Dieldrin | 60571 | 3.71 | 5.40 | 3.30 | 3.87 | 3.99 | 3.66 | 3.46 | 3.37 | 3.88 | 6.87 |

| | | | | | | | | | | | |
|---|---|---|---|---|---|---|---|---|---|---|---|
| Di(2-ethylhexyl) phthalate | 117817 | 2.34 | 7.45 | 2.49 | 5.49 | 3.42 | 6.65 | 3.37 | 4.45 | 5.16 | 8.83 |
| Dimethyl phthalate | 131113 | 1.76 | 1.60 | 0.53 | 0.86 | 0.67 | 0.33 | 0.78 | 1.37 | 2.32 | 3.27 |
| Diphenylamine | 122394 | 1.48 | 3.50 | 2.00 | 2.37 | 2.40 | 1.48 | 1.91 | 2.37 | 2.77 | 3.56 |
| 4-Dodecylphenol | 104438 | 3.78 | 7.91 | 2.06 | 5.85 | 2.96 | 6.22 | 3.03 | 4.69 | 5.08 | 6.54 |
| Fenvalerate | 51630581 | 2.79 | 6.20 | 4.07 | 4.50 | 4.23 | 4.52 | 3.73 | 3.79 | 5.14 | 8.07 |
| 2,2',3,4,4',5',6-Heptachlorobiphenyl | 52663691 | 5.84 | 7.92[a] | 3.69 | 5.86 | 2.94 | 6.38 | 3.02 | 4.70 | 5.27 | 6.03 |
| 1,2,3,4,6,7,8-Heptachlorodibenzofuran | 67562394 | 3.62 | 7.92 | 3.55 | 5.86 | 2.94 | 5.86 | 3.02 | 4.70 | 5.28 | 6.23 |
| 2,2',4,4',6,6'-Hexabromobiphenyl | 59261084 | 3.96 | 8.09[a] | 2.56 | 5.99 | 2.77 | 7.01 | 2.88 | 4.79 | 4.58 | 7.60 |
| Hexachlorobenzene | 118741 | 4.26 | 5.73 | 3.71 | 4.13 | 4.16 | 4.48 | 3.62 | 3.55 | 4.18 | 4.01 |
| 2,2',3,3',4,4'-Hexachlorobiphenyl | 38380073 | 5.77 | 7.31 | 5.01 | 5.37 | 3.55 | 5.58 | 3.45 | 4.38 | 5.14 | 5.65 |
| 2,2',3,4,5,5'-Hexachlorobiphenyl | 52712046 | 5.81 | 7.19 | 5.17 | 5.28 | 3.66 | 5.67 | 3.52 | 4.31 | 5.16 | 5.64 |
| 2,2',4,4',6,6'-Hexachlorobiphenyl | 33979032 | 4.93 | 7.55 | 4.68 | 5.56 | 3.32 | 6.06 | 3.30 | 4.50 | 5.20 | 5.64 |
| 1,2,3,4,7,8-Hexachlorodibenzo-p-dioxin | 39227286 | 3.54 | 7.80 | 3.71 | 5.76 | 3.07 | 4.52 | 3.12 | 4.63 | 5.22 | 5.97 |
| Hydroquinone | 123319 | 1.60 | 0.59 | 0.50 | 0.07 | -0.25 | 0.59 | 0.55 | 0.84 | 0.71 | 1.94 |
| Isofenphos | 25311711 | 2.17 | 4.12 | 2.47 | 2.85 | 2.96 | 2.85 | 2.45 | 2.70 | 4.81 | 7.98 |
| p,p'-Methoxychlor | 72435 | 3.10 | 5.08 | 3.21 | 3.61 | 3.77 | 4.11 | 3.25 | 3.20 | 5.28 | 6.01 |
| 9-Methylanthracene | 779022 | 3.66 | 5.07 | 3.20 | 3.61 | 3.76 | 3.18 | 3.25 | 3.20 | 3.99 | 4.28 |
| 2-Methylnaphthalene | 91576 | 3.20 | 3.86 | 2.27 | 2.65 | 2.72 | 2.26 | 2.22 | 2.56 | 2.76 | 3.16 |
| Molinate | 2212671 | 1.41 | 3.21 | 1.77 | 2.14 | 2.13 | 0.46 | 1.68 | 2.22 | 4.10 | 4.52 |
| Naphthalene | 91203 | 1.64 | 3.30 | 1.77 | 2.21 | 2.22 | 1.83 | 1.75 | 2.27 | 4.10 | 4.52 |
| 3-Nitroaniline | 99092 | 0.92 | 1.37 | 0.35 | 0.68 | 0.46 | 0.46 | 0.71 | 1.25 | 1.28 | 2.28 |
| 2-Nitrophenol | 88755 | 1.60 | 1.79 | 0.68 | 1.01 | 0.84 | 0.54 | 0.85 | 1.47 | 1.10 | 2.23 |
| 3-Nitrophenol | 554847 | 1.40 | 2.00 | 0.84 | 1.18 | 1.03 | 1.03 | 0.93 | 1.58 | 1.15 | 2.22 |
| 4-Nonylphenol | 104405 | 2.45 | 5.76 | 2.74 | 4.15 | 4.17 | 4.80 | 3.63 | 3.56 | 4.49 | 5.36 |
| 2,2',3,3',4,5,5',6-Octachlorobiphenyl | 68194172 | 5.88 | 8.22[a] | 2.81 | 6.09 | 2.63 | 6.95 | 2.77 | 4.86 | 5.28 | 6.41 |
| Octachloronaphthalene | 2234131 | 3.44 | 8.24 | 3.73 | 6.11 | 2.61 | 6.37 | 2.76 | 4.87 | 5.25 | 5.90 |
| Octachlorostyrene | 29082744 | 4.52 | 7.40[a] | 4.18 | 5.45 | 3.46 | 5.85 | 3.40 | 4.42 | 5.12 | 5.30 |
| Pentachlorobenzene | 608935 | 3.86 | 5.17 | 3.28 | 3.68 | 3.84 | 3.73 | 3.32 | 3.25 | 3.78 | 3.63 |
| 2,2',3,4,5'-Pentachlorobiphenyl | 38380028 | 5.38 | 6.85 | 5.19 | 5.01 | 3.94 | 5.05 | 3.66 | 4.14 | 4.97 | 5.26 |
| 2,2',4,4',6-Pentachlorobiphenyl | 39685831 | 3.37 | 6.84[a] | 5.29 | 5.00 | 3.94 | 5.36 | 3.66 | 4.13 | 5.02 | 5.26 |
| 1,2,3,7,8-Pentachlorodibenzo-p-dioxin | 40321764 | 4.50 | 6.64 | 4.41 | 4.85 | 4.07 | 3.82 | 3.71 | 4.02 | 2.06 | 2.83 |
| Permethrin | 52645531 | 3.39 | 6.50 | 2.66 | 4.74 | 4.14 | 4.77 | 3.73 | 3.95 | 5.02 | 7.36 |
| Phenanthrene | 85018 | 3.42 | 4.46 | 2.73 | 3.12 | 3.26 | 2.72 | 2.76 | 2.88 | 3.55 | 3.95 |
| Phenol | 108952 | 1.24 | 1.46 | 0.42 | 0.75 | 0.54 | 0.75 | 0.73 | 1.30 | 0.38 | 1.83 |
| N-Phenyl-2-naphthylamine | 135886 | 2.17 | 4.38 | 2.67 | 3.06 | 3.19 | 2.40 | 2.69 | 2.84 | 4.09 | 4.66 |
| 2-Phenyldodecane | 2719611 | 2.65 | 8.19 | 1.68 | 6.07 | 2.66 | 6.30 | 2.80 | 4.84 | 5.07 | 6.37 |
| 2,2',5,5'-Tetrabromobiphenyl | 59080374 | 4.80 | 6.82[a] | 5.00 | 4.99 | 3.96 | 5.14 | 3.67 | 4.12 | 5.28 | 6.19 |
| 2,3,4,5-Tetrachloroaniline | 634833 | 2.69 | 4.27 | 2.59 | 2.97 | 3.09 | 2.79 | 2.59 | 2.78 | 3.46 | 3.41 |
| 1,2,3,4-Tetrachlorobenzene | 634662 | 3.77 | 4.60 | 2.84 | 3.23 | 3.38 | 3.02 | 2.88 | 2.95 | 3.23 | 3.25 |

**Table 15.2 (continued)　Measured Log BCF Values and Log BCF Values Predicted by Various Methods for 100-Compound Test Set**

| Compound Name | CAS No. | Measured log BCF | Measured log $K_{ow}$ | BCFWIN ver. 2.14 | Veith and Kosian (1983) | Bintein et al. (1993) | Park and Lee (1993) | Dimitrov et al. (2002) | Dearden (this work) | Sabljić (2001) | Koch (1983) |
|---|---|---|---|---|---|---|---|---|---|---|---|
| 1,2,4,5-Tetrachlorobenzene | 95943 | 3.76 | 4.64 | 2.87 | 3.27 | 3.41 | 3.02 | 2.91 | 2.97 | 3.32 | 3.24 |
| 2,2',3,5'-Tetrachlorobiphenyl | 41464395 | 4.84 | 5.81 | 4.39 | 4.19 | 4.19 | 4.44 | 3.65 | 3.59 | 4.70 | 4.88 |
| 2,3',4',5-Tetrachlorobiphenyl | 32598111 | 4.77 | 6.23 | 4.72 | 4.52 | 4.22 | 4.44 | 3.74 | 3.81 | 4.74 | 4.88 |
| 1,3,6,8-Tetrachlorodibenzo-p-dioxin | 33423926 | 3.36 | 7.10 | 4.67 | 5.21 | 3.74 | 3.29 | 3.56 | 4.27 | 4.96 | 5.19 |
| 2,3,7,8-Tetrachlorodibenzo-p-dioxin | 1746016 | 4.06 | 6.80 | 4.54 | 4.97 | 3.97 | 3.12 | 3.68 | 4.11 | 4.93 | 5.19 |
| 2,3,7,8-Tetrachlorodibenzofuran | 51207319 | 3.53 | 6.53 | 4.33 | 4.76 | 4.13 | 4.10 | 3.73 | 3.97 | 4.93 | 5.07 |
| Tetrachloroethene | 127184 | 1.74 | 3.40 | 1.92 | 2.29 | 2.31 | 1.79 | 1.83 | 2.32 | 2.16 | 2.13 |
| Tetrachloroguaiacol | 2539175 | 2.71 | 4.59 | 2.43 | 3.23 | 3.37 | 2.98 | 2.87 | 2.95 | 3.70 | 3.78 |
| 1,2,3,4-Tetrachloronaphthalene | 20020024 | 4.10 | 5.75 | 4.35 | 4.14 | 4.17 | 3.94 | 3.63 | 3.56 | 4.31 | 4.36 |
| 2,3,5,6-Tetrachloronitrobenzene | 117180 | 3.20 | 4.38 | 2.67 | 3.06 | 3.19 | 3.28 | 2.69 | 2.84 | 3.69 | 3.65 |
| Toluene | 108883 | 1.12 | 2.73 | 1.40 | 1.76 | 1.70 | 1.29 | 1.33 | 1.97 | 0.95 | 2.05 |
| 1,3,5-Tribromobenzene | 626391 | 3.85 | 4.51 | 2.77 | 3.16 | 3.30 | 3.08 | 2.80 | 2.90 | 4.33 | 3.84 |
| 2,4,5-Trichloroaniline | 636306 | 2.61 | 3.45 | 1.96 | 2.33 | 2.35 | 2.04 | 1.87 | 2.35 | 2.96 | 3.02 |
| 2,4,6-Trichloroanisole | 87401 | 2.94 | 4.11 | 2.46 | 2.85 | 2.95 | 2.25 | 2.45 | 2.69 | 3.14 | 3.28 |
| 1,3,5-Trichlorobenzene | 108703 | 3.38 | 4.19 | 2.53 | 2.91 | 3.02 | 2.59 | 2.52 | 2.74 | 2.82 | 2.86 |
| 2,2',5-Trichlorobiphenyl | 37680652 | 4.27 | 5.55 | 4.19 | 3.98 | 4.08 | 4.00 | 3.54 | 3.45 | 4.35 | 4.50 |
| 2,4',5-Trichlorobiphenyl | 16606023 | 3.75 | 5.69 | 4.30 | 4.10 | 4.15 | 4.00 | 3.60 | 3.52 | 4.39 | 4.49 |
| 1,2,4-Trichlorodibenzo-p-dioxin | 39227582 | 2.36 | 6.35 | 4.19 | 4.62 | 4.20 | 2.59 | 3.74 | 3.87 | 4.57 | 4.82 |
| 2,4,5-Trichlorodiphenyl ether | 52322802 | 4.18 | 5.44 | 3.49 | 3.90 | 4.02 | 3.46 | 3.48 | 3.39 | 4.33 | 4.62 |
| 1,1,1-Trichloroethane | 71556 | 0.95 | 2.49 | 1.22 | 1.57 | 1.48 | 1.49 | 1.19 | 1.84 | 3.68 | 1.88 |
| Trichloroethene | 79016 | 1.59 | 2.42 | 1.16 | 1.51 | 1.42 | 1.29 | 1.15 | 1.80 | 0.90 | 1.79 |
| 3,4,5-Trichloroguaiacol | 57057837 | 2.41 | 3.77 | 1.80 | 2.58 | 2.64 | 2.28 | 2.14 | 2.52 | 3.20 | 3.39 |
| 1,3,7-Trichloronaphthalene | 55720371 | 4.43 | 5.35 | 4.04 | 3.83 | 3.96 | 3.34 | 3.43 | 3.35 | 4.07 | 3.97 |
| 2,4,6-Trichloronitrobenzene | 18708708 | 2.88 | 3.69 | 2.14 | 2.52 | 2.57 | 2.66 | 2.07 | 2.47 | 3.23 | 3.26 |
| Xanthene | 92831 | 3.62 | 4.23 | 2.56 | 2.94 | 3.05 | 1.66 | 2.55 | 2.76 | 3.53 | 3.91 |
| 1,3-Xylene | 108383 | 1.27 | 3.20 | 1.76 | 2.13 | 2.13 | 1.71 | 1.67 | 2.22 | 1.77 | 2.37 |
| 2,4-Xylenol | 105679 | 2.18 | 2.30 | 1.07 | 1.42 | 1.31 | 1.59 | 1.08 | 1.74 | 1.97 | 2.48 |

[a] No measured value available. Value given is the mean of values calculated by KOWWIN ver. 1.66, ClogP ver. 1.0.0 (www.biobyte.com) and ChemSilico (www.logp.com).

**Table 15.3 Analysis of Log BCF Predictions Shown in Table 15.2**

| Method | Percentage of Compounds with Errors in Range | | | Correlation Statistics | | |
|---|---|---|---|---|---|---|
| | <0.5 | ≥0.5–1.0 | >1.0 | R² | s | F |
| BCFWIN ver. 2.14 | 56% | 27% | 17% | 0.725 | 0.722 | 258.4 |
| Veith & Kosian (1983) | 49% | 32% | 19% | 0.645 | 0.820 | 178.1 |
| Bintein et al. (1993) | 49% | 33% | 18% | 0.566 | 0.907 | 128.0 |
| Park & Lee (1993) | 47% | 27% | 26% | 0.618 | 0.851 | 158.7 |
| Dimitrov et al. (2002) | 46% | 30% | 24% | 0.676 | 0.784 | 204.5 |
| Dearden (this work) | 38% | 43% | 19% | 0.645 | 0.820 | 178.4 |
| Sabljić (2001) | 39% | 26% | 35% | 0.419 | 1.049 | 70.8 |
| Koch (1983) | 28% | 25% | 47% | 0.324 | 1.132 | 46.9 |

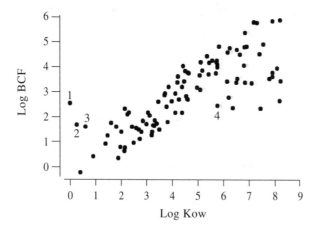

**Figure 15.1** Correlation of measured log BCF with log $K_{ow}$ for 100-compound test-set listed in Table 15.2.

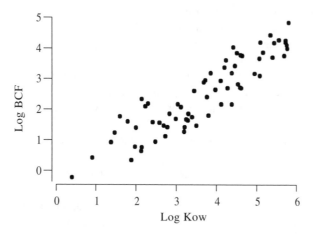

**Figure 15.2** Correlation of log BCF with log $K_{ow}$ for those compounds in 100-compound test-set with log $K_{ow}$ < 6, and excluding also acrolein (1), acrylonitrile (2), hydroquinone (3), and 4-nonylphenol (4).

The less than satisfactory modeling of log BCF by log $K_{ow}$ has encouraged numerous workers to examine the potential of other descriptors to model BCF values, the most widely used of which are undoubtedly molecular connectivities. These are topological descriptors (see Chapter 5 and Kier and Hall, 1976; 1986) calculated from knowledge of the bonds between atoms in a molecule, and they have been found to have wide applicability. The descriptors are information-rich, but difficulty is encountered in comprehending their physicochemical significance. Dearden et al. (1988)

have shown that the lower order molecular connectivity terms correlate well with molecular size, and it is essentially as surrogates for size that molecular connectivities have been used to model BCF values.

In a series of papers, Sabljić and coworkers (Sabljić and Protić, 1982; Sabljić, 1987; Sabljić and Piver, 1992; Sabljić, 2001) carried out studies leading to the development of the following QSAR for chlorinated compounds:

$$\log \text{BCF} = 2.28 \; {}^2\chi^v - 0.17 \; ({}^2\chi^v)^2 - 2.36 \qquad (15.16)$$

$$n = 20, \; R^2 = 0.939, \; s = 0.293$$

where ${}^2\chi^v$ = second-order valence molecular connectivity, which is calculated across all two-bond (three-atom) paths in a molecule, and also takes account of heteroatoms and multivalent bonds. Sabljić (1987) validated this QSAR with a set of 67 diverse chemicals.

Koch (1983) developed a rectilinear QSAR using first-order valence molecular connectivity for a small diverse set of chemicals:

$$\log \text{BCF} = 0.789 \; {}^1\chi^v + 0.147 \qquad (15.17)$$

$$n = 21, \; R^2 = 0.916, \; s \text{ not given}$$

The molecular connectivity model proposed by Lu et al. (1999) for a diverse group of non-polar chemicals incorporated five different $\chi$ values:

$$\log \text{BCF} = 0.757 \; {}^0\chi^v - 2.650 \; {}^1\chi + 3.372 \; {}^2\chi - 1.186 \; {}^2\chi^v - 1.807 \; {}^3\chi_c + 0.770 \qquad (15.18)$$

$$n = 80, \; R^2 = 0.907, \; s = 0.364$$

where ${}^3\chi_c$ = third-order cluster molecular connectivity.

This QSAR must be regarded as of dubious validity, because there is a high level of collinearity between $\chi$ values of various orders (Dearden et al., 1988). Such collinearity renders the statistics meaningless (Livingstone, 1995).

One problem with the use of molecular connectivities is that they cannot model well such properties as polarity and hydrogen bonding. Lu et al. (2000) attempted to allow for this by the incorporation of polarity correction factors for the presence of polar functional groups:

$$\log \text{BCF} = 1.564 \; {}^0\chi^v - 0.041 \; ({}^0\chi^v)^2 - 5.809 \; ({}^1\chi^v)^{0.5} + 0.615 \; {}^2\chi - 0.785 \; {}^3\chi_c + \Sigma F_j n_j + 3.179 \quad (15.19)$$

$$n = 239, \; R^2 = 0.810, \; s = 0.615$$

where $F_j$ is the correction factor for, and $n_j$ is the number of, group $j$ present in the molecule. Again, one has to question the validity of a QSAR that contains several terms likely to possess high collinearity.

Ferreira (2001) used first-order molecular connectivity, molecular volume ($V_m$), and aqueous solubility ($S_{aq}$) to improve the *Daphnia magna* log BCF-log $K_{ow}$ correlation for a series of polycyclic aromatic hydrocarbons:

$$\log \text{BCF} = 0.246 \log K_{ow} - 0.238 \log S_{aq} + 0.246 \; {}^1\chi + 0.225 \; V_m + \text{constant (not given)} \quad (15.20)$$

$$n = 11, \; R^2 = 0.870, \; s = 0.295$$

Although there is a strong negative correlation between partition coefficient and aqueous solubility (Hansch et al., 1968; Chiou et al., 1977), and a strong positive correlation between $^1\chi$ and molecular volume (Dearden et al., 1988), the use of the partial least squares (PLS) method in this study allows the simultaneous use of intercorrelated descriptors. Nevertheless, the use of four descriptors to model the bioconcentration factor of only 11 compounds contravenes the Topliss and Costello (1972) rule, and renders the QSAR of dubious validity.

Banerjee and Baughman (1991) used solubility in octanol ($S_o$) and melting point (MP, $^\circ$C) as descriptors to improve the ability of $K_{ow}$ to model the BCF of a series of diverse chemicals in fish:

$$\log BCF = 1.02 \log K_{ow} + 0.84 \log S_o + 0.0004 \, (MP - 25) - 1.13 \qquad (15.21)$$

$$n = 36, \, R^2 = 0.90, \, s \text{ not given}$$

The octanol solubility term is included to compensate for the difference in lipid solubility and octanol solubility of large, hydrophobic molecules, and the melting point term is intended to allow octanol solubilities for both solids and liquids to be included in the same equation. Octanol solubility is, however, highly correlated with melting point (Yalkowsky et al., 1983).

Another attempt to circumvent the failure of $K_{ow}$ to model BCF well was made by Escuder-Gilabert et al. (2001), who used bio-partitioning micellar chromatography retention factor (k):

$$\log BCF = -1.1 \log k + 1.3 \, (\log k)^2 + 0.5 \qquad (15.22)$$

$$n = 66, \, R^2 = 0.78, \, s = 0.499, \, F = 112$$

In fact, the correlation is not as good as that between log BCF and log $K_{ow}$ for the same compounds (see Table 15.1). That is not to say that other surrogates for partitioning into lipid may not be better than octanol (Nendza, 1991). Gobas et al. (1987) suggested that the use of membrane-water partition coefficients might prove useful in this respect, but none appears to have been investigated to date. It may be mentioned here that Mailhot (1987) found that of 10 physicochemical properties used to model log BCF, those reflecting hydrophobicity (i.e., log $K_{ow}$ and high performance liquid chromatography [HPLC] capacity factor) gave better correlations than did those reflecting molecular size (including first-order molecular connectivity).

The so-called solvatochromic or linear solvation energy relationship (LSER) descriptors developed by Abraham and coworkers (Kamlet et al., 1983) have proved valuable in correlating a wide variety of biological endpoints and physicochemical properties, and two studies have utilized them to model BCF. Park and Lee (1993) found the following QSAR for the fish BCF values of a set of diverse chemicals:

$$\log BCF = 0.88 \, \alpha - 4.39 \, \beta + 4.74 \, V_I/100 - 0.95 \qquad (15.23)$$

$$n = 51, \, R^2 = 0.897, \, s = 0.42$$

where $\alpha$ = hydrogen bond donor ability, $\beta$ = hydrogen bond acceptor ability, and $V_I$ = intrinsic molecular volume.

From the above QSAR, it can be seen that solute hydrogen bond donor ability increases BCF, as does molecular volume, while solute hydrogen bond acceptor ability decreases BCF. The LSER descriptors can now be calculated using the computer program Absolv (www.sirius-analytical.com).

Ivanciuc (1998), using the same data set as Park and Lee (1993), applied a neural network method to assess how well LSER descriptors could model BCF. He observed better correlation between observed and predicted log BCF values than did Park and Lee, but found that the neural network (NN) model gave large errors for compounds whose input and/or output values were outside the range of the corresponding values in the training set.

A number of studies (Chiou et al., 1977; Kenaga and Goring, 1980, Geyer et al., 1982; Davies and Dobbs, 1984; Isnard and Lambert, 1988; Jørgensen et al., 1998, and references cited therein) have correlated BCF with aqueous solubility. Geyer et al. (1982) found the correlation slightly poorer ($R^2 = 0.889$) than the correlation with partition coefficient ($R^2 = 0.912$), although it must be pointed out that they incorrectly used weight rather than mole concentrations for solubility values. It should also be noted that aqueous solubility determinations are generally less accurate than are those of partition coefficients, and that aqueous solubility predictions are less accurate than are those of partition coefficients (Dearden et al., 2002a, 2002b).

The Weighted Holistic Invariant Molecular (WHIM) descriptors developed by Todeschini and co-workers have been found to correlate well with the BCF of chlorobenzenes ($n = 13$, $R^2 = 0.840$, $Q^2 = 0.780$ with a single size descriptor and $Q^2 = 0.836$ with two descriptors) (Todeschini and Gramatica, 1997a; 1997b). Grammatica and Papa (2003) have used WHIM descriptors to model a large, very diverse data-set with good results. They found that a five-descriptor QSAR gave considerably better results than did the BCFWIN software (*vide infra*) (www.epa.gov/oppt/exposure/docs/episuitedl.htm). For 238 chemicals, they found $R^2 = 0.828$, root mean square error = 0.57, whereas BCFWIN yielded, for the same chemicals, $R^2 = 0.686$, RMS error = 0.77. The software to calculate WHIM descriptors can be downloaded free of charge from www.disat.unimib.it/chm/.

It is perhaps surprising, given the current interest in quantum chemical descriptors, that these have been little used to model BCF. Wei et al. (2001) found a good correlation for a series of polychlorinated organic compounds:

$$\log BCF = 0.748\ E_{HOMO} + 0.047\ E_{LUMO} + 0.020\ CCR + 10.857 \qquad (15.24)$$

$$n = 27, R^2 = 0.931, s = 0.137, F = 103.0$$

where $E_{HOMO}$ and $E_{LUMO}$ = energies of the highest occupied and lowest unoccupied molecular orbitals, and CCR = core-core repulsion energy.

Tao et al. (2000, 2001) have developed QSARs based on fragment constants and structural correction factors. For a large, diverse data set they obtained:

$$\log BCF = \Sigma\ n_i f_i + \Sigma\ m_j F_j \qquad (15.25)$$

$$n = 337, R^2_{adj} = 0.980, \text{ mean absolute error} = 0.315$$

where $f_i$ is the fragment constant for fragment $i$, $n_i$ is the number of fragments $i$, $F_j$ is the correction factor for fragment $j$, $m_j$ is the number of fragments $j$, and $R^2_{adj}$ is the square of the correlation coefficient adjusted for degrees of freedom (see Chapter 7).

Concern must be expressed over the very high $R^2_{adj}$ value reported by Tao et al. It has already been mentioned that there can be considerable error on BCF values, and examination of the other QSARs given herein suggests that an $R^2$ value of about 0.9 is the best that can be expected. This is, in fact, very good considering that BCF values are measured in whole organisms. Thus, it is likely that the QSAR developed by Tao et al. has overfitted the data.

## C.  Expert Systems

So far as can be ascertained, there is only one commercially available expert system for BCF prediction. Named BCFWIN and developed by Syracuse Research Corporation, it can be downloaded free of charge from the U.S. Environmental Protection Agency (EPA) website (www.epa.gov/oppt/exposure/docs/episuitedl.htm). It uses the QSARs developed by Meylan et al. (1999). BCF predictions have been made using BCFWIN ver. 2.14 for the 100 compounds listed in Table 15.2, and the predicted log BCF values are given there. The results are rather disappointing, with only 56% of values being predicted within ±0.5 log units of the measured log BCF, 27% being between 0.5 and 1.0 log units in error, and 17% having errors >1.0 log units.

Jørgensen et al. (1998) included in their book a diskette containing a program called WINTOX, which gives published QSARs and QSPRs for a number of endpoints and properties, including BCF. The user inputs structures in compound (CMP) format, and selects the most relevant QSARs. (Mekenyan et al. [1994] have explained the CMP format.) The program then calculates the log BCF value for each compound from each QSAR selected. No guidance is given on how to select the best QSAR to use. For example, the program lists 10 QSARs for the prediction of fish log BCF from log $K_{ow}$, and reports predicted log BCF values for aniline ranging from –3.14 to 1.92 (i.e., a range of 5 orders of magnitude). The program also lists 9 QSARs for the prediction of fish log BCF from aqueous solubility, and the predicted values for aniline range from –0.76 to 3.56 (i.e., covering 4.3 orders of magnitude). This clearly emphasises the need to select a QSAR carefully. Factors that need to be considered include whether the BCF data are all determined in the same laboratory using the same protocol on the same species; the number and classes of chemicals tested; and the source of the log $K_{ow}$, solubility, or other property used to model BCF.

## D. Comparison of Some BCF Prediction Models

Space does not permit of validation of all of the QSARs mentioned above for the prediction of BCF. A number of models have been tested by using them to predict the log BCF values of the chemicals listed in Table 15.2, the log $K_{ow}$ values of which range from –0.01 to 8.24. The criteria for selection were that each model must be applicable to heterogeneous data sets, must have given good results in its authors' hands, and must use descriptors that are readily available and easily calculated. A number of models (namely BCFWIN version 2.14 and those of Bintein et al. [1993], Sabljić [2001], and Dimitrov et al. [2002]) attempt to allow for the fact that apparent BCF values of some (but not all) very hydrophobic compounds are low by using a biphasic relationship to model bioconcentration. The other models (Veith and Kosian, 1983; Koch, 1983; Park and Lee, 1993; and Dearden [this work]) use rectilinear models. The Absolv program was used to obtain the Abraham descriptors needed for the Park and Lee equation.

The biphasic models based on log $K_{ow}$ perform better on the whole (with the exception of that of Bintein et al. [1993]); BCFWIN version 2.14 (which includes structural correction factors) is clearly the best. The rectilinear model of Veith and Kosian (1983) is almost as good as the biphasic model of Dimitrov et al. (2002), while the models that use molecular connectivity perform badly.

There is widespread acknowledgment that it is extremely difficult to predict BCF values accurately for very hydrophobic compounds. Macdonald et al. (2002) have commented that for compounds with log $K_{ow}$ values greater than about 6, "fitting any equation in this region is very adventurous." For compounds with log $K_{ow} > 8$, they recommend the use of expert judgment.

## E. Vegetation-Water and Vegetation-Air Bioconcentration

The bioconcentration of chemicals in plants, especially plants that are important links in the food chain, is of evident concern. Chemicals may be taken up in aqueous solution via roots or leaves (the latter particularly if chemicals are present in rainfall), or absorbed directly from the atmosphere via leaves if they are in the vapor phase. However, in contrast to the large number of QSAR studies on bioconcentration in aquatic species, there have been relatively few QSAR investigations of bioconcentration by plants. Several mechanistic models have been propounded (e.g., Paterson et al., 1994, Trapp and Mathies, 1995; Hung and Mackay, 1997).

Kerler and Schönherr (1988) observed a good correlation of plant cuticle-water partition coefficient ($K_{cw}$) with log $K_{ow}$ for a diverse set of chemicals:

$$\log K_{cw} = 0.97 \log K_{ow} + 0.045 \tag{15.26}$$

$$n = > 40, R^2 = 0.970, s \text{ not given}$$

A very similar correlation was obtained by Schönherr and Riederer (1989), who also found a good negative correlation ($R^2 = 0.956$) between $\log K_{cw}$ and $\log S_{aq}$ ($S_{aq}$ = aqueous solubility) based on 50 data-points from 13 diverse chemicals. In this study (and that described below) the authors used all published data, so if they found, for example, four separate results for a compound, they put all the data in as four separate data points.

Sabljić et al. (1990), using 50 data points for 15 diverse organic chemicals, found a good correlation of $K_{cw}$ with third-order valence molecular connectivity ($^3\chi^v$) and the number of aliphatic hydroxyl groups ($N_{OHaliphatic}$):

$$\log K_{cw} = 1.32 \ ^3\chi^v - 1.43 \ N_{OHaliphatic} - 0.26 \tag{15.27}$$

$$n = 50, R^2 = 0.970, s = 0.345, F = 754$$

Sabljić et al. (1990) commented that plant cuticle-water partition coefficients are influenced primarily by molecular size, with the number of hydroxyl groups (presumably a surrogate for hydrogen bonding) having a "fine-tuning" effect.

Platts and Abraham (2000) used their LSER or linear free energy relationship (LFER) descriptors to model plant cuticular matrix-water partitioning ($K_{mxw}$):

$$\log K_{mxw} = 0.596 \ R_2 - 0.413 \ \pi_2^H - 0.508 \ \Sigma\alpha_2^H - 4.096 \ \Sigma\beta_2^H + 3.908 \ V_x - 0.415 \tag{15.28}$$

$$n = 62, R^2 = 0.981, s = 0.236, F = 566$$

where $R_2$ = excess molar refraction (a measure of polarisability), $\pi_2^H$ = a polarity/polarizability term, $\Sigma\alpha_2^H$ = sum of hydrogen bond donor ability, $\Sigma\beta_2^H$ = sum of hydrogen bond acceptor ability, and $V_x$ = McGowan characteristic volume. The coefficients are approximately autoscaled, and so indicate the relative contributions of each term. It can be seen that, in contrast to the comment of Sabljić et al. (1990), plant cuticle-water partition coefficients are not influenced primarily by molecular size; hydrogen bond acceptor ability in particular appears to be as important as molecular size. Schönherr and Riederer (1989) have commented that it is primarily the polarity of a solute (in which they implicitly included hydrogen bonding) that controls plant cuticle-water partitioning.

Travis and Arms (1988) modeled bioconcentration from soil to vegetation ($K_{vs}$) for a series of pesticides:

$$\log K_{vs} = -0.578 \log K_{ow} + 1.588 \tag{15.29}$$

$$n = 29, R^2 = 0.53, s \text{ not given}$$

This weak correlation indicates that bioconcentration decreases as $\log K_{ow}$ increases. Travis and Arms interpret this as meaning that the controlling factor is aqueous solubility, which is inversely related to hydrophobicity. A more likely explanation is that the hydrophobicity of the organic matter of the soil is greater than the hydrophobicity of the vegetation.

A similar QSAR was obtained by Dowdy and McKone (1997) for vegetation-soil partitioning of a data set comprising mostly pesticides:

$$\log K_{vs} = -0.419 \log K_{ow} + 0.840 \tag{15.30}$$

$$n = 30, \ R^2 = 0.53, \ s = 0.896$$

Use of a polarity-corrected first-order molecular connectivity ($^1\chi_{pc}$) gave a correlation with a better $R^2$ value but a slightly worse standard error. Interestingly, the same authors found a positive correlation between log $K_{vs}$ for roots and log $K_{ow}$.

The computer program Absolv, has, *inter alia*, a log $K_{cw}$ prediction module. For the chemicals used by Sabljić et al. (1990), Absolv gave reasonably good predictions. For example, for partitioning into the cuticular membrane of the tomato (*Lycopersicon esculentum*), for which data for 13 chemicals were available, 8 Absolv predictions were within 0.5 log units of the measured values, 4 were between 0.5 and 1.0 units in error, and 1 was more than 1 log unit in error. The measured and Absolv-predicted values correlated well ($R^2 = 0.954$, $s = 0.425$, $F = 230.2$).

Vegetation-air partitioning has received relatively little attention in general (Schönherr and Riederer, 1989; Welke et al., 1998). It is to be hoped that this situation will be remedied, because the high affinity of plant cuticular material for organic compounds, and the extensive leaf area of many types of vegetation, indicate that substantial quantities of volatile substances are taken up from the atmosphere by vegetation. Simonich and Hites (1995) have estimated that 44 ± 18% of polyaromatic hydrocarbons emitted into the atmosphere in the northeastern U.S. is removed by vegetation.

Müller et al. (1994) developed from theoretical concepts a multi-term expression for plant-air partitioning ($K_{pa}$), the essence of which is that $K_{pa}$ should be proportional to $K_{ow}/H$, where $H$ = Henry's law constant (air-water partition coefficient); this means that $K_{pa}$ should therefore be proportional to $K_{oa}$, the octanol-air partition coefficient. They found a reasonable correlation between measured and predicted log $K_{pa}$ values; for example, for 20 chlorohydrocarbons they found $R^2 = 0.79$. Bacci et al. (1990) also used log $K_{ow}$ and log $H$ to correlate *Azalea indica* leaf-air partitioning of 10 chemicals (mostly pesticides) with $R^2 = 0.92$.

Most QSAR studies of plant-air partitioning have involved the use of $K_{oa}$ (Paterson et al., 1991; Tolls and McLachlan, 1994; Dowdy and McKone, 1997; Keymeulen, et al., 1997; Brown et al., 1998). Unfortunately, from a statistical viewpoint, they have also used small data sets. For example, Keymeulen et al. (1997), using *Hedera helix* (ivy) plant cuticles, found the following QSAR for alkylbenzenes:

$$\log K_{ca} = 0.96 \log K_{oa} - 0.57 \tag{15.31}$$

$$n = 5, \ R^2 = 0.994, \ s \text{ not given}$$

Brown et al. (1998), using whole needles of *Pinus sylvestris* (Scots pine), obtained a QSAR with very different coefficients for a very small series of chlorinated alkanes and alkenes:

$$\log K_{pa} = 3.09 \log K_{oa} - 7.24 \tag{15.32}$$

$$n = 4, \ R^2 = 0.96, \ s \text{ not given}$$

Dowdy and McKone (1997) compared the ability of log $K_{ow}$, log $K_{oa}$, and a polarity-corrected first-order simple molecular connectivity ($^1\chi_{pc}$) to model log $K_{pa}$ of unspecified vegetation for a series of 14 chemicals comprising mostly chlorinated hydrocarbons, with the following results: log $K_{ow}$, $R^2 = 0.74$, $s = 3.08$; log $K_{oa}$, $R^2 = 0.73$, $s = 0.72$; $^1\chi_{pc}$, $R^2 = 0.78$, $s = 1.44$. Clearly, log $K_{oa}$ models the data best, with by far the lowest standard error.

Welke et al. (1998) investigated the correlation of cuticular matrix-air partition coefficients (log $K_{mxa}$) of a diverse group of chemicals with a number of physicochemical properties, including boiling point (BP) in °C:

$$\log K_{mxa} = 0.017 \text{ BP} + 1.343 \tag{15.33}$$

$$n = 50, R^2 = 0.829, s = 0.260, F = 232.8$$

vapor pressure (VP) in pascal:

$$\log K_{mxa} = -0.892 \log \text{VP} + 6.290 \tag{15.34}$$

$$n = 50, R^2 = 0.899, s = 0.199, F = 428.7$$

and octanol-air partition coefficient ($K_{oa}$) (cf. Müller et al. [1994]):

$$\log K_{mxa} = 0.668 \log K_{oa} + 0.820 \tag{15.35}$$

$$n = 38, R^2 = 0.819, s = 0.237, F = 163.3$$

In addition, they found a good correlation with first-order simple molecular connectivity, but only for a congeneric series of aliphatic alcohols. They also investigated the approach of Magee (1991) in developing a mechanism-based QSAR, using hydrogen bond donation, molar refractivity, and components of log $K_{ow}$; this approach resulted in a correlation no better than that with log $K_{oa}$ ($n = 49$, $R^2 = 0.812$, $s = 0.271$, $F = 37.1$).

Platts and Abraham (2000) used their LSER approach to model plant cuticular matrix-air partitioning, and developed a four-term QSAR:

$$\log K_{mxa} = 1.310 \, \pi_2^H + 3.116 \, \Sigma\alpha_2^H + 0.793 \, \Sigma\beta_2^H + 0.877 \log L^{16} - 0.641 \tag{15.36}$$

$$n = 62, R^2 = 0.994, s = 0.230, F = 2361$$

Most of the descriptors have been explained earlier; $L^{16}$ is the hexadecane-gas partition coefficient. The extremely high correlation coefficient is of some concern, and may indicate overfitting of the data. From this QSAR, it can be deduced that hydrogen bond donor ability and polarity/polarizability are most important in controlling $K_{mxa}$.

The Absolv prediction program includes a module for plant-air partitioning, and the present author has tested its predictive ability. Unfortunately most of the QSAR studies carried out on plant-air partitioning have used small data sets, and since data from different plant genera cannot safely be combined, the only reasonably large data set that could be found was that used by Platts and Abraham. This was in all probability the data set used to develop the Absolv prediction module, and so the Absolv predictions would be expected to be good, as in fact they are. For the 50-chemical test set, 88% had predicted log $K_{mxa}$ values within 0.5 log units of the measured values, and the remaining 12 per cent were within 1 log unit. Observed and predicted values correlated well: $R^2 = 0.865$, $s = 0.230$, $F = 307.6$. As pointed out above, in using any prediction method for plant-air partitioning, it should be remembered that such partitioning is very dependent on the genus of plant used.

As mentioned earlier, many plant-air partitioning QSARs use log $K_{oa}$ as a descriptor. Unfortunately, there are very few measured values of log $K_{oa}$ available, and so most workers have used values calculated as the difference between log $K_{ow}$ and log $K_{aw}$ (log H) (air-water partition coefficient). Harner and Mackay (1995) reported measured log $K_{oa}$ values for 12 chlorinated aromatic hydrocarbons, and so a check can be made of how well log $K_{oa}$ values can be predicted. Absolv includes modules for the prediction of both dry octanol-air and wet octanol-air partition coefficients. The Syracuse Research Corporation software, freely downloadable from the EPA website (www.epa.gov/oppt/exposure/docs/episuitedl.htm), includes modules for the prediction of log $K_{ow}$ (KOWWIN) and log H (HENRYWIN). HENRYWIN offers two predictions, namely, a group contribution prediction and a bond contribution prediction, but the group contribution method is not able to make predictions for all compounds. Dearden and Schüürmann (2003) found, using a test set of 700 chemicals, that the group contribution method could be used for only 392 of the 700 compounds, while the bond contribution method worked for all of them.. The Absolv methods and the KOWWIN/HENRYWIN method all gave very good predictions of log $K_{oa}$, with the best being from the Absolv wet octanol-air module ($R^2 = 0.970$, $s = 0.311$, $F = 326.3$; 11 out of the 12 predictions were within 0.26 log units of the measured values. The prediction for 2,2′,4,4′,6,6′-hexachlorobiphenyl was in error by 1.07 log units).

## III. BIOMAGNIFICATION

As stated in the introduction, biomagnification (BMF), defined in Equation 15.2, is the accumulation of a xenobiotic *via* the food chain. There have been only a few QSAR studies of biomagnification, and some examples are discussed below. Kenaga (1980) found a rather weak correlation of BMF in beef fat with hydrophobicity:

$$\log BMF = 0.500 \log K_{ow} - 1.476 \qquad (15.37)$$

$$n = 23, R^2 = 0.624, s \text{ not given}$$

Travis and Arms (1988) developed QSARs for biomagnification of pesticides in beef and milk from cattle food:

$$\log BMF(beef) = 1.033 \log K_{ow} - 7.735 \qquad (15.38)$$

$$n = 36, R^2 = 0.66, s \text{ not given}$$

$$\log BMF(milk) = 0.992 \log K_{ow} - 8.056 \qquad (15.39)$$

$$n = 28, R^2 = 0.55, s \text{ not given}$$

Dowdy et al. (1996) used molecular connectivity to model biomagnification of pesticides in beef meat and cattle milk from cattle food:

$$\log BMF(beef) = 0.525 \, {}^1\chi_{pc} - 5.904 \qquad (15.40)$$

$$n = 35, R^2 = 0.903, s = 0.49$$

$$\log BMF (milk) = 0.421 \, {}^1\chi_{pc} - 5.879 \qquad (15.41)$$

$$n = 34, R^2 = 0.887, s = 0.43$$

where ${}^1\chi_{pc}$ = first-order simple molecular connectivity corrected with appropriate polarity factors.

The QSARs in Equation 15.40 and Equation 15.41 were much better than those in which log $K_{ow}$ was used as the independent variable (Equation 15.38 and Equation 15.39), $R^2 = 0.65$, $s = 0.92$ and $R^2 = 0.54$, $s = 0.92$, respectively). The authors offered no explanation of this; it is possible that molecular size dominates the biomagnification process here.

van Bavel et al. (1996) carried out a principal components analysis of the biomagnification of 18 PCBs in the 3-spined stickleback (*Gasterosteus aculeatus*). Using three components representing various physicochemical properties, they were able to explain 91%of the variance in BMF.

The uptake of pesticides by earthworms from soil has been modeled by Magee (1991):

$$\log K_{\text{worm-soil}} = 0.455 \log K_{ow} + 1.15 \tag{15.42}$$

$$n = 19, R^2 = 0.766, s = 0.446, F = 55.6$$

The correlation is not very good, and could perhaps be improved by the inclusion of additional descriptors.

Voutsas et al. (2002) have presented QSARs for bioaccumulation in each of four trophic levels of an aquatic food web: plankton, benthic invertebrates, planktivorous fish, and piscivorous fish. All are based on log $K_{ow}$ and (log $K_{ow}$)$^2$. They conclude that biomagnification becomes important only at log $K_{ow}$ values greater than 6, in agreement with the findings of other workers (Mackay, 1991; Metcalfe and Metcalfe, 1997). Compared to the first level, bioaccumulation is a factor of 5 higher in the third level, for log $K_{ow}$ = 8, and a factor of 30 higher in the fourth level.

## IV. RECOMMENDATIONS

### A. Bioconcentration in Aquatic Species

Devillers et al. (1996) have commented that most QSARs for the prediction of BCF perform similarly up to log $K_{ow}$ ~ 6. In view of the fact that the computer program BCFWIN version 2.14 is freely available from the EPA website (www.epa.gov/oppt/exposure/docs/episuitedl.htm), it is recommended that this be used for BCF prediction for chemicals with log $K_{ow} \leq 6$; the proviso is that highly reactive chemicals will probably have a higher than predicted BCF, perhaps by up to two orders of magnitude. BCFWIN requires that the chemical structure be input using Simplified Molecular Line Entry System (SMILES) notation (Weininger, 1988) or as a Chemical Abstracts Service (CAS) number.

For chemicals with log $K_{ow}$ values between 6 and 8, BCF values of non-polar chemicals (hydrocarbons and halogenated hydrocarbons) should be estimated using one of the rectilinear log BCF-log $K_{ow}$ QSARs such as that of Veith and Kosian (1983). BCFWIN should be used for polar chemicals with log $K_{ow}$ values in this range. It should also be used for all chemicals with log $K_{ow}$ values > 8. If measured log $K_{ow}$ values are not available, it is recommended that a consensus value be used, being the average of two to three calculated values (e.g., KOWWIN, ClogP [www.biobyte.com], ChemSilico [www.logp.com]).

### B. Vegetation-Water Bioconcentration

Inasmuch as the Absolv software includes a plant cuticle-water partitioning module, which has been found to perform well, it is recommended that this be used if possible. If this is not possible, the simple log $K_{cw}$-log $K_{ow}$ of Kerler and Schönherr (1988) can be used.

### C. Vegetation-Air Bioconcentration

The Absolv software includes a plant cuticle-air partitioning module that yields good predictions. It is recommended that this be employed, if possible, for the calculation of plant-air partitioning. If Absolv is not available, one of the QSARs developed by Welke et al. (1998) for a reasonably large diverse data set ($n = 38$), using either boiling point, vapor pressure, or log $K_{oa}$ as the descriptor should be used.

## D. Biomagnification

Unfortunately, there are too few QSAR analyses of biomagnification available for any firm recommendations to be made. If a prediction is required, the user should check whether any of the published biomagnification QSARs is appropriate.

## REFERENCES

Abraham, M.H., Chadha, H.S., and Mitchell, R.C., The factors that influence skin penetration of solutes, *J. Pharm. Pharmacol.*, 47, 8–16, 1995.

Bacci, E., Calamari, D., Gaggi, C., and Vighi, M., Bioconcentration of organic chemical vapors in plant leaves: experimental measurements and correlation, *Environ. Sci. Tech.*, 24, 885–889, 1990.

Banerjee, S. and Baughman, G.L., Bioconcentration factors and lipid solubility, *Environ. Sci. Tech.*, 25, 536–539, 1991.

Bintein, S., Devillers, J., and Karcher, W., Nonlinear dependence of fish bioconcentration on *n*-octanol/water partition coefficient, *SAR QSAR Environ. Res.*, 1, 29–39, 1993.

Brown, R.H.A., Cape, J.N., and Farmer, J.G., Partitioning of chlorinated solvents between pine needles and air, *Chemosphere*, 36, 1799–1810, 1998.

Chiou, C.T., Freed, V.H., Schmedding, D.W., and Kohnert, R.L., Partition coefficient and bioaccumulation of selected organic chemicals, *Environ. Sci. Tech.*, 11, 475–478, 1977.

Connell, D.W., *Bioaccumulation of Xenobiotic Compounds*, CRC Press, Boca Raton, FL, 1990.

Connell, D.W., Prediction of bioconcentration and related lethal and sublethal effects with aquatic organisms, *Mar. Pollut. Bull.*, 31, 201–205, 1995.

Connell, D.W. and Hawker, D.W., Use of polynomial expressions to describe the bioconcentration of hydrophobic chemicals by fish, *Ecotoxicology Environ. Saf.*, 16, 242–257, 1988.

Davies, R.P. and Dobbs, A., The prediction of bioconcentration in fish, *Water Res.*, 18, 1253–1262, 1984.

Dearden, J.C., Bradburne, S.J.A., Cronin, M.T.D., and Solanki, P., The physical significance of molecular connectivity, in *QSAR 88*, Turner, J.E., Williams, M.W., Schultz, T.W., and Kwaak, N.J., Eds., U.S. Department of Energy, Oak Ridge, TN, 1988, pp. 43–50.

Dearden, J.C., Netzeva, T.I., and Bibby, R., Comparison of a number of commercial software programs for the prediction of octanol-water partition coefficients, *J. Pharm. Pharmacol.*, 54 (Suppl.) S65–S66, 2002a.

Dearden, J.C., Netzeva, T.I., and Bibby, R., Comparison of a number of commercial software programs for the prediction of aqueous solubility, *J. Pharm. Pharmacol.*, 54 (Suppl.) S66, 2002b.

Dearden, J.C. and Schüürmann, G., Quantitative structure-property relationships for predicting Henry's law constant from molecular structure, *Environ. Toxicol. Chem.*, 22, 1755–1770, 2003.

Deneer, J.W., Sinnige, T.L., Seinen, W., and Hermens, J.L.M., Quantitative structure-activity relationships for the toxicity and bioconcentration factor of nitrobenzene derivatives towards the guppy (*Poecilia reticulata*), *Aquatic Toxicol.*, 10, 115–129, 1987.

Devillers, J., Bintein, S., and Domine, D., Comparison of BCF models based on log P, *Chemosphere*, 33, 1047–1065, 1996.

Dimitrov, S.D., Dimitrova, N.C., Walker, J.D., Veith, G.D., and Mekenyan, O.G., Predicting bioconcentration factors of highly hydrophobic chemicals. Effect of molecular size, *Pure Appl. Chem.*, 74, 1823–1830, 2002.

Dimitrov, S.D., Dimitrova, N.C., Walker, J.D., Veith, G.D., and Mekenyan, O.G., Bioconcentration potential predictions based on molecular attributes — an early warning approach for chemicals found in humans, birds, fish and wildlife, *QSAR Comb. Sci.*, 22, 58–68, 2003.

Dimitrov, S.D., Mekenyan, O.G., and Walker, J.D., Non-linear modelling of bioconcentration using partition coefficients for narcotic chemicals, *SAR QSAR Environ. Res.*, 13, 177–184, 2002.

Dowdy, D.L. and McKone, T.E., Predicting plant uptake of organic chemicals from soil or air using octanol/water and octanol/air partition ratios and a molecular connectivity index, *Environ. Toxicol. Chem.*, 16, 2448–2456, 1997.

Dowdy, D.L., McKone, T.E., and Hsieh, D.P.H., Prediction of chemical biotransfer of organic chemicals from cattle diet into beef and milk using the molecular connectivity index, *Environ. Sci. Tech.*, 30, 984–989, 1996.

Ellgehausen, H., Guth, J.A., and Esser, H.O., Factors determining the bioaccumulation potential of pesticides in the individual compartments of aquatic food chains, *Ecotoxicology Environ. Saf.*, 4, 134–157, 1980.

Escuder-Gilabert, L., Martín-Biosca, Y., Sagrado, S., Villanueva-Camañas, R.M., and Medina-Hernández, M.J., Biopartitioning micellar chromatography to predict ecotoxicity, *Analytica Chimica Acta*, 448, 173–185, 2001.

European Centre for Ecotoxicology and Toxicology of Chemicals, *The Role of Bioaccumulation in Environmental Risk Assessment: The Aquatic Environment and Related Food Webs*, Technical Report 67, Brussels, Belgium, 1996.

Ferreira, M.M.C., Polycyclic aromatic hydrocarbons: a QSPR study, *Chemosphere*, 44, 125–146, 2001.

Geyer, H.J., Politzki, G., and Freitag, D., Prediction of ecotoxicological behaviour of chemicals: relationship between n-octanol/water partition coefficient and bioaccumulation of organic chemicals by algae *Chlorella*, *Chemosphere*, 13, 269–284, 1984.

Geyer, H.J., Scheunert, I., Brüggemann, R., Steinberg, C., Korte, F., and Kettrup, A., QSAR for organic chemical bioconcentration in *Daphnia*, algae, and mussels, *Sci. Total Environ.*, 109/110, 387–394, 1991.

Geyer, H., Sheehan, D., Kotzias, D., Freitag, D., and Korte, F., Prediction of ecotoxicological behaviour of chemicals: relationship between physicochemical properties and bioaccumulation of organic chemicals in the mussel, *Chemosphere*, 11, 1121–1134, 1982.

Gobas, F.A.P.C., Clark, K.E., Shiu, W.Y., and Mackay, D., Bioconcentration of polybrominated benzenes and biphenyls and related superhydrophobic chemicals in fish: role of bioavailability and elimination into the feces, *Environ. Toxicol. Chem.* 8, 231–245, 1989.

Gobas, F.A.P.C., Shiu, W.Y., and Mackay, D., Factors determining partitioning of hydrophobic organic chemicals in aquatic organisms, in *QSAR in Environmental Toxicology — II*, Kaiser, K.L.E., Ed., Dordrecht, The Netherlands: D. Reidel, 1987, pp. 107–123.

Gobas, F.A.P.C. and Zhang, X., Measuring bioconcentration factors and rate constants of chemicals in aquatic organisms under conditions of variable water concentrations and short exposure time, *Chemosphere*, 25, 1961–1971, 1995.

Grammatica, P. and Papa, E., QSAR modelling of bioconcentration factor by theoretical molecular descriptors, *QSAR Comb. Sci.*, 22, 374–385, 2003.

Hansch, C., Quinlan, J.E., and Lawrence, G.L., The linear free-energy relationship between partition coefficients and the aqueous solubility of organic liquids, *J. Org. Chemistry*, 33, 347–350, 1968.

Harner, T. and Mackay, D., Measurement of octanol-air partition coefficients for chlorobenzenes, PCBs, and DDT, *Environ. Sci. Tech.*, 29, 1599–1606, 1995.

Hawker, D.W. and Connell, D.W., Prediction of bioconcentration factors under non-equilibrium conditions, *Chemosphere*, 14, 1835–1843, 1985.

Hawker, D.W. and Connell, D.W., Bioconcentration of lipophilic compounds by some aquatic organisms, *Ecotoxicology Environ. Saf.*, 11, 184–197, 1986.

Hermens, J., Prediction of environmental toxicity based on structure-activity relationships using mechanistic information, *Sci. Total Environ.*, 171, 235–242, 1995.

Hung, H. and Mackay, D., A novel and simple model of the uptake of organic chemicals by vegetation from air and soil, *Chemosphere*, 35, 959–977, 1997.

Isnard, P. and Lambert, S., Estimating bioconcentration factors from *n*-octanol/water partition coefficient and aqueous solubility, *Chemosphere*, 17, 21–34, 1988.

Ivanciuc, O., Artificial neural networks applications. Part 7. Estimation of bioconcentration factors in fish using solvatochromic parameters, *Revue Roumaine de Chimie*, 43, 347–354, 1988.

Jørgensen, S.E., Halling-Sørensen, B., and Mahler, H., *Handbook of Estimation Methods in Ecotoxicology and Environmental Chemistry*, Lewis Publishers, Boca Raton, FL, 1998.

Kamlet, M.J., Abboud, J.-L.M., Abraham, M.H., and Taft, R.W., Linear solvation energy relationships. 23. A comprehensive collection of the solvatochromic parameters π*, α and β, and some methods for simplifying the generalized solvatochromic equation, *J. Org. Chem.*, 48, 2877–2887, 1983.

Kenaga, E.E., Correlation of bioconcentration factors of chemicals in aquatic and terrestrial organisms with their physical and chemical properties, *Environ. Sci. Tech.*, 14, 553–556, 1980.

Kenaga, E.E. and Goring, C.A.I., Relationship between water solubility and soil sorption, octanol-water partitioning and bioconcentration of chemicals in biota, in *Aquatic Toxicology*, Special Technical Publication 707, Eaton, J.G., Parrish, P.R.P. and Hendricks, A.C., Eds., American Society for Testing and Materials, Philadelphia, PA, 1980, pp. 78–115.

Kerler, F. and Schönherr, J., Accumulation of lipophilic chemicals in plant cuticles: prediction from 1-octanol/water partition coefficient, *Arch. Environ. Contamination Toxicol.*, 17, 1–6, 1988.

Keymeulen, R., De Bruyn, G., and Van Langenhove, H., Headspace gas chromatographic determination of the plant cuticle-air partition coefficients for monocyclic aromatic hydrocarbons as environmental compartment, *J. Chromatogr. A*, 774, 213–221, 1997.

Kier, L.B. and Hall, L.H., *Molecular Connectivity in Chemistry and Drug Research*, Academic Press, New York, 1976.

Kier, L.B. and Hall, L.H., *Molecular Connectivity in Structure-Activity Analysis*, Wiley, London, 1986.

Koch, R., Molecular connectivity index for assessing ecotoxicological behavior of organic compounds, *Toxicol. Environ. Chem.*, 6, 87–96, 1983.

Könemann, H. and van Leeuwen, C., Toxicokinetics in fish: accumulation and elimination of six chloroben-zenes in guppies, *Chemosphere*, 9, 3–19, 1980.

Kubinyi, H., Quantitative structure-activity relationships. IV. Non-linear dependence of biological activity on hydrophobic character: a new model, *Arzneimittel-Forschung – Drug Res.*, 26, 1991–1997, 1976.

Lassiter, R.R., Modeling Dynamics of Biological and Chemical Components on Aquatic Ecosystems, U.S. Environmental Protection Agency Report No. EPA-660/3-75-012, Washington, D.C., 1975.

Livingstone, D., *Data Analysis for Chemists: Applications to QSAR and Chemical Product Design*, Oxford University Press, Oxford, 1995.

Lu, X., Tao, S., Cao, J., and Dawson, R.W., Prediction of fish bioconcentration factors of nonpolar organic pollutants based on molecular connectivity indices, *Chemosphere*, 39, 987–999, 1999.

Lu, X., Tao, S., Hu, H., and Dawson, R.W., Estimation of bioconcentration factors of non-ionic organic compounds in fish by molecular connectivity indices and polarity correction factors, *Chemosphere*, 41, 1675–1688, 2000.

Macdonald, D., Breton, R., Sutcliffe, R., and Walker, J., Uses and limitations of quantitative structure-activity relationships (QSARs) to categorize substances on the Canadian Domestic Substance List as persistent and/or bioaccumulative, and inherently toxic to non-human organisms, *SAR QSAR Environ. Res.*, 13, 43–55, 2002.

Mackay, D., Correlation of bioconcentration factors, *Environ. Sci. Tech.*, 16, 274–276, 1982.

Mackay, D., *Multimedia Environmental Models: The Fugacity Approach*, Lewis Publishers, Chelsea, MI, 1991.

Mackay, D., Paterson, S., and Shiu, W.Y., Generic models for evaluating the regional fate of chemicals, *Chemosphere*, 24, 695–717, 1992.

Magee, P.S., Complex factors in hydrocarbon/water, soil/water and fish/water partitioning, *Sci. Total Environ.*, 109/110, 155–178, 1991.

Mailhot, H., Prediction of algal bioaccumulation and uptake rate of nine organic compounds by ten physic-ochemical properties, *Environ. Sci. Tech.*, 21, 1009–1013, 1987.

Mekenyan, O.G., Karabunarliev, S.H., Ivanov, J.M., and Dimitrov, D.N., A new development of the OASIS computer-system for modelling molecular properties, *Comput. Chem.*, 18, 173–187, 1994.

Metcalfe, T.L. and Metcalfe, C.D., The trophodynamics of PCBs, including mono- and non-ortho congeners in the food web of north-central Lake Ontario, *Sci. Total Environ.*, 201, 245–272, 1997.

Meylan, W.M., Howard, P.H., Boethling, R.S., Aronson, D., Printup, H., and Gouchie, S., Improved method for estimating bioconcentration/bioaccumulation factor from octanol/water partition coefficient, *Environ. Toxicol. Chem.*, 18, 664–672, 1999.

Müller, J.F., Hawker, D.W., and Connell, D.W., Calculation of bioconcentration factors of persistent hydro-phobic compounds in the air/vegetation system, *Chemosphere*, 29, 623–640, 1994.

Neely, W.B., Branson, D.R., and Blau, G.E., Partition coefficients to measure bioconcentration potential of organic chemicals in fish, *Environ. Sci. Tech.*, 8, 1113–1115, 1974.

Nendza, M., QSARs of bioconcentration: validity assessment of log $P_{ow}$/log BCF correlations, in *Bioaccumulation in Aquatic Systems*, Nagel, R. and Loskill, R., Eds., VCH, Weinheim, Germany, 1991, pp. 43–66.

Nendza, M., *Structure-Activity Relationships in Environmental Sciences*, Chapman & Hall, London, 1998.

Organization for Economic Cooperation and Development, Guide-Line for Testing of Chemicals: Bioaccumulation, Paris, France, 1981.

Park, J.H. and Lee, H.J., Estimation of bioconcentration factor in fish, adsorption coefficient for soils and sediments and interfacial tension with water for organic nonelectrolytes based on the linear solvation energy relationships, *Chemosphere*, 26, 1905–1916, 1993.

Paterson, S., Mackay, D., Bacci, E., and Calamari, D., Correlation of the equilibrium and kinetics of leaf-air exchange of hydrophobic organic chemicals, *Environ. Sci. Tech.*, 25, 866–871, 1991.

Paterson, S., Mackay, D., and McFarlane, C., A model of organic chemical uptake by plants from soil and the atmosphere, *Environ. Sci. Tech.*, 28, 2259–2266, 1994.

Platts, J.A. and Abraham, M.H., Partition of volatile organic compounds from air and from water into plant cuticular matrix: an LFER analysis, *Environ. Sci. Tech.*, 34, 318–323, 2000.

Saarikoski, J. and Viluksela, M., Relation between physicochemical properties of phenols and their toxicity and accumulation in fish, *Ecotoxicology Environ. Saf.* 6, 501–512, 1982.

Sabljić, A., The prediction of fish bioconcentration factors of organic pollutants from the molecular connectivity model, *Zeitschrift Für die Gesamte Hygiene und ihre Grenzgebiete*, 33, 493-496, 1987.

Sabljić, A., QSAR models for estimating properties of persistent organic pollutants required in evaluation of their environmental fate and risk, *Chemosphere*, 43, 363–375, 2001.

Sabljić, A., Güsten, H., Schönherr, J., and Riederer, M., Modeling plant uptake of airborne organic chemicals. Plant cuticle/water partitioning and molecular connectivity, *Environ. Sci. Tech.*, 24, 1321–1326, 1990.

Sabljić, A. and Piver, W.T., Quantitative modeling of environmental fate and impact of commercial chemicals, *Environ. Toxicol. Chem.*, 11, 961–972, 1992.

Sabljić, A. and Protić, M., Molecular connectivity: a novel method for prediction of bioconcentration factor of hazardous chemicals, *Chemico-Biol. Interactions*, 42, 301–310, 1982.

Schönherr, J. and Riederer, M., Foliar penetration and accumulation of organic chemicals in plant cuticles, *Rev. Environ. Contam. Toxicol.*, 108, 1–70, 1989.

Schüürmann, G. and Klein, W., Advances in bioconcentration prediction, *Chemosphere*, 17, 1551–1574, 1988.

Simonich, S.L. and Hites, R.A., Organic pollutant accumulation in vegetation, *Environ. Sci. Tech.*, 29: 2905-2914, 1995.

Streit, B., Bioaccumulation processes in ecosystems, *Experientia*, 48: 955-970, 1992.

Tao, S., Hu, H.Y., Dawson, R., and Lu, X.X., Fragment constant method for prediction of fish bioconcentration factors of nonpolar chemicals, *Chemosphere*, 41, 1563–1568, 2000.

Tao, S., Hu, H., Xu, F., Dawson, R., Li, B., and Cao, J., QSAR modeling of bioconcentration factors in fish based on fragment constants and structural correction factors, *J. Environ. Sci. Health Part B — Pest. Food Contaminants Agric. Wastes*, 36, 631-649, 2001.

Todeschini, R. and Gramatica, P., The WHIM theory: new 3D molecular descriptors for QSAR in environmental modelling, *SAR QSAR Environ. Res.,* 7, 89–115, 1997a.

Todeschini, R. and Gramatica, P., 3D-modelling and prediction by WHIM descriptors. Part 5. Theory development and chemical meaning of WHIM descriptors, *Quant. Struct.-Act. Relat.*, 16, 113–119, 1997b.

Tolls, J. and McLachlan, M.S., Partitioning of semivolatile organic compounds between air and *Lolium multiflorum* (Welsh ray grass), *Environ. Sci. Tech.*, 28, 159–166, 1994.

Topliss, J.G. and Costello, R.J., Chance correlations in structure-activity studies using multiple regression analysis, *J. Med. Chem.*, 15, 1066–1069, 1972.

Trapp, S. and Matthies, M., Generic one-compartment model for uptake of organic chemicals by foliar vegetation, *Environ. Sci. Tech.*, 29, 2333–2338, 1995.

Travis, C.C. and Arms, A.D., Bioconcentration of organics in beef, milk and vegetation, *Environ. Sci. Tech.*, 22, 271–274, 1988.

U.S. Environmental Protection Agency, Fish BCF, OPPTS 850.1730, EPA/712-C-96-129 (draft), Washington, D.C., 1996.

van Bavel, B., Andersson, P., Wingfors, H., Åhgren, J., Bergqvist, P.-A., Norrgren, L., Rappe, C. and Tysklind, M., Multivariate modeling of PCB bioaccumulation in three-spined stickleback (*Gasterosteus aculeatus*), *Environ. Toxicol. Chem.*, 15, 947–954, 1996.

Veith, G.D. and Kosian, P., Estimating bioconcentration potential from octanol/water partition coefficients, in *Physical Behaviour of PCBs in the Great Lakes*, Mackay, D., Paterson, S., Eisenreich, S.J., and Simmons, M.S., Eds., Ann Arbor Science, Ann Arbor, MI, 1983, pp. 269–282.

Voutsas, E., Magoulas, K., and Tassios, D., Prediction of the bioaccumulation of persistent organic pollutants in aquatic food webs, *Chemosphere*, 48, 645–651, 2002.

Wei, D., Zhang, A., Wu, C., Han, S., and Wang, L., Progressive study and robustness test of QSAR model based on quantum chemical parameters for predicting BCF of polychlorinated organic compounds, *Chemosphere*, 44, 1421–1428, 2001.

Weininger, D., SMILES, a chemical language and information system. 1. Introduction to methodology and encoding rules, *J. Chem. Inf. Comput. Sci.*, 28, 31–36, 1998.

Welke, B., Ettlinger, K., and Riederer, M., Sorption of volatile organic chemicals in plant surfaces, *Environ. Sci. Tech.*, 32, 1099–1104, 1998.

Yalkowsky, S.H., Valvani, S.C., and Roseman, T.J., Solubility and partitioning. 6. Octanol solubility and octanol-water partition coefficients, *J. Pharm. Sci.*, 72, 866–870, 1983.

Zok, S., Görge, G., Kalsch, W., and Nagel, R., Bioconcentration, metabolism and toxicity of substituted anilines in the zebrafish (*Brachydanio rerio*), *Sci. Total Environ.*, 109/110, 411–421, 1991.

# CHAPTER 16

# QSAR Modeling of Soil Sorption

John C. Dearden

## CONTENTS

## I. INTRODUCTION

Environmental pollution from chemicals is a global problem, and poses many risks. If we are to quantify those risks, it is necessary to have knowledge of the ways in which a chemical can be distributed in the environment. The fugacity model (Mackay, 1991; Mackay et al., 1992) incorporates the factors responsible for such distribution, one of which is soil sorption. Delle Site (2001) and Doucette (2000; 2003) have reviewed soil sorption in detail.

Sorption of a chemical (usually from water) by soil prevents the chemical, albeit generally temporarily, from being transported further in the environment. The umbrella term sorption is used to cover absorption, adsorption, and chemisorption, as it is often difficult to know the precise mechanism by which a chemical is taken up by soil. It is, however, known (Karickhoff, 1984) that virtually all soil sorption takes place on or in the organic layer that coats soil particles, particularly for soils with high organic matter content, and that pure silica has negligible sorptive ability. For this reason the sorption coefficient is written as $K_{oc}$, where "oc" stands for organic carbon, or sometimes as $K_{om}$ ("om" = organic matter); the relationship between the two is: $\log K_{oc} = \log K_{om} + 0.2365$ (Nendza, 1998). $K_{oc}$ is defined as:

$$K_{oc} = \frac{\text{concentration of chemical sorbed to soil}}{\text{concentration of chemical in water}} \tag{16.1}$$

Soil sorption is a partition phenomenon, as the above equation indicates. A common surrogate for partitioning of a chemical between an organism and water is its octanol-water partition coefficient, $K_{ow}$ (sometimes written simply as P) (Nendza, 1998). It will be seen later that this property (in the form of its common logarithm, log $K_{ow}$) in many cases correlates well with soil sorption.

As will be seen, the large amount of quantitative structure-activity relationship (QSAR) modeling that has been carried out for soil sorption has almost exclusively involved nonionic organic compounds. For strongly ionizing and inorganic chemicals, no QSARs are available. However, Bintein and Devillers (1994) developed a soil sorption QSAR that incorporated correction factors for ionization of weak acids and bases.

## II. METHODOLOGY

Soil sorption coefficients are most often determined using a batch technique (Organization for Economic Cooperation and Development, 1983; American Society for Testing and Materials, 1993; 2001), whereby a small quantity of the soil is agitated for a period of time (e.g., 18 hours) with an aqueous solution of the chemical under investigation; the phases are then separated and the concentration of chemical measured in one or both of the phases. While in principle the method is simple, problems can arise due to, for example, incomplete phase separation, lack of time for equilibration, volatilization loss, and chemical instability.

The packed column technique is also widely used (Lee et al., 1991; Macintyre et al., 1991), in which a solution of the chemical under investigation is pumped through a soil column, with the effluent concentration being monitored over time. Good agreement between batch and column methods has generally been observed (Doucette, 2003), although some appreciable differences have been noted (Maraqua et al., 1998).

Other methods that have been used to determine $K_{oc}$ values are the so-called box method (Macintyre et al., 1991), the continuous stirred reactor method (de Jonge et al., 1999) and headspace methods, which are especially useful for volatile chemicals (Garbarini and Lion, 1985). Delle Site (2001) has recently reviewed the available methods for the determination of $K_{oc}$ values.

Nendza (1998) and Doucette (2003) have discussed reasons for the variability of measured $K_{oc}$ values. These include soil inhomogeneity, lack of attainment of equilibrium, sorption of the test chemical onto vessel walls, volatilization of the test chemical, instability of the test chemical in both water and soil, pH and buffer capacity of the aqueous phase, presence of solubilizing agents and co-solutes, and incomplete phase separation prior to analysis.

It must also be pointed out that, because there is no such thing as a standard soil, reproducibility of results, especially between laboratories, is difficult to achieve. Nendza (1998) has reported that measured $K_{oc}$ values for a given chemical can vary by up to three orders of magnitude; for example, Doucette (2003) reported log $K_{oc}$ values of 2.85 and 4.13 for 1,3,5-trichlorobenzene. On the other hand, Hassett et al. (1980) found that $K_{oc}$ values measured on 13 different soils and sediments were indistinguishable, whilst Chiou et al. (1979) observed that the makeup of the organic matter of soils is not critical.

## III. QSAR MODELS

As mentioned above, soil sorption can be regarded as essentially a partitioning phenomenon, at least so far as adsorption and absorption are concerned. (Chemisorption clearly additionally involves reactivity.) It is thus not surprising that many QSARs have employed partition coefficient to model soil sorption. However, Bintein and Devillers (1994) and Baker et al. (1997) have pointed out that very hydrophilic chemicals (log $K_{ow}$ < 1.7) may not fit such a model, as they may not bind by a hydrophobic mechanism. Gawlik et al. (1997), Nendza (1998) and Doucette (2000; 2003) have comprehensively reviewed QSAR modeling of soil sorption.

Most QSAR studies have been carried out on specific chemical classes, and Nendza (1998) has drawn attention to the marked differences in both slopes and intercepts of such models. This is clearly shown by the work of Sabljić et al. (1995), who developed QSARs for a large number of chemical classes, some of which are given below.

1. Hydrocarbons and halogenated hydrocarbons

$$\log K_{oc} = 0.81 \log K_{ow} + 0.10 \tag{16.2}$$

$$n = 81, R^2 = 0.887, s = 0.451, F = 629$$

2. Alcohols

$$\log K_{oc} = 0.39 \log K_{ow} + 0.50 \tag{16.3}$$

$$n = 13, R^2 = 0.747, s = 0.397, F = 36$$

3. Anilines

$$\log K_{oc} = 0.62 \log K_{ow} + 0.85 \tag{16.4}$$

$$n = 20, R^2 = 0.808, s = 0.341, F = 81$$

4. Dinitroanilines

$$\log K_{oc} = 0.38 \log K_{ow} + 1.92 \tag{16.5}$$

$$n = 20, R^2 = 0.817, s = 0.242, F = 86$$

The pronounced differences between Equations 16.2 to 16.5 indicate that it might not be possible to obtain good soil sorption QSARs for diverse data sets; Nendza (1998) has suggested that the differences may reflect class-specific sorption mechanisms. However, some sorption QSARs based on diverse chemicals have been developed. Baker et al. (1997) found a remarkably good correlation:

$$\log K_{oc} = 0.903 \log K_{ow} + 0.094 \tag{16.6}$$

$$n = 72, R^2 = 0.91, s = 0.397$$

Gerstl (1990) used a very large diverse data set to obtain the following QSAR:

$$\log K_{oc} = 0.679 \log K_{ow} + 0.663 \tag{16.7}$$

$$n = 419, R^2 = 0.831, s \text{ not given}$$

It would be expected, in view of the differences in slopes and intercepts found for class-specific QSARs, that the correlation observed for diverse data sets would be appreciably lower than those found for class-specific data sets. The evidence from Equations 16.6 and 16.7 does not support that. Furthermore, Bintein and Devillers (1994), in a successful attempt to model both the variation

**Table 16.1    Some Soil Sorption QSARs Based on Log $K_{ow}$**

| QSAR | Chemical Class | n | $R^2$ | Reference |
|---|---|---|---|---|
| $\log K_{om} = 0.53 \log K_{ow} + 0.64$ | Pesticides | 105 | 0.90 | Briggs (1981) |
| $\log K_{oc} = 0.402 \log K_{ow} + 1.071$ | Pesticides | 15 | 0.69 | Kanazawa (1989) |
| $\log K_{oc} = 0.57 \log K_{ow} + 1.08$ | Phenols and benzonitriles | 24 | 0.76 | Sabljić et al. (1995) |
| $\log K_{oc} = 0.545 \log K_{ow} + 0.943$ | Ureas | 57 | 0.712 | Gerstl (1990) |
| $\log K_{oc} = 0.827 \log K_{ow} + 0.293$ | Aromatic hydrocarbons | 20 | 0.897 | Hodson and Williams (1988) |
| $\log K_{oc} = 1.03 \log K_{ow} - 0.61$ | Diverse | 117 | 0.90 | Seth et al. (1999) |
| $\log K_{oc} = 0.47 \log K_{ow} + 1.09$ | Diverse agricultural | 216 | 0.681 | Sabljić et al. (1995) |
| $\log K_{oc} = 0.544 \log K_{ow} + 1.377$ | Diverse (mostly pesticides) | 45 | 0.74 | Kenaga and Goring (1980) |

of organic content of soils and ionization of chemicals, developed the following QSAR based on 229 data points for 53 diverse chemicals:

$$\log K_p = 0.93 \log K_{ow} + 1.09 f_{oc} + 0.32 \, CFa - 0.55 \, CFb' + 0.25 \tag{16.8}$$

$$n = 229, \, R^2 = 0.933, \, s = 0.433, \, F = 786.1$$

where $K_p$ = sorption coefficient uncorrected for organic content, $f_{oc}$ = fraction of organic carbon in soil, $CFa$ = correction factor for acid ionization, and $CFb'$ = correction factor for base ionization.

Another QSAR involving correction for ionization was developed by van Gestel et al. (1991; 1993). Their 1991 paper reported the following correlation for a series of chlorobenzenes, chlorophenols, and chloroanilines, in which the $\log K_{ow}$ values of the chlorophenols were corrected for ionization:

$$\log K_{om} = 0.89 \log K_{ow} - 0.32 \tag{16.9}$$

$$n = 34, \, R^2 = 0.87, \, s \text{ not given}$$

A selection of published $\log K_{oc}$-$\log K_{ow}$ correlations is listed in Table 16.1. Others can be found in the comprehensive compilations of Gawlik et al. (1997), Nendza (1998), and Doucette (2000, 2003).

The author has selected a very diverse 100-compound test set of chemicals (see Table 16.2) from the compilation of Doucette (2003). Many of these chemicals are chlorinated compounds, reflecting the fact that a great number of environmental pollutants are chlorine-containing. For the 100-chemical set the correlation between soil sorption and hydrophobicity is found to be:

$$\log K_{oc} = 0.745 \log K_{ow} + 0.414 \tag{16.10}$$

$$n = 100, \, R^2 = 0.845, \, s = 0.521, \, F = 536.0$$

This QSAR is quite similar to that of Gerstl (1990) shown in Equation 16.7. The correlation is depicted in Figure 16.1, from which it can be seen that there is a greater spread of $\log K_{oc}$ values at higher $\log K_{ow}$ values, a phenomenon also observed by Sabljić et al. (1995). Whether this reflects inherently less accuracy in the determination of $\log K_{oc}$ for hydrophobic chemicals, or a greater dependency of $K_{oc}$ values of hydrophobic chemicals on soil type, is difficult to say.

In contrast to log BCF- (bioconcentration factor) $\log K_{ow}$ correlations (Meylan et al., 1999), no falloff in $\log K_{oc}$ values appears to be observed for highly hydrophobic chemicals, and no biphasic QSARs based on $\log K_{ow}$ appear to have been published. This suggests that equilibrium, or at least quasi-equilibrium, is generally achieved within the timescale of the determination of $K_{oc}$. In addition,

**Table 16.2 Measured Log K_oc values and Log K_oc Values Predicted by Various Methods for 100-Compound Test Set**

| Compound Name | CAS No. | Measured logK$_{oc}$ | Measured log K$_{ow}$ | PCKOCWIN ver. 1.66 | Magee (1991) | Poole & Poole (1999) | Dearden (this work) | Absolv | Gerstl (1990) | Baker et al. (1997) | Gerstl & Helling (1987) |
|---|---|---|---|---|---|---|---|---|---|---|---|
| Acridine | 260946 | 4.11 | 3.40 | 4.31 | 2.97 | 3.24 | 2.95 | 3.25 | 2.97 | 3.16 | 3.72 |
| 2-Aminoanthracene | 613138 | 4.45 | 3.43[a] | 4.52 | 2.73 | 3.93 | 2.97 | 3.94 | 2.99 | 3.19 | 3.78 |
| 6-Aminochrysene | 2642980 | 5.21 | 4.99 | 5.59 | 3.59 | 5.08 | 4.13 | 5.10 | 4.05 | 4.60 | 5.47 |
| Anisole | 100663 | 1.54 | 2.11 | 2.07 | 2.09 | 1.82 | 1.99 | 1.80 | 2.10 | 2.00 | 1.81 |
| Anthracene | 120127 | 4.42 | 4.54 | 4.31 | 3.44 | 4.20 | 3.80 | 4.22 | 3.75 | 4.19 | 3.94 |
| Benzene | 71432 | 1.92 | 2.13 | 2.22 | 1.98 | 1.90 | 2.00 | 1.90 | 2.11 | 2.02 | 1.47 |
| Benzoic acid | 65850 | 1.99 | 1.87 | 1.16 | 1.62 | 1.48 | 1.81 | 1.48 | 1.93 | 1.78 | 1.65 |
| Benzoic acid methyl ester | 93583 | 2.32 | 2.12 | 1.89 | 2.18 | 1.83 | 1.99 | 1.85 | 2.10 | 2.01 | 2.00 |
| Benzoic acid phenyl ester | 93992 | 3.53 | 3.59 | 3.23 | 3.06 | 3.15 | 3.09 | 3.16 | 3.10 | 3.34 | 3.23 |
| 2,2'-Biquinoline | 119915 | 4.02 | 4.31 | 5.89 | 3.69 | 4.10 | 3.62 | 4.11 | 3.59 | 3.99 | 5.22 |
| Bromacil | 314409 | 1.86 | 2.11 | 2.02 | 2.15 | 1.77 | 1.99 | 1.77 | 2.10 | 2.00 | 3.53 |
| 4-Bromonitrobenzene | 586787 | 2.42 | 2.55 | 2.49 | 2.40 | 2.78 | 2.31 | 2.77 | 2.39 | 2.40 | 1.47 |
| 4-Bromophenol | 106412 | 2.41 | 2.59 | 2.64 | 1.93 | 2.13 | 2.34 | 2.13 | 2.42 | 2.43 | 1.21 |
| 3-(3-Bromophenyl)-1-methyl-1-methoxyurea | 3060897 | 2.02 | 2.38 | 2.33 | 2.22 | 2.11 | 2.19 | 2.11 | 2.28 | 2.24 | 2.24 |
| Butralin | 33629479 | 3.91 | 4.87[a] | 3.55 | 3.95 | 3.98 | 4.04 | 3.98 | 3.97 | 4.49 | -0.50 |
| Butyl benzyl phthalate | 85687 | 4.23 | 4.91 | 3.97 | 4.07 | 4.39 | 4.07 | 4.38 | 4.00 | 4.53 | 5.25 |
| Butylbenzene | 104518 | 3.39 | 4.38 | 3.25 | 3.13 | 3.07 | 3.68 | 3.05 | 3.64 | 4.05 | 2.94 |
| Butyl-N-phenylcarbamate | 1538745 | 2.26 | 3.30 | 2.45 | 2.53 | 2.88 | 2.87 | 2.86 | 2.90 | 3.07 | 3.09 |
| Carbaryl | 63252 | 2.02 | 2.36 | 2.38 | 2.23 | 2.87 | 2.17 | 2.88 | 2.27 | 2.23 | 3.35 |
| 2-Chloroacetanilide | 533175 | 1.58 | 1.28 | 1.80 | 1.60 | 2.07 | 1.37 | 2.08 | 1.53 | 1.25 | 1.77 |
| 3-Chloroacetanilide | 588078 | 1.86 | 2.15 | 1.79 | 1.92 | 2.12 | 2.02 | 2.12 | 2.12 | 2.04 | 1.36 |
| Chlorobenzene | 108907 | 2.41 | 2.89 | 2.43 | 2.34 | 2.30 | 2.57 | 2.29 | 2.63 | 2.70 | 1.34 |
| Chloroform | 67663 | 1.65 | 1.97 | 1.54 | 1.46 | 1.69 | 1.88 | 1.69 | 2.00 | 1.87 | -3.21 |
| 3-Chloro-4-methoxyaniline | 5345540 | 1.93 | 1.85 | 1.72 | 1.78 | 1.96 | 1.79 | 1.95 | 1.92 | 1.76 | 1.76 |
| 3-(3-Chlorophenyl)-1,1-dimethylurea | 587348 | 1.79 | 2.00 | 1.92 | 2.01 | 2.14 | 1.90 | 2.14 | 2.02 | 1.90 | 1.35 |
| p,p'-DDT | 50293 | 5.38 | 6.91 | 5.34 | 4.76 | 5.95 | 5.56 | 5.95 | 5.35 | 6.33 | -2.19 |
| Diallate | 2303164 | 3.28 | 4.49 | 3.02 | 3.60 | 3.16 | 3.76 | 3.15 | 3.71 | 4.15 | 2.82 |
| 1,2,5,6-Dibenzanthracene | 53703 | 6.31 | 6.50 | 6.42 | 4.73 | 6.53 | 5.26 | 6.54 | 5.08 | 5.96 | 6.68 |
| 1,2,7,8-Dibenzocarbazole | 239645 | 6.11 | 6.40 | 6.16 | 4.21 | 5.34 | 5.18 | 5.34 | 5.01 | 5.87 | 6.31 |
| 1,2-Dibromo-3-chloropropane | 96128 | 2.11 | 2.96 | 2.12 | 2.45 | 2.76 | 2.62 | 2.76 | 2.67 | 2.77 | 4.49 |
| 1,2-Dibromoethane | 106934 | 1.64 | 1.96 | 1.64 | 1.93 | 2.12 | 1.87 | 2.12 | 1.99 | 1.86 | 3.94 |
| Dibutyl phthalate | 84742 | 3.14 | 4.72 | 3.16 | 3.82 | 3.63 | 3.93 | 3.65 | 3.87 | 4.36 | 4.78 |
| Dichlobenil | 1194656 | 2.37 | 2.74 | 2.43 | 2.48 | 2.85 | 2.46 | 2.83 | 2.52 | 2.57 | 1.94 |
| 3,4-Dichloroacetanilide | 2150938 | 2.34 | 3.00 | 2.01 | 2.31 | 2.49 | 2.65 | 2.49 | 2.70 | 2.80 | 1.98 |

**Table 16.2 (continued)**    Measured Log $K_{oc}$ values and Log $K_{oc}$ Values Predicted by Various Methods for 100-Compound Test Set

| Compound Name | CAS No. | Measured logK$_{oc}$ | Measured log K$_{ow}$ | PCKOCWIN ver. 1.66 | Magee (1991) | Poole & Poole (1999) | Dearden (this work) | Absolv | Gerstl (1990) | Baker et al. (1997) | Gerstl & Helling (1987) |
|---|---|---|---|---|---|---|---|---|---|---|---|
| 1,2-Dichlorobenzene | 95501 | 2.50 | 3.43 | 2.65 | 2.63 | 2.69 | 2.97 | 2.69 | 2.99 | 3.19 | 1.94 |
| 1,3-Dichlorobenzene | 541731 | 2.47 | 3.53 | 2.64 | 2.66 | 2.75 | 3.04 | 2.75 | 3.06 | 3.28 | 1.11 |
| Dichloromethane | 75092 | 1.44 | 1.25 | 1.38 | 1.49 | 1.34 | 1.35 | 1.36 | 1.51 | 1.22 | 0.17 |
| 3,4-Dichloronitrobenzene | 99547 | 2.53 | 3.12 | 2.71 | 2.64 | 3.00 | 2.74 | 3.00 | 2.78 | 2.91 | 2.03 |
| 2,3-Dichlorophenol | 576249 | 2.39 | 2.84 | 2.86 | 2.05 | 2.38 | 2.53 | 2.39 | 2.59 | 2.66 | 2.08 |
| 3-(3,4-Dichlorophenyl)-1-methylurea | 3567622 | 2.46 | 2.94 | 2.06 | 2.35 | 2.35 | 2.60 | 2.36 | 2.66 | 2.75 | 2.50 |
| Di-(2-ethylhexyl) phthalate | 117817 | 4.94 | 7.45 | 5.22 | 5.45 | 5.88 | 5.96 | 5.88 | 5.72 | 6.82 | 8.08 |
| Diethyl phthalate | 84662 | 2.58 | 2.47 | 2.10 | 2.68 | 2.49 | 2.25 | 2.49 | 2.34 | 2.32 | 3.00 |
| N,N-Dimethylaniline | 121697 | 2.63 | 2.31 | 1.89 | 2.29 | 1.85 | 2.13 | 1.85 | 2.23 | 2.18 | 1.46 |
| Dinoseb | 88857 | 2.09 | 3.56 | 3.55 | 2.73 | 3.74 | 3.07 | 3.75 | 3.08 | 3.31 | 3.12 |
| Diphenylamine | 122394 | 2.78 | 3.50 | 3.28 | 2.59 | 2.91 | 3.02 | 2.91 | 3.04 | 3.25 | 3.09 |
| Fluchloralin | 33245395 | 3.56 | 4.63 | 3.99 | 3.82 | 4.39 | 3.86 | 4.38 | 3.81 | 4.27 | 3.91 |
| Fluorene | 86737 | 3.68 | 4.18 | 4.05 | 3.24 | 3.73 | 3.53 | 3.72 | 3.50 | 3.87 | 4.06 |
| 2,2',3,4,5,5',6-Heptachlorobiphenyl | 52712057 | 5.95 | 7.93 | 5.32 | 5.13 | 6.50 | 6.32 | 6.51 | 6.05 | 7.25 | 5.63 |
| Hexachlorobenzene | 118741 | 3.59 | 5.73 | 3.53 | 3.80 | 4.65 | 4.68 | 4.64 | 4.55 | 5.27 | 4.58 |
| 2,2',3,3',4,4'-Hexachlorobiphenyl | 38380073 | 5.05 | 7.31 | 5.11 | 4.82 | 5.98 | 5.86 | 5.98 | 5.63 | 6.69 | 5.67 |
| Hexanoic acid | 142621 | 1.38 | 1.92 | 0.88 | 1.61 | 1.33 | 1.84 | 1.33 | 1.97 | 1.83 | 1.85 |
| Isocil | 314421 | 2.11 | 1.52[a] | 1.73 | 1.85 | 1.54 | 1.55 | 1.54 | 1.70 | 1.47 | 2.30 |
| Lindane | 58899 | 2.96 | 3.72 | 3.53 | 3.05 | 3.93 | 3.19 | 3.93 | 3.19 | 3.45 | 6.36 |
| 4-Methoxyacetanilide | 51661 | 1.40 | 1.05 | 1.43 | 1.55 | 1.58 | 1.20 | 1.59 | 1.38 | 1.04 | 1.91 |
| p,p'-Methoxychlor | 72435 | 4.90 | 5.08 | 4.63 | 4.15 | 4.95 | 4.20 | 4.97 | 4.11 | 4.68 | -1.25 |
| 3-Methylacetanilide | 537928 | 1.45 | 1.68 | 1.79 | 1.75 | 1.95 | 1.67 | 1.97 | 1.80 | 1.61 | 1.37 |
| 3-Methylaniline | 108441 | 1.65 | 1.40 | 1.86 | 1.50 | 1.90 | 1.46 | 1.90 | 1.61 | 1.36 | 1.15 |
| 9-Methylanthracene | 779022 | 4.81 | 5.07 | 4.54 | 3.72 | 4.50 | 4.19 | 4.50 | 4.11 | 4.67 | 4.36 |
| 4-Methylbenzoic acid ethyl ester | 94086 | 2.84 | 2.88[a] | 2.36 | 2.63 | 2.41 | 2.56 | 2.42 | 2.62 | 2.69 | 2.05 |
| 4-Methylbenzoic acid | 99945 | 2.28 | 2.27 | 1.37 | 1.86 | 1.77 | 2.11 | 1.76 | 2.20 | 2.14 | 1.52 |
| 3-Methyl-4-bromoaniline | 6933104 | 2.26 | 2.53 | 2.08 | 2.05 | 2.50 | 2.30 | 2.50 | 2.38 | 2.38 | 2.11 |
| Methyl-N-phenylcarbamate | 2603103 | 1.73 | 1.76 | 1.65 | 1.72 | 1.98 | 1.73 | 1.99 | 1.86 | 1.68 | 2.02 |
| Metribuzin | 21087649 | 1.98 | 1.70 | 3.08 | 2.08 | -0.82 | 1.68 | -0.83 | 1.82 | 1.63 | -1.50 |
| Monolinuron | 1746812 | 1.84 | 2.30 | 2.33 | 2.14 | 1.91 | 2.13 | 1.89 | 2.22 | 2.17 | 2.25 |
| Naphthalene | 91203 | 3.11 | 3.30 | 3.26 | 2.70 | 3.07 | 2.87 | 3.06 | 2.90 | 3.07 | 2.74 |
| 1-Naphthylamine | 134327 | 3.26 | 2.25 | 3.48 | 2.01 | 2.77 | 2.09 | 2.78 | 2.19 | 2.13 | 2.77 |

| Compound | CAS | | | | | | | | | | |
|---|---|---|---|---|---|---|---|---|---|---|---|
| Nitrobenzene | 98953 | 1.94 | 1.85 | 2.28 | 2.01 | 2.14 | 1.79 | 2.15 | 1.92 | 1.76 | 1.61 |
| 4-Nitrobenzoic acid | 62237 | 2.07 | 1.89 | 1.22 | 1.76 | 1.74 | 1.82 | 1.72 | 1.95 | 1.80 | 1.79 |
| 4-Nitrobenzoic acid ethyl ester | 997774 | 2.67 | 2.33 | 2.22 | 2.47 | 2.39 | 2.15 | 2.39 | 2.25 | 2.20 | 2.32 |
| 2,2',3,5',6-Pentachlorobiphenyl | 38379996 | 5.55 | 6.55 | 4.88 | 4.46 | 5.54 | 5.29 | 5.53 | 5.11 | 6.01 | 4.00 |
| Pebulate | 1114712 | 2.80 | 3.84 | 2.68 | 3.20 | 2.49 | 3.27 | 2.48 | 3.27 | 3.56 | 4.69 |
| 2,2',3',4,5-Pentachlorobiphenyl | 41464511 | 5.69 | 6.67 | 4.88 | 4.50 | 5.54 | 5.38 | 5.53 | 5.19 | 6.12 | 4.42 |
| 2,2',4,5,5'-Pentachlorobiphenyl | 37680732 | 4.63 | 6.85 | 4.87 | 4.57 | 5.54 | 5.52 | 5.53 | 5.31 | 6.28 | 3.68 |
| Phenanthrene | 85018 | 4.36 | 4.46 | 4.32 | 3.41 | 4.20 | 3.74 | 4.22 | 3.69 | 4.12 | 4.11 |
| Phenol | 108952 | 1.43 | 1.46 | 2.43 | 1.38 | 1.57 | 1.50 | 1.55 | 1.65 | 1.41 | 1.35 |
| 3-Phenyl-1-cyclohexylurea | 886599 | 2.07 | 2.77 | 2.27 | 2.49 | 2.92 | 2.48 | 2.91 | 2.54 | 2.60 | 4.86 |
| 3-Phenyl-1-cyclopentylurea | 13140891 | 1.93 | 2.65 | 2.01 | 2.37 | 2.63 | 2.39 | 2.62 | 2.46 | 2.49 | 4.37 |
| Phenylacetic acid ethyl ester | 101973 | 2.11 | 2.28 | 2.41 | 2.40 | 2.41 | 2.11 | 2.43 | 2.21 | 2.15 | 2.38 |
| Phenylurea | 64108 | 1.35 | 0.83 | 1.47 | 1.33 | 1.39 | 1.03 | 1.40 | 1.23 | 0.84 | 1.63 |
| Picloram | 1918021 | 1.23 | 1.77[a] | 1.26 | 1.89 | 1.78 | 1.73 | 1.77 | 1.86 | 1.69 | 2.53 |
| Propazine | 139402 | 2.56 | 2.93 | 2.55 | 2.54 | 2.89 | 2.60 | 2.90 | 2.65 | 2.74 | 0.15 |
| Propyl N-phenylcarbamate | 5532901 | 2.06 | 2.80 | 2.19 | 2.26 | 2.57 | 2.50 | 2.57 | 2.56 | 2.62 | 2.52 |
| Simazine | 122349 | 2.33 | 2.18 | 2.17 | 2.11 | 2.42 | 2.04 | 2.41 | 2.14 | 2.06 | 2.14 |
| Tetracene | 92240 | 5.81 | 5.90 | 5.36 | 4.22 | 5.37 | 4.81 | 5.38 | 4.67 | 5.42 | 5.14 |
| 1,2,4,5-Tetrachlorobenzene | 95943 | 4.27 | 4.64 | 3.07 | 3.24 | 3.60 | 3.87 | 3.60 | 3.81 | 4.28 | 2.25 |
| 2,3,7,8-Tetrachlorodibenzo-p-dioxin | 1746016 | 6.66 | 6.80 | 5.17 | 4.49 | 4.54 | 5.48 | 4.54 | 5.28 | 6.23 | 4.13 |
| 1,1,2,2-Tetrachloroethane | 79345 | 1.90 | 2.39 | 2.03 | 1.78 | 2.24 | 2.19 | 2.24 | 2.29 | 2.25 | 0.08 |
| Tetrachloroethene | 127184 | 2.42 | 3.40 | 2.03 | 2.53 | 2.45 | 2.95 | 2.44 | 2.97 | 3.16 | 0.10 |
| Tetrachloroguaiacol | 1539175 | 2.85 | 4.59 | 3.17 | 2.97 | 3.28 | 3.83 | 3.30 | 3.78 | 4.24 | 3.68 |
| 2,3,4,5-Tetrachlorophenol | 4901513 | 4.12 | 4.21 | 3.30 | 2.72 | 3.28 | 3.55 | 3.29 | 3.52 | 3.90 | 3.07 |
| Toluene | 108883 | 2.06 | 2.73 | 2.43 | 2.29 | 2.18 | 2.45 | 2.18 | 2.52 | 2.56 | 1.34 |
| 1,1,1-Trichloroethane | 71556 | 2.26 | 2.49 | 1.69 | 2.12 | 2.01 | 2.27 | 2.00 | 2.35 | 2.34 | -10.49 |
| 1,1,2-Trichloroethane | 79005 | 1.89 | 1.89 | 1.83 | 1.89 | 1.96 | 1.82 | 1.95 | 1.95 | 1.80 | 0.12 |
| 3,4,5-Trichlorophenol | 609198 | 3.48 | 4.01 | 3.07 | 2.57 | 2.79 | 3.40 | 2.80 | 3.39 | 3.72 | 2.22 |
| 3-Trifluoromethylaniline | 98168 | 2.36 | 2.29 | 2.50 | 1.84 | 2.02 | 2.12 | 2.02 | 2.22 | 2.16 | 1.28 |
| Trifluralin | 1582098 | 4.49 | 5.34 | 3.99 | 4.08 | 4.25 | 4.39 | 4.27 | 4.29 | 4.92 | 3.85 |
| 1,3,5-Trimethylbenzene | 108678 | 2.82 | 3.42 | 2.85 | 2.72 | 2.75 | 2.96 | 2.75 | 2.99 | 3.18 | 0.83 |
| 1,2-Xylene | 95476 | 2.11 | 3.12 | 2.65 | 2.52 | 2.46 | 2.74 | 2.47 | 2.78 | 2.91 | 1.80 |
| 1,3-Xylene | 108383 | 2.22 | 3.20 | 2.64 | 2.55 | 2.46 | 2.80 | 2.47 | 2.84 | 2.98 | 1.13 |
| 1,4-Xylene | 106423 | 2.31 | 3.15 | 2.64 | 2.53 | 2.46 | 2.76 | 2.47 | 2.80 | 2.94 | 1.21 |

[a] No measured value available. Value given is the mean of values calculated by KOWWIN (ver. 1.66), ClogP (ver. 1.0.0) (www.biobyte.com) and ChemSilico (www.logp.com)

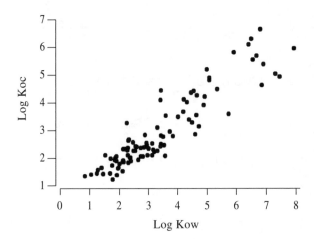

**Figure 16.1**   Correlation of measured log $K_{oc}$ with log $K_{ow}$ for 100-compound test-set listed in Table 16.2.

as one would expect for sorption on to non-living organic matter, there is little or no degradation of the chemical. Sabljić et al. (1995) have commented that one of the main reasons for outliers to log $K_{oc}$-log $K_{ow}$ correlations is specific binding to appropriate soil constituents.

Magee (1991) used an interesting approach to correct for the fact that log $K_{ow}$ is not an ideal surrogate for soil sorption. Log $K_{ow}$ has components of molecular size, polarity/polarizability, and hydrogen bonding (Abraham et al., 1994), but the relative contributions of each do not match those of the organic matter of soil. Magee included molar refractivity (MR), which itself has components of both molecular size and polarizability (Hansch et al., 2003), and an indicator variable (HBD) for hydrogen bond donor ability, to improve the correlation. (He found that hydrogen bond acceptor ability was not significant.) For a large, diverse data set he obtained:

$$\log K_{om} = 0.365 \log K_{ow} + 0.0175 \text{ MR} - 0.385 \text{ HBD} + 0.513 \qquad (16.11)$$

$$n = 128, \text{ R}^2 = 0.874, \, s = 0.276, \, F = 237.5$$

In addition to the partition coefficient, aqueous solubility has also been used extensively to model soil sorption, and Doucette (2000; 2003) has given comprehensive listings of published QSARs. It is pointed out, however, that although solubility is inversely related to partition coefficient (Hansch et al., 1968; Chiou et al., 1977), aqueous solubility is more difficult both to measure and to calculate accurately than is partition coefficient (Dearden et al., 2002a; 2002b). Furthermore, whilst partition coefficient is dimensionless, aqueous solubility is not, so that note must be taken of the units used. There are even a few examples of the wrong units being used (Kenaga and Goring, 1980; Kanazawa, 1989); it is essential in QSAR studies always to use mole and not weight concentrations.

Two examples of solubility-based soil sorption QSARs are that of Gerstl (1990) for ureas ($S_{aq}$ in mol/L):

$$\log K_{oc} = -0.381 \log S_{aq} + 1.177 \qquad (16.12)$$

$$n = 57, \text{ R}^2 = 0.616, \, s \text{ not given}$$

and that of Chiou et al. (1979) for chlorinated hydrocarbons ($S_{aq}$ in μmol/L):

$$\log K_{oc} = -0.557 \log S_{aq} + 4.040 \qquad (16.13)$$

$$n = 15, \text{ R}^2 = 0.988, \, s \text{ not given}$$

Note that a change in units from mol/L to μmol/L does not alter the slope, but changes the intercept by an increment of 6.

Reversed-phase high performance liquid chromatography (RP HPLC) works by eluates partitioning between mobile and stationary phases, and the retention factor log $k'$ is proportional to log $K_{ow}$. Numerous studies have correlated $K_{oc}$ with log $k'$, and Gawlik et al. (1997) list 35 such correlations, most of which relate to specific chemical classes. Note that the correlations vary with the stationary phase used. Some examples are:

1. Aromatic hydrocarbons (Pussemier et al., 1990)
   a. C18 column

$$\log K_{oc} = 1.332 \log k' - 0.261 \tag{16.14}$$

$$n = 11, R^2 = 0.829, s = 0.409$$

   b. Physically immobilized humic acid column

$$\log K_{oc} = 0.893 \log k' + 1.803 \tag{16.15}$$

$$n = 11, R^2 = 0.971, s = 0.169$$

2. Pesticides (Kördel et al., 1993)
   a. C18 column

$$\log K_{oc} = 1.40 \log k' + 2.5 \tag{16.16}$$

$$n = 30, R^2 = 0.76, s \text{ not given}$$

   b. Trimethylammoniumpropylic column

$$\log K_{oc} = 2.0 \log k' + 3.0 \tag{16.17}$$

$$n = 46, R^2 = 0.85, s \text{ not given}$$

The topological indices known as molecular connectivities (see Chapter 5 and Kier and Hall [1976; 1986]) have been extensively used to model soil sorption, and Gawlik et al. (1997) and Doucette (2000; 2003) have presented an extensive tabulation of published QSARs involving these descriptors. They are calculated from knowledge of the bonds between atoms in a molecule, and have been found to have wide applicability. The descriptors are information-rich, but their physicochemical significance is difficult to understand. Dearden et al. (1988) have shown that the lower order molecular connectivity terms correlate well with molecular size, and it is in effect as surrogates for size that molecular connectivities have been used to model $K_{oc}$ values. Some examples of soil sorption QSARs involving molecular connectivities are given below.

Bahnick and Doucette (1988) devised a polarity term, $\Delta^1\chi^v$, calculated as the difference between a modified $^1\chi^v$ (calculated assuming that polar atoms have been replaced by carbon) and the normal $^1\chi^v$ (i.e., 1st order valence molecular connectivity). Note that this is not the same as $\Delta^1\chi$ devised by Kier and Hall (1986).

$$\log K_{oc} = 0.53 \,^1\chi - 2.09 \,\Delta^1\chi^v + 0.64 \tag{16.18}$$

$$n = 56, R^2 = 0.94, s = 0.34$$

Gerstl and Helling (1987) developed a QSAR based on two 3rd order molecular connectivities for a heterogeneous set of non-pesticides:

$$\log K_{oc} = 1.953 \,^3\chi_p{}^v - 4.010 \,^3\chi_c{}^v + 0.17 \tag{16.19}$$

$$n = 23, R^2 = 0.905, s \text{ not given}$$

where $^3\chi_p{}^v$ = third-order valence path molecular connectivity, and $^3\chi_c{}^v$ = third-order valence cluster molecular connectivity. The use of valence connectivity terms means that there is some electronic component to the descriptors.

Sabljić et al. (1995) developed a molecular connectivity model for hydrophobic chemicals (i.e., hydrocarbons and halogenated hydrocarbons) based on first-order $\chi$ values:

$$\log K_{oc} = 0.52 \,^1\chi + 0.70 \tag{16.20}$$

$$n = 81, R^2 = 0.961, s = 0.264, F = 1993$$

This model is better than that based on log $K_{ow}$ which Sabljić et al. (1995) derived for the same chemicals:

$$\log K_{oc} = 0.81 \log K_{ow} + 0.10 \tag{16.21}$$

$$n = 81, R^2 = 0.887, s = 0.451, F = 629$$

Clearly, $^1\chi$ encodes more relevant information (probably size) than does log $K_{ow}$, which does contain a size component, but also contains hydrogen bonding and polarity/polarizability components (Dearden and Bentley, 2002). Log $K_{ow}$ would, however, be expected to be a better descriptor for polar chemicals. In connection with this, Gerstl and Helling (1987) commented that the ability of molecular connectivities to predict log $K_{oc}$ was rather limited for diverse data sets. Baker et al. (2001) included two cluster connectivity terms to improve the correlation of soil sorption of a small hydrophobic data set, yielding $R^2 = 0.806$ and $s = 0.302$.

Sabljić (1987) used polarity correction factors together with $^1\chi$ to model log $K_{om}$ values of 160 diverse chemicals with $R^2 = 0.945$ and $s = 0.276$. Meylan et al. (1992) made a successful attempt to circumvent the limitations of molecular connectivity indices by the inclusion of fragmental contributions. Their method forms the basis of the PCKOCWIN software, which can be downloaded free of charge from the U.S. Environmental Protection Agency (EPA) website (www.epa.gov/oppt/exposure/docs/episuitedl.htm).

Given the apparent importance of molecular size in modeling soil sorption of hydrophobic chemicals, it is not surprising to find various size descriptors having been used for that purpose, including parachor (Briggs, 1981), molecular weight (Kanazawa, 1989), molar refractivity (Koch and Nagel, 1988) and total surface area (TSA) (Hansen et al., 1999). The QSAR developed by Hansen et al. for a series of polychlorinated biphenyls (PCBs) was:

$$\log K_{oc} = 0.0252 \, TSA - 0.677 \tag{16.22}$$

$$n = 48, R^2 = 0.92, \text{rms error} = 0.17$$

Park and Lee (1993) applied the Abraham linear solvation energy relationship (LSER) descriptors (Kamlet et al., 1983) to the prediction of log $K_{oc}$ values of a diverse data set:

$$\log K_{oc} = 4.84 \ V_I/100 - 0.50 \ \pi^* - 0.59 \ \delta - 1.11 \ \beta \qquad (16.23)$$

$$n = 42, \ R^2 = 0.937, \ s = 0.36$$

where $V_I$ = intrinsic molecular volume, $\pi^*$ = polarity/polarizability descriptor, $\delta$ = polarizability correction factor, and $\beta$ = hydrogen bond acceptor ability. They reported that the correlation was comparable to that of models using molecular connectivity.

Baker et al. (1997) also used the Abraham descriptors to model $K_{oc}$ of a diverse data set, but with rather poor results:

$$\log K_{oc} = -2.09 \ \beta + 3.36 \ V_i/100 + 0.93 \qquad (16.24)$$

$$n = 68, \ R^2_{adj} = 0.72, \ s \ \text{not given}$$

Poole and Poole (1999) used more up-to-date LSER descriptors to model soil sorption of a large diverse data set:

$$\log K_{oc} = 0.74 \ R_2 - 0.31 \ \Sigma\alpha_2^H - 2.27 \ \Sigma\beta_2^O + 2.09 \ V_X + 0.21 \qquad (16.25)$$

$$n = 131, \ R^2 = 0.955, \ s = 0.245, \ F = 655$$

where $R_2$ = a measure of polarizability, $\Sigma\alpha_2^H$ = hydrogen bond donor ability, $\Sigma\beta_2^O$ = hydrogen bond acceptor ability of oxygen, and $V_X$ = McGowan molecular volume. Polarizability and molecular size both increase soil sorption, while solute hydrogen bonding ability decreases it. Since the LSER descriptors are approximately autoscaled, one can also obtain an indication of the relative contribution of each type of interaction; in the above example, solute hydrogen bond acceptor ability and molecular size are very important, whereas solute hydrogen bond donor ability plays only a small part.

Several workers (Von Oepen et al., 1991; Reddy and Locke, 1994; Dai et al., 1999; 2000) have utilized quantum chemical descriptors to model soil sorption. Dai et al. (2000) found a good QSAR for a series of benzaldehydes, using the Austin Model 1 (AM1) Hamiltonian:

$$\log K_{oc} = -1.02 \ \mu - 9.24 \ QH^+ - 0.49 \ Q^- + 5.99 \qquad (16.25)$$

$$n = 14, \ R^2 = 0.864, \ s = 0.129, \ F = 28.5$$

where $\mu$ = dipole moment, $QH^+$ = most positive charge on hydrogen, and $Q^-$ = most negative atomic charge. Reddy and Locke (1994), however, found that the incorporation of $\mu$ and energy of the highest occupied molecular orbital ($E_{HOMO}$) gave no improvement over the use of a size term alone (van der Waals volume) to model the soil sorption of 71 herbicides.

Mention has already been made of the method of Meylan et al. (1992) which uses first-order molecular connectivity and fragmental contributions to model soil sorption. A similar approach was developed by Tao and Lu (1999), who developed a QSAR containing 3 molecular connectivity terms and 15 fragmental constants for a data set of 543 compounds. They found $R^2 = 0.860$ and a mean error of 0.346 log units.

Tao et al. (1999) used fragmental constants alone to model soil sorption, for a data set of 592 chemicals. Using 74 fragmental constants and 24 structural factors, they found a very high $R^2$ value of 0.967, with a mean error of 0.366 log units. Tao et al. (2001) compared the fragment constant and molecular connectivity models for the estimation of soil sorption, using the same 592 chemicals.

**Table 16.3    Analysis of Log $K_{oc}$ Predictions Shown in Table 16.2**

| Method | Percentage of Compounds with Errors in Range | | | Correlation Statistics | | |
|---|---|---|---|---|---|---|
| | <0.5 | ≥0.5–1.0 | >1.0 | $R^2$ | s | F |
| PCKOCWIN ver. 1.66 | 82% | 13% | 5% | 0.863 | 0.490 | 619.4 |
| Magee (1991) | 75% | 12% | 13% | 0.830 | 0.547 | 477.6 |
| Poole & Poole (1999) | 72% | 24% | 4% | 0.815 | 0.570 | 431.9 |
| Dearden (this work) | 72% | 20% | 8% | 0.845 | 0.521 | 536.4 |
| Absolv | 70% | 26% | 4% | 0.815 | 0.569 | 433.1 |
| Gerstl (1990) | 70% | 22% | 8% | 0.846 | 0.521 | 536.7 |
| Baker et al. (1997) | 69% | 21% | 10% | 0.845 | 0.521 | 535.6 |
| Gerstl & Helling (1987) | 47% | 14% | 39% | 0.209 | 1.179 | 25.9 |

They found the former considerably better, since their molecular connectivity model had $R^2 = 0.765$ with a mean error of 0.440 log units. It should be noted that the $R^2$ value of 0.967 for their fragmental constant method is so high that overfitting has probably occurred, since the error on measured log $K_{oc}$ values is quite high (Nendza, 1998).

Another group contribution method that has been applied to the prediction of soil sorption is the UNIquac Functional-group Activity Coefficient (UNIFAC, where UNIQUAC = Universal Quasichemical) approach (Fredenslund et al., 1977). Ames and Grulke (1995) applied the method to a small diverse set of chemicals, with rather poor results. They did not report any correlations, but from their results it can be shown that the correlation of observed and predicted log $K_{oc}$ values using the Bondi method was: $n = 17$, $R^2 = 0.571$, $s = 0.524$, and $F = 20.0$; eight chemicals were predicted with an error of < 0.5 log units, 7 chemicals were predicted with an error between 0.5 and 1.0 log units, and 2 chemicals were predicted with an error of > 1.0 log units.

## IV. COMMERCIAL SOFTWARE FOR THE PREDICTION OF LOG $K_{OC}$

So far as the author is aware, there are two commercial software packages currently available for the prediction of soil sorption coefficients; these are PCKOCWIN from Syracuse Research Corporation, freely downloadable from the EPA website, and Absolv, developed by Sirius Analytical Ltd., Forest Row, Surrey, U.K. (www.sirius-analytical.com), and based on the Abraham LSER descriptors. Their predictions for a 100-chemical test set are given in Table 16.2 and Table 16.3, and are commented on in the next section.

## V. COMPARISON OF SOME $K_{OC}$ PREDICTION MODELS

A number of models have been selected for validation using the 100-chemical test set listed in Table 16.2, the criteria being that each must be applicable to heterogeneous data sets, must have given good results in their authors' hands, and must use descriptors that are readily available and easily calculated. The methods examined are the two commercially available software programs, PCKOCWIN (version 1.66) and Absolv, and the published methods of Magee (1991), Poole and Poole (1999), Gerstl (1990), Baker et al. (1997), Gerstl and Helling (1987), and Dearden (this work). Note that the QSAR developed by Magee (1991) used log $K_{om}$ and not log $K_{oc}$. As Nendza (1998) has pointed out, log $K_{om}$ values can be converted to log $K_{oc}$ values by the addition of 0.2365. The Abraham descriptors for the Poole and Poole equation include $\Sigma\beta_2^O$, the hydrogen bond acceptor ability of oxygen; this descriptor is not available in Absolv, which the present author used to obtain the Abraham descriptors; $\Sigma\beta_2^H$ (total hydrogen bond acceptor ability), which is very similar to $\Sigma\beta_2^O$, was used instead. The predictions are listed in Table 16.2, and the statistics are given in Table 16.3.

It is clear that PCKOCWIN, which uses molecular connectivities and fragmental constants, gives by far the best predictions of log $K_{oc}$. The methods based on LSER descriptors (Poole and Poole [1999] and Absolv) are quite good, although they both underpredicted two compounds (metribuzin and 2,3,7,8-tetrachlorodibenzo-$p$-dioxin) by more than two log units. Interestingly, the method of Magee, which uses hydrogen bonding and molecular size/polarizability as (in effect) correction terms for log $K_{ow}$, gives better results than the two LSER methods. The three methods based on log $K_{ow}$ alone (those of Dearden, Gerstl, and Baker et al.) are also of similar predictive ability to the previous two. The methods based solely on molecular connectivities give poor predictions. This is probably because molecular connectivities are really unable to model polarity and hydrogen bonding, while they are probably able to predict soil sorption of hydrophobic compounds, they should not be used for polar compounds.

## VI. RECOMMENDATIONS

PCKOCWIN (version 1.66) has been found (see Table 16.3) to give the best predictions of log $K_{oc}$, and it is recommended that this be used, as it is freely downloadable from the EPA website (www.epa.gov/oppt/exposure/docs/episuitedl.htm). PCKOCWIN requires that the chemical structure be input using Simplified Molecular Line Entry System (SMILES) notation (Weininger, 1988) or as a Chemical Abstract Service (CAS) number.

## REFERENCES

Abraham, M.H., Chadha, H.S., and Mitchell, R.C., Hydrogen bonding. 32. An analysis of water-octanol and water-alkane partitioning and the Δlog P parameter of Seiler, *J. Pharm. Sci.*, 83, 1085–1100, 1994.

Ames, T.T. and Grulke, E.A., Group contribution method for predicting equilibria of nonionic organic compounds between soil organic matter and water, *Environ. Sci. Tech.*, 29, 2273–2279, 1995.

American Society for Testing and Materials, *Standard Test Method for Determining a Sorption Constant ($K_{oc}$) for an Organic Chemical in Soils and Sediments,* Philadelphia, PA, 1993.

American Society for Testing and Materials, *Standard Test Method for 24-Hour Batch-Type Measurement of Containment Sorption by Soils and Sediments,* Philadelphia, PA, 2001.

Bahnick, D.A. and Doucette, W.J., Use of molecular connectivity indices to estimate soil sorption coefficients for organic chemicals, *Chemosphere*, 17, 1703–1715, 1988.

Baker, J.R., Mihelcic, J.R., Luehrs, D.C., and Hickey, J.P., Evaluation of estimation methods for organic carbon normalized sorption coefficients, *Water Environ. Res,*, 69, 136–144, 1997.

Baker, J.R., Milhelcic, J.R., and Sabljić, A., Reliable QSAR for estimating $K_{oc}$ for persistent organic pollutants: correlation with molecular connectivity indices, *Chemosphere*, 45, 213–221, 2001.

Bintein, S. and Devillers, J., QSAR for organic chemical sorption in soils and sediments, *Chemosphere* 28, 1171–1188, 1994.

Briggs, G.G., Theoretical and experimental relationships between soil adsorption, octanol-water partition coefficients, water solubilities, bioconcentration factors and the parachor, *J. Agric. Food Chem.*, 29, 1050–1059, 1981.

Chiou, C.T., Peters, L.J., and Freed, V.H., A physical concept of soil-water equilibria for nonionic organic compounds, *Science*, 206, 831–832, 1979.

Chiou, C.T., Freed, V.H., Schmedding, D.W., and Kohnert, R.L., Partition coefficient and bioaccumulation of selected organic chemicals, *Environ. Sci. Tech.*, 11, 475–478, 1977.

Dai, J., Sun, C., Han, S., and Wang, L., QSAR for polychlorinated organic compounds (PCOCs). I. Prediction of partition properties for PCOCs using quantum chemical parameters, *Bull. Environ. Contamination Toxicol.*, 62, 530–538, 1999.

Dai, J.Y., Xu, M., and Wang, L.S., Prediction of octanol/water partitioning coefficient and sediment sorption coefficient for benzaldehydes by various molecular descriptors, *Bull. Environ. Contamination Toxicol.*, 65, 190–199, 2000.

Dearden, J.C. and Bentley, D., The components of the "critical quartet" log $K_{ow}$ values assessed by four commercial software packages, *SAR QSAR Environ. Res.*, 13, 185–197, 2002.

Dearden, J.C., Bradburne, S.J.A., Cronin, M.T.D., and Solanki, P., The physical significance of molecular connectivity, in *QSAR 88*, Turner, J.E., Williams, M.W. Schultz, T.W., and Kwaak, N.J., Eds., U.S. Department of Energy, Oak Ridge, TN, 1988, pp. 43–50.

Dearden, J.C., Netzeva, T.I., and Bibby, R., Comparison of a number of commercial software programs for the prediction of octanol-water partition coefficients, *J. Pharm. Pharmacol.*, 54 (Suppl.), S65–S66, 2002a.

Dearden, J.C., Netzeva, T.I., and Bibby, R., Comparison of a number of commercial software programs for the prediction of aqueous solubility, *J. Pharm. Pharmacol.*, 54 (Suppl.), S66, 2002b.

de Jonge, H., Heimovaara, T.J., and Verstraten, J.M., Naphthalene sorption to organic soil materials studied with continuous stirred flow experiments, *Soil Sci. Soc. Am. J.*, 63, 297–306, 1999.

Delle Site, A., Factors affecting sorption of organic compounds in natural sorbent/water systems and sorption coefficients for selected pollutants. A review, *J. Phys. Chem. Ref. Data*, 30, 187–439, 2001.

Doucette, W.J., Soil and sediment sorption coefficients, in *Handbook of Property Estimation Methods for Chemicals: Environmental and Health Sciences*, Boethling, R.S. and Mackay, D., Eds., Lewis Publishers, Boca Raton, FL, 2000, pp. 141–188.

Doucette, W.J., Quantitative structure-activity relationships for predicting soil/sediment sorption coefficients for organic chemicals, *Environ. Toxicol. Chem.*, 22, 1771–1788, 2003.

Fredenslund, Aa., Gmehling, J., and Rasmussen, P., *Vapor-Liquid Equilibria using UNIFAC*, Elsevier Scientific, New York, NY, 1977.

Garbarini, D.R. and Lion, L.W., Evaluation of sorptive partitioning of non-ionic pollutants in closed systems by headspace analysis, *Environ. Sci. Tech.*, 19, 1122–1128, 1985.

Gawlik, B.M., Sotiriou, N., Feicht, E.A., Schulte-Hostede, S., and Kettrup, A., Alternatives for the determination of the soil adsorption coefficient, $K_{oc}$, of non-ionic organic compounds — a review, *Chemosphere*, 34, 2525–2551, 1997.

Gerstl, Z., Estimation of organic chemical sorption by soils, *J. Contaminant Hydrology*, 6, 357–375, 1990.

Gerstl, Z. and Helling, C.S., Evaluation of molecular connectivity as a predictive method for the adsorption of pesticides by soils, *J. Environ. Sci. Health Part B – Pest. Food Contaminants Agric. Wastes*, 22, 55–69, 1987.

Hansch, C., Quinlan, J.E., and Lawrence, G.L., The linear free-energy relationship between partition coefficients and the aqueous solubility of organic liquids, *J. Org. Chem.*, 33, 347–350, 1968.

Hansch, C., Steinmetz, W.E., Leo, A.J., Mekapati, S.B., Kurup, A., and Hoekman, D., On the role of polarizability in chemical-biological interactions, *J. Chem. Inf. Comput. Sci.*, 43, 120–125, 2003.

Hansen, B.G., Paya-Perez, A.B., Rahman, M., and Larsen, B.R., QSARs for $K_{ow}$ and $K_{oc}$ of PCB congeners: a critical examination of data, assumptions and statistical approaches, *Chemosphere*, 39, 2209–2228, 1999.

Hassett, J.J., Means, J.C., Banwart, W.L., and Wood, S.J., *Sorption properties of sediments and energy-related pollutants,* U.S. Environmental Protection Agency Report No. EPA-600/3-80-041, Washington, D.C., 1980.

Hodson, J. and Williams, N.A., The estimation of the adsorption coefficient ($K_{oc}$) for soils by high performance liquid chromatography, *Chemosphere*, 17, 67–77, 1988.

Kamlet, M.J., Abboud, J.-L.M., Abraham, M.H., and Taft, R.W., Linear solvation energy relationships. 23. A comprehensive collection of the solvatochromic parameters $\pi^*$, $\alpha$ and $\beta$, and some methods for simplifying the generalized solvatochromic equation, *J. Org. Chem.*, 48, 2877–2887, 1983.

Kanazawa, J., Relationship between the soil sorption constants for pesticides and their physicochemical properties, *Environ. Toxicol. Chem.*, 8, 477–484, 1989.

Karickhoff, S.W., Organic pollutant sorption in aquatic systems, *J. Hydraul. Eng.*, 110, 707–735, 1984.

Kenaga, E.E. and Goring, C.A.I., Relationship between water solubility and soil sorption, octanol-water partitioning and bioconcentration of chemicals in biota, in *Aquatic Toxicology*, Special Technical Publication 707, Eaton, J.G., Parrish, P.R.P., and Hendricks, A.C., Eds., American Society for Testing and Materials, Philadelphia, PA, 1980, pp. 78–115.

Kier, L.B. and Hall, L.H., *Molecular Connectivity in Chemistry and Drug Research*, Academic Press, New York, NY, 1976.

Kier, L.B. and Hall, L.H., *Molecular Connectivity in Structure-Activity Analysis*, Wiley, London, 1986.

Koch, R. and Nagel, M., Quantitative structure-activity relationships in soil ecotoxicology, *Toxicol. Environ. Chem.*, 77, 269–276, 1988.

Kördel, W., Stutte, J., and Kotthoff, G., HPLC-screening method for the determination of the adsorption-coefficient on soil — comparison of different stationary phases, *Chemosphere*, 27, 2341–2352, 1993.

Lee, L.S, Rao, P.S.C., and Brusseau, M.L., Nonequilibrium sorption and transport of neutral and ionized chlorophenols, *Environ. Sci. Tech.*, 25, 722–729, 1991.

Macintyre, W.G., Stauffer, T.B., and Antworth, C.P., A comparison of sorption coefficients determined by batch, column, and box methods on a low organic-carbon aquifer material, *Ground Water*, 29, 908–913, 1991.

Mackay, D., *Multimedia Environmental Models: The Fugacity Approach*, Lewis Publishers, Chelsea, MI, 1991.

Mackay, D., Paterson, S., and Shiu, W.Y., Generic models for evaluating the regional fate of chemicals, *Chemosphere*, 24, 695–717, 1992.

Magee, P.S., Complex factors in hydrocarbon/water, soil/water and fish/water partitioning, *Sci. Total Environ.*, 109/110, 155–178, 1991.

Maraqa, M.A., Zhao, X., Wallace, R.B., and Voice T.C., Retardation coefficients of non-ionic compounds determined by batch and column techniques, *Soil Sci. Soc. Am. J.*, 62, 142–152, 1998.

Meylan, W., Howard, P.H., and Boethling, R.S., Molecular topology/fragment contribution method for predicting soil sorption coefficients, *Environ. Sci. Tech.*, 26, 1560–1567, 1992.

Meylan, W.M., Howard, P.H., Boethling, R.S., Aronson, D., Printup, H., and Gouchie, S., Improved method for estimating bioconcentration/bioaccumulation factor from octanol/water partition coefficient, *Environ. Toxicol. Chem.*, 18, 664–672, 1999.

Nendza, M., *Structure-Activity Relationships in Environmental Sciences*, Chapman & Hall, London, 1998.

Organization for Economic Cooperation and Development, *Guide-Line for Testing of Chemicals: Adsorption/Desorption*, Paris, France, 1983.

Park, J.H. and Lee, H.J., Estimation of bioconcentration factor in fish, adsorption coefficient for soils and sediments and interfacial tension with water for organic nonelectrolytes based on the linear solvation energy relationships, *Chemosphere*, 26, 1905–1916, 1993.

Poole, S.K. and Poole, C.F., Chromatographic models for the sorption of neutral organic compounds by soil from air and water, *J. Chromatogr. A*, 845, 381–400, 1999.

Pussemier, L., Szabó, G., and Bulman, R.A., Prediction of the soil adsorption coefficient $K_{oc}$ for aromatic pollutants, *Chemosphere*, 21, 1199–1212, 1990.

Reddy, K.N. and Locke, M.A., Prediction of soil sorption ($K_{oc}$) of herbicides using semiempirical molecular properties, *Weed Sci.*, 42, 453–461, 1994.

Sabljić, A., On the prediction of soil sorption coefficients of organic pollutants from molecular structure: application of molecular topology model, *Environ. Sci. Tech.*, 21, 358–366, 1987.

Sabljić, A., Güsten, H., Verhaar, H., and Hermens, J., QSAR modelling of soil sorption. Improvements and systematics of log $K_{oc}$ *vs.* log $K_{ow}$ correlations, *Chemosphere*, 31, 4489–4514, 1995.

Seth, R., Mackay, D., and Muncke, J., Estimating the organic carbon partition coefficient and its variability for hydrophobic chemicals, *Environ. Sci. Tech.*, 33, 2390–2394, 1999.

Tao, S. and Lu, X.X., Estimation of organic carbon normalized sorption coefficient ($K_{oc}$) for soils by topological indices and polarity factors, *Chemosphere*, 39, 2019–2034, 1999.

Tao, S., Lu, X.X., Cao, J., and Dawson, R., A comparison of the fragment and molecular connectivity indices models for normalized sorption coefficient estimation, *Water Environ. Res.*, 73, 307–313, 2001.

Tao, S., Piao, H., Dawson, R., Lu, X., and Hu, H., Estimation of organic carbon normalized sorption coefficient ($K_{oc}$) using the fragment constant method, *Environ. Sci. Tech.*, 33, 2719–2725, 1999.

van Gestel, C.A.M. and Ma, W.-C., Development of QSARs in soil ecotoxicology: earthworm toxicity and soil sorption of chlorophenols, chlorobenzenes and chloroanilines, *Water, Air Soil Pollut.*, 69, 265–276, 1993.

van Gestel, C.A.M., Ma, W.-C., and Smit, C.E., Development of QSARs in terrestrial ecotoxicology: earthworm toxicity and soil sorption of chlorophenols, chlorobenzenes and dichloroaniline, *Sci. Total Environ.*, 109/110, 589–604, 1991.

Von Oepen, B., Kördel, W., Klein, W., and Schüürmann, G., Predictive QSPR models for estimating soil sorption coefficients: potential and limitations based on dominating processes, *Sci. Total Environ.*, 109/110, 343–354, 1991.

Weininger, D., SMILES, a chemical language and information system. 1. Introduction to methodology and encoding rules, *J. Chem. Inf. Comput. Sci.*, 28, 31–36, 1988.

# CHAPTER 17

# Application of Catabolic-Based Biosensors to Develop QSARs for Degradation

Graeme I. Paton, Jacob G. Bundy, Colin D. Campbell, and Helena Maciel

## CONTENTS

# I. INTRODUCTION

The prediction of biodegradation is less well established than the prediction of toxicity (Damborsky and Schultz, 1997). The majority of quantitative structure-biodegradability relationships (QSBRs) are based on data derived from the Organisation for Economic Co-operation and Development (OECD) biodegradability test (301), as producing a standardized inoculum poses significant difficulties. Biodegradation is measured (as biomass production, $O_2$ depletion, or $CO_2$ production) over a 28-day period, and a chemical is then classified semiquantitatively as readily biodegradable, partially biodegradable, or non-biodegradable. The results generated produce QSBRs that are frequently semi-quantitative and are based on a summation of structural fragments (Howard et al. 1991; Degner et al., 1991). More sophisticated methods use multivariate analysis and neural networks to classify chemicals or to predict rates.

The ideal predictive tool for degradation will link an appropriate metabolic pathway with a response that indicates actual degradation. Biological tools capable of fulfilling this requirement have become readily available in recent years, though few researchers have made the linkage with QSARs.

## A. Reporter Genes and Biosensors

Bioluminescence is widely accepted as an excellent reporter mechanism for microbial biosensors (Atlas and Bartha, 1992; Meighen, 1988). By coupling the genes for bioluminescence to genes for specific promoters, biosensors can be developed that are activated by a particular biological response or activity (Barkay et al., 1995). Biosensors of this type have been constructed to report on heavy metal resistance, nitrogen and phosphorus starvation, and a range of stresses (Tescione and Belfort, 1993; Selifonova et al., 1993; Kragelund et al., 1995; Prest et al., 1997; Belkin et al., 1997). The largest group, however, report on hydrocarbon degradation pathways: catabolic biosensors exist for naphthalene (*nah*) and salicylate, toluene (*tod*), and other monocyclic aromatic hydrocarbons; isopropylbenzene (*ipb*); octane (*oct*); and biphenyl (King et al., 1990; Applegate et al., 1998; Ikariyama et al., 1997; Willardson et al., 1998; Selifonova and Eaton, 1996; Sticher et al., 1997; Layton et al., 1998). These catabolic sensors use genes regulating degradation as promoters for bioluminescence, and respond in a sensitive, rapid, and quantitative way to hydrocarbons.

## B. Hydrocarbon Degradation

The biochemistry and genetics of aerobic bacterial hydrocarbon degradation have been extensively studied. Bacterial pathways for hydrocarbon degradation typically have a low specificity, allowing microbes to deal with a large number of different, but related, compounds (Zylstra and Gibson, 1991). In practice, this means that a microbial catabolic sensor will respond to a range of structurally related compounds. There can even be a response to chemicals surprisingly different from the archetype. For example, Selifonova and Eaton (1996) found that an isopropylbenzene biosensor was induced by a range of compounds including halogenated chemicals (aliphatic and aromatic) and even heterocycles. It follows that *lux*-marked biosensors can be used to study the regulation of hydrocarbon degradation pathways. Data can be produced rapidly, easily making them ideal candidates for developing quantitative structure-activity relationships (QSARs).

## C. Overview of Metabolic Pathways for Consideration

During the construction of hydrocarbon-based catabolic biosensors, extensive characterization is conducted. It is not the intention of this chapter to report these findings, but a short summary of the key catabolic factors will be given here. A general observation may be made at the outset that hydrocarbon degradation lacks specificity, so individual genes have the ability to degrade a wide

**Table 17.1    Summary of the Biosensors Used in this Study**

| Biosensor | Parent Strain | Constraint Name | Induced Catabolic Ability | Typical Degradation Product | Reference |
|-----------|---------------|-----------------|---------------------------|-----------------------------|-----------|
| A | *P. putida* F1 | *P. putida* TVA8 | *Tod* | Toluene | Applegate et al., 1998 |
| B | *P. fluorescens* | *P. fluorescens* HK44 | *Nah* | Naphthalene | King et al., 1990 |
| C | *P. putida* RE204 | *E. coli* HMS174 | *Ipb* | Isopropylbenzene | Selfinova and Eaton, 1996 |
| D | *P. oleovorans* | *E.coli* DH5α | *Oct* | Octane | Sticher et al., 1997 |

range of compounds. As such the catabolic activity is governed by the induction of the genes which is a reflection more on the bioavailability of the analyte rather than the enzymatic specificity. The bioavailability of the molecule dictates degradation rather than the perceived specificity of the enzymatic activity. The biosensors used are listed in Table 17.1 and described in more detail below.

## 1.  Tod Pathway (Biosensor A)

*Pseudomonas putida* F1 (which contains a chromosomally encoded tod operon for toluene degradation) was genetically modified by the introduction of *tod-lux* CDABE into the chromosome yielding the biosensor *Pseudomonas putida* TVA8 (Applegate et al., 1998). The *tod* operon is induced when exposed to several compounds such as benzene, toluene, ethylbenzene and xylene (Shingleton et al., 1998).

*P. putida* F1, oxidizes toluene to *cis*-toluene 2,3-dihydrodiol through a dihydrodiol pathway (Gibson et al., 1970). This reaction is catalyzed by a multicomponent enzyme system designated toluene dioxygenase (Yeh et al., 1977). The genes encoding toluene dioxygenase and the other enzymes involved in the degradation of toluene are located on the chromosome of *P. putida* F1 and have been given the designation tod (Wang et al., 1995; Lau et al., 1997).

## 2.  Nah Pathway (Biosensor B)

The *lux*-marked catabolic biosensor *Pseudomonas fluorescens* HK44 (pUTK21) responds sensitively and quantitatively to naphthalene (Heitzer et al., 1992). The biosensor is metabolite-controlled and an ideal sensor for biotransformation. This is made possible by the lack of specificity in microbial hydrocarbon degradation pathways: the *nah* operon has been reported to mediate the biotransformation of the three-ring polyaromatic hydrocarbons (PAHs) anthracene and phenanthrene (Menn et al., 1993; Sanseverino et al., 1993). The *nah* operon is arranged in two pathways, upper and lower: the upper encodes for the proteins that degrade naphthalene to salicylate, and the lower encodes for the proteins that degrade salicylate. The inducer of the pathway is salicylate, so if naphthalene is present a small amount will be degraded by the upper pathway to salicylate. The *lux*CDABE cassette is inserted behind *nahG* in the lower pathway and is controlled by *nahR*, hence luminescence is induced only in the presence of salicylate. *P. fluorescens* HK44 (pUTK21) is likely to respond to a wider range of aromatic compounds than naphthalene alone (LeBlond et al., 2000; 2001).

## 3.  Degradation of Isopropylbenzene (Biosensor C)

*Escherichia coli* HMS174 (pOS25) was developed with a plasmid pRE4 from *Pseudomonas putida* RE204 (involved in the regulation of isopropylbenzene catabolism operon, *ipb*) and the *Vibrio fischeri* luciferase gene, *lux*CDABE. *E. coli* HMS174 produces light in the presence of inducers of the *ipb* operon (Selifonova et al., 1996). This specific biosensor is induced by several compounds: mono-alkylbenzene, substituted benzenes and toluenes, single alkanes and cyclo-alkanes, chlorinated solvents, naphthalenes, gasoline, diesel fuel, some jet fuels, and creosote (Selifonova et al., 1996).

Isopropylbenzene catabolism is analogous to the degradation of toluene by the *tod* pathway (Eaton and Timmis, 1986). The initial step involves attack by a dioxygenase to form a *cis*-dihydrodiol. The *ipb* operon is regulated at the transcriptional level and the regulator is thought to be a protein of the XylS family (Berendes et al., 1998).

### 4. Oct Plasmid (Biosensor D)

The *E. coli* DH5α (pGEc74, pJAMA7) biosensor carries the regulatory gene alkS from *Pseudomonas oleovorans* and a transcriptional fusion of $P_{alkB}$ from the same strain with the promoterless luciferase luxAB genes from *Vibrio harveyi* on two separately introduced plasmids (Sticher et al., 1997). *E.coli* DH5α is induced by octane and middle-chain-length alkanes and some related compounds (Sticher et al., 1997). Octane (or other similar *n*-alkanes) is oxidized by an alkane hydroxylase to the corresponding monoterminal alcohol. This initial product is then further oxidised to an aldehyde, then to the corresponding carboxylic acid. The carboxylic acid can then be fully metabolized by standard biochemical pathways. The product of the regulatory alkS gene is the AlkS protein. In the presence of octane (or similar *n*-alkane), the Pb promoter is induced controlling the degradative genes.

### D. Summary and Aims of this Chapter

Microbial biosensors have been developed based on key hydrocarbon degradation pathways. These biosensors can be used as tools to examine the regulation of the degradation pathways. In addition to the use of empirically derived data from laboratory studies, a selective meta-analysis of published data from literature sources using catabolic hydrocarbon biosensors was conducted. The aim of this work was to demonstrate that biosensor specific QSARs may be developed to first assess the specificity of degradation pathways and then to assess the possibility of predicting analyte-specific degradation characteristics.

## II. MATERIALS AND METHODS

### A. Sources of Data

The data used came mainly from empirical data generated at the School of Biological Sciences, University of Aberdeen, Scotland. In the case of biosensors C and D we have compared the response of the biosensors with literature values. Logarithms of the induction data (in molar units) for the various biosensors are listed in Table 17.2.

### B. Data Handling and Analysis

For the biotransformation assays, values were normalized to a percentage response relative to a control (apart from biosensor A, where the data were normalized against the concentration that produced maximum induction). Percent response was also plotted against exposure time. The area under the time-response curves was then normalized by dividing by the total time and plotted against concentration to give dose-response curves. The data used in the QSAR analyses were obtained by taking the maximum increase in light output and normalizing this relative to the concentration at which the maximum luminescence was observed.

**Table 17.2    Calculated and Reported Physicochemical Parameters Used in Deriving Regression Equations**

| | log $K_{ow}$ | (log $K_{ow}$)$^2$ | $E_{LUMO}$ (eV) | $E_{HOMO}$ (eV) | Molar Volume (cm$^3$ mol$^{-1}$) | Equation[a] |
|---|---|---|---|---|---|---|
| 1,2,4-Trimethylbenzene | 3.70 | 13.69 | 3.45 | −11.95 | | 17.1, 17.2 |
| Naphthalene | 3.30 | 10.89 | 1.93 | −11.11 | | 17.1, 17.2, 17.3, 17.4 |
| 1-Methylnaphthalene | 3.87 | 14.98 | 1.87 | −10.79 | | 17.3, 17.4 |
| 2-Methylnaphthalene | 3.86 | 14.90 | 1.90 | −10.91 | | 17.3, 17.4 |
| 1-Ethylnaphthalene | 4.39 | 19.27 | 1.92 | −10.72 | | 17.3, 17.4 |
| Ethylbenzene | 3.15 | 9.92 | 3.72 | −12.80 | | 17.3, 17.4 |
| n-Butylbenzene | 4.38 | 19.18 | 3.74 | −12.80 | | 17.3, 17.4 |
| Isopropylbenzene | 3.66 | 13.40 | 3.73 | −12.80 | | 17.3 |
| Propylbenzene | 3.69 | | | | | 17.3 |
| Pyridine | 0.63 | 0.40 | 3.99 | −12.46 | | 17.1, 17.2 |
| Phenol | 0.65 | 0.42 | 3.87 | −12.41 | | 17.1 |
| Indole | 2.14 | 4.58 | 3.35 | −10.54 | | 17.2 |
| Toluene | 2.73 | 7.45 | 3.74 | −12.89 | | 17.1, 17.3, 17.4 |
| Benzene | 2.13 | 4.54 | 4.07 | −13.89 | | 17.1, 17.2, 17.3, 17.4 |
| 4-Xylene | 3.15 | 9.92 | 3.54 | −12.14 | | 17.3, 17.4 |
| 2-Chlorotoluene | 3.42 | 11.70 | 3.34 | −12.49 | | 17.3, 17.4 |
| Pentane | | | | | 115.30 | 17.5 |
| Hexane | | | | | 126.70 | 17.5 |
| Heptane | | | | | 146.50 | 17.5 |
| Octane | | | | | 162.50 | 17.5 |
| Nonane | | | | | 207.20 | 17.5 |
| Decane | | | | | 229.40 | 17.5 |
| Pyrrole | 0.75 | | 5.51 | −11.36 | | 17.1, 17.2 |
| Furan | 1.34 | | 4.70 | −11.45 | | 17.1, 17.2 |
| n-Pentylbenzene | 4.92 | | | | | 17.1 |
| n-Hexylbenzene | 5.52 | | | | | 17.1 |
| m-Cresol | 1.96 | | | | | 17.1 |
| Anisole | 2.11 | | | | | 17.1 |
| Thiophene | 1.81 | | 3.89 | −11.56 | | 17.2 |
| 2-Methylthiphene | 2.33 | | 3.65 | −10.89 | | 17.2 |
| Benzothiophene | 3.12 | | 3.03 | −10.54 | | 17.2 |
| Benzofuran | 2.67 | | 3.14 | −10.93 | | 17.2 |
| Dibenzothiophene | 4.59 | | 2.59 | −10.47 | | 17.2 |
| Dibenzofuran | 4.12 | | 2.57 | −11.10 | | 17.2 |
| Quinoline | 2.03 | | 1.92 | −11.55 | | 17.2 |
| Isoquinoline | 2.08 | | 1.87 | −11.24 | | 17.2 |
| Quinaldine | 2.59 | | 1.97 | −11.37 | | 17.2 |
| Acridine | 3.41 | | 0.59 | −9.98 | | 17.2 |

[a] Equation column lists which regression equations a chemical has been used in.

The data for biosensors A, B, and C were log-transformed; this was found not to be necessary for biosensor D. The data for biosensor C were also normalized because they were reported at different concentrations. The biological response is referred to as log(*ind*) or *ind*, where *ind* is the level of induction.

Octanol-water partition coefficient (log $K_{ow}$) values were taken from Hansch et al. (1995). The energy of the highest unoccupied molecular orbital ($E_{HOMO}$) and the energy of the lowest unoccupied molecular orbital ($E_{LUMO}$) values were calculated using HyperChem (Hypercube, FL, USA). The lowest-energy configuration was first obtained for each compound within a molecular force field; frontier orbital energies were then calculated using the CNDO/2 (Complete Neglect of Diatomic Overlap) option. Physicochemical descriptors are reported in Table 17.2.

**Table 17.3   Biological Data Used to Derive the QSARs**

| Biosensor: | A<br>log($ind$/mM)[a] | B<br>log($ind$/mM)[a] | C<br>log($ind$/μM)[a] | D<br>100$ind$/$ind_{oct}$[b] |
|---|---|---|---|---|
| 1,2,4-Trimethylbenzene | | | 0.68 | |
| Naphthalene | | 1.71 | 2.23 | |
| 1-Methylnaphthalene | | | 1.08 | |
| 2-Methylnaphthalene | | | 1.04 | |
| 1-Ethylnaphthalene | | | −0.01 | |
| Ethylbenzene | | | 1.68 | |
| $n$-Butylbenzene | | | 1.30 | |
| Isopropylbenzene | | | 1.69 | |
| Pyridine | −2.3237 | −0.51 | −2.70 | |
| Phenol | −1.5298 | | −2.40 | |
| Indole | | * | | |
| Toluene | −0.7093 | | 1.36 | |
| Benzene | −0.6596 | 0.27 | 0.61 | |
| 4-Xylene | | | 1.30 | |
| 2-Chlorotoluene | | | 1.95 | |
| Pentane | | | | 13 |
| Hexane | | | | 44 |
| Heptane | | | | 81 |
| Octane | | | | 100 |
| Nonane | | | | 100 |
| Decane | | | | 69 |
| 3-Methylheptane | | | | 36 |
| Pyrrole | −1.2403 | * | | |
| Furan | −1.8660 | −1.05 | | |
| $n$-Pentylbenzene | 0.7730 | | | |
| $n$-Hexylbenzene | 0.9540 | | | |
| 3-Cresol | −1.1879 | | | |
| Anisole | −0.5712 | | | |
| Thiophene | | 0.03 | | |
| Benzothiophene | | 1.33 | | |
| 2-Methylthiophene | | * | | |
| Benzofuran | | 0.34 | | |
| Dibenzothiophene | | * | | |
| Dibenzofuran | | * | | |
| Quinoline | | 1.43 | | |
| Isoquiniline | | * | | |
| Quinaldine | | 1.24 | | |
| Acridine | | 2.94 | | |

[a] Data are log-transformed induction values, $ind$, where $ind$ represents increase in signal relative to control, normalized by concentration (μM).
[b] Data are induction values normalized to response against octane.

For the biotransformation assay results, concentration-normalized maximum induction values were modeled. Stepwise multiple linear regression analysis was used to select the most suitable parameters from: log $K_{ow}$, $E_{HOMO}$, $E_{LUMO}$, and the difference in $E_{HOMO}$ and $E_{LUMO}$ ($E_{LUMO} - E_{HOMO}$). The QSARs obtained were then tested by cross validation and visual examination of plots of fitted values against residuals. Cross-validation was performed by leave-one-out (LOO) testing. Each data point was omitted in turn from a regression, and the actual value of the omitted point compared to the value predicted by the revised model. The difference was referred to as a deletion residual. $Q^2$ values (an analog of the summary statistic $R^2$) were then calculated from the sum of squares of the deletion residuals. The $Q^2$ statistic provides a measure of the predictive power of a regression, and is therefore more relevant for QSAR modeling than the $R^2$ statistic (Damborsky and Schultz, 1997).

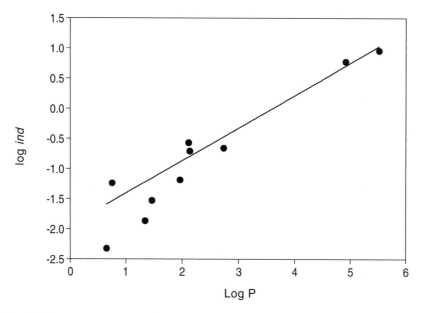

**Figure 17.1** QSAR model for the induction of biosensor A against log Kow (Equation 17.1).

## III. RESULTS

### A. Physical and Chemical Data

The values for log $K_{ow}$ and the frontier orbital energies are reported in Table 17.2. Induction data are presented in Table 17.3 in the format used in the equations, (i.e., normalized and log transformed where appropriate).

### 1. Biosensor A

Data were reported as induction levels (calculated as a percentage relative to the maximum induction). Data are given for ten compounds (dissolved at aqueous solubility limit). No co-solvents were used. The following QSAR was developed:

$$\log \textit{ind} = 0.54 \log K_{ow} - 1.94 \tag{17.1}$$

$$n = 10, R^2 = 0.90$$

Figure 17.1 shows the relationship between induction and hydrophobicity.

### 2. Biosensor B

The biosensors did not respond to aromatic compounds with more than two rings hence they have been excluded from the data analysis. Figure 17.2 shows the best fit of induction values is explained by $E_{LUMO}$ values:

$$\log \textit{ind} = -0.841\ E_{LUMO} + 3.27 \tag{17.2}$$

$$n = 10, R^2 = 0.90$$

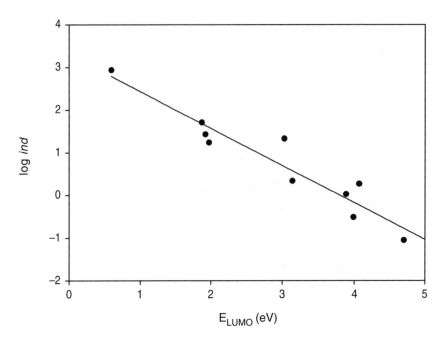

**Figure 17.2**  Relationship between the induction of *P. fluorescens* HK44 and $E_{LUMO}$ (Equation 17.2).

## 3.  Biosensor C

This sensor did not respond to the selected compounds as predicted, and in the first appraisal of data no significant relationship was found for the set of hydrocarbons tested. The biosensor response could be better explained for the set of aromatic hydrocarbons, insofar as these were more similar to isopropylbenzene. The data were remodeled, resulting in an improved regression model:

$$\log ind = 7.17 \log K_{ow} - 1.23 \, (\log K_{ow})^2 - 0.433 \, E_{LUMO} - 7.3 \tag{17.3}$$

$$n = 12, \, R^2 = 0.95$$

Examination of the residuals plot shows isopropylbenzene as an outlier, in that the predicted value is too low. Recalculation of the data after removing isopropylbenzene resulted in the model:

$$\log ind = 6.20 \log K_{ow} - 1.09 \, (\log K_{ow})^2 - 0.544 \, E_{LUMO} - 5.34 \tag{17.4}$$

All the parameters are highly significant ($p < 0.001$, except for the constant: $p = 0.007$). A plot of fitted values against residuals shows slight clustering of adjacent points (not shown); no problem with the regression model is evident upon examination of the plot of actual against predicted values (Figure 17.3). It should be noted however that the use of 3 descriptors in Equation 17.4 is at the limit of statistical acceptability for this number of compounds.

## 4.  Biosensor D

Induction data for 24 compounds (23 hydrocarbons and dicyclopropylketone) that have been normalized to the response for octane (100%) are reported. Only seven of these compounds (*n*-alkanes from pentane to decane, and 3-methylheptane) showed induction significantly above

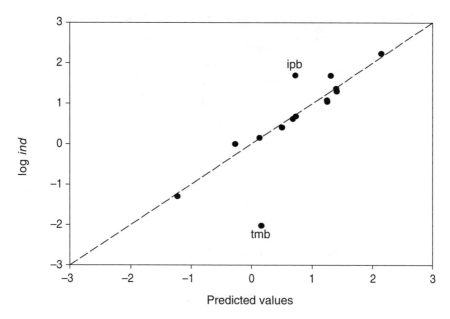

**Figure 17.3**   QSAR model for biosensor C, aromatic hydrocarbon compounds only (Equation 17.4).
*Note:* Dashed line indicates model of perfect fit (x = y). Isopropylbenzene (ipb) and 1,2,4,5-tetramethylbenzene (tmb) were not included in regression model: data are labelled on graph for sake of comparison.

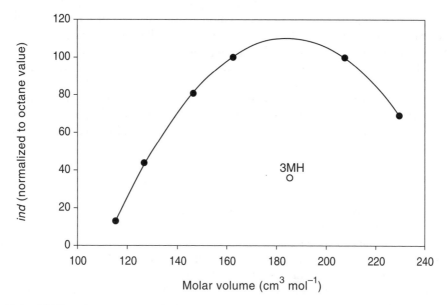

**Figure 17.4**   QSAR model for biosensor D (Equation 17.5), where 3MH is 3-methylheptane.

background. These are too few data points to attempt to derive a QSAR, but in this case the biological response indicates a dependence on carbon chain length, with maximum induction reached at a chain length of eight to nine and a rapid decline in luminescence thereafter. A slightly better fit was obtained by modeling the data (without 3-methylheptane) against molar volume, rather than simple chain length (Figure 17.4):

**Table 17.4    Summary of Significant QSAR Regression Models for the Four Biosensors Considered**

| Sensor | QSAR | $R^2$ | $Q^2$ | Equation |
|---|---|---|---|---|
| Biosensor A | Log ($ind$) = 0.54 log $K_{ow}$ − 1.94 | 0.90 | | 17.1 |
| Biosensor B | Log ($ind$) = −0.841 $E_{LUMO}$ + 3.27 | 0.90 | 0.87 | 17.2 |
| Biosensor C | Log ($ind$) = 6.20 log $K_{ow}$ − 1.90 (log $K_{ow}$)$^2$ − 0.544 $E_{LUMO}$ − 5.34 | 0.95 | 0.90 | 17.4 |
| Biosensor D | $Ind$ = 7.46 vol − 0.0202 (vol)$^2$ − 578 | 1 | 1 | 17.5 |

$$ind = 7.46 \text{ vol} - 0.0202 \text{ (vol)}^2 - 578 \tag{17.5}$$

$$n = 6, R^2 = 1$$

where vol is the molar volume in $cm^3 \text{ mol}^{-1}$

The significant QSARs are summarized in Table 17.4.

## IV. DISCUSSION

### A. Individual Biosensor Responses

Biodegradation or biotransformation is usually quantified by incubating a compound with a degrader and measuring disappearance of oxygen or production of $CO_2$. Measuring the appearance of metabolites is an alternative approach that is likely to be more sensitive in detecting biotransformation, and also provides direct confirmation of biotransformation. Such tests typically involve sophisticated chemical analysis or the use of isotopically labelled compounds, and therefore few chemicals are tested. Traditional biochemical testing is not ideal to provide new data for developing QSARs, as large data sets including many different chemicals are a prerequisite for good QSAR modeling (Degner et al., 1991). Environmental biodegradation is mediated by many different organisms under different conditions. As a result, the QSAR for biotransformation reported can be considered valid only for organisms containing key degradation genes. In addition, the assay utilizes active, pregrown cells added to a substrate. These data must be carefully evaluated prior to relating the response to environmental scenarios.

### 1. Biosensor A

Examining the QSAR for *P. putida* TVA8 for the chemicals tested (which represent different classes of compounds), a highly predictive QSAR equation (Equation 17.1, $R^2 = 0.90$) was obtained that described induction by hydrophobicity alone. The ten chemicals tested all significantly induced the *tod* operon. This increase in induction correlated with the log $K_{ow}$ values. Other aromatic compounds, such as naphthalene, failed to induce TVA8.

Ehtylbenzene and 3-xylene (log $K_{ow}$ values of 3.15 and 3.20, respectively) were removed because they were outliers ($R^2 = 0.73$ when included in the analysis) (data not shown). Their removal can be justified by the fact that for the highest concentrations tested, these compounds were found to be toxic to a luminescent metabolic biosensor of *P. putida* F1, and it may be assumed that they were toxic to *P. putida* TVA8 accordingly. As a consequence of the removal of these data, there is a significant gap of chemicals for octanol-water partition coefficients between 2.73 and 4.90.

*P. putida* TVA8 was expected to respond to benzene, toluene, ethylbenzene, 3- and 4-xylenes, and phenol (Applegate et al., 1998) since it contains a chromosomally encoded *tod* operon for toluene degradation. *P. putida* TVA8 has also been used to sense for trichloroethylene because the

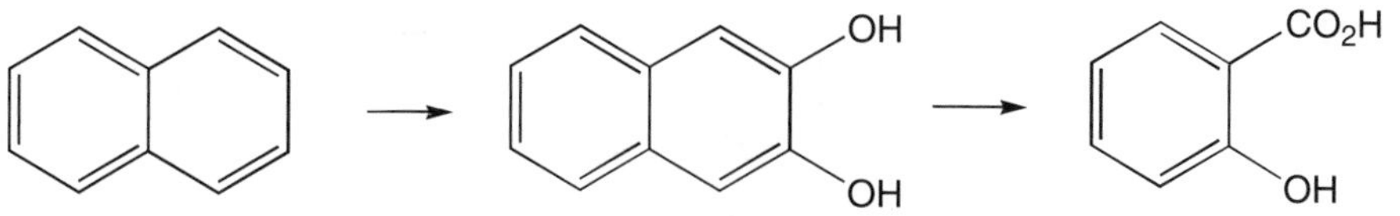

**Figure 17.5**  Illustration that salicylate, or an analogous carbon-substituted monoaromatic ring, cannot be derived from an unsubstituted monoaromatic compound.

*lux* and *tod* operons are under the same regulation (toluene dioxygenase catabolises trichloroethylene) (Shingleton et al., 1998).

Toluene dioxygenase (the first enzyme of the *tod* pathway) has been reported as being capable of the co-metabolism of n-alkylbenzenes ($C_3$-$C_7$), biphenyl, styrene, and cumene (Cho et al., 2000). This explains the response to n-pentylbenzene and n-hexylbenzene.

Following the application of 3-cresol and anisole, induction of *P. putida* TVA8 was achieved and there are some studies (e.g., Boyd and Sheldrake, 1998), suggesting several substrates and mono- and poly-cyclic aromatic rings may be cleaved by toluene dioxygenase. Other cyclic compounds (e.g., pyrrole, pyridine, and furan) also induce *P. putida* TVA8, but the specific mechanism and enzymes involved are unknown. Further studies on this aspect are required to comprehensively address this question.

## 2.  Biosensor B

The stepwise multiple linear regression demonstrated $E_{LUMO}$ alone as the most significant parameter. The model is highly statistically significant ($p < 0.001$), and has high $R^2$ and $Q^2$ values. A plot of induction versus $E_{LUMO}$ shows a good fit is obtained (Figure 17.2). It should be pointed out that the QSAR is not predictive. Only the compounds causing induction were included in the model, with no *a priori* method of predicting whether a compound would cause induction.

The model's dependence on $E_{LUMO}$ and not on log $K_{ow}$ is reasonable because frontier orbital energies control chemical reactivity and $E_{LUMO}$ energy can be considered as a measure of a compound's susceptibility to nucleophilic attack (Lynam et al., 1998). Previous relationships have been shown to be more influenced by electronic properties than by hydrophobicity (Damborsky and Schultz, 1997).

Salicylate, or an analogous carbon-substituted mono-aromatic ring, could not be derived from an unsubstituted monoaromatic compound (see Figure 17.5). The question must be addressed whether the biosensor response for compounds other than naphthalene is genuinely related to gene induction, or whether some other non-specific mechanism may be a cause. It has been observed that *P. fluorescens* HK44 (pUTK21) increases light output on exposure to a wide range of organic solvents, inducing toluene, alcohols, and methyl tertiary butyl ether. Toxic organic compounds affect the cell membrane and may therefore lead to a release of membrane-derived fatty acids into the cytoplasm. These can then be reduced by the *lux* proteins to form a supplementary substrate to tetradecanal for the luciferase enzyme.

There is strong evidence for making the assumption that the increase in luminescence observed is caused by induction of a metabolite for most of the compounds tested. First, outliers in QSAR regressions can be used to determine the limits of applicability of a QSAR (Lipnick, 1991). If the biosensor response to all compounds other than naphthalene was a non-specific response with no relationship to biotransformation, then it would be expected that the value for naphthalene would be a clear outlier. Instead, the value for naphthalene is close to the predicted value, as shown in Figure 17.3. A dose-response behavior is indicative of a specific mechanism.

The isomers quinoline and isoquinoline elicit different responses: quinoline induces light production up to a concentration of approximately 160 μM, and beyond that exerts a toxic effect. Isoquinoline by contrast shows no induction, and light output is reduced relative to the control at

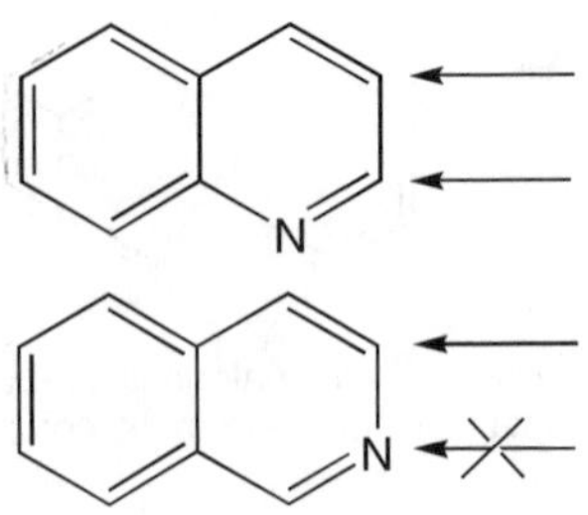

**Figure 17.6**  Illustration that the presence of the N atom in isoquinoline blocks the site of attack of the naphthalene dioxygenase enzyme.

all concentrations. Quinoline and isoquinoline have similar levels of toxicity to bacterial biosensors; it is therefore improbable that the differences are the result of a non-specific mechanism. One explanation is that the compounds have different biodegradability relative to the *nah* pathway, because the presence of the N atom in isoquinoline blocks the site of attack of the naphthalene dioxygenase enzyme as shown in Figure 17.6.

## 3.  Biosensor C

The removal of 1,2,4,5-tetramethylbenzene and isoropylbenzene from Equation 17.3 as outliers can be justified. Inspection of the data reveals that the bioluminescence response for 1,2,4,5-tetramethylbenzene is very low (–2.02), and much lower than for similar compounds (e.g. 1,2,3,4-tetramethylbenzene has a light level of 0.15). This implies that the compound is not an inducer of the *ipb* pathway and it is not part of the homologous series of compounds on which the model is based. It is fundamental to the development of QSARs that they are based on a genuinely homologous series (i.e., the chemicals elicit a biological response through the same mode of action).

It is reasonable that Equation 17.3 is improved by removal of isopropylbenzene, giving Equation 17.4. Isopropylbenzene is the archetypal substrate for this degradation pathway, and would be expected to be more strongly inducing than its analogs.

## 4.  Biosensor D

This sensor responds to octane. This compound is significantly more volatile than the other test componds, proving to be problematic in the development of a significant QSAR. The compound 3-methylheptane clearly does not fall into the same group as the *n*-alkanes. The value predicted for induction (111%) using Equation 17.5 is different from the actual value of 36%. Further testing with a wider range of branched alkanes would be necessary to determine if, like *n*-alkanes, they formed a QSAR that was dependent on molar volume.

## B.  General Discussion

In addition to the sensors tested in this study other catabolic biosensors have been widely applied. Abril et al. (1989) used *lacZ* reporter gene biosensors to study the specificity of the regulatory proteins of the *tol* pathway. They found that the XylR protein had a broad effector molecule specificity, responding to a range of mono-, di-, and tri-substituted alkyl and chlorobenzenes. Abril and coworkers (1989) concluded that substitution of the benzene ring is necessary for activation of XylR. Further chloro- or alkyl substituents also led to activation. In addition, benzaldehydes did not cause activation except for 4-chlorobenzaldehyde. These functional groups and the corresponding induction could be well defined by the hydrophobicity of the test compounds (Bundy, 2000).

Used effectively, QSARs can give clues as to the underlying factors controlling specificity. Hydrophobicity is shown to be the most important descriptor for *tol-* and *tod*-based biosensors. It is not surprising that the response can be predicted using a single chemical descriptor as the chemicals tested were selected because they were known inducers of the target promoter, and hence form a homologous set. The linear free energy relationship (LFER) hypothesis states that small changes in the physical and chemical properties of chemicals within a set cause linear changes in the free energy of a reaction (such as binding to a protein), enabling a linear relationship to be derived (Okey and Stensel, 1996). Because the chemicals form a known series, trends within that series are governed by simple properties. Conversely, a set including chemicals from outside of that series might require additional descriptors and more complicated models to predict the response.

The response of the *tod* (this study) and *tol* plasmid-based (Abril et al., 1989) biosensors can be compared to the response of the ipb-based biosensor C. Equation 17.4 shows that hydrophobicity and $E_{LUMO}$ are both highly significant for the set of aromatic hydrocarbons. Similar factors appear to regulate specificity. Hydrocarbon degradation pathways are both evolutionarily and functionally conserved (Williams and Sayers, 1994), although *ipb* genes show greater homology with *tod* than *xyl* genes (Aoki et al., 1996). It is interesting that the induction promoted by aromatic hydrocarbons has a quadratic relationship with log $K_{ow}$, whereas the non-hydrocarbons show a linear relationship. This could be because the log $K_{ow}$ values (i.e., the domain of the model) for the hydrocarbons are mostly higher than those of the non-hydrocarbons. Compounds with low hydrophobicity may be expected to diffuse slowly across cell membranes. It is possible that the response of low log $K_{ow}$ compounds is limited by the rate of cell uptake not regulator protein specificity, resulting in a positive relationship with log $K_{ow}$.

QSARs can also be used to examine the biological behavior of individual chemicals, or even chemical classes. Outliers can indicate the limits of applicability of a QSAR (Lipnick, 1991): 3-methylheptane is shown to fall into a different biological set from the *n*-alkanes for induction of *alkB* (Equation 17.5). 1,2,3,5-Tetramethylbenzene stimulated too little bioluminescence to be considered part of the set of compounds recognized by the regulatory elements of the *ipb* pathway. In contrast, the archetypal substrate isopropylbenzene stimulates too much. Three differently acting groups of chemicals can be distinguished, apart from the outliers. Non-hydrocarbons are dependent on log $K_{ow}$ alone, however aromatic hydrocarbons require $E_{LUMO}$ as well as $(\log K_{ow})^2$. Aliphatic hydrocarbons are further shown to have a clearly different biological response from aromatic hydrocarbons.

## C.  Relationship to Biodegradation Potential

An important point is determining to what extent these results relate to actual biodegradation potential of the compounds concerned. In other words, could the regression models be considered QSBRs. This is of particular significance because a lack of reproducible quantitative data on biodegradation is one of the factors that limits the development of QSBRs (Degner et al., 1991).

For the response of a biosensor to truly reflect QSBRs there are two points that must be addressed:

1. We must be familiar with the metabolic pathway for the compounds selected and the relative specificity at that pathway. The more specific the pathway, the more likely the derived QSBR reflects biodegradation. To fully appreciate these parameters, we need a sound knowledge both of the chemical compound and indeed the process level biochemistry.
2. The genetic marking of the biosensor must be sympathetic to the question posed. This is particularly pertinent when *P. fluorescens* HK44 (pUTK21) is compared with *P. putida* TVA8. The first sensor responds to the transformation of naphthalene to salicylate (and corresponding analogs). The second biosensor is able to both induce and follow the toluene dioxygenase pathway and reflects the relative activity through the expression of bioluminescence. The second biosensor, a far superior

environmental tool, reflects a decade of advancement in microbial molecular biology. It may be concluded therefore that the tool kit required to comprehensively address QSBRs are only now becoming available; the biosensors of the next decade will realize the required potential.

The biosensors that underpin this chapter have genes that code for the regulatory proteins that control hydrocarbon degradation fused to luminescence genes. The biosensors will respond to gratuitous inducers — compounds that induce gene expression, but are not actually substrates of the initial enzyme. Strictly speaking, the QSARs derived are not QSBRs. In particular, biosensor D, based on genes from the *oct* plasmid, showed an extremely narrow specificity, responding only to *n*-alkanes from $C_5$ to $C_{10}$ and to 3-methylheptane. The oct genes are known to hydroxylate *n*-alkanes up to $C_{12}$, and similar gene systems can hydroxylate a wide range of compounds including cyclic aliphatics and alkylbenzenes (van Beilen et al., 1994), so there is a discrepancy between the known degradative ability of the pathway and the bioluminescence response. This could be experimentally defined and limited. In addition, the biosensor did not respond to dicyclopropylketone, which is a known gratuitous inducer.

## V. CONCLUSIONS

Structure-activity relationships can be generated that predict the specificity of regulatory proteins for hydrocarbon degradation pathways. Hydrophobicity is the most important parameter in these relationships (possibly because it affects initial uptake by the cell); $E_{LUMO}$ energy is also significant in several of the models that have been generated. These relationships do not, however, predict biodegradation, because the set of inducer compounds for a pathway is not identical to the set of possible substrates. The sensors do prove to be useful in increasing understanding of the mechanisms of hydrocarbon degradation and assessing enzymatic specificity.

Microbial biosensors are well suited to producing rapid and highly reliable data for simple modeling, allowing large numbers of chemicals to be tested. This is ideal for the requirements of QSAR modeling.

## REFERENCES

Abril, M.A., Michan, C., Timmis, K.N., and Ramos, J.L., Regulator and enzyme specificities of the TOL plasmid-encoded upper pathway for degradation of aromatic hydrocarbons and expansion of the substrate range of the pathway, *J. Bacteriol.*, 171, 6782–6790, 1989.

Aoki, H., Kimura, T., Habe, H., Yamane, H., Kodama, T., and Omori, T., Cloning, nucleotide sequence, and characterization of the genes encoding enzymes involved in the degradation of cumene to 2-hydroxy-6-oxo-7-methylocta-2,4-dienoic acid in *Pseudomonas fluorescens* IP01, *J. Fermentation Bioeng.*, 81, 187–196, 1996.

Applegate, B.M., Kehrmeyer, S.R., and Sayler, G.S., A chromosomally-based *tod-luxCDABE* whole-cell reporter for benzene, toluene, ethylbenzene, and xylene (BTEX) sensing, *App. Environ. Microbiol.*, 64, 2730–2735, 1998.

Atlas, R.M. and Bartha, R., Hydrocarbon biodegradation and oil-spill bioremediation, *Adv. Microbial Ecol.*, 12, 287–338, 1992.

Barkay, T., Nazaret, S., and Jeffrey, W., Degradative genes in the environment, in *Microbial Transformation and Degradation of Toxic Organic Chemicals*, Young, L.Y. and Cerniglia. C.E., Eds., Wiley-Liss, New York, 1995, pp. 545–577.

Belkin, S., Smulski, D.R., Dadon, S., Vollmer, A.C., Van Dyk, T.K., and LaRossa, R.A., A panel of stress-responsive luminous bacteria for the detection of selected classes of toxicants, *Water Res.*, 31, 3009–3016, 1997.

Berendes, F., Sabarth, N., Averhoff, B., and Gottschalk, G., Construction and use of an *ipb* DNA module to generate *Pseudomonas* strains with constitutive trichloroethene and isopropylbenzene oxidation activity, *Appl. Environ. Microbiol.*, 64, 2454–2462, 1998.

Boyd, D.R. and Sheldrake, G.N., The dioxygenase-catalysed formation of vicinal *cis*-diols, *Nat. Prod. Rep.*, 15-3, 309–323, 1998.

Bundy, J.G., *The Use of Biological Methods for the Assessment of Oil Contamination and Bioremediation*, Ph.D. thesis, University of Aberdeen, Scotland, 2000.

Cho, M.C., Kang, D-O., Yoon, B.D., and Lee K., Toluene degradation pathway from Pseudomonas putida F1: substrate specificity and gene induction by 1-substituted benzenes, *J. Ind. Microbiol. Biotechnol.*, 25, 163–170, 2000.

Damborsky, J. and Schultz, T.W., Comparison of the QSAR models for toxicity and biodegradability of anilines and phenols, *Chemosphere*, 34, 429–446, 1997.

Degner, P., Nendza, M., and Klein, W., Predictive QSAR models for estimating biodegradation of aromatic compounds, *Sci. Total Environ.*, 109, 253–259, 1991.

Eaton, R.W. and Timmis, K.N., Characterization of a plasmid-specified pathway for catabolism of isopropylbenzene in *Pseudomonas putida* RE204, *J. Bacteriol.*, 168, 123–131, 1986.

Gibson, D.T., Hensley, M., Yoshioka, H., and Mabry, T.J., Formation of (+)-cis-2,3-dihydroxy-1-methylcyclohexa-4,6-diene from toluene by *Pseudomonas putida, Biochemistry*, 9, 1626–1630, 1970.

Hansch, C., Leo, A., and Hoekman, D., *Exploring QSAR: Hydrophobic, Electronic, and Steric Constants*, American Chemical Society, Washington, D.C., 1995.

Heitzer, A., Webb, O.F., Thonnard, J.E., and Sayler, G.S., Specific and quantitative assessment of naphthalene and salicylate bioavailability by using a bioluminescent catabolic reporter bacterium, *Appl. Environ. Microbiol.*, 58, 1839–1846, 1992.

Howard, P.H., Boethling, R.S., Stiteler, W., Meylan, W., and Beauman, J., Development of a predictive model for biodegradability based on BIODEG, the evaluated biodegradation data base, *Sci. Total Environ.*, 109/110, 635–641, 1991.

Ikariyama, Y., Nishiguchi, S., Koyama, T., Kobatake, E., and Aisawa, M., Fiber-optic-based biomonitoring of benzene derivatives by recombinant *E. coli* bearing luciferase gene-fused TOL-plasmid immobilized on the fiber-optic end, *Anal. Chem.*, 69, 2600–2605, 1997.

King, J.M.H., DiGrazia, P.M., Applegate, B., Burlage, R., Sanseverino, J., Dunbar, P., Larimer, F., and Sayler, G.S., Rapid, sensitive bioluminescence reporter technology for naphthalene exposure and biodegradation, *Science*, 249, 778–781, 1990.

Kragelund, L., Christoffersen, B., Nybroe, O., and de Bruijn, F.J., Isolation of *lux* reporter gene fusions in *Pseudomonas fluorescens* DF57 inducible by nitrogen or phosphorus starvation, *FEMS Microbiol. Ecol.*, 17, 95–106, 1995.

Layton, A.C., Muccini, M., Ghosh, M.M., and Sayler, G.S., Construction of a bioluminescent reporter strain to detect polychlorinated biphenyls, *Appl. Environ. Microbiol.*, 64, 5023–5026, 1998.

Lau, P.C.K., Wang, Y., and Patel, A., A bacterial basic region leucine zipper histidine regulating toluene degradation, *Proc. Natl. Acad. Sci. USA*, 94, 1453–1458, 1997.

LeBlond, J.D., Applegate, B.M., Menn, F.M., Schultz, T.W., and Sayler, G.S., Structure-toxicity assessment of metabolites of the aerobic bacterial transformation of substituted naphthalenes, *Environ. Toxicol. Chem.*, 19, 1235-1246, 2000.

LeBlond, J.D., Schultz, T.W., and Sayler, G.S., Observations on the preferential biodegradation of selected components of polyaromatic hydrocarbon mixtures, *Chemosphere*, 42, 333-343, 2001.

Lipnick, R.L., Outliers: their origin and use in the classification of molecular mechanisms of toxicity, *Sci. Total Environ.*, 109/110, 131–153, 1991.

Lynam, M.M., Kuty, M., Damborsky, J., Koca, J., and Adriaens, P., Molecular orbital calculations to describe microbial reductive dechlorination of polychlorinated dioxins, *Environ. Toxicol. Chem.*, 17, 988–997, 1998.

Meighen, E.A., Enzymes and genes from the *lux* operons of bioluminescent bacteria, *Ann. Rev. Microbiol.*, 42, 151–176, 1988.

Menn, F.M., Applegate, B.M., and Sayler, G.S., NAH plasmid-mediated catabolism of anthracene and phenanthrene to naphthoic acids, *Appl. Environ. Microbiol.*, 59, 1938–1942, 1993.

Okey, R.W. and Stensel, H.D., A QSAR-based biodegradability model: a QSBR, *Water Res.*, 30, 2206–2214, 1996.

Prest, A.G., Winson, M.K., Hammond, J.R.M., and Stewart, G.S.A.B., Construction and application of a *lux*-based nitrate biosensor, *Lett. Appl. Microbiol.*, 24, 355–360, 1997.

Sanseverino, J., Applegate, B.M., King, J.M.H., and Sayler, G.S., Plasmid-mediated mineralization of naphthalene, phenanthrene, and anthracene, *Appl. Environ. Microbiol.*, 59, 1931–1937, 1993.

Selifonova, O. and Eaton, R.W., Use of an *ipb-lux* fusion to study regulation of the isopropylbenzene catabolism operon of *Pseudomonas putida* RE204 and to detect hydrophobic pollutants in the environment, *Appl. Environ. Microbiol.*, 62, 778–783, 1996.

Selifonova, O., Burlage, R., and Barkay, T., Bioluminescent sensors for detection of bioavailable Hg(II) in the environment, *Appl. Environ. Microbiol.*, 59, 3083–3090, 1993.

Shingleton, J.T., Applegate, B.M., and Nagel, A.C., Induction of the *tod* operon by trichlorethylene in *Pseudomonas putida* TVA8, *Appl. Environ. Microbiol.*, 64, 5049–5052, 1998.

Sticher, P., Jaspers, M.C.M., Stemmler, K., Harms, H., Zehnder, A.J.B., and Van der Meer, J.R., Development and characterization of a whole-cell bioluminescent sensor for bioavailable middle-chain alkanes in contaminated groundwater samples, *Appl. Environ. Microbiol.*, 63, 4053–4060, 1997.

Tescione, L. and Belfort, G., Construction and evaluation of a metal-ion biosensor, *Biotech. Bioeng.*, 42, 945–952, 1993.

van Beilen, J.B., Wubbolts, M.G., and Witholt, B., Genetics of alkane oxidation by *Pseudomonas oleovorans*, *Biodegradation*, 5, 161–174, 1994.

Wang, Y., Rawlings, M., and Gibson, D.T., Identification of a membrane- protein and a truncated lysr-type regulator associated with the toluene degradation pathway in *Pseudomonas putida* F1, *Molecular and Gen. Genet.*, 246, 570–579, 1995.

Willardson, B.M., Wilkins, J.F., Rand, T.A., Schupp, J.M., Hill, K.K., Keim, P., and Jackson, P.J., Development and testing of a bacterial biosensor for toluene-based environmental contaminants, *Appl. Environ. Microbiol.*, 64, 1006–1012, 1998.

Williams, P.A. and Sayers, J.R., The evolution of pathways for aromatic hydrocarbon oxidation in *Pseudomonas*, *Biodegradation*, 5, 195–217, 1994.

Yeh, W.K., Gibson, D.T., and Liu, T-N., Toluene dioxygenase: a multi-component enzyme system, *Biochem. Biophys. Res. Commun.*, 78, 401–410, 1977.

Zylstra, G.J. and Gibson, D.T., Aromatic hydrocarbon degradation: a molecular approach, *Genet. Eng.*, 13, 183–203, 1991.

# SECTION 5

# Application

CHAPTER **18**

# The Tiered Approach to Toxicity Assessment Based on the Integrated Use of Alternative (Non-Animal) Tests

**Andrew P. Worth**

## CONTENTS

## I. INTRODUCTION

### A. Alternative Methods to Animal Testing

In the context of laboratory animal use, alternative methods include all procedures that can completely replace the need for animals (replacement alternatives), reduce the number of animals required (reduction alternatives), or diminish the amount of distress or pain suffered by animals (refinement alternatives), in meeting the essential needs of man and other animals (Smyth, 1978).

The concept of the three Rs (replacement, reduction, and refinement), attributed to Russell and Burch (1959), is now enshrined in the laws of many countries and in Directive 86/609/EEC on the protection of animals used for experimental and other scientific purposes (European Commission, 1986). This directive requires that replacement alternatives, reduction alternatives, and refinement alternatives should be used wherever and whenever possible.

Alternative methods include: (1) computer-based methods (mathematical models and expert systems); (2) physicochemical methods, in which physical or chemical effects are assessed in systems lacking cells; and, most typically, (3) *in vitro* methods, in which biological effects are observed in cell cultures, tissues, or organs.

Alternative methods for the safety and toxicity testing of chemicals and products (e.g., cosmetics, medicines, and vaccines) are particularly important, since regulations exist at both the national and international levels to ensure that such chemicals and products can be manufactured, transported and used without adversely affecting human health or the environment. Traditionally, safety and toxicity testing has been conducted on animals. However, animal tests have been criticized not only on ethical grounds, but also on scientific and economic grounds. There has been a considerable effort to develop and validate alternative tests, with a view to increasing their use for regulatory purposes. Validation is a crucial stage in the evolution of any alternative test from its development to its routine application. It consists of the independent assessment of the relevance and reliability of the test, and therefore forms the scientific basis on which regulators can decide whether to incorporate the alternative test into legislation or into a test guideline. A number of successfully validated alternative tests have already been accepted by regulatory authorities at national and international levels, and incorporated into various regulations and test guidelines (European Commission, 2000; Organization for Economic Co-operation and Development, 2002a; 2002b; 2002c). A comprehensive review of the current status of alternative tests has recently been produced by European Center for the Validation of Alternative Methods (ECVAM) (Worth and Balls, 2002).

## B. Prediction Models and Structure-Activity Relationships

To make predictions of toxic potential by using a physicochemical or an *in vitro* test system, it is necessary to have a means of extrapolating the physicochemical or *in vitro* data to the *in vivo* level. To achieve this, Bruner et al. (1996) introduced the concept of the prediction model (PM), which has been defined as an unambiguous decision rule that converts the results of one or more alternative methods into the prediction of an *in vivo* pharmacotoxicological endpoint (Worth and Balls, 2001). A PM could be a classification model (CM) for predicting toxic potential, or it could be a regression model for predicting toxic potency.

The usefulness of an alternative method for regulatory purposes is formally assessed by performing an interlaboratory validation study. The alternative method is judged valid for a specific purpose (e.g., the classification of chemicals on the basis of skin corrosivity) if it meets predefined criteria of reliability and relevance (Balls and Karcher, 1995). In this context, reliable means that the data generated by the alternative method are reproducible (within and between laboratories). Relevant means that the method has a sound scientific basis (mechanistic relevance) and is associated with a PM of sufficient predictive ability (predictive relevance).

In addition to using PMs, predictions of toxic hazard can also be made by using structure-activity relationships (SARs). A quantitative structure-activity relationship (QSAR) can be defined as any mathematical model for predicting biological activity from the structure or physicochemical properties of a chemical. In this chapter, the premodifer quantitative is used in accordance with the recommendation of Livingstone (1995) to indicate that a quantitative measure of chemical structure is used. In contrast, a SAR is simply a (qualitative) association between a specific molecular (sub)structure and biological activity.

A subtle distinction can be made between QSARs and the PMs associated with physicochemical tests. The distinction is that while any PM (associated with a physicochemical test) could also be called a QSAR, not all QSARs could also be called PMs. For example, QSARs can also be based on theoretical descriptors (e.g., topological indices) or on experimental properties that are themselves more easily predicted than measured (e.g., the octanol-water partition coefficient). Furthermore, QSARs developed for the prediction of physicochemical and *in vitro* end points would not be regarded as PMs.

## C. Tiered Testing Strategies

Because of the limitations of individual alternative (non-animal) methods for predicting toxicological hazard, there is a growing emphasis on the use of integrated approaches that combine the use of two or more alternative tests. This has led to the concept of the integrated testing strategy, which has been defined as follows (Blaauboer et al., 1999):

> An integrated testing strategy is any approach to the evaluation of toxicity which serves to reduce, refine or replace an existing animal procedure, and which is based on the use of two or more of the following: physicochemical, *in vitro*, human (e.g., epidemiological, clinical case reports), and animal data (where unavoidable), and computational methods, such as (quantitative) structure-activity relationships ([Q]SAR) and biokinetic models.

Since integrated testing strategies are based on the use of different types of information, they are expected to be particularly successful at predicting *in vivo* end points that are too complex in biochemical and physiological terms for any single method to reproduce.

A particular type of integrated testing strategy is the so-called tiered (stepwise or hierarchical) testing strategy. This is based on the sequential use of existing information and data derived from alternative methods, before any animal testing is performed. The outlines of tiered testing strategies have been proposed for a variety of human health end points (Worth and Balls, 2002).

An important principle in the design of many strategies for hazard classification is that chemicals that are predicted to be toxic in an early step are classified without further assessment. Conversely, chemicals that are predicted to be non-toxic proceed to the next step for further assessment. In this way, it is intended that toxic chemicals will be identified by non-animal methods, while the animal tests performed at the end of the stepwise procedure will merely serve to confirm predictions of non-toxicity made in previous steps.

At the regulatory level, a stepwise approach for classifying skin irritants and corrosives has been based on this principle, and is included in a supplement to Organization for Economic Co-operation and Development (OECD) Test Guideline 404 (Organization for Economic Cooperation and Development, 2001). This testing strategy is an adaptation of a testing strategy adopted by the OECD in November 1998 (Organization for Economic Co-operation and Development, 1998).

## D. Statistical Assessment of Classification Models

QSARs, PMs based on physicochemical data, and PMs based on *in vitro* data can all be used to make predictions on a categorical scale. Such CMs are often developed and evaluated on the basis that they will be applied as stand-alone alternatives to animal experiments, but in practice they are more likely to be used in the context of a tiered testing strategy.

The predictive performance of a CM is often expressed in terms of a contingency table (Table 18.1) containing the numbers of true and false positive and negative predictions made by the CM, and in terms of the CM's Cooper statistics, which are derived from the contingency table. Definitions of the Cooper statistics are provided in Table 18.2.

**Table 18.1    A 2 × 2 Contingency Table**

| | | Predicted Class | | |
| --- | --- | --- | --- | --- |
| | | Non-toxic | Toxic | Marginal Totals |
| Observed (*in vivo*) | Non-toxic | a | b | a + b |
| Class | Toxic | c | d | c + d |
| | Marginal totals | a + c | b + d | a + b + c + d |

**Table 18.2    Definitions of the Cooper Statistics**

| Statistic | Definition: "The Proportion (or Percentage) of the ... | |
| --- | --- | --- |
| Sensitivity | Toxic chemicals (chemicals that give positive results *in vivo*) which the CM predicts to be toxic." | $= d/(c + d)$ |
| Specificity | Non-toxic chemicals (chemicals that give negative results *in vivo*) which the CM predicts to be non-toxic." | $= a/(a + b)$ |
| Concordance or accuracy | Chemicals which the CM classifies correctly." | $= (a + d)/(a + b + c + d)$ |
| Positive predictivity | Chemicals predicted to be toxic by the CM that give positive results *in vivo*." | $= d/(b + d)$ |
| Negative predictivity | Chemicals predicted to be non-toxic by the CM that give negative results *in vivo*." | $= a/(a + c)$ |
| False positive (overclassification) rate | Non-toxic chemicals that are falsely predicted to be toxic by the CM." | $= b/(a + b)$ <br> $= 1 - $ specificity |
| False negative (under-classification) rate | Toxic chemicals that are falsely predicted to be non-toxic by the CM." | $= c/(c + d)$ <br> $= 1 - $ sensitivity |

## E.  Purpose of this Chapter

The objectives of this chapter are to illustrate:

1.  The development of a tiered testing strategy for predicting a particular kind of toxic potential, skin corrosion, based on the sequential use of a QSAR; a PM based on physicochemical (pH) data; and a PM based on *in vitro* data obtained with the EPISKIN™ test, a particular type of human skin model
2.  A method for evaluation of the tiered testing strategy in terms of its predictive capacity and its ability to reduce and refine the use of laboratory animals

## II.  DEVELOPMENT OF A TIERED APPROACH TO HAZARD CLASSIFICATION

To develop a tiered approach to hazard classification, it is first necessary to use existing data to develop the CMs that will serve as the individual steps of the tiered strategy. The example presented in this chapter used existing data on skin corrosion, and represents a development of earlier work (Worth et al., 1998).

## A.  Development of a Quantitative Structure-Activity Relationship

Before developing a QSAR for skin corrosion, a data set of 277 organic chemicals (Table 18.3) was constructed from a variety of literature sources (Barratt, 1995; 1996a; 1996b; European Centre for Ecotoxicology and Toxicology of Chemicals, 1995; National Institutes of Health, 1999; Whittle et al., 1996). Chemicals taken from the European Centre for Ecotoxicology and Toxicology of Chemicals (ECETOC) data bank (European Centre for Ecotoxicology and Toxicology of Chemicals, 1995) were classified for skin corrosion potential according to European Union (EU) classification criteria; in the case of the chemicals taken from the other sources, the published classifications of corrosion potential were used.

**Table 18.3 Skin Corrosion Data for 277 Organic Chemicals**

| | Chemical | Source | C/NC | MP | MW |
|---|---|---|---|---|---|
| 1 | 1-Naphthoic acid | Barratt (1996a) | NC | 106.7 | 172.2 |
| 2 | 1-Naphthol | Barratt (1996a) | NC | 67.7 | 144.2 |
| 3 | 2,3-Lutidine | Barratt (1996a) | NC | −7.6 | 107.2 |
| 4 | 2,3-Xylenol | Barratt (1996a) | C | 25.4 | 122.2 |
| 5 | 2,4,6-Trichlorophenol | Barratt (1996a) | NC | 63.8 | 197.5 |
| 6 | 2,4-Dichlorophenol | Barratt (1996a) | NC | 46.8 | 163.0 |
| 7 | 2,4-Dinitrophenol | Barratt (1996a) | NC | 118.5 | 184.1 |
| 8 | 2,4-Xylenol | Barratt (1996a) | C | 25.4 | 122.2 |
| 9 | 2,5-Dinitrophenol | Barratt (1996a) | NC | 118.5 | 184.1 |
| 10 | 2,5-Xylenol | Barratt (1996a) | C | 25.4 | 122.2 |
| 11 | 2,6-Xylenol | Barratt (1996a) | C | 25.4 | 122.2 |
| 12 | 2-Bromobenzoic acid | Barratt (1995b) | NC | 81.6 | 201.0 |
| 13 | 2-Butyn-1,4-diol | Barratt (1996b) | C | 29.0 | 86.1 |
| 14 | 2-Chlorobenzaldehyde | Barratt (1996b) | C | 8.7 | 140.6 |
| 15 | 2-Chloropropanoic acid | Barratt (1996a) | C | 8.1 | 108.5 |
| 16 | 2-Ethylphenol | Barratt (1996a) | NC | 27.1 | 122.2 |
| 17 | 2-Hydroxyethyl acrylate | Barratt (1996b) | C | −15.9 | 116.1 |
| 18 | 2-Mercaptoethanoic acid | Barratt (1996a) | C | 18.8 | 92.1 |
| 19 | 2-Naphthoic acid | Barratt (1996a) | NC | 106.7 | 172.2 |
| 20 | 2-Naphthol | Barratt (1996a) | NC | 67.7 | 144.2 |
| 21 | 2-Nitrophenol | Barratt (1996a) | NC | 70.8 | 139.1 |
| 22 | 2-Phenylphenol | Barratt (1996a) | NC | 86.6 | 170.2 |
| 23 | 3-Methylbutanal | Barratt (1996b) | NC | −79.3 | 86.1 |
| 24 | 3-Nitrophenol | Barratt (1996a) | NC | 70.8 | 139.1 |
| 25 | 3-Picoline | Barratt (1996a) | NC | −25.9 | 93.1 |
| 26 | 3-Toluidine | Barratt (1995b) | NC | 11.6 | 107.2 |
| 27 | 4-Ethylbenzoic acid | Barratt (1996a) | NC | 73.5 | 150.2 |
| 28 | 4-Methoxyphenol | Barratt (1996a) | NC | 25.2 | 124.1 |
| 29 | 4-Nitrophenol | Barratt (1996a) | NC | 70.8 | 139.1 |
| 30 | 4-Nitrophenylacetic acid | Barratt (1996a) | NC | 124.3 | 181.2 |
| 31 | 4-Picoline | Barratt (1996a) | NC | −25.9 | 93.1 |
| 32 | Acridine | Barratt (1995b) | NC | 100.3 | 179.2 |
| 33 | Acrolein | Barratt (1995b) | C | −94.6 | 56.1 |
| 34 | Acrylic acid | Barratt (1995b) | C | −36.5 | 74.1 |
| 35 | Amino*tris*(methylphosphonic acid) | Barratt (1996a) | C | 90.3 | 299.1 |
| 36 | Barbituric acid | Barratt (1996a) | NC | 199.0 | 128.1 |
| 37 | Benzoic acid | Barratt (1996a) | NC | 48.9 | 122.1 |
| 38 | Benzylamine | Barratt (1996a) | C | −6.2 | 93.1 |
| 39 | Butyric acid | Barratt (1996a) | C | 3.0 | 88.1 |
| 40 | Catechol | Barratt (1996a) | NC | 45.7 | 110.1 |
| 41 | Citric acid | Barratt (1995b) | NC | 169.2 | 192.1 |
| 42 | Cocoamine (dodecylamine) | Barratt (1995b) | C | 35.1 | 185.4 |
| 43 | Cyanoacetic acid | Barratt (1996a) | C | 38.0 | 85.1 |
| 44 | Cyclopropane carboxylic acid | Barratt (1996a) | C | 13.0 | 86.1 |
| 45 | Decanoic acid | Barratt (1995b) | NC | 62.7 | 172.3 |
| 46 | Formaldehyde | Barratt (1996b) | C | −110.9 | 30.0 |
| 47 | Fumaric acid | Barratt (1996a) | NC | 84.1 | 116.1 |
| 48 | Glycolic acid | Barratt (1996a) | NC | 23.3 | 76.1 |
| 49 | Glyoxylic acid | Barratt (1996a) | C | 16.1 | 74.0 |
| 50 | Hexylcinnamic aldehyde | Barratt (1996b) | NC | 44.4 | 216.3 |
| 51 | Hydrogenated tallow amine (hexadecylamine) | Barratt (1996a) | NC | 75.6 | 241.5 |
| 52 | Hydroquinone | Barratt (1996a) | NC | 45.7 | 110.1 |
| 53 | Imidazole | Barratt (1995b) | NC | 18.5 | 68.1 |
| 54 | Iodoacetic acid | Barratt (1996a) | C | 29.6 | 186.0 |
| 55 | *Iso*butanal | Barratt (1996b) | NC | −80.2 | 72.1 |
| 56 | *Iso*butyric acid | Barratt (1996a) | C | −8.3 | 88.1 |
| 57 | *Iso*eugenol | Barratt (1996a) | NC | 61.9 | 164.2 |
| 58 | *Iso*quinoline | Barratt (1995b) | NC | 37.6 | 129.2 |

**Table 18.3 (continued)   Skin Corrosion Data for 277 Organic Chemicals**

| Chemical | | Source | C/NC | MP | MW |
|---|---|---|---|---|---|
| 59 | Kojic acid | Barratt (1996a) | NC | 96.2 | 142.1 |
| 60 | Lactic acid | Barratt (1995b) | C | 22.7 | 90.1 |
| 61 | Malic acid | Barratt (1996a) | NC | 112.7 | 134.1 |
| 62 | Malonic (propanedioic) acid | Barratt (1996a) | NC | 73.3 | 104.1 |
| 63 | 3-Cresol | Barratt (1995b) | C | 15.7 | 108.1 |
| 64 | Methoxyacetic acid | Barratt (1996a) | C | 8.7 | 90.1 |
| 65 | Methyl *iso*thiocyanate | Barratt (1996b) | C | −63.3 | 73.1 |
| 66 | Morpholine | Barratt (1995b) | C | −15.2 | 87.1 |
| 67 | Myristic (tetradecanoic) acid | Barratt (1995b) | NC | 99.7 | 228.4 |
| 68 | 2-Cresol | Barratt (1995b) | C | 15.7 | 108.1 |
| 69 | Oxalic (ethanedioic) acid | Barratt (1995b) | C | 63.0 | 90.0 |
| 70 | 4-Cresol | Barratt (1995b) | C | 15.7 | 108.1 |
| 71 | Propargyl alcohol | Barratt (1996b) | C | −49.0 | 56.1 |
| 72 | Propylphosphonic acid | Barratt (1996a) | C | 28.3 | 124.1 |
| 73 | Pyridine | Barratt (1995b) | NC | -44.5 | 79.1 |
| 74 | Pyruvic acid | Barratt (1996a) | C | 28.2 | 88.1 |
| 75 | Quinoline | Barratt (1995b) | NC | 37.6 | 129.2 |
| 76 | Salicylic acid | Barratt (1995b) | NC | 93.8 | 138.1 |
| 77 | Succinic acid | Whittle (1996) | NC | 83.3 | 118.1 |
| 78 | Thymol | Barratt (1996a) | C | 38.1 | 150.2 |
| 79 | *trans*-Cinnamic acid | Barratt (1995b) | NC | 69.5 | 148.2 |
| 80 | 3-Methoxyphenol | Barratt (1996a) | NC | 25.2 | 124.1 |
| 81 | 4-Ethylphenol | Barratt (1996a) | NC | 27.1 | 122.2 |
| 82 | Phenol | Barratt (1995b) | C | −2.3 | 94.1 |
| 83 | 1,1,1-Trichloroethane | ECETOC (1995) | NC | −72.0 | 133.4 |
| 84 | 1,13-Tetradecadiene | ECETOC (1995) | NC | −1.2 | 194.4 |
| 85 | 1,3-Dibromopropane | ECETOC (1995) | NC | −27.0 | 201.9 |
| 86 | 1,5-Hexadiene | ECETOC (1995) | NC | −96.7 | 82.2 |
| 87 | 1,6-Dibromohexane | ECETOC (1995) | NC | 7.9 | 244.0 |
| 88 | 1,9-Decadiene | ECETOC (1995) | NC | −46.8 | 138.3 |
| 89 | 10-Undecenoic Acid | ECETOC (1995) | NC | 71.5 | 184.3 |
| 90 | 1-Bromo-2-chloroethane | ECETOC (1995) | NC | −58.0 | 143.4 |
| 91 | 1-Bromo-4-chlorobutane | ECETOC (1995) | NC | −33.6 | 171.5 |
| 92 | 1-Bromo-4-fluorobenzene | ECETOC (1995) | NC | −19.1 | 175.0 |
| 93 | 1-Bromohexane | ECETOC (1995) | NC | −41.6 | 165.1 |
| 94 | 1-Bromopentane | ECETOC (1995) | NC | −53.8 | 151.1 |
| 95 | 1-Decanol | ECETOC (1995) | NC | 7.9 | 158.3 |
| 96 | 1-Formyl-1-methyl-4(4-methyl-3-penten-1-yl)-3-cyclohexane | ECETOC (1995) | NC | 46.5 | 208.4 |
| 97 | 2,3-Dichloroproprionitrile | ECETOC (1995) | NC | −21.2 | 124.0 |
| 98 | 2,4-Decadienal | ECETOC (1995) | NC | 6.0 | 154.3 |
| 99 | 2,4-Dimethyl-3-cyclohexene-1-carboxaldehyde | ECETOC (1995) | NC | −10.1 | 138.2 |
| 100 | 2,4-Dimethyltetrahydrobenzaldehyde | ECETOC (1995) | NC | −10.1 | 138.2 |
| 101 | 2,4-Dinitromethylaniline | ECETOC (1995) | NC | 108.9 | 197.2 |
| 102 | 2,4-Hexadienal | ECETOC (1995) | NC | −56.2 | 96.1 |
| 103 | 2,4-Xylidine | ECETOC (1995) | NC | 34.7 | 135.2 |
| 104 | 2,5-Methylene-6-propyl-3-cyclo-hexen-carbaldehyde | ECETOC (1995) | NC | 15.2 | 164.3 |
| 105 | 2,6-Dimethyl-2,4,6-octatriene | ECETOC (1995) | NC | −21.2 | 134.2 |
| 106 | 2,6-Dimethyl-4-heptanol | ECETOC (1995) | NC | −38.1 | 144.3 |
| 107 | 2-Bromobutane | ECETOC (1995) | NC | −78.1 | 137.0 |
| 108 | 2-Bromopropane | ECETOC (1995) | NC | −91.0 | 123.0 |
| 109 | 2-Chloronitrobenzene | ECETOC (1995) | NC | 48.8 | 157.6 |
| 110 | 2-Ethoxyethyl methacrylate | ECETOC (1995) | NC | −25.2 | 158.2 |
| 111 | 2-Ethylhexanal | ECETOC (1995) | NC | −42.3 | 128.2 |
| 112 | 2-Ethylhexylpalmitate | ECETOC (1995) | NC | 117.2 | 368.7 |
| 113 | 2-Fluorotoluene | ECETOC (1995) | NC | −54.2 | 110.1 |
| 114 | 2-Methoxyethyl acrylate | ECETOC (1995) | C | −56.2 | 128.2 |
| 115 | 2-Methoxyphenol (guaiacol) | Barratt (1996a) | NC | 25.2 | 124.1 |

**Table 18.3 (continued)** **Skin Corrosion Data for 277 Organic Chemicals**

| Chemical | | Source | C/NC | MP | MW |
|---|---|---|---|---|---|
| 116 | 2-Methyl-4-phenyl-2-butanol | ECETOC (1995) | NC | 30.4 | 164.3 |
| 117 | 2-Methylbutyric acid | ECETOC (1995) | C | 3.6 | 102.1 |
| 118 | 2-Phenylethanol (phenylethylalcohol) | ECETOC (1995) | NC | 5.8 | 122.2 |
| 119 | 2-Phenylpropanal (2-phenylpropionaldehyde) | ECETOC (1995) | NC | −10.0 | 134.2 |
| 120 | 2-*tert*-Butylphenol | ECETOC (1995) | C | 36.9 | 150.2 |
| 121 | 3,3′-Dithiopropionic acid | ECETOC (1995) | NC | 141.5 | 210.3 |
| 122 | 3,7-Dimethyl-2,6-nonadienal | ECETOC (1995) | NC | −3.9 | 180.3 |
| 123 | 3-Chloro-4-fluoronitrobenzene | ECETOC (1995) | NC | 44.2 | 175.6 |
| 124 | 3-Diethylaminopropionitrile | ECETOC (1995) | NC | −0.4 | 126.2 |
| 125 | 3-Mercapto-1-propanol | ECETOC (1995) | NC | −33.6 | 92.2 |
| 126 | 3-Methoxypropylamine | ECETOC (1995) | NC | −40.4 | 89.1 |
| 127 | 3-Methylphenol | ECETOC (1995) | NC | 15.7 | 108.1 |
| 128 | 3-Methylbutyraldehyde | ECETOC (1995) | NC | −79.3 | 86.1 |
| 129 | 4-(Methylthio)-benzaldehyde | ECETOC (1995) | NC | 28.6 | 152.2 |
| 130 | 4,4′-Methylene-*bis*-(2,6-di*tert*-butylphenol) | ECETOC (1995) | NC | 208.5 | 424.7 |
| 131 | 4-Amino-1,2,4-triazole | ECETOC (1995) | NC | 31.0 | 84.1 |
| 132 | 4-Tricyclo-decylindene-8-butanal | ECETOC (1995) | NC | 233.9 | 494.9 |
| 133 | 6-Butyl-2,4-dimethyldihydropyrane | ECETOC (1995) | NC | −2.3 | 168.3 |
| 134 | α-Hexyl cinnamic aldehyde | ECETOC (1995) | NC | 44.4 | 216.3 |
| 135 | α-Ionol | ECETOC (1995) | NC | 45.2 | 194.3 |
| 136 | Allyl bromide | ECETOC (1995) | C | −80.5 | 121.0 |
| 137 | Allyl heptanoate | ECETOC (1995) | NC | −10.8 | 170.3 |
| 138 | Allyl phenoxyacetate | ECETOC (1995) | NC | 36.5 | 192.2 |
| 139 | α-Terpineol | ECETOC (1995) | NC | 12.4 | 154.3 |
| 140 | α-Terpinyl acetate | ECETOC (1995) | NC | 21.5 | 196.3 |
| 141 | Benzyl acetate | ECETOC (1995) | NC | −0.5 | 150.2 |
| 142 | Benzyl acetone | ECETOC (1995) | NC | 12.8 | 148.2 |
| 143 | Benzyl alcohol | ECETOC (1995) | NC | −5.4 | 108.1 |
| 144 | Benzyl benzoate | ECETOC (1995) | NC | 70.8 | 212.3 |
| 145 | Benzyl salicylate | ECETOC (1995) | NC | 115.5 | 228.3 |
| 146 | β-Ionol | ECETOC (1995) | NC | 54.5 | 194.3 |
| 147 | Butyl propanoate | ECETOC (1995) | NC | −44.6 | 130.2 |
| 148 | Carvacrol | ECETOC (1995) | C | 38.1 | 150.2 |
| 149 | Cinnamaldehyde | ECETOC (1995) | NC | 0.0 | 132.2 |
| 150 | Cinnamyl alcohol | ECETOC (1995) | NC | 15.8 | 134.2 |
| 151 | *cis*-Cyclooctene | ECETOC (1995) | NC | −58.8 | 110.2 |
| 152 | *cis*-Jasmone | ECETOC (1995) | NC | 40.2 | 164.3 |
| 153 | Citrathal | ECETOC (1995) | NC | 4.8 | 226.4 |
| 154 | Cyclamen aldehyde | ECETOC (1995) | NC | 29.1 | 190.3 |
| 155 | Diacetyl | ECETOC (1995) | NC | −41.7 | 86.1 |
| 156 | Dichloromethane | ECETOC (1995) | NC | −89.5 | 84.9 |
| 157 | Diethyl phthalate | ECETOC (1995) | NC | −1.7 | 222.2 |
| 158 | Diethylaminopropylamine | ECETOC (1995) | C | 0.7 | 130.2 |
| 159 | Dihydromercenol | ECETOC (1995) | NC | −10.6 | 156.3 |
| 160 | Dimethyl disulphide | ECETOC (1995) | NC | −69.7 | 94.2 |
| 161 | Dimethylbenzylcarbinyl acetate | ECETOC (1995) | NC | 28.3 | 192.3 |
| 162 | Dimethyldipropylenetriamine | ECETOC (1995) | C | 40.4 | 159.3 |
| 163 | Dimethylisopropylamine | ECETOC (1995) | C | −95.4 | 87.2 |
| 164 | Dimethyl butylamine | ECETOC (1995) | C | −70.6 | 101.2 |
| 165 | Dipropyl disulphide | ECETOC (1995) | NC | −21.8 | 150.3 |
| 166 | Dipropylene glycol | ECETOC (1995) | NC | 6.1 | 134.2 |
| 167 | dl-Citronellol | ECETOC (1995) | NC | −12.2 | 156.3 |
| 168 | d-Limonene | ECETOC (1995) | NC | −40.8 | 136.2 |
| 169 | Dodecanoic (lauric) acid | ECETOC (1995) | NC | 81.9 | 200.3 |
| 170 | Erucamide | ECETOC (1995) | NC | 183.4 | 337.6 |
| 171 | Ethyl thioethyl methacrylate | ECETOC (1995) | NC | −8.5 | 174.3 |
| 172 | Ethyl tiglate | ECETOC (1995) | NC | −53.9 | 128.2 |

**Table 18.3 (continued)  Skin Corrosion Data for 277 Organic Chemicals**

| Chemical | | Source | C/NC | MP | MW |
|---|---|---|---|---|---|
| 173 | Ethyl triglycol methacrylate | ECETOC (1995) | NC | 51.3 | 246.3 |
| 174 | Ethyl trimethyl acetate | ECETOC (1995) | NC | −68.4 | 116.2 |
| 175 | Eucalyptol | ECETOC (1995) | NC | 8.1 | 154.3 |
| 176 | Eugenol | ECETOC (1995) | NC | 60.6 | 164.2 |
| 177 | Fluorobenzene | ECETOC (1995) | NC | −73.0 | 96.1 |
| 178 | Geraniol | ECETOC (1995) | NC | −10.8 | 154.3 |
| 179 | Geranyl dihydrolinalol | ECETOC (1995) | NC | 60.0 | 292.5 |
| 180 | Geranyl linalool | ECETOC (1995) | NC | 58.5 | 290.5 |
| 181 | Glycol bromoacetate | ECETOC (1995) | C | 1.2 | 303.9 |
| 182 | Heptanal | ECETOC (1995) | NC | −43.0 | 114.2 |
| 183 | Heptyl butyrate | ECETOC (1995) | NC | 1.7 | 186.3 |
| 184 | Heptylamine | ECETOC (1995) | C | −21.6 | 115.2 |
| 185 | Hexyl salicylate | ECETOC (1995) | NC | 99.7 | 222.3 |
| 186 | Hydroxycitronellal | ECETOC (1995) | NC | 23.4 | 172.3 |
| 187 | *Iso*bornyl acetate | ECETOC (1995) | NC | 34.1 | 196.3 |
| 188 | *Iso*butyraldehyde | ECETOC (1995) | NC | −92.1 | 72.1 |
| 189 | *Iso*propanol | ECETOC (1995) | NC | −89.2 | 60.1 |
| 190 | *Iso*propyl isostearate | ECETOC (1995) | NC | 80.6 | 326.6 |
| 191 | *Iso*propyl myristate | ECETOC (1995) | NC | 44.4 | 270.5 |
| 192 | *Iso*propyl palmitate | ECETOC (1995) | NC | 72.0 | 298.5 |
| 193 | *Iso*stearic acid | ECETOC (1995) | NC | 125.2 | 284.5 |
| 194 | *Iso*stearyl alcohol | ECETOC (1995) | NC | 77.3 | 270.5 |
| 195 | Lilestralis lilial | ECETOC (1995) | NC | 46.3 | 204.3 |
| 196 | Linalol | ECETOC (1995) | NC | −11.4 | 154.3 |
| 197 | Linalol oxide | ECETOC (1995) | NC | 31.1 | 170.3 |
| 198 | Linalyl acetate | ECETOC (1995) | NC | −2.1 | 196.3 |
| 199 | Methacrolein | ECETOC (1995) | C | −90.6 | 70.1 |
| 200 | Methyl 2-methylbutyrate | ECETOC (1995) | NC | −68.4 | 116.2 |
| 201 | Methyl caproate | ECETOC (1995) | NC | −44.6 | 130.2 |
| 202 | Methyl laurate | ECETOC (1995) | NC | 23.2 | 214.4 |
| 203 | Methyl lavender ketone (1-hydroxy-3-decanone) | ECETOC (1995) | NC | 42.7 | 172.3 |
| 204 | Methyl linoleate | ECETOC (1995) | NC | 70.8 | 294.5 |
| 205 | Methyl palmitate | ECETOC (1995) | NC | 63.2 | 270.5 |
| 206 | Methyl stearate | ECETOC (1995) | NC | 81.6 | 298.5 |
| 207 | Methyl trimethyl acetate | ECETOC (1995) | NC | −62.5 | 116.2 |
| 208 | Decylidene methyl anthranilate | ECETOC (1995) | NC | 99.9 | 289.4 |
| 209 | *N,N*-Dimethylbenzylamine | ECETOC (1995) | NC | −12.8 | 135.2 |
| 210 | Nonanal | ECETOC (1995) | NC | −19.5 | 142.2 |
| 211 | Octanoic acid | ECETOC (1995) | C | 48.4 | 144.2 |
| 212 | Oleyl propylene diamine dioleate | ECETOC (1995) | NC | 142.1 | 324.6 |
| 213 | Phenethyl bromide | ECETOC (1995) | NC | 2.5 | 185.1 |
| 214 | 4-*Iso*propylphenylacetaldehyde | ECETOC (1995) | NC | 18.4 | 162.2 |
| 215 | 4-Mentha-1,8-dien-7-ol | ECETOC (1995) | NC | 11.1 | 152.2 |
| 216 | 4-*tert*-Butyl dihydrocinnamaldehyde | ECETOC (1995) | NC | 46.3 | 190.3 |
| 217 | Salicylaldehyde | ECETOC (1995) | NC | 42.6 | 122.1 |
| 218 | Tetrachloroethylene | ECETOC (1995) | NC | −60.6 | 165.8 |
| 219 | Tetrahydrogeranial | ECETOC (1995) | NC | −30.0 | 156.3 |
| 220 | Tonalid | ECETOC (1995) | NC | 98.7 | 244.4 |
| 221 | Trichloroethylene | ECETOC (1995) | NC | −60.6 | 165.8 |
| 222 | 1-(2-Aminoethyl)piperazine | NIH (1999) | C | 53.7 | 129.2 |
| 223 | 1,2-Diaminopropane | NIH (1999) | C | −22.9 | 74.1 |
| 224 | 1,4-Diaminobutane | NIH (1999) | C | 0.9 | 88.2 |
| 225 | 2,3-Dimethylcyclohexylamine | NIH (1999) | C | −11.1 | 127.2 |
| 226 | 2-Ethylhexylamine | NIH (1999) | C | −21.0 | 129.3 |
| 227 | 2-Mercaptoethanol | NIH (1999) | C | −45.6 | 78.1 |
| 228 | 3-Diethylaminopropylamine | NIH (1999) | C | 0.7 | 130.2 |
| 229 | Acetic acid | NIH (1999) | C | −21.3 | 60.1 |

**Table 18.3 (continued)  Skin Corrosion Data for 277 Organic Chemicals**

| Chemical | | Source | C/NC | MP | MW |
|---|---|---|---|---|---|
| 230 | Acetic anhydride | NIH (1999) | C | −95.1 | 102.1 |
| 231 | Acetyl bromide | NIH (1999) | C | −53.0 | 123.0 |
| 232 | Benzene sulphonyl chloride | NIH (1999) | C | 61.2 | 176.6 |
| 233 | Benzyl chloroformate | NIH (1999) | C | 11.6 | 170.6 |
| 234 | Bromoacetic acid | NIH (1999) | C | 29.2 | 139.0 |
| 235 | Bromoacetyl bromide | NIH (1999) | C | −1.7 | 201.9 |
| 236 | Butanoic acid | NIH (1999) | C | 3.0 | 88.1 |
| 237 | Butylamine | NIH (1999) | C | −58.8 | 73.1 |
| 238 | Butylbenzene | NIH (1999) | NC | −23.3 | 134.2 |
| 239 | Butyric anhydride | NIH (1999) | C | −44.6 | 158.2 |
| 240 | Chloroacetic acid | NIH (1999) | C | 10.9 | 94.5 |
| 241 | Crotonic acid | NIH (1999) | C | 2.4 | 86.1 |
| 242 | Cyanuric chloride | NIH (1999) | C | 68.8 | 184.4 |
| 243 | Cyclohexylamine | NIH (1999) | C | −27.1 | 99.2 |
| 244 | Dichloroacetic acid | NIH (1999) | C | 24.2 | 128.9 |
| 245 | Dichloroacetyl chloride | NIH (1999) | C | −32.5 | 147.4 |
| 246 | Dichlorophenyl phosphine | NIH (1999) | C | −4.9 | 179.0 |
| 247 | Dicyclohexylamine | NIH (1999) | C | 27.7 | 181.3 |
| 248 | Diethylamine | NIH (1999) | C | −79.7 | 73.1 |
| 249 | Diethylene triamine | NIH (1999) | C | 17.8 | 103.2 |
| 250 | Dimethylcarbamyl chloride | NIH (1999) | C | -15.9 | 107.5 |
| 251 | Dodecyl trichlorosilane | NIH (1999) | C | 51.0 | 303.8 |
| 252 | Ethanolamine | NIH (1999) | C | −27.6 | 61.1 |
| 253 | Ethylene diamine | NIH (1999) | C | −23.8 | 60.1 |
| 254 | Formic acid | NIH (1999) | C | −25.0 | 46.0 |
| 255 | Fumaryl chloride | NIH (1999) | C | 6.8 | 153.0 |
| 256 | Hexanoic acid | NIH (1999) | C | 26.2 | 116.2 |
| 257 | Hexanol | NIH (1999) | NC | −37.9 | 102.2 |
| 258 | Maleic acid | NIH (1999) | NC | 84.1 | 116.1 |
| 259 | Maleic anhydride | NIH (1999) | C | −51.6 | 98.1 |
| 260 | Mercaptoacetic acid | NIH (1999) | C | 18.8 | 92.1 |
| 261 | Nonanol | NIH (1999) | NC | −3.2 | 144.3 |
| 262 | 2-Anisoyl chloride | NIH (1999) | C | 36.7 | 170.6 |
| 263 | Octadecyl trichlorosilane | NIH (1999) | C | 107.7 | 387.9 |
| 264 | Octyl trichlorosilane | NIH (1999) | C | 8.1 | 247.7 |
| 265 | Pentanoyl (valeryl) chloride | NIH (1999) | C | −42.4 | 120.6 |
| 266 | Phenyl acetyl chloride | NIH (1999) | C | 13.7 | 154.6 |
| 267 | Phenyl trichlorosilane | NIH (1999) | C | 5.8 | 211.6 |
| 268 | Propanoic acid | NIH (1999) | C | −9.0 | 74.1 |
| 269 | Pyrrolidine | NIH (1999) | C | −36.0 | 71.1 |
| 270 | Tetraethylenepentamine | NIH (1999) | C | 112.7 | 189.3 |
| 271 | Tributylamine | NIH (1999) | NC | 0.8 | 185.4 |
| 272 | Trichloroacetic acid | NIH (1999) | C | 26.7 | 163.4 |
| 273 | Trichlorotoluene | NIH (1999) | NC | 10.4 | 195.5 |
| 274 | Triethanolamine | NIH (1999) | NC | 83.3 | 149.2 |
| 275 | Triethylene tetramine | NIH (1999) | C | 68.2 | 146.2 |
| 276 | Trifluoroacetic acid | NIH (1999) | C | −24.0 | 114.0 |
| 277 | Undecanol | NIH (1999) | NC | 18.7 | 172.3 |

*Note:*  C = corrosive; MP = melting point (°C); MW = molecular weight (g/mol); NC = non-corrosive.

The following physicochemical properties, which were considered to be possible predictors of acute skin toxicity, were calculated for the 277 chemicals in Table 18.3:

1. Molecular weight (MW), surface area (MSA), and volume (MV)
2. Log $K_{ow}$
3. Melting point (MP)

4. Surface tension (ST)
5. Dermal permeability coefficient ($K_p$)
6. Dipole moment (DM)
7. Energies of the lowest unoccupied molecular orbital and the highest occupied molecular orbital ($E_{LUMO}$ and $E_{HOMO}$, respectively)

Log $K_{ow}$, MP, and $K_p$ values were calculated with the Syracuse Research Corporation (SRC, Syracuse, NY) KOWWIN, MPBPWIN, and DERMWIN software packages (SRC, Syracuse, NY, USA), respectively, using the SMILES codes of the chemicals as the input. Values of MW and ST were calculated with the Advanced Chemistry Development (ACD) ChemSketch software.

A two-step decision rule was envisaged. In the first step, it was hypothesized that discrimination could be based on MP alone. This hypothesis was based on the grounds that chemicals existing as solids at skin temperature are not expected to be corrosive, whereas chemicals existing as liquids may or may not be, depending on other factors that could be assessed in a second step. The 277 chemicals were separated into two groups. One group contained 88 chemicals having predicted MPs greater than 37°C, and the other contained 189 chemicals having predicted MPs less than or equal to 37°C, revealed that 74 of the 88 predicted solids (84%) are non-corrosive, as expected, whereas 14 of them (16%) are corrosive, contrary to expectation.

To identify the best variable for discriminating between corrosive and non-corrosive liquids, classification tree (CT) analysis was applied to the values of MW, log $K_{ow}$, log Kp, ST, DM, $E_{LUMO}$, and $E_{HOMO}$ for the 189 liquids. CT analysis was performed by using the CART (Classification and Regression Tree) algorithm (Breiman et al., 1984) in STATISTICA 5.5 for Windows (Statsoft Inc., Tulsa, OK). Equal prior probabilities were set for the two classes (C/NC), the Gini index was used as the measure of node homogeneity, and a minimum node size of five observations was used as the stopping rule (i.e., a node would only be split if it contained more than five observations).

The best discriminating variable was found to be log $K_{ow}$. However, the resulting CT predicted liquids with log $K_{ow}$ values greater than 1.32 to be non-corrosive, and liquids with log $K_{ow}$ values less than or equal to 1.32 to be corrosive. The direction of this inequality is contrary to expectation, since corrosive chemicals are generally expected to be more hydrophobic than non-corrosive chemicals and have higher, not lower, values of log $K_{ow}$. A possible explanation for this finding is that log $K_{ow}$ is significantly correlated with MW ($r = 0.69$, $p < 0.001$), meaning it is the smaller chemicals that are more likely to be corrosive, not the less hydrophobic ones. Log $K_{ow}$ was removed from the set of input variables, and CT analysis was applied again. This time, CT analysis identified MW as the best discriminating variable, with an optimal cutoff value of 123 g/mol. On this basis of this finding, CM 18.1 was formulated for predicting the corrosion potential of organic liquids, and the variable selection procedure was stopped:

$$\text{If MW} \leq 123 \text{ g/mol, predict as C; otherwise predict as NC.} \qquad \text{(CM 18.1)}$$

## B. Development of a Prediction Model Based on pH Data

To develop a PM based on measured pH values for skin corrosion potential, a training set of 44 organic and inorganic chemicals (Table 18.4) was taken from a data set of 60 chemicals used in the ECVAM validation study on alternative methods for skin corrosion (Barratt et al., 1998; Fentem et al., 1998). For the purposes of the current investigation, 44 chemicals were chosen from the full set of 60 on the basis that: (1) they are water-soluble, do not decompose, and do not react with water (as indicated in Fentem et al., 1998); and (2) they have unambiguous identities (for example, 20/80 coconut/palm soap was omitted from the training set). The pH data for 10% solutions of these chemicals had been obtained using a pH meter (Accumet 15, Fisher Scientific Ltd., Loughborough, U.K.) by BIBRA International (Croydon, U.K.) under the terms of an ECVAM contract. The chemicals were classified as skin corrosives (C) or non-corrosives (NC) by applying

**Table 18.4    Skin Corrosion Classifications and pH Data for 44 Chemicals**

| Chemical | | Known (*in vivo*) Classification | pH | Predicted Classification |
|---|---|:---:|:---:|:---:|
| 1 | Hexanoic acid | C | 2.57 | C |
| 2 | 1,2-Diaminopropane | C | 12.02 | C |
| 3 | Carvacrol | C | 4.91 | NC |
| 4 | Methacrolein | C | 4.18 | NC |
| 5 | Phenethyl bromide | NC | 5.40 | NC |
| 6 | *Iso*propanol | NC | 5.86 | NC |
| 7 | 2-Methoxyphenol (Guaiacol) | NC | 4.86 | NC |
| 8 | 2,4-Xylidine (2,4-Dimethylaniline) | NC | 8.73 | NC |
| 9 | 2-Phenylethanol (phenylethylalcohol) | NC | 5.31 | NC |
| 10 | 3-Methoxypropylamine | C | 11.78 | C |
| 11 | Allyl bromide | C | 3.15 | C |
| 12 | Dimethyldipropylenetriamine | C | 11.38 | C |
| 13 | Methyl trimethylacetate | NC | 4.96 | NC |
| 14 | Dimethylisopropylamine | C | 11.81 | C |
| 15 | Potassium hydroxide | C | 13.76 | C |
| 16 | Tetrachloroethylene | NC | 7.13 | NC |
| 17 | Ferric (iron [III]) chloride | C | 1.11 | C |
| 18 | Butyl propanoate | NC | 4.57 | NC |
| 19 | 2-*tert*-Butylphenol | C | 8.17 | NC |
| 20 | Sulfuric acid | C | 0.33 | C |
| 21 | *Iso*stearic acid | NC | 4.78 | NC |
| 22 | Methyl palmitate | NC | 5.69 | NC |
| 23 | 65/35 Octanoic/decanoic acids | C | 3.72 | C |
| 24 | 2-Bromobutane | NC | 3.89 | NC |
| 25 | 4-(Methylthio)-benzaldehyde | NC | 6.38 | NC |
| 26 | 70/30 Oleine/octanoic acid | NC | 3.26 | C |
| 27 | 2-Methylbutyric acid | C | 2.81 | C |
| 28 | 2-Ethoxyethyl methacrylate | NC | 9.52 | NC |
| 29 | Octanoic acid (caprylic acid) | C | 3.67 | C |
| 30 | Benzyl acetone | NC | 4.81 | NC |
| 31 | Heptylamine | C | 11.88 | C |
| 32 | Cinnamaldehyde | NC | 4.03 | NC |
| 33 | 60/40 Octanoic/decanoic acids | C | 3.77 | C |
| 34 | Eugenol | NC | 3.68 | C |
| 35 | 55/45 Octanoic/decanoic acids | C | 3.80 | C |
| 36 | Methyl laurate | NC | 5.67 | NC |
| 37 | Sodium bicarbonate | NC | 7.89 | NC |
| 38 | Sulfamic acid | NC | 0.70 | C |
| 39 | Sodium bisulphite | NC | 3.85 | NC |
| 40 | 1-(2-Aminoethyl)piperazine | C | 11.67 | C |
| 41 | 1,9-Decadiene | NC | 4.15 | NC |
| 42 | Phosphoric acid | C | 1.63 | C |
| 43 | 10-Undecenoic acid | NC | 3.88 | NC |
| 44 | 4-Amino-1,2,4-triazole | NC | 5.92 | NC |

*Note:*  C = corrosive (EU risk phrases R34 and R35); NC = non-corrosive. The pH data were
provided by BIBRA International (Surrey, U.K.) and refer to measurements made on
a 10% solution. The data in this table constitute the training set for CM 18.2.

Predictions are based on the PM. If pH < 3.9 or pH > 10.5, predict C; otherwise predict NC.

EU classification criteria (European Commission, 1983) to the animal data (European Centre for
Ecotoxicology and Toxicology of Chemicals, 1993).

The application of CT analysis to the pH data in Table 18.4 generated a CT (Figure 18.1). The
CT is interpreted by reading from the root node (node 1) at the top of the tree to the terminal nodes
(nodes 3, 4, and 5) at the bottom. The nodes are numbered in the top left corner. Before the splitting

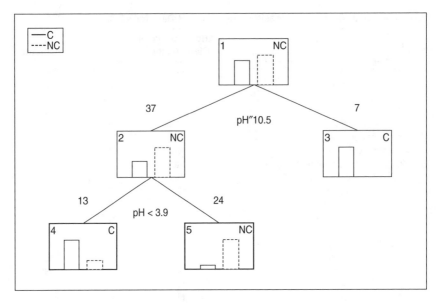

**Figure 18.1** Classification tree for distinguishing between corrosive and non-corrosive chemicals on the basis of pH measurements.

process begins, all 44 observations are placed in node 1. According to the first decision rule, which is applied to all observations, 7 observations with pH values greater than 10.5 are placed in node 3 and are predicted to be corrosive. The remaining 37 observations are placed in node 2 and subjected to a second decision rule. Application of the second rule leads to 13 observations with pH values less than 3.9 being placed in node 4 and being predicted to be corrosive. The remaining 24 observations are placed in node 5 and are predicted to be noncorrosive. The numbers above each node show how many observations (chemicals) are sent to each node, and the histograms illustrate the relative proportions of C and NC chemicals in each node. The CT for skin corrosion potential can be summarized in the form of CM 18.2.

If pH < 3.9 or if pH > 10.5, then predict as C; otherwise, predict NC.    (CM 18.2)

In CM 18.2, pH is measured for a 10% solution (w/v in the case of liquids, and w/w in the case of solids). Because of the identities of the chemicals in the training set (Table 18.4), the domain of the model is expected to cover organic acids, inorganic acids, organic bases, inorganic bases, mixtures, neutral organics (such as alcohols, ketones and esters), phenols, and electrophiles (such as aldehydes and alkyl halides). It is important to note that the domain of CM 18.2 excludes insoluble chemicals and chemicals that react with water.

## C. Development of a Prediction Model Based on EPISKIN Data

PMs based on the EPISKIN *in vitro* end point were developed from the data obtained during the ECVAM Skin Corrosivity Validation Study (Barratt et al., 1998; Fentem et al., 1998). The EPISKIN data are cell viabilities, measured following treatment for 3 minutes, 1 hour, and 4 hours.

The application of CT analysis to the EPISKIN data for 60 chemicals (Table 18.5) produced the following CM:

If the EPISKIN viability after 4-h exposure < 36%, predict C; otherwise, predict NC.  (CM 18.3)

Table 18.5    EPISKIN Data for the 60 Chemicals Tested in the ECVAM Skin Corrosivity Study

| Chemical | | Classification | EPISKIN 3 min | EPISKIN 1 h | EPISKIN 4 h |
|---|---|---|---|---|---|
| 1 | Hexanoic acid | C | 15.49 | 2.30 | 2.64 |
| 2 | 1,2-Diaminopropane | C | 63.78 | 49.76 | 25.36 |
| 3 | Carvacrol | C | 126.04 | 16.17 | 16.28 |
| 4 | Boron trifluoride dihydrate | C | 2.26 | 3.79 | 3.43 |
| 5 | Methacrolein | C | 105.75 | 36.21 | 21.07 |
| 6 | Phenethyl bromide | NC | 140.42 | 148.54 | 156.59 |
| 7 | 3,3′-Dithiodipropionic acid | NC | 98.31 | 108.10 | 99.79 |
| 8 | Isopropanol | NC | 96.05 | 99.27 | 89.67 |
| 9 | 2-Methoxyphenol (Guaiacol) | NC | 143.83 | 16.27 | 8.87 |
| 10 | 2,4-Xylidine (2,4-Dimethylaniline) | NC | 117.38 | 54.50 | 39.40 |
| 11 | 2-Phenylethanol (phenylethylalcohol) | NC | 115.65 | 106.72 | 69.14 |
| 12 | Dodecanoic (lauric) acid | NC | 106.41 | 121.01 | 112.69 |
| 13 | 3-Methoxypropylamine | C | 50.87 | 41.83 | 17.60 |
| 14 | Allyl bromide | C | 117.84 | 28.23 | 22.46 |
| 15 | Dimethyldipropylenetriamine | C | 79.23 | 38.39 | 17.07 |
| 16 | Methyl trimethylacetate | NC | 112.89 | 96.97 | 70.56 |
| 17 | Dimethylisopropylamine | C | 78.22 | 22.30 | 13.99 |
| 18 | Potassium hydroxide (10% aq.) | C | 64.19 | 15.30 | 11.11 |
| 19 | Tetrachloroethylene | NC | 104.55 | 91.00 | 55.96 |
| 20 | Ferric [iron (III)] chloride | C | 91.04 | 66.28 | 33.24 |
| 21 | Potassium hydroxide (5% aq.) | NC | 44.42 | 12.11 | 9.85 |
| 22 | Butyl propanoate | NC | 102.89 | 117.33 | 77.09 |
| 23 | 2-tert-Butylphenol | C | 93.16 | 8.09 | 8.17 |
| 24 | Sodium carbonate (50% aq.) | NC | 115.11 | 112.61 | 88.39 |
| 25 | Sulfuric acid (10% wt.) | C | 94.95 | 22.08 | 2.12 |
| 26 | Isostearic acid | NC | 103.39 | 94.21 | 108.30 |
| 27 | Methyl palmitate | NC | 104.63 | 104.59 | 99.06 |
| 28 | Phosphorus tribromide | C | 3.55 | 14.34 | 11.87 |
| 29 | 65/35 Octanoic/decanoic acids | C | 108.83 | 6.16 | 4.25 |
| 30 | 4,4′-Methylene-bis-(2,6-ditert-butylphenol) | NC | 120.08 | 99.06 | 103.29 |
| 31 | 2-Bromobutane | NC | 132.81 | 121.21 | 55.02 |
| 32 | Phosphorus pentachloride | C | 43.34 | 15.25 | 3.28 |
| 33 | 4-(Methylthio)-benzaldehyde | NC | 134.39 | 176.61 | 161.64 |
| 34 | 70/30 Oleine/octanoic acid | NC | 107.55 | 104.67 | 59.56 |
| 35 | Hydrogenated tallow amine | NC | 98.63 | 101.41 | 104.20 |
| 36 | 2-Methylbutyric acid | C | 44.05 | 3.58 | 3.82 |
| 37 | Sodium undecylenate (34% aq.) | NC | 122.64 | 34.37 | 12.22 |
| 38 | Tallow amine | C | 94.44 | 110.84 | 113.97 |
| 39 | 2-Ethoxyethyl methacrylate | NC | 135.96 | 200.93 | 178.77 |
| 40 | Octanoic acid (caprylic acid) | C | 80.19 | 4.20 | 3.17 |
| 41 | 20/80 Coconut/palm soap | NC | 112.49 | 116.04 | 136.18 |
| 42 | 2-Mercaptoethanol, Na salt (46% aq.) | C | 101.39 | 104.44 | 101.49 |
| 43 | Hydrochloric acid (14.4% wt.) | C | 73.57 | 3.29 | 3.55 |
| 44 | Benzyl acetone | NC | 139.35 | 158.09 | 151.71 |
| 45 | Heptylamine | C | 97.22 | 355.39 | 348.54 |
| 46 | Cinnamaldehyde | NC | 129.99 | 118.92 | 64.14 |
| 47 | 60/40 Octanoic/decanoic acids | C | 108.75 | 7.01 | 5.12 |
| 48 | Glycol bromoacetate (85%) | C | 95.55 | 3.80 | 3.98 |
| 49 | Eugenol | NC | 136.42 | 91.06 | 49.02 |
| 50 | 55/45 Octanoic/decanoic acids | C | 122.38 | 7.72 | 4.67 |
| 51 | Methyl laurate | NC | 110.30 | 101.50 | 115.60 |
| 52 | Sodium bicarbonate | NC | 101.19 | 109.76 | 102.33 |
| 53 | Sulfamic acid | NC | 101.44 | 25.02 | 2.53 |
| 54 | Sodium bisulfite | NC | 116.31 | 103.18 | 107.76 |
| 55 | 1-(2-Aminoethyl)piperazine | C | 117.32 | 88.34 | 53.89 |

**Table 18.5 (continued)    EPISKIN Data for the 60 Chemicals Tested in the ECVAM Skin Corrosivity Study**

| Chemical | | Classification | EPISKIN 3 min | EPISKIN 1 h | EPISKIN 4 h |
|---|---|---|---|---|---|
| 56 | 1,9-Decadiene | NC | 116.16 | 116.28 | 129.68 |
| 57 | Phosphoric acid | C | 87.02 | 36.78 | 3.98 |
| 58 | 10-Undecenoic acid | NC | 104.41 | 72.22 | 53.90 |
| 59 | 4-Amino-1,2,4-triazole | NC | 107.66 | 106.11 | 107.30 |
| 60 | Sodium lauryl sulfate (20% aq.) | NC | 114.14 | 109.82 | 71.59 |

*Note:*   C = corrosive (EU risk phrases R34 and R35); NC = non-corrosive. The data in this table constitute the training set for CM 18.3. The EPISKIN data refer to percentage viabilities.

**Table 18.6    Performance of the CMs for Skin Corrosion**

| Model | Sensitivity | Specificity | Concordance | False Positive Rate | False Negative Rate |
|---|---|---|---|---|---|
| CM 18.1[a] | 70 | 68 | 69 | 32 | 30 |
| CM 18.1[b] | 66 | 62 | 64 | 38 | 34 |
| CM 18.2[c] | 85 | 88 | 86 | 12 | 15 |
| CM 18.2[d] | 70 | 75 | 73 | 25 | 30 |
| CM 18.3[e] | 88 | 85 | 86 | 15 | 12 |
| CM 18.3[f] | 88 | 79 | 83 | 21 | 12 |

[a] Statistics based on the application of CM 18.1 to its training set of 189 organic liquids.
[b] Cross-validated statistics based on the three-fold cross-validation of CM 18.1.
[c] Statistics based on the application of CM (18.2) to its training set of 44 chemicals.
[d] Cross-validated statistics based on the the three-fold cross-validation of CM 18.2.
[e] Statistics based on the application of CM 18.3 to its training set of 60 chemicals.
[f] Cross-validated statistics based on the three-fold cross-validation of CM 18.3.

## D.  Assessment of the Classification Models

CM 18.1 to CM 18.3 were assessed in terms of their Cooper statistics, which define an upper limit to predictive performance. In addition, cross-validated Cooper statistics, which provide a more realistic indication of a model's capacity to predict the classifications of independent data, were obtained by applying the threefold cross-validation procedure to the best-sized CTs. In the threefold cross-validation procedure, the data set is randomly divided into three approximately equal parts, the CT is re-parameterized using two thirds of the data, and predicted classifications are made for the remaining third of the data. The cross-validated Cooper statistics are the mean values of the usual Cooper statistics, taken over the three iterations of the cross-validation procedure. The Cooper statistics for CM 18.1 to CM 18.3 are summarized in Table 18.6.

## E.  Incorporation of the Classification Models into a Tiered Testing Strategy

The three CMs (the QSAR based on MW, the PM based on pH data, and the PM based on EPISKIN data) were arranged into a three-step sequence to represent a simple three-step testing strategy (Figure 18.2). The ordering of the three steps was based on the relative ease of applying the models. The first step was based on the application of the QSAR, since QSARs are the easiest CMs to apply, not being based on experimental data, and the subsequent steps were based on PMs, using physicochemical (pH) data before *in vitro* (EPISKIN) data.

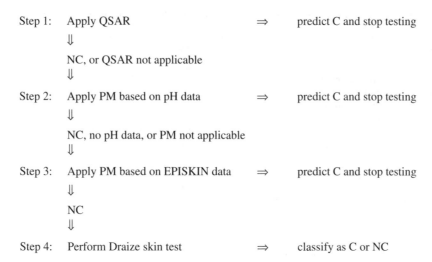

Step 1:  Apply QSAR  ⇒  predict C and stop testing
⇓
NC, or QSAR not applicable
⇓
Step 2:  Apply PM based on pH data  ⇒  predict C and stop testing
⇓
NC, no pH data, or PM not applicable
⇓
Step 3:  Apply PM based on EPISKIN data  ⇒  predict C and stop testing
⇓
NC
⇓
Step 4:  Perform Draize skin test  ⇒  classify as C or NC

**Figure 18.2**  A tiered testing strategy for skin corrosion based on the OECD approach to hazard classification. C = corrosive; NC = non-corrosive. Step 1: If MP ≤ 37°C and MW ≤ 123 g/mol, predict C; otherwise predict NC. Step 2: If pH < 3.9 or pH > 10.5, predict C; otherwise predict NC. Step 3: If EPISKIN viability at 4 h < 36%, predict C; otherwise predict NC.

## III. EVALUATION OF THE TIERED APPROACH TO HAZARD CLASSIFICATION

### A. Evaluation Method

The tiered approach to hazard classification was evaluated by simulating possible outcomes obtained when a stepwise strategy comprising three alternative tests and one animal test (Figure 18.2) is applied to a heterogeneous set of 51 chemicals (Table 18.7). The decision rules in steps 1 to 3 are based on the CMs for skin corrosion developed above.

The 51 chemicals in Table 18.7 form a subset of the 60 test chemicals in Table 18.5 that were used in the ECVAM Skin Corrosivity Validation Study (Barratt et al., 1998; Fentem et al., 1998). The subset of 51 chemicals was chosen in the interests of consistency, on the basis that each chemical had been tested neat, rather than as a dilution.

A number of simulations were performed to assess the effects of applying the different combinations of the three alternative tests before the Draize skin corrosion test. Each combination of alternative tests is referred to hereafter as a different sequence. Specifically, assessments were made of the sequences applied before the Draize test:

Sequence 1 — A QSAR, a PM based on pH data, and a PM based on EPISKIN data
Sequence 2 — A QSAR and a PM based on pH data
Sequence 3 — A QSAR and a PM based on EPISKIN data
Sequence 4 — A PM based on pH data and a PM based on EPISKIN data

The outcome of each simulation was used to compare the ability of each stepwise sequence to predict EU classifications and to reduce and refine the use of animals, with the corresponding ability of the EPISKIN test, when used as a stand-alone alternative method.

The predicted and known classifications of skin corrosion potential are given in Table 18.7. Predictions of corrosion potential made by the QSAR in step 1 are made only for the 36 single chemicals that are organic liquids, since the domain of the QSAR excludes inorganic substances,

**Table 18.7    Data Set of 51 Chemicals Used to Evaluate a Tiered Testing Strategy for Skin Corrosion**

| Chemical | Step 1 | Step 2 | Step 3 | Draize Test |
|----------|--------|--------|--------|-------------|
| 1   Hexanoic acid | C | C | C | C |
| 2   1,2-Diaminopropane | C | C | C | C |
| 3   Carvacrol | NC | NC | C | C |
| 4   Boron trifluoride dehydrate | np | np | C | C |
| 5   Methacrolein | C | NC | C | C |
| 6   Phenethyl bromide | NC | NC | NC | NC |
| 7   3,3′-Dithiodipropionic acid | NC | np | NC | NC |
| 8   *Iso*propanol | C | NC | NC | NC |
| 9   2-Methoxyphenol (Guaiacol) | NC | C | C | NC |
| 10  2,4-Xylidine (2,4-Dimethylaniline) | NC | NC | NC | NC |
| 11  2-Phenylethanol (phenylethylalcohol) | C | NC | NC | NC |
| 12  Dodecanoic (lauric) acid | NC | C | NC | NC |
| 13  3-Methoxypropylamine | C | C | C | C |
| 14  Allyl bromide | C | C | C | C |
| 15  Dimethyldipropylenetriamine | NC | C | C | C |
| 16  Methyl trimethylacetate | C | NC | NC | NC |
| 17  Dimethylisopropylamine | C | C | C | C |
| 18  Tetrachloroethylene | NC | NC | NC | NC |
| 19  Ferric (iron [III]) chloride | np | C | C | C |
| 20  Butyl propanoate | NC | NC | NC | NC |
| 21  2-*tert*-Butylphenol | NC | NC | C | C |
| 22  *Iso*stearic acid | np | NC | C | NC |
| 23  Methyl palmitate | NC | NC | NC | NC |
| 24  Phosphorus tribromide | NC | np | NC | C |
| 25  65/35 Octanoic/decanoic acids | np | C | C | C |
| 26  4,4′-Methylene-bis-(2,6-di*tert*-butylphenol) | NC | np | NC | NC |
| 27  2-Bromobutane | NC | NC | NC | NC |
| 28  Phosphorus pentachloride | np | np | C | C |
| 29  4-(Methylthio)-benzaldehyde | NC | NC | NC | NC |
| 30  70/30 Oleine/octanoic acid | np | C | NC | NC |
| 31  Hydrogenated tallow amine | np | np | NC | NC |
| 32  2-Methylbutyric acid | C | C | C | C |
| 33  Tallow amine | np | np | NC | C |
| 34  2-Ethoxyethyl methacrylate | NC | NC | NC | NC |
| 35  Octanoic acid (caprylic acid) | NC | C | C | C |
| 36  20/80 Coconut/palm soap | np | np | NC | NC |
| 37  Benzyl acetone | NC | NC | NC | NC |
| 38  Heptylamine | C | C | NC | C |
| 39  Cinnamaldehyde | NC | NC | NC | NC |
| 40  60/40 Octanoic/decanoic acids | np | C | C | C |
| 41  Eugenol | NC | C | NC | NC |
| 42  55/45 Octanoic/decanoic acids | np | C | C | C |
| 43  Methyl laurate | NC | NC | NC | NC |
| 44  Sodium bicarbonate | np | NC | NC | NC |
| 45  Sulfamic acid | np | C | C | NC |
| 46  Sodium bisulphite | np | NC | NC | NC |
| 47  1-(2-Aminoethyl)piperazine | NC | C | NC | C |
| 48  1,9-Decadiene | NC | NC | NC | NC |
| 49  Phosphoric acid | np | C | C | C |
| 50  10-Undecenoic acid | NC | NC | NC | NC |
| 51  4-Amino-1,2,4-triazole | C | NC | NC | NC |

*Note:*    C = corrosive (R34 or R35); NC = non-corrosive; np = no prediction (chemical outside domain of CM).
Step 1: If MP ≤ 37°C and MW ≤ 123 g/mol, predict C; otherwise predict NC.
Step 2: If pH < 3.9 or pH > 10.5, predict C; otherwise predict NC.
Step 3: If EPISKIN viability at 4 h < 36%, predict C; otherwise predict NC.
Shading indicates the step where a classification of corrosive potential (C) is assigned to a given chemical, and testing is stopped.

**Table 18.8    Possible Outcomes of Tiered Testing Strategies for Skin Corrosion**

| Sequence of Steps | No. of Chemicals Entering Step | Known Corrosion Potential | | No. of Positive Predictions | No. of True Positives | No. of False Positives |
|---|---|---|---|---|---|---|
| Step 1 | 51 | 22 C | 29 NC | 12 | 8 | 4 |
| Step 2 | 39 | 14 C | 25 NC | 13 | 8 | 5 |
| Step 3 | 26 | 6 C | 20 NC | 5 | 4 | 1 |
| Draize test | 21 | 2 C | 19 NC | — | — | — |
| Step 1 | 51 | 22 C | 29 NC | 12 | 8 | 4 |
| Step 2 | 39 | 14 C | 25 NC | 13 | 8 | 5 |
| Draize test | 26 | 6 C | 20 NC | — | — | — |
| Step 1 | 51 | 22 C | 29 NC | 12 | 8 | 4 |
| Step 3 | 39 | 14 C | 25 NC | 14 | 11 | 3 |
| Draize test | 25 | 3 C | 22 NC | — | — | — |
| Step 2 | 51 | 22 C | 29 NC | 20 | 15 | 5 |
| Step 3 | 31 | 7 C | 24 NC | 6 | 5 | 1 |
| Draize test | 25 | 2 C | 23 NC | — | — | — |
| Step 3 | 51 | 22 C | 29 NC | 21 | 18 | 3 |
| Draize test | 30 | 4 C | 26 NC | — | — | — |

*Note:*   C = corrosive; NC = non-corrosive.

**Table 18.9    Contingency Tables for the Predictive Abilities of Four Stepwise Sequences for Skin Corrosion and the Stand-Alone Use of the EPISKIN Test**

| Test or Stepwise Sequence | True Positives | True Negatives | False Positives | False Negatives |
|---|---|---|---|---|
| Steps: 1 → 2 → 3 | 20 | 19 | 10 | 2 |
| Steps: 1 → 2 | 16 | 20 | 9 | 6 |
| Steps: 1 → 3 | 19 | 22 | 7 | 3 |
| Steps: 2 → 3 | 20 | 23 | 6 | 2 |
| Step 3: EPISKIN test | 18 | 26 | 3 | 4 |

solids, and mixtures. For the 15 chemicals in Table 18.7 that lie outside the domain of the QSAR, no prediction (np) is made. In such cases, it is necessary to proceed to step 2, to continue the assessment of toxic hazard. Similarly, in step 2, no prediction is made for 8 chemicals that fall outside the domain of the PM based on pH data.

The possible outcomes obtained when the 4 sequences of alternative tests and the Draize test are applied to the data for the 51 chemicals are summarized in Table 18.8, along with the outcome of applying just one *in vitro* test (the EPISKIN test) before the Draize test. For each sequence, Table 18.8 gives the number of chemicals that enter each step, the distribution of these chemicals in terms of their known corrosion potential (C or NC), the number of positive predictions (i.e., chemicals for which no further assessment is made), and the numbers of true and false positives.

Contingency tables for the four sequences of CMs and for the stand-alone use of the EPISKIN PM are given in Table 18.9. The number of true positives for a given sequence was obtained by adding the numbers of true positives obtained by applying the individual steps in the sequence (Table 18.8). Similarly, the number of false positives for each sequence was obtained by summing the numbers of false positives for the individual steps (Table 18.8). The numbers of true negatives in Table 18.9 were calculated by Equation 18.1, since it was known that the sum of true negatives and false positives should equal the total number of non-corrosive chemicals in the data set, (i.e., 29):

$$\text{Number of true negatives} = 29 - \text{number of false positives} \qquad (18.1)$$

**Table 18.10  Predictive Abilities of Stepwise Sequences of Alternative Tests for Skin Corrosion Compared with the Stand-Alone Use of the EPISKIN Test**

| Test or Stepwise Sequence | Sensitivity | Specificity | Concordance |
|---|---|---|---|
| Steps: 1 → 2 → 3 | 91 | 66 | 76 |
| Steps: 1 → 2 | 73 | 69 | 71 |
| Steps: 1 → 3 | 86 | 76 | 80 |
| Steps: 2 → 3 | 91 | 79 | 84 |
| Step 3: EPISKIN test | 82 | 90 | 86 |

Step 1: If MP ≤ 37°C and MW ≤ 123 g/mol, predict C; otherwise predict NC.
Step 2: If pH < 3.9 or pH > 10.5, predict C; otherwise predict NC.
Step 3: If EPISKIN viability at 4 h < 36%, predict C; otherwise predict NC.

The most predictive test or sequence for each end point is shaded. All performance measures are expressed as percentages.

Similarly, since the numbers of true positives and false negatives should equal the total number of corrosive chemicals in the data set (i.e., 22), the numbers of false negatives were calculated by using Equation 18.2:

$$\text{Number of false negatives} = 22 - \text{number of true positives} \qquad (18.2)$$

The numbers of false negatives should also equal the number of chemicals identified as corrosive by the Draize test (Table 18.8).

Finally, Cooper statistics for the application of 4 sequences of CMs and for the application of the EPISKIN PM alone are given in Table 18.10. The statistics in Table 18.10 were calculated from the data in Table 18.9, using the definitions of the Cooper statistics given in Table 18.2.

## B.  Results of the Evaluation

The Cooper statistics in Table 18.10 show that the stand-alone use of the EPISKIN test gives the best overall predictive performance (concordance of 86%), and provides for the best identification of NC chemicals (specificity of 90%). However, it is the sequential application of the pH test and the EPIKSIN test (steps 2→3) that results in the best identification of corrosive chemicals (sensitivity of 91%) associated with a high overall predictive performance (concordance of 84%). The use of all three alternative tests (steps 1→2→3) also enables 91% of the corrosive chemicals to be correctly identified, but the overall concordance of 76% is lower because the specificity is also lower (66%). There is no scientific advantage in using the QSAR, which lowers the overall concordance of the testing strategy because of its relatively high false positive and negative rates (Table 18.6).

Having considered the predictive performance of each sequence of alternative tests in comparison with the EPISKIN test, it is also important to examine the effect of each sequence in terms of the extent to which it reduces and refines the use of the Draise rabbit skin test in comparison with the stand-alone use of the EPISKIN test.

The application of all three alternative tests before the Draize skin test would result in 21 chemicals being tested on rabbits, of which just two would be corrosive. The effect of applying the most predictive two-step sequence (steps 2→3) would be the testing of 25 chemicals on rabbits, of which just two chemicals would be found corrosive. If only one alternative test, the EPISKIN test, were applied before the Draize test, then 30 chemicals would be tested on rabbits, of which four would be corrosive. The best stepwise strategy that can be constructed from the CMs reported in this study is the two-step combination of the pH test and the EPISKIN test, applied before the Draize test. This combination maximizes predictive performance, while at the same time reducing and refining animal testing as much as possible.

## IV. CONCLUSIONS

It is concluded that:

1. Testing strategies based on the sequential use of alternative methods prior to the use of animal methods provide an effective means of reducing and refining the use of animals, without compromising the ability to classify chemicals on the basis of toxic hazard.
2. A CM of high sensitivity (but low specificity) could be combined with a CM of high specificity (but low sensitivity) to exploit the strengths and compensate for the weaknesses, of the two models.

## V. DISCUSSION

### A. Interpretation of the Classification Models

The importance of MP can be related to the physical state of the substance under the conditions of Draize test. In this study, it was assumed that chemicals with a MP less than or equal to 37°C would exist as liquids in the test procedure and that, in general, liquids would be more likely than solids to cause corrosion and irritation. The results confirm that there is indeed a relationship between physical state and the potential for acute skin toxicity. The fact that some solids are corrosive or irritant may relate to the fact that their MPs are not much higher than 37°C and that they exist as wax-like substances, which are more capable of penetrating into the skin than are solids with higher MPs. For example, carvacrol, and thymol, which are both irritant and corrosive, have predicted MPs of 38°C and 38.1°C, respectively. In the case of other solids, such as benzene sulfonyl chloride (MP = 61°C), the corrosive response may be due to a more toxic derivative (e.g., benzene sulfonic acid).

The importance of MW is probably related to the fact that small molecules are more likely to penetrate into the skin and cause corrosion than are larger chemicals. An alternative explanation could be that chemicals with lower MWs are applied in greater molar amounts than chemicals with higher MWs, since a fixed volume (or weight) of test substance is applied in the Draize test. This could be regarded as a limitation in the protocol of the Draize test, which could be improved by adopting a fixed molar dose of the test substance.

Log $K_{ow}$ was also found to discriminate between corrosive and non-corrosive chemicals, but the direction of the separation was contrary to expectation. Low log $K_{ow}$ values were associated with the presence of corrosion, whereas high log $K_{ow}$ values were associated with the absence of corrosion. An inverse relationship between corrosion potential and log $K_{ow}$ also emerged (but was not commented upon) in several PCA studies (Barratt, 1996b; Barratt et al., 1998). It was decided that the apparent importance of log $K_{ow}$ may be a reflection of the importance of MW, resulting from the collinearity between log $K_{ow}$ and MW. It has been argued (e.g., Barratt, 1995) that log $K_{ow}$ plays a role in skin corrosion on the basis that hydrophobic chemicals are more likely than hydrophilic ones to diffuse across the stratum corneum. If it is assumed that the rate-limiting step in the production of a corrosive response is the transfer of the applied chemical from its bulk phase (solid or liquid) into the skin, one could question the importance of log $K_{ow}$ on the grounds that this provides a measure of the ability of a chemical to partition between octanol and water, rather than between the neat substance and the stratum corneum; this would be the more appropriate partitioning process to model, given that in the Draize skin test, most liquids and solids are applied neat, rather than as aqueous solutions. In other words, the octanol-water partition coefficient may be a poor substitute for the liquid-stratum corneum partition coefficient.

The acidity/basicity descriptor pH provides a useful means of identifying substances that are corrosive to the skin by disrupting its pH balance (away from a physiological value of about 5.5). In Table 18.4, three chemicals (2-bromobutane, sodium bisulfite and 4-amino-1,2,4-triazole) have borderline predictions because of the proximity of their pH values to the cut-offs (3.4 and 10.5) of the PM.

The PM for skin corrosion based on EPISKIN measurements (CM 18.3) is similar to the one evaluated in the ECVAM Skin Corrosivity Validation Study, in which a cut-off value of 35% viability was evaluated (Barratt et al., 1998; Fentem et al., 1998).

## B. Comments of the Design of Tiered Testing Strategies

The results of this study and of a previous study (Worth et al., 1998) show that stepwise approaches to hazard classification, in which alternative methods are applied before animal tests, provide a promising means of reducing and refining the use of animals. In these approaches, fewer animal experiments need to be conducted, and of those chemicals tested *in vivo*, the majority are found to be non-toxic.

The validity of this conclusion depends on the adequate performance of each alternative method included in the stepwise sequence. Methods that overpredict toxic potential will tend to compromise the performance of strategies in which they are incorporated, since (according to the approach evaluated) chemicals found to be toxic do not undergo further testing. When designing a tiered testing strategy, it is important that the models included should have low false positive rates (i.e., high specificities). For example, it might be decided that false positive rates should not exceed 10%. In general, models with lower false positive rates tend to have lower sensitivities. Even models with sensitivities less than or equal to 50% may be useful in the context of a tiered testing strategy; it is not important that any single model is capable of identifying a majority of the toxic chemicals in a test set, as long as there is a high degree of certainty associated with the positive predictions.

An alternative approach to the design of a tiered testing strategy would be that models of high specificity (low false positive rate) could be used to identify toxic substances, whereas models of high sensitivity (low false negative rate) would be used to identify non-toxic substances. Again, chemicals predicted to be toxic would not undergo further testing, whereas chemicals predicted to be non-toxic would be tested directly in animals. In this approach, models that identify toxic chemicals would be used to terminate the testing process, whereas models that identify non-toxic substances would be used to expedite the process (by skipping intermediate steps based on more time-consuming and expensive alternative methods).

The rationale behind the alternative approach is that some models are better suited for identifying toxic chemicals, whereas others are better suited for identifying non-toxic chemicals, because of the inescapable overlap between toxic and non-toxic chemicals along certain variables. Although such models may be unacceptable as stand-alone alternatives to animal experiments, their combined use should provide a means of exploiting their strengths and compensating for their weaknesses. In particular, it is foreseen that highly specific methods could be successfully combined with highly sensitive ones. The main difference between the alternative approach and the conventional approach evaluated in this chapter concerns the consequence of negative predictions. In the conventional approach, further tests are conducted to confirm predictions of non-toxicity, which means that there is no useful role to be played by a model that only identifies non-toxic chemicals.

Finally, it is important to note that the approaches to hazard classification described in this chapter represent just two possible ways of integrating the use of different CMs; other designs are conceivable. For example, if each prediction of toxic and non-toxic potential were associated with a probability (e.g., a 70% probability of being corrosive), thresholds other than 50% could be chosen for the identification of toxic and non-toxic chemicals. In fact, models derived by logistic regression or linear discriminant analysis can be used to assign probabilities, but these are likely to be misleading if the assumptions of the methods are not obeyed. Alternatively, the identification of toxic potential could proceed according to a majority voting system, in which predictions were made by several models, with classifications being assigned when a majority of models made the same prediction.

# REFERENCES

Balls, M. and Karcher, W., The validation of alternative test methods, *Alt. Lab. Anim. (ATLA)*, 23, 884–886, 1995.

Barratt, M.D., Quantitative structure-activity relationships for skin corrosivity: appendix A to the report of ECVAM Workshop 6, *Alt. Lab. Anim. (ATLA)*, 23, 219–255, 1995.

Barratt, M.D., Quantitative structure-activity relationships for skin irritation and corrosivity of neutral and electrophilic organic chemicals, *Toxicol. in vitro*, 10, 247–256, 1996a.

Barratt, M.D., Quantitative structure-activity relationships (QSARs) for skin corrosivity of organic acids, bases and phenols: principal components and neural network analysis of extended datasets, *Toxicol. in vitro*, 10, 85–94, 1996b.

Barratt, M.D., Brantom, P.G., Fentem, J.H., Gerner, I., Walker, A.P., and Worth, A.P., The ECVAM international validation study on *in vitro* tests for skin corrosivity. 1. Selection and distribution of the test chemical, *Toxicol. in vitro*, 12, 471–482, 1998.

Blaauboer, B.J., Barratt, M.D., and Houston, J.B., The integrated use of alternative methods in toxicological risk evaluation. ECVAM integrated testing strategies Task Force report 1, *Alt. Lab. Anim. (ATLA)*, 27, 229–237, 1999.

Breiman, L., Friedman, J.H., Olshen, R.A., and Stone, C.J., *Classification and Regression Trees*, Wadsworth, Monterey, CA, 1984.

Bruner, L.H., Carr, G.J., Chamberlain, M., and Curren, R.D., Validation of alternative methods for toxicity testing, *Toxicol. in vitro*, 10, 479–501, 1996.

European Commission (EC), Council Directive 86/609/EEC of 24 November 1986 on the approximation of laws, regulations and administrative provisions of the Member States regarding the protection of animals used for experimental and other scientific purposes, *Off. J. Eur. Communities*, L358, 1-29, 18.12.1986, 1986.

European Commission (EC), Commission Directive 2000/33/EC of 25 April 2000 adapting to technical progress for the 27th time Council Directive 67/548/EEC on the approximation of laws, regulations and administrative provisions relating to the classification, packaging and labelling of dangerous substances, *Off. J. Eur. Communities*, L136A, 90–97, 2000.

European Centre for Ecotoxicology and Toxicology of Chemicals (ECETOC), Skin Irritation and Corrosion: Reference Chemicals Data Bank, ECETOC Technical Report No. 66, Brussels, Belgium, 1995.

Fentem, J.H., Archer, G.E.B., Balls, M., Botham, P.A., Curren, R.D., Earl, L.K., Esdaile, D.J., Holzhütter, H.G., and Liebsch, M., The ECVAM international validation study on *in vitro* tests for skin corrosivity. 2. Results and evaluation by the management team, *Toxicol. in vitro*, 12, 483–524, 1998.

Livingstone, D., *Data Analysis for Chemists: Applications to QSAR and Chemical Product Design*, Oxford University Press, Oxford, 1995.

National Institutes of Health (NIH), Corrositex: an *in vitro* test method for assessing dermal corrosivity potential of chemicals, NIH Publication No. 99-4495, National Toxicology Program (NTP) Inter-agency Center for the Evaluation of Alternative Toxicological Methods (NICEATM), Research Triangle Park, North Carolina, 1999.

Organization for Economic Co-operation and Development, Harmonized Integrated Hazard Classification System for Human Health and Environmental Effects of Chemical Substances, Paris, France, 1998.

Organization for Economic Co-operation and Development (OECD), Revised Proposals for Updated Test Guidelines 404 and 405: Dermal and Eye Corrosion/Irritation Studies. ENV/JM/TG(2001)2, 112 pp. Paris, France, 2001.

Organization for Economic Co-operation and Development (OECD), OECD Guidelines for the Testing of Chemicals No. 430: *in vitro* Skin Corrosion — Transcutaneous Electrical Resistance Test (TER), Paris, France, 2002a.

Organization for Economic Co-operation and Development (OECD), OECD Guidelines for the Testing of Chemicals No. 431: *In vitro* Skin Corrosion — Human Skin Model Test, Paris, France, 2002b.

Organization for Economic Co-operation and Development (OECD), OECD Guidelines for the Testing of Chemicals No. 432: *In vitro* 3T3 NRU Phototoxicity, Paris, France, 2002c.

Russell, W.M.S. and Burch, R.L., *The Principles of Humane Experimental Technique*, London: Methuen, London, 1959.

Smyth, D.H., *Alternatives to Animal Experiments*, Scolar Press-Royal Defence Society, London, 1978.

Whittle, E.G., Barratt, M.D., Carter, J.A., Basketter, D.A., and Chamberlain, M., The skin corrosivity potential of fatty acids: *in vitro* rat and human skin testing and QSAR studies, *Toxicol. in vitro*, 10, 95–100, 1996.

Worth A.P., Fentem J.H., Balls M., Botham P.A., Curren R.D., Earl L.K., Esdaile D.J., and Liebsch M., An evaluation of the proposed OECD testing strategy for skin corrosion, *Alt. Lab. Anim. (ATLA)*, 26: 709-720, 1998.

Worth, A.P. and Balls, M., The importance of the prediction model in the development and validation of alternative tests, *Alt. Lab. Anim. (ATLA)*, 29, 135–143, 2001.

Worth, A.P. and Balls, M., Alternative (non-animal) methods for chemicals testing: current status and future prospects: a report prepared by ECVAM and the ECVAM Working Group on Chemicals, *Alt. Lab. Anim. (ATLA)*, 30 (Suppl 1), 1–125, 2002.

# CHAPTER 19

# The Use by Governmental Regulatory Agencies of Quantitative Structure-Activity Relationships and Expert Systems to Predict Toxicity

Mark T.D. Cronin

## CONTENTS

0-415-27180-0/04/$0.00+$1.50
© 2004 by CRC Press LLC

# I. INTRODUCTION

An obvious area of application of quantitative structure activity relationships QSARs is by governmental regulatory agencies. There are a number of reasons for the use of methods to predict toxicity by national and international agencies. There are clearly considerable savings in cost and time for the assessment of chemical hazard, and more importantly for the filling of data gaps. In addition, the open use of structure-based methods by regulatory agencies also allows for industrial producers of chemicals to know how their product will be assessed by the relevant agency.

There are acknowledged to be three main areas where QSARs may be applied by governmental regulatory agencies:

1.  Prioritization of existing chemicals for further testing or assessment
2.  Classification and labeling of new chemicals
3.  Risk assessment of new and existing chemicals

This chapter provides an overview of the use of QSARs by regulatory agencies worldwide. This is an ever-changing topic that is driven more by the requirements of national and international legislation, rather than advances in the scientific basis of QSAR. This chapter first addresses some factors affecting the use of QSARs and expert systems by regulatory authorities and then provides examples of their application by relevant regulatory authorities. This is a detailed and complex field; for more information regarding the use of QSARs by regulatory agencies, the reader is referred to the detailed reviews of Cronin et al. (2003a; 2003b) and Walker et al. (2002).

# II. FACTORS AFFECTING THE USE OF QSARS BY REGULATORY AGENCIES

## A.  Regulatory Guidance

Currently there is relatively little guidance for the use of QSARs to predict the toxicity and fate (especially in the environment) of chemicals. Some guidance is provided within the European Union (EU) where a comprehensive technical guidance document (TGD) was produced to support the Directive on New Substances and the Regulation on Existing Substances (European Economic Community, 1996). This document includes a substantial chapter providing guidance on the use of QSARs in environmental risk assessments.

The general tenet of advice provided by regulatory agencies is that precautionary and conservative use of QSAR is recommended. On occasion a predicted value may be accepted for an endpoint, if it suggests the worst possible scenario. For example, a number of QSARs for biodegradability exist (see Chapter 14). On occasions a prediction that a compound is nonreadily degradable will be accepted, without the requirement for testing. A prediction of readily degradable is less likely to be accepted (or not at all).

In the future, the use of QSARs may be more comprehensive. Using the above example, predictions of biodegradability may be accepted for both nonreadily and readily degradable compounds. The comprehensive use of QSAR will depend on the endpoint being modeled and the model itself. Much will depend on the quality of model and the original data on which it is based, the philosophy of its development, and the process of validation (see Chapters 18 and 20), and confidence associated with it. Another endpoint specific issue is the acceptability of a false prediction. Returning to the previous example, in terms of environmental risk assessment, a readily degradable compound that is predicted to be nondegradable is not a problem, but a nondegradable compound predicted to be degradable is of greater concern. These issues are discussed below and the reader is also referred to Walker et al. (2003a) for more details.

**Table 19.1    A Summary of Selected Validation Studies of QSARs and Expert Systems Performed By Regulatory Agencies**

| Regulatory Agency | End Points Assessed | QSAR/Expert System Studied | Reference |
|---|---|---|---|
| U.S. Environmental Protection Agency/European Union | Various physicochemical properties, environmental fate parameters, ecotoxicological and human health endpoints. | EPA and EU regulatory methods | U.S. Environmental Protection Agency (1994) |
| German Federal Institute for Health Protection of Consumers and Veterinary Medicine | Skin sensitization | DEREK expert system | Zinke et al. (2002) |
| Danish Environmental Protection Agency | Mutagenicity, carcinogenicity, skin sensitization, acute toxicity, aquatic toxicity | M-CASE and TOPKAT models (see Table 19.5) | Tyle et al. (2001) |
| Environment Canada | Fathead minnow acute toxicity | ECOSAR, TOPKAT, probabilistic neural network, computational neural network, ASTER, OASIS | Moore et al. (2003) |

## B. Evaluation and Validation of QSARs for Application by Regulatory Authorities

In order for a regulatory authority to make a decision using a QSAR or other computer-based method, there needs to some assessment of the quality of the model and the likely accuracy of the prediction. This will enable the regulator and the regulated to assign some level of confidence to the prediction. Assignment of quality to any object or process tends to be a subjective evaluation. However, certain criteria may be assigned to high-quality components of QSARs (i.e., the biological data, physicochemical properties, and statistical modeling process itself [Cronin and Schultz, 2003; Eriksson et al., 2003; Schultz and Cronin, 2003]). These components are summarized below. In addition, a useful set of guidelines for developing and using QSARs in a regulatory setting has been described by Walker et al. (2003c).

In order to assess how predictive QSARs or commercial expert systems are, a number of validation studies have been performed. Typically these have involved using the QSAR or expert system to make a prediction for the biological activity of a chemical not included in the original model. Some of the major independent validation studies performed by regulatory agencies are summarized in Table 19.1, the reader is also referred to the recent work of Hulzebos et al. (2003). The stumbling block in many of these validation studies is the availability of biological data to form the test set. By the nature of the development of many commercial expert systems, (nearly) all literature data are often included in the models. This means there may be few (if any) openly available data with which to validate models. Regulatory agencies (as well as private companies with historical databases) are well placed to perform validations using data from regulatory submissions. It has also been proposed that data for the high production volume (HPV) chemicals could be utilized to assist in QSAR validation (Walker et al., 2003b). While regulatory submissions and HPV data may often provide an adequate source of information, the confidential nature of the toxicity values usually means that the predictions themselves are not reported openly. Instead, summaries of the predictions are reported (e.g., concordance, accuracy, etc.). These statistics are useful, but do not allow for detailed analysis of the precise strengths and weaknesses of expert systems. A formal framework for the validation of QSARs and computer-based models for toxicity predicted is proposed in Chapter 20.

## C. Indicators of the Quality of QSARs and Expert Systems

The use of high-quality biological and physicochemical data is a prerequisite for the development of high-quality QSARs and expert systems. Criteria to define high-quality biological data are given by Cronin and Schultz (2003) and Schultz and Cronin (2003). It is appreciated that any biological measurement will contain error. When the activity is in the form of a potency value (i.e., $EC_{50}$, $LC_{50}$, etc.) the error may be quantifiable. Ideally, for the development of predictive models, biological data with as little error in them as possible should be used. To achieve this, biological data should preferably be sourced from a reliable and well-standardized assay. The most reliable data sets are also those measured within a single laboratory by a single protocol and preferably even the same workers. It should be remembered that even within a test protocol error will vary among the different chemicals tested. In an acute ecotoxicity assay, error will be lower for a relatively hydrophilic chemical acting by nonpolar narcosis, than for a more hydrophobic reactive chemical.

There are also issues relating to the quality of physicochemical and structural descriptors of chemicals used in QSARs or computer-based predictive approaches. All measured or calculated descriptors are subject to error (see Chapters 4 and 6). Some estimate of this error must be taken into account in the modeling process. For risk assessment procedures, the guidance from regulatory authorities is generally that preference should be given to measured (where possible), over-calculated values. In most circumstances, descriptors with a direct physicochemical meaning or interpretation are preferred over those with less defined meaning. With regard to the modeling process itself, much credence is placed on the transparency of QSARs by regulatory authorities. Some statistical approaches, such as regression analysis, are considered to be intrinsically more transparent than others, such as the use of neural networks (see Chapter 7). Methods for the assessment of the reliability of toxicological QSARs are described by Eriksson et al. (2003) and placed into context by Jaworska et al. (2003).

## III. USE OF QSAR BY REGULATORY AGENCIES IN THE U.S.

In the U.S., the Interagency Testing Committee (ITC) acts as an independent advisory committee to the Environmental Protection Agency (EPA) Administrator. The ITC has representatives from 16 U.S. governmental agencies on it. The ITC was created to identify chemicals that should be regulated under the Toxic Substances Control Act (TSCA) and chemical classes in need of testing, such that these may be added to the Priority Testing List provided to the EPA Administrator. The ITC's statutory mandate is to discover suspect toxic chemicals and chemical classes that are in need of testing. There are a number of agencies on the ITC that use QSAR relatively broadly. These include: the Agency for Toxic Substances and Disease Registry (ATSDR), EPA, Food and Drug Administration (FDA), and to a limited extent the National Institute for Occupational Safety and Health (NIOSH) (Walker, 2003). A summary of the use of QSARs by each of these U.S. regulatory agencies is provided below.

## A. EPA

Of all governmental regulatory agencies worldwide, the EPA probably has the most widespread application of, and expertise in, QSAR. The use is underpinned by TSCA, which provides the framework for the regulation of new industrial chemicals by the EPA. Within TSCA, Section 5 of the Act specifies the premanufacture reporting requirements for new chemicals. Within TSCA the burden of proof is placed on the EPA to demonstrate risk and exposure for a new chemical. Since there is no requirement at the premanufacture notification (PMN) stage for the producer to provide toxicological and fate information, the EPA may need to predict over 150 attributes for a chemical during an assessment (a full case study of the regulatory assessment of a new PMN chemical is provided in Zeeman et al. [1999]). At the time of writing, since 1979, the EPA has received about 38,000 PMNs and currently there are about 2000 notifications per year. Of these, less than 20% of current notices have human health toxicity test data, less than 10% have environmental fate/transport test data, and less than 5% have environmental toxicity test data. The role of the Office of Pollution Prevention and Toxics (OPPT) to process PMNs within the EPA is described by Zeeman (1995) and Zeeman et al. (1995; 1999).

## 1. *Carcinogenicity*

Within the EPA the OPPT utilizes the OncoLogic system to assess the carcinogenic potential of substances (Woo et al., 1995). The OncoLogic system is marketed by LogiChem Inc., and provides a series of structure-activity relationships (SARs) to predict the potential of chemicals to cause cancer. It is developed by experts in the field of carcinogenicity evaluation in cooperation with the EPA. The system incorporates detailed knowledge regarding mechanisms of action and human epidemiological studies. It differs from many expert systems for toxicity by providing a line of reasoning for the prediction of toxicity, as well as being able to evaluate a much broader variety of compounds, including the following chemical classes: general organic chemicals, fibres, polymers, metals, metalloids, and metal-containing compounds.

## 2. *Skin Absorption*

QSARs utilized by the U.S. EPA for the prediction of the dermal uptake (absorption through the skin) of compounds are well described by Walker et al. (2002). Predictions of the ability of chemicals to be absorbed across the skin allowed for the potential of dermal toxicity to be assessed. Typically simple regression-based QSARs, which were based either on hydrophobicity and molecular size, or hydrophobicity alone, were utilised.

## 3. *Environmental and Ecological Effects*

The EPA uses QSARs to predict a large number of ecological effects, as well as for environmental fate within the PMN process. The EPA's website (www.epa.gov) provides a valuable source of further information on all these predictive methods, as well as a database and aquatic toxicity values and detailed information on how the models have been validated. Many of the predictive models have been brought together into the EPISUITE software (see Table 19.2 for a listing of the models available). This includes the OPPT's models used for the prediction of physical and chemical properties for new chemical substances. The EPISUITE software is downloadable free of charge (www.epa.gov/oppt/exposure/docs/episuitedl.htm). This provides not only an excellent resource for the development of QSARs, but also a transparent mechanism for the assessment of PMNs.

At the time of writing, the Internet-based PBT Profiler has become available. This is a web-based tool to help identify chemicals that potentially may persist, bioaccumulate, and be toxic to aquatic life (i.e., PBT chemicals). It uses computerized methods, such as SARs and standard scenarios, to predict risk-related data (physical/chemical properties, bioconcentration, environmental fate, carcinogenicity, toxicity to aquatic organisms, worker and general population exposure, and other information) on chemicals lacking experimental data. The Pollution Prevention (P2) Framework website (www.epa.gov/oppt/p2framework/) has information on the models and how to use them. The PBT Profiler is available from www.pbtprofiler.net as well as www.epa.gov/ oppt/pbt-profiler/ and is likely to see increasing widespread usage.

**Table 19.2   Summary of the Main Models Available in the EPA's EPISUITE Software**

| Name of and Model | Properties Predicted |
| --- | --- |
| EPIWIN | This is an interface program that transfers a single SMILES notation to separate structure estimation programs listed below (excluding DERMWIN). The EPIWIN interface program is a convenience for users because it automatically executes each program in succession without user interaction. In addition, the interface program executes the WVOLWIN (Volatilization Rate from Water), STPWIN (Sewage Treatment Plant Fugacity Model), and LEVEL3NT (Level III Fugacity Model) programs. |
| AOPWIN | The rate constant for the atmospheric, gas-phase reaction between photochemically produced hydroxyl radicals and organic chemicals and the rate constant for the gas-phase reaction between ozone and olefinic/acetylenic compounds. The rate constants estimated by the program are then used to calculate atmospheric half-lives for organic compounds. |
| BCFWIN | Bioconcentration factor (BCF) from log $K_{ow}$. |
| BIOWIN | Probability for the rapid aerobic biodegradation of an organic chemical in the presence of mixed populations of environmental microorganisms. |
| DERMWIN | Dermal permeability coefficient and the dermally absorbed dose per event. |
| ECOSAR | Aquatic toxicity of chemicals based on their similarity of structure to chemicals for which the aquatic toxicity has been previously measured (see Table 19.3 for a list of endpoints) |
| HENRYWIN | Henry's Law Constant at 25°C |
| HYDROWIN | Aqueous hydrolysis rate constants for esters, carbamates, epoxides, halomethanes, and selected alkyl halides. Acid- and base-catalyzed rate constants are estimated, but not neutral hydrolysis rate constants. |
| KOWWIN | Logarithm of the octanol-water partition coefficient (log $K_{ow}$). |
| MPBPWIN | Boiling point (at 760 mm Hg), melting point, and vapor pressure. |
| PCKOCWIN | Soil adsorption coefficecient ($K_{oc}$). |
| WSKOWWIN | Water solubility from log $K_{ow}$. |

## 4.  Physicochemical Properties

A knowledge of physicochemical properties underpins many of the predictions of toxicity made by the EPA. In addition, properties are required for the estimation of fate. The logarithm of the octanol-water partition coefficient (log $K_{ow}$) is one of the values for which predictions are most widely accepted (many agencies worldwide will accept predicted values for log $K_{ow}$). The KOWWIN program as well as the other programs in the EPISUITE software noted in Table 19.2 are used routinely by the EPA to calculate the physicochemical properties of new chemical substances for PMNs.

## 5.  Acute and Chronic Toxicity

The OPPT uses the ECOSAR program to calculate the acute and chronic toxicity of chemicals to aquatic species. ECOSAR contains over 150 SARs that have been developed for more than 50 chemical classes. Most SAR calculations in the ECOSAR program are based upon log $K_{ow}$ alone. Various surfactant SAR calculations are based on the average length of carbon chains or the number of ethoxylate units. A summary of the endpoints covered is provided in Table 19.3 and more details may be obtained from the OPPT SAR manual (Clements, 1996). A useful assessment of the applicability of ECOSAR has been made recently (Hulzebos and Posthumus, 2003).

## 6.  Environmental Fate

Estimates of the persistence of compounds in the environment are made from a variety of programs. These include those from the EPISUITE software described in Table 19.2, such as BIOWIN for biodegradation, BCFWIN for bioconcentration, and PCKOCWIN for soil sorption.

**Table 19.3    Major Toxicity End Points Predicted
by the ECOSAR Program**

**A. Freshwater Organisms**

1. Fish acute toxicity
2. Daphnid acute toxicity
3. Green algal toxicity
4. Fish Chronic Value (ChV)
5. Daphnid ChV
6. Green algal ChV

**B. Saltwater Organisms**

1. Fish acute toxicity
2. Mysid shrimp acute toxicity
3. Green algal toxicity
4. Fish ChV
5. Mysid ChV
6. Green algal ChV

**C. Benthic or Sediment-Dwelling Organisms**

**D. Terrestrial Plants**

**E. Terrestrial Soil**

1. Earthworms
2. Insects

**F. Birds**

1. Mallards
2. Quail
3. Raptors

**G. Wild Mammals**

1. Marine
2. Terrestrial

**H. Terrestrial Insects**

## 7.  Endocrine Disruption

The prediction of the ability of chemicals to cause endocrine disruption has also been addressed by the EPA. In 1996, the OPPTs established the Endocrine Disruptors Screening and Testing Advisory Committee (EDSTAC). The EDSTAC developed a conceptual framework to screen and test chemicals for endocrine disruption. As part of this, the Endocrine Disruptor Priority Setting Database (EDPSD1) was developed. Chemical formulae, molecular weights, Simplified Molecular Input Line Entry System (SMILES) notations, and structures for 57,811 chemicals were entered into EDPSD1. More details regarding the prediction of endocrine disruption are available in Chapter 13.

Within EDPSD1, three QSAR-based approaches to predict the relative binding affinities (RBAs) of estrogens have been used. These were: Holographic QSAR (HQSAR), Comparative Molecular Field Analysis (CoMFA), and COmmon REactivity PAttern (COREPA). The use and success of these approaches is described in more detail by Schmieder et al. (2003). Following the development and implementation of EDPSD1, EDPSD2 was created with substantially more resources. EDPSD2 is a decision support tool that is used to select chemicals for endocrine disruption screening assays.

COREPA (Mekenyan et al., 2002) and 4-phase (Shi et al., 2002; Chapter 13) QSAR-based approaches were used to screen the 623 inert high production volume chemicals for estrogen relative binding affinity. The Endocrine Disruptor Knowledge Base (EDKB) has subsequently been used to predict the estrogen receptor binding of 58,000 chemicals (Hong et al., 2002; Chapter 13).

## B.  Agency for Toxic Substances and Disease Registry

The ATSDR use of QSAR and models to predict toxicity is well described by El-Masri et al. (2002). In 1998, the ATSDR established a computational toxicology laboratory and initiated efforts to use Physiologically Based PharmacoKinetic (PBPK) models, BenchMark Dose (BMD) models, and QSARs. PBPK models are used by the ATSDR to:

1. Establish connections between exposure scenarios and biological indicators such as tissue dose or endpoint response.
2. Identify the significance of exposure routes in producing tissue levels of possible contaminants for people living near hazardous waste sites.
3. Provide a credible tool for extrapolating route-to-route health effects endpoints.

BMD models are used to estimate human health guidance values for environmental substances. QSARs are used to provide data estimates for chemicals that lack adequate experimental documentation. The ATSDR uses two commercial computational toxicology models to make toxicity predictions based on QSARs. To increase confidence in the models' predictions, ATSDR used the models similarity search features and established a minimum threshold similarity distance value of 0.25 to increase the probability that predicted toxicity values are close to nearest analog chemicals.

## C.  Food and Drug Administration

The FDA's Center for Drug Evaluation and Research (CDER) has recently considered applications of QSARs to support regulatory decisions when toxicology data are unavailable or limited. Further, to initiate the regulatory applications of QSARs for drugs, CDER is developing an electronic toxicology database that includes rodent carcinogenicity data; plans to extend it to acute, chronic, reproductive, developmental toxicity and genotoxicity databases are being developed. Additional details are available on CDER's website (www.fda.gov/cder) and from Matthews et al. (2000).

### 1.  Carcinogenicity

The FDA's CDER has evaluated the ability of several QSAR-based commercial computational toxicology models (such as TOPKAT [TOxicity Prediction by Komputer-Assisted Technology] and Multi-CASE [Multiple Computer Automated Structure Evaluation]) to make carcinogenicity predictions for about 400 pharmaceuticals that had been tested in 2-year carcinogenicity studies. The results of a trial using TOPKAT to predict the carcinogenicity of chemicals tested by the National Toxicology Program were disappointing, with a low rate of successful prediction (Prival, 2001). It should be emphasised however that TOPKAT was the only system evaluated in this manner, and it is unlikely that other expert systems would have performed better.

More importantly, the FDA has been instrumental in the release of data and information from regulatory submissions. Matthews and Contrera (1998) report the development of MultiCASE for the prediction of carcinogenicity using data released from the FDA under a Cooperative Research and Development Agreement (CRADA). The model developed with the FDA data had greatly improved predictivity. This is a novel contribution to the science and illustrates the usefulness of such an initiative.

## D. National Toxicology Program and Associated Agencies

The National Toxicology Program (NTP) was established in 1978 by the U.S. Department of Health and Human Services. Its function is to coordinate toxicological testing programs within the Department; strengthen the science base in toxicology; develop and validate improved testing methods; and provide information about potentially toxic chemicals to health regulatory and research agencies, the scientific and medical communities, and the public. The NTP is an inter-agency program consisting of relevant toxicology activities of the National Institute of Environmental Health Sciences (NIEHS), NIOSH, and the FDA's National Center for Toxicological Research (NCTR). The National Cancer Agency was a charter agency and remains active in the NTP.

Within the NTP, the Interagency Committee for Chemical Evaluation and Coordination (ICCEC) accepts chemicals nominated for review due to their potential for hazard. The nominations may come from the FDA, NCI, NIEHS, NIOSH, other federal agencies, state agencies, private citizens, labor and industry groups, and academia. Acceptable nominations are then reviewed. The ICCEC have used SARs to compare the toxic potential of many structurally related chemicals. The general role of ICCEC and a few examples of chemicals it has reviewed (chlorinated butenes, chlorinated alkyl ethers and halogenated propanes) are described by Walker (2003). The NTP has also been instrumental in organizing two blind trials for the prediction of carcinogenicity (see Chapter 8 for more details). These have demonstrated that many of the commercial softwares for carcinogenicity prediction (see Chapter 9) provide only a poor prediction, or may not even be predictive at all. Further the trials demonstrated that the best models tend to be those that can integrate mechanism-based reasoning with biological data (Richard and Benigni, 2002).

SARs are also used by NIOSH to develop Current Intelligence Bulletins. These are issued by NIOSH to disseminate new scientific information about occupational hazards. A Current Intelligence Bulletin may draw attention to a hazard previously unrecognized or may report new data suggesting that a known hazard is either more or less dangerous than was previously thought. NIOSH posts the Current Intelligence Bulletins on its web site (www.cdc.gov/niosh). Examples of Current Intelligence Bulletins on SAR-based chemical groups are discussed by Walker (2003).

## IV. USE OF QSAR BY REGULATORY AGENCIES IN CANADA

The use of QSARs by regulatory agencies in Canada has been stimulated by the Canadian Environmental Protection Act in 1999. This requires that the 23,000 substances on the Domestic Substance List (DSL) should be categorized and screened for persistence or bioconcentration and inherent toxicity. To assist in the categorization process, a technical advisory group (TAG) was constituted. This comprised a range of experts to advise on matters relating to structure-based prediction of toxicity and fate. Further, an international workshop on QSARs hosted by Environment Canada was held in Philadelphia in November 1999 to establish rules of thumb for predicting properties and effects of the structurally diverse DSL chemicals. The participants of the work agreed that mode of action QSARs (e.g., Assessment Tools for the Evaluation of Risk (ASTER) [Russom et al., 1991] and Optimized Approach based on Structural Indices Set (OASIS) [Mekenyen et al., 1997]) should be used for the prediction of ecological toxicity endpoints (MacDonald et al., 2002). Environment Canada also recently funded a study to assess and evaluate six modeling packages to predict acute toxicity, with particular application to prioritizing chemicals within the Canadian DSL (Moore et al., 2003). The six packages assessed were ECOSAR, TOPKAT, a probabilistic neural

**Table 19.4    Summary of the Main Areas Covered by Chapter 4 of the European Union Technical Guidance Document**

| Area | Guidance Provided |
|---|---|
| QSAR selection and evaluation | Criteria underpinning the acceptability of QSARs |
| Guidance on the use of (Q)SARs in environmental risk assessment | Toxicity of nonpolar and polar narcosis to aquatic species — recommended QSARs |
| | Calculation of log $K_{ow}$ — recommended methods and limitations |
| | Calculation of soil and sediment sorption — recommended QSARs |
| | Calculation of Henry's law constant — recommended QSARs |
| | Calculation of bioconcentration factor (aquatic and terrestrial organisms) — recommended QSARs |
| | Calculation of biodegradation — recommended QSARs |
| | Calculation of photolysis in air and water, hydrolysis — recommended QSARs |
| Guidance on the use of (Q)SARs in human health risk assessment | No (Q)SARs recommended |

network (PNN), a computational neural network (CNN), the QSAR components of the ASTER system, and the OASIS system. Of these, the PNN model was found to provide the best predictions on the basis of an external test set of compounds. The TOPKAT model was also found to provide excellent predictions, but only for compounds classified as falling within the optimum prediction space (OPS) of the model.

## V. USE OF QSAR BY REGULATORY AGENCIES IN THE EUROPEAN UNION

In the European Union (EU), risk assessment of chemical substances is driven by the requirements of European Commission Directives. In order to ensure consistency of application of the Environmental Risk Assessment (ERA) process, in 1996 the EU produced a comprehensive technical guidance document (TGD) to support the Directive on New Substances and the Regulation on Existing Substances (European Economic Community, 1996). Chapter 4 of this document provides guidance on the use of QSARs in the ERA process in terms of where they should be used, how they should be used, and which ones should be used. A summary of the areas covered is provided in Table 19.4. From the table it is obvious that while considerable information is provided in the TGD regarding the prediction of ecological effects and environmental fate, no formal recommendations are given on the use of QSARs for the prediction of human health effects. The TGD can be downloaded from the Internet (ecb.jrc.it/existing-chemicals).

For toxicity, data obtained from QSARs can be used according to the guidance on the use of QSARs for specific groups of substances found in Part IV of the TGD. The TGD provides recommendations for the use of QSARs to predict acute toxicity to fish (96-h $LC_{50}$), *Daphnia* (48-h $EC_{50}$), and algae (72–96-h $EC_{50}$). In particular, QSARs are provided for chemicals acting by nonpolar narcosis and polar narcosis mechanisms of action. No QSARs have been recommended for substances that act by more specific modes of action.

The TGD provides recommendations for the use of QSARs to predict long term toxicity to fish (no observed effect concentration [NOEC], 28 days) and to *Daphnia* (NOEC, 21 days). In particular QSARs are provided for chemicals acting by non-polar narcosis and polar narcosis mechanisms of action. No QSARs have been recommended for substances that act by more specific modes of action. For persistence, the TGD recommends two of the SRC BIOWIN models, namely the BIOWIN2 nonlinear model and the BIOWIN3 survey model for ultimate biodegradation. The exact cutoff points for these models have been calibrated on the basis of the model score for 1,2,4-trichlorobenzene — a substance that is known to be relatively persistent under environmental conditions. In model terms the cutoff values for identifying potentially persistent substances are: [BIOWIN2] < 0.5 and [BIOWIN3] < 2.2. For bioaccumulation, BCF values may be estimated from the octanol/water partition coefficient using QSAR models where experimental data are not available. For highly hydrophobic substances (e.g., with log $K_{ow}$ > 6), the available BCF models can lead to very different results. Hence an assessment must be done on a case-by-case basis taking into account what is known about the BCF QSAR-models and the specific properties of the substance, in particular what is known to affect uptake and the potential for metabolism in aquatic organisms.

The TGD discusses a number of QSARs to predict soil and sediment sorption coefficients. In total 19 QSARs, based solely on log $K_{ow}$, are reported. These QSARs are for distinct chemical classes and were taken from the article by Sabljić et al. (1995).

According to the current EU system of chemicals legislation, new and existing substances are not subject to the same testing requirements, which means that there is a lack of knowledge about the potential danger represented by many existing substances. Existing substances make up about 99% of the total volume of chemicals on the EU market. To address this problem, and other shortcomings of the current EU system, the European Commission has proposed a new policy on chemicals, in which new and existing substances will be subject to the same information requirements. In addition, the new proposals place the burden of performing hazard and risk assessments on industry, rather than the regulatory authorities (European Commission, 2002). The proposed system is called REACH (Registration, Evaluation, and Authorization of CHemicals).

When the REACH system is introduced, it is possible that additional human health and ecotoxicological information could be required for up to 30,000 existing chemicals, which are currently marketed in volumes greater than 1 t/year (t.p.a.). Therefore, QSAR and other computer-based methods for predicting toxicity are expected to play an increasingly important role, not only for the priority setting of chemicals that need further assessment, but also for hazard assessment purposes. As yet no formal procedures have been put in place for the use of QSAR in the REACH system.

Within the EU, member states apply QSAR differently according to national legislation. A small number of the uses of QSAR to predict toxicological endpoints by member states are described below.

## A. Use of QSAR by Regulatory Agencies in Denmark

The Danish Environmental Protection Agency (EPA) has used QSARs to classify dangerous substances. This has enabled the Danish EPA to draw up an advisory list of these dangerous substances — the so-called process of self-classification. Approximately 47,000 substances were examined, and of these 20,624 substances were identified as requiring classification for one or more of the following dangerous properties: acute oral toxicity, sensitization by skin contact, mutagenicity, carcinogenicity, and danger to the aquatic environment. Table 19.5 lists the main models that were applied by the Danish EPA to predict toxicity. Further, the Danish EPA attempted to validate many of the models which they applied for the self-classification process.

The validation process enabled the Danish EPA to state that, "the (QSAR) models used here are now so reliable that they are able to predict whether a given substance has one or more of the properties selected with an accuracy of approximately 70–85%." In addition to the use described above, the Danish EPA has developed a QSAR database that contains predicted data on more than 166,000 substances (OSPAR Commission, 2000). The Danish EPA used a suite of commercially available and proprietary QSARs for environmental and human health endpoints (see those listed in Table 19.5). The predictions were made off-line and were stored in a database (derived from the CHEM-X software). The database was searchable by Chemical Abstract Service (CAS) number or chemical name. Only discrete organic chemicals can be stored in the database.

**Table 19.5    Summary of the Main Models That Were Applied by the Danish EPA for Self-Classification Purposes**

| Endpoint | Model |
|---|---|
| Acute Oral Toxicity | TOPKAT Mouse Oral LD$_{50}$ |
| Sensitization by skin contact | TOPKAT — sensitizer vs. nonsensitizer |
| | TOPKAT strong vs. weak/moderate sensitiser |
| | M-CASE Model A33: Allergic contact dermatitis |
| Mutagenicity | M-CASE Model A2E: Structural alerts for DNA reactivity |
| | M-CASE Model A62: Induction of micronuclei |
| | TOPKAT: Salmonella (Ames) mutagenicity |
| | M-CASE Model A2H: *Salmonella* (Ames) mutagenicity |
| | M-CASE Model A61: Chromosomal aberrations |
| | M-CASE Model A2F: Mutations in mouse lymphoma |
| Carcinogenicity | TOPKAT NTP Carcinogenicity: Male rat |
| | TOPKAT NTP Carcinogenicity: Female rat |
| | TOPKAT NTP Carcinogenicity: Male mouse |
| | TOPKAT NTP Carcinogenicity: Female mouse |
| | TOPKAT FDA Carcinogenicity: Male rat |
| | TOPKAT FDA Carcinogenicity: Female rat |
| | TOPKAT FDA Carcinogenicity: Male mouse |
| | TOPKAT FDA Carcinogenicity: Female mouse |
| | M-CASE Carcinogenic Potency Database model: Rat (Danish EPA version of A0D) |
| | M-CASE Carcinogenic Potency Database model: Mouse (Danish EPA version of A0E) |
| Danger to the aquatic environment | M-CASE Fathead Minnow Acute Aquatic LC$_{50}$ |
| | SRC BIOWIN |
| | SRC BCFWIN |

*Source:* Adapted from Tyle, H., Larsen, H.S., Wedebye, E.B., and Niemela, J., Identification of Potential PBTs and vPvBs by Use of QSARs, Danish EPA, Copenhagen, Denmark, Nov. 11, 2001. With permission.

## B.  Use of QSAR by Regulatory Agencies in Germany

In Germany, new chemicals are notified to the Federal Institute for Health Protection of Consumers and Veterinary Medicine (BgVV). To provide a tool for the evaluation of physicochemical properties and probable toxic effects of notified substances, the BgVV has developed a computerized database from data sets containing physicochemical and toxicological properties. The database has been used to develop specific SAR models for predicting skin and eye irritation/corrosion, which have been incorporated into a decision support system (DSS) (Gerner et al., 2000a; 2000b; Zinke et al., 2000).

The BgVV has developed the database from regulatory test results that have been used to develop specific SAR models for predicting the following acute toxicities: oral LD$_{50}$, dermal LD$_{50}$, and inhalation LD$_{50}$. These models have been incorporated into the DSS (Gerner et al., 2000a; 2000b; Zinke et al., 2000). The DSS is mainly a rule based approach, the rules being developed not only on substructural molecular features, but also on physicochemical properties such as molecular weight, aqueous solubility, and log K$_{ow}$. The DSS is designed to predict EU risk phrases and has undergone a rigorous process of validation.

## VI.  RECOMMENDATIONS FROM THE ORGANISATION FOR ECONOMIC CO-OPERATION AND DEVELOPMENT FOR THE USE OF QSARS

While not a regulatory body, the Organization for Economic Co-operation and Development (OECD) has increasingly become involved in the use of QSARs and expert systems to predict toxicity. In particular, the OECD has organized a freely available database on methods to predict chemical and physical properties of molecules (available from www.oecd.org or webdomino1.oecd.org/comnet/env/models.nsf)

## VII. CONCLUSIONS

There is widespread use by regulatory authorities of QSARs and computer-based approaches to predict toxicity and fate. This is especially true in the U.S., but also increasingly in Canada and the EU. Due to desire to obtain more information regarding the deleterious effects of chemicals to the environment and man, it is likely that these approaches will be used more often. Most regulatory use of QSARs is cautionary (i.e., a prediction of toxicity may be accepted for classification, while a prediction of no toxicity may be met with a requirement for further testing). Most often, commercial expert systems are applied, a notable exception being the use of the EPISUITE collection of programs that have been developed with the backing of the U.S. EPA. As yet, little validation of QSARs and expert systems for regulatory use has been attempted. Considerably greater efforts to validate predictions of toxicity and fate further will be required to increase confidence in the use of these techniques.

## ACKNOWLEDGMENTS

I am indebted to many scientists across the globe for their input into the chapter, provision of information and references, and general stimulation of my interest in this application of QSAR. In particular, I would like to thank Mike Comber, Joanna Jaworska, John Walker, Chris Watts, Andrew Worth, and Maurice Zeeman. Further, I acknowledge the hard work, effort, and dedication of many scientists and civil servants worldwide who have endeavored to apply QSAR methods to regulatory decision-making processes.

## REFERENCES

Clements, R.G., Ed., *Estimating Toxicity of Industrial Chemicals to Aquatics Organisms Using Structure-Activity Relationships*, Office of Pollution Prevention and Toxics, U.S. Environmental Protection Agency, Washington, D.C., 1996, p. 404. Available from: www.epa.gov/opptintr/newchems/sarman.pdf

Cronin, M.T.D. and Schultz, T.W., Pitfalls in QSAR, *J. Mol. Struct. (Theochem)*, 622, 39–51, 2003.

Cronin, M.T.D., Jaworska, J.S., Walker, J.D., Comber, M.H.I., Watts, C.D., and Worth, A.P., Use of QSARs in international decision-making frameworks to predict health effects of chemical substances, *Environ. Health Perspect.*, 111, 1391–1401, 2003a.

Cronin, M.T.D., Walker, J.D., Jaworska, J.S., Comber, M.H.I., Watts, C.D., and Worth, A.P., Use of QSARs in international decision-making frameworks to predict ecologic effects and environmental fate of chemical substances, *Environ. Health Perspect.*, 111, 1376–1390, 2003b.

El-Masri, H.A., Mumtaz, M.M., Choudhary, G., Cibulas, W., and De Rosa, C.T., Applications of computational toxicology methods at the Agency for Toxic Substances and Disease Registry, *International J. Hyg. Environ. Health*, 205, 63–69, 2002.

Eriksson, L., Jaworska, J., Worth, A.P., Cronin, M.T.D., McDowell, R.M., and Gramatica, P., Methods for reliability and uncertainty assessment and applicability evaluations of classification- regression-based and QSARs, *Environ. Health Perspect.*, 111, 1361–1375, 2003.

European Commission, *White Paper on the Strategy for a Future Chemicals Policy*, European Commission, Brussles, Belgium, 2002, available at europa.eu.int/comm/environment/chemicals/whitepaper.htm.

European Economic Community, Technical Guidance Document in Support of Commission Directive 93/67/EEC on Risk Assessment for New Notified Substances and Commission Regulation (EC) No 1488/94 on Risk Assessment for Existing Substances, European Commission, Office for Official Publications of the European Communities, Luxembourg, 1996.

Gerner, I., Graetschel, G., Kahl, J., and Schlede, E., Development of a decision support system for the introduction of alternative methods into local irritancy/corrosivity testing strategies. Development of a relational database, *Alt. Lab. Anim. (ATLA)*, 28, 11–28, 2000a.

Gerner, I., Zinke, S., Graetschel, G., and Schlede, E., Development of a decision support system for the introduction of alternative methods into local irritancy/corrosivity testing strategies: creation of fundamental rules for a decision support system, *Alt. Lab. Anim. (ATLA)*, 28, 665–698, 2000b.

Hong, H.W., Tong, H., Fang, L., Shi, R., Xie, Q., Wu, J., Perkins, R., Walker, J.D., Branham, W., and Sheehan, D., Prediction of estrogen receptor binding for 58,000 chemicals using an integrated system of a tree-based model with structural alerts, *Environ. Health Perspect.*, 110: 29-36, 2002.

Hulzebos, E.M., Maslankiewicz, L., and Walker, J.D., Verification of literature-derived SARs for skin irritation and corrosion, *QSAR Combinatorial Sci.*, 22, 351–363, 2003.

Hulzebos, E.M. and Posthumus, R., (Q)SARs: gatekeepers against risk on chemicals?, *SAR QSAR Environ. Res.*, 14, 285–316, 2003.

Jaworska, J.S., Comber, M., Auer, C., and Van Leeuwen, C.J., Summary of a workshop on regulatory acceptance of (Q)SARS for human health and environmental endpoints, *Environ. Health Perspect.*, 111, 1358–1360, 2003.

MacDonald, D., Breton, R., Sutcliffe, R., and Walker, J., Uses and limitations of quantitative structure-activity relationships (QSARs) to categorize substances on the Canadian Domestic Substance List as persistent and/or bioaccumulative, and inherently toxic to non-human organisms, *SAR QSAR Environ. Res.*, 13, 43–55, 2002.

Matthews, E.J., Benz, R.D., and Contrera, J.F., Use of toxicological information in drug design, *J. Mol. Graphics Modelling*, 18, 605–614, 2000.

Matthews, E.J. and Contrera, J.F., A new highly specific method for predicting the carcinogenic potential of pharmaceuticals in rodents using enhanced MCASE QSAR-ES software, *Regul. Toxicol. Pharmacol.*, 28, 242–264, 1998.

Mekenyan, O., Ivanov, J., Karabunarliev, S., Bradbury, S.P., Ankley, G.T., and Karcher, W., A computationally-based hazard identification algorithm that incorporates ligand flexibility. 1. Identification of potential androgen receptor ligands, *Environ. Sci. Tech.*, 31, 3702–3711, 1997.

Mekenyan, O., Kamenska, V., Serafimova, R., Poellinger, L., Brower, A., and Walker, J., Development and validation of an average mammalian estrogen receptor-based QSAR model, *SAR QSAR Environ. Res.*, 13, 579–595, 2002.

Moore, D.R.J., Breton, R.L., and MacDonald, D.B., A comparison of model performance for six quantitative structure-activity relationship packages that predict acute toxicity to fish, *Environ. Toxicol. Chem.*, 22, 1799–1809, 2003.

OSPAR Commission, *Briefing Document on the Work of DYNAMEC and the DYNAMEC Mechanism for the Selection and Prioritisation of Hazardous Substances*, OSPAR Commission, London, 2000.

Prival, M.J., Evaluation of the TOPKAT system for predicting the carcinogenicity of chemicals, *Environ. Mol. Mutagenesis*, 37, 55–69, 2001.

Richard, A.M. and Benigni, R., AI and SAR approaches for predicting chemical carcinogenicity: survey and status report, *SAR QSAR Environ. Res.*, 13, 1–19, 2002.

Russom, C.L., Anderson, E.B., Greenwood, B.E., and Pilli, A., ASTER: an integration of the AQUIRE data base and the QSAR system for use in ecological risk assessments, *Sci. Total Environ.*, 109/110, 667–670, 1991.

Sabljić, A., Gusten, H., Verhaar, H. and Hermens, J., QSAR modeling of soil sorption — improvements and systematics of log K-oc vs Log K-ow correlations, *Chemosphere*, 31, 4489–4514, 1995.

Schmieder, P., Ankley, G., Mekenyan, O., Walker, J.D., and Bradbury, S., Quantitative structure-activity relationship models for prediction of estrogen receptor binding affinity of structurally diverse chemicals, *Environ. Toxicol. Chem.*, 22, 1844–1854, 2003.

Schultz, T.W. and Cronin, M.T.D., Essential and desirable characteristics of ecotoxicity quantitative structure-activity relationships, *Environ. Toxicol. Chem.*, 22, 599–607, 2003.

Shi, L., Tong, W., Fang, H., Xie, Q., Hong, H., Perkins, R., Wu, J., Tu, M., Blair, R.M., Branham, W.S., Waller, C., Walker, J., and Sheehan, D.M., An integrated "4-phase" approach for setting endocrine disruption screening priorities — phase I and II predictions of estrogen receptor binding affinities, *SAR QSAR Environ. Res.*, 13, 69–88, 2002.

Tyle, H., Larsen, H.S., Wedebye, E.B., and Niemela, J., Identification of Potential PBTs and vPvBs by Use of QSARs, Danish EPA, Copenhagen, Denmark, Nov. 11, 2001.

U.S. Environmental Protection Agency, U.S. EPA/EC Joint Project on the Evaluation of (Quantitative) Structure-Activity Relationships (EPA 743-R-94-001), Office of Prevention, Pesticides, and Toxic Substances, Washington D.C., 1994, available electronically from: www.epa.gov/opptintr/newchems/sar_report.pdf

Walker, J.D., Applications of QSARs in toxicology: a U.S. Government perspective, *J. Mol. Struct. (Theochem)*, 622, 167–184, 2003.

Walker, J.D., Carlsen. L., Hulzebos, E., and Simon-Hettich, B., Global government applications of analogues, SARs and QSARs to predict aquatic toxicity, chemical or physical properties, environmental fate parameters and health effects of organic chemicals, *SAR QSAR Environ. Res.*, 13, 607–616, 2002.

Walker, J.D., Carlsen. L., and Jaworska, J., Improving opportunities for regulatory acceptance of QSARs: the importance of model domain, uncertainty, validity and predictability, *QSAR Combinatorial Sci.*, 22, 346–350, 2003a.

Walker, J.D., Dimitrov, S., and Mekenyan, O., Using HPV chemical data to develop QSARs for non-HPV chemicals: opportunities to promote more efficient use of chemical testing resources, *QSAR Combinatorial Sci.*, 22, 386–395, 2003b.

Walker, J.D., Jaworska, J., Comber, M.H.I., Schultz, T.W., and Dearden, J.C., Guidelines for developing and using quantitative structure-activity relationships, *Environ. Toxicol. Chem.*, 22, 1653–1665, 2003c.

Woo, Y.T., Lai, D.Y., Argus, M.F., and Arcos, J.C., Development of structure-activity relationship rules for predicting carcinogenic potential of chemicals, *Toxicol. Lett.*, 79, 219–228, 1995.

Zeeman, M., EPA's framework for ecological effects assessment in *Screening and Testing Chemicals in Commerce* (OTA-BP-ENV-166), U.S. Congress, Office of Technology Assessment, Washington, D.C., 1995, pp. 169–178.

Zeeman, M., Fairbrother, A., and Gorsuch, J.W., Environmental toxicology: testing and screening, in *Screening and Testing Chemicals in Commerce* (OTA-BP-ENV-166), U.S. Congress, Office of Technology Assessment, Washington, D.C., 1995, pp. 159–167.

Zeeman, M., Rodier, D., and Nabholz J.V., Ecological risks of a new industrial chemical under TSCA, in *Ecological Risk Assessment in the Federal Government*, a report from the U.S. White House Committee on Environment and Natural Resources (CENR) (CENR/5-99/001), Washington, D.C., 1999, pp. 32–61.

Zinke, S., Gerner, I., Graetschel, G., and Schlede, E., Local irritation/corrosion testing strategies: development of a decision support system for the introduction of alternative methods, *Alt. Lab. Anim. (ATLA)*, 28, 29–40, 2000.

Zinke, S., Gerner, I., and Schlede, E., Evaluation of a rule base for identifying contact allergens by using a regulatory database: comparison of data on chemicals notified in the European union with "structural alerts" used in the DEREK expert system, *Alt. Lab. Anim. (ATLA)*, 30, 285–298, 2002.

# CHAPTER 20

# A Framework for Promoting the Acceptance and Regulatory Use of (Quantitative) Structure-Activity Relationships

**Andrew P. Worth, Mark T.D. Cronin, and Cornelius J. Van Leeuwen***

## CONTENTS

---

* This manuscript expresses the views of the authors, and does not necessarily reflect the position of the European Commission.

# I. INTRODUCTION

Structure-activity relationships (SARs) and quantitative structure-activity relationships (QSARs), referred to collectively as QSARs, can be used for the prediction of physicochemical properties, environmental fate parameters (e.g., accumulation and biodegradation), human health effects, and ecotoxicological effects. A SAR is a (qualitative) association between a chemical substructure and the potential of a chemical containing the substructure to exhibit a certain physical or biological effect. A QSAR is a mathematical model that relates a quantitative measure of chemical structure (e.g., a physicochemical property) to a physical property or to a biological effect (e.g., a toxicological endpoint).

QSARs are potentially useful components of strategies for the regulatory assessment of chemicals. At present, the use of QSARs for regulatory purposes is limited (mostly to the priority setting of substances to be tested), and there is an ongoing debate over the extent to which the predictions made by QSARs can be relied upon. One of the reasons for the widespread caution about, and sometimes opposition to, the use of QSARs is that these models are generally proposed in the scientific literature by their developers. Unfortunately, they are seldom subjected to an independent validation and peer-review process, in the way that other test methods, and alternative (non animal) test methods in particular, have been (Cronin and Schultz, 2003). A possible reason for this is that, in the QSAR field, the term validation is generally understood to be a statistical exercise performed by the model developer, with considerable freedom in the choice of statistical approach and acceptability criteria. Conversely, in the field of *in vitro* toxicology, validation is widely understood as an independent process by which the relevance and reliability of a method are established for a particular purpose (Balls et al., 1995). The principle of independence is paramount in this concept of validation, and is a key element in establishing the scientific credibility of the method.

A powerful way of overcoming the credibility problem associated with QSARs is to establish a formal framework for promoting the use of QSARs. Ideally, this framework should be managed by an organization that is independent of national and sectoral interests, which is close to the regulatory process, and has no commercial interests associated with the use of QSARs. The framework should not only facilitate the development and validation of QSARs, but should also provide a means for the independent peer review of the validation process and initiate the regulatory acceptance process wherever appropriate.

This chapter describes in detail how the overall process of QSAR development, validation, peer review, and acceptance could be implemented, by means of an imaginary case study in which a QSAR for predicting acute toxicity to fish is proposed for regulatory use.

# II. A PROCESS FOR PROMOTING THE ACCEPTANCE AND USE OF QSARS

The need to develop and apply acceptability criteria for the use of QSARs has been acknowledged for at least 10 years. Proposals for acceptability criteria were made in the European Union (EU) Technical Guidance Document (TGD) on Risk Assessment (European Economic Community, 1996). These criteria were subsequently expressed in the context of the European Centre for the Validation of Alternative Methods' (ECVAM) acceptability criteria for *in vitro* methods (Worth et al., 1998), and further developed and agreed at the Workshop on Regulatory Acceptance of QSARs for Human Health and Environmental Endpoints, organized and funded by CEFIC/ICCA (European Chemical Industry Council [CEFIC]/International Council of Chemical Associations [ICCA]) and held in Setubal, Portugal, in March 2002 (Cronin et al., 2003a, 2003b; Eriksson et al., 2003; Jaworska et al., 2003). These criteria were presented at the 4th World Congress on Alternatives and Animal Use in the Life Sciences, held in New Orleans in August 2002, along with the outline of a proposal for the practical validation of computer models, such as QSARs (Worth and Cronin, 2004).

While there is widespread agreement on these generic criteria, it is also widely felt that detailed guidance, possibly including more specific criteria, should be developed. It is foreseen that the development of such guidance will be overseen by an Organization for Economic Cooperation and Development (OECD) expert working group. Any independent process for promoting the acceptance and use of QSARs should ultimately take account of internationally agreed acceptability criteria. Since such criteria have not yet been agreed, it should be noted that the criteria adopted in this chapter are presented as suggestions, developed for the purposes of illustrating a general framework that is independent of specific criteria.

It is suggested that the independent process should be based on six steps that cover the:

1. Selection of QSARs
2. Confirmation of model parameters and goodness-of-fit properties for selected QSARs
3. Assessment of data quality and development of predictivity criteria
4. Assessment of predictive capacity
5. Independent peer review of the validation process
6. Progression toward regulatory acceptance

These steps are described in more detail below.

## A. Independent Selection of Candidate QSARs

In the first stage of the process, the independent body selects QSARs that are potentially useful in the context of a specific regulatory system, such as the future Registration, Evaluation and Authorization of Chemicals (REACH) system in the EU (European Commission, 2001). The selection could be based on a review of the scientific literature, or on the assessment of dossiers submitted by the proponents of QSAR models, such as industry, academia, or regulatory bodies.

All QSARs selected for validation should meet a minimum set of criteria, which could be called QSAR development criteria:

1. The QSAR should have a clearly defined scientific and regulatory purpose. In other words, it should make predictions for a well-defined physical effect or biological endpoint that is of regulatory relevance. Reference should be made to the test method upon which the QSAR is based, including details of the endpoint determined, the test species, and any other experimental conditions that may be important determinants of the endpoint being modeled. Ideally, the test method should follow an internationally accepted protocol, such as an OECD Test Guideline or a test method listed in Annex V of Council Directive 67/548/EEC. Where appropriate, the units of the endpoint being modeled should also be given.
2. The proposed application of the QSAR should be explicitly defined. It should be clear whether the QSAR is intended to:
   a. Fully replace the use of an experimental system or animal.
   b. Act as a partial replacement in the context of a testing strategy (i.e., for particular classes of chemicals or a particular range of the spectrum of endpoint values).
   c. Contribute solely to the priority setting of chemicals for further testing, without making predictions that are directly used in the assessment itself.
   d. Guide the selection of additional tests, without making predictions that are directly used in the assessment itself.
   Depending on the proposed application of the QSAR, the specific acceptability criteria used to evaluate it are likely to be more or less stringent.
3. The domain of applicability of the QSAR should be explicitly defined. The QSAR should be associated with a description of the chemical classes for which it is applicable (inclusion rules) or inapplicable (exclusion rules). For QSARs, there should be an indication of the range of descriptor values for which reliable predictions can be made.

4. The statistical method used to develop the QSAR should be specified (including details of any software packages used).

5. Full details of the QSAR training set should be provided, including details of chemical structure (names and structural formulae, and chemical abstract service [CAS] numbers if available) and, in the case of a QSAR, data for all descriptor and response variables. If the data used to develop the model were based upon the processing of raw data (e.g., the averaging of replicate values), it is preferable if all of the raw data are supplied.

6. A QSAR should be associated with basic statistics for its goodness-of-fit to the training set (e.g., r and $R^2$ values in the case of regression models).

The prioritization of QSAR models for validation is likely to take account of regulatory needs. The selection of QSARs could also take into account the mechanistic basis of the QSAR. In the case of QSARs, this would involve a qualitative assessment of the relevance of descriptor variables to the endpoint being modeled. A strong mechanistic basis could be regarded as a desirable criterion, rather than an essential criterion, since mechanistic relevance cannot always be established; this is especially the case since the fundamental processes underlying the expression of biological endpoints are often unknown.

As an additional consideration, the degree of confidence in a QSAR would be increased by the knowledge that similar QSARs had already been validated. Such QSARs could be similar in terms of their mechanism of action or their domain of applicability, but they could have been developed for an equivalent endpoint in a different species.

## B. Independent Confirmation of Candidate QSARs

Using the information obtained (from the literature) or provided (by the proponent of the QSAR) in step 1, there should be an independent confirmation of the QSAR. This means that someone with statistical expertise should apply the specified statistical method to the training set of data, to check that the same QSAR model and (goodness-of-fit) statistics are obtained. If the QSAR model is confirmed, step 3 should be initiated. Alternatively, the proposed QSAR could be modified, so that the modified QSAR is subjected to independent assessment.

## C. Independent Assessment of Data Quality and Development of Predictivity Criteria

In this step, one or more independent experts should evaluate the quality of the training set data along with any other available data for the endpoint predicted by the QSAR. This should enable an evaluation to be made of the maximal predictive capacity that could be expected for the QSAR. For QSARs, the inevitable variability in descriptor and response variable data should be taken into consideration when defining criteria for predictive capacity. For example, in the case of a regression-based QSAR, it might be decided that its predictions should fall within a specified prediction interval, and that the $R^2$ value for predictions of independent data should exceed a specified value. Issues relating to the quality of data for use in QSARs are discussed in Cronin and Schultz (2003) and Schultz and Cronin (2003).

When defining predictivity criteria, the proposed application of the QSAR should also be taken into account, to accommodate the consequences of making incorrect (or inaccurate) predictions. In principle, QSARs can be used for the priority setting of chemicals, or as components in hazard and risk assessment strategies. In such strategies, QSARs could serve as (partial) replacements of animal experiments, or as screening methods for selecting the experimental tests that need to be conducted.

## D.  Independent Assessment of Predictive Capacity

In this step, it is necessary to identify or compile an independent test set of data. In other words, it is necessary to obtain a data set containing values of the descriptors and response variables for chemical structures that were not used in the training set. The selection of chemicals for the test set should take into account the predefined domain of applicability for the QSAR.

Predictions should then be obtained for the test set of data, and appropriate statistical measures of QSAR predictivity should be derived. These measures should then be compared with the predefined predictivity criteria to assess the predictive performance of the model.

## E.  Independent Peer Review of the Validation Process

The outcome of step 2 to step 4 should be documented in the form of a report, which should be submitted to one or more independent experts for peer review. The peer review should be conducted by experts who were not involved in the previous steps, and who were not involved in the development of the QSAR. The goal of the peer review is to ensure that acceptability criteria were adequately developed and applied, and to consider the appropriateness of the test set.

The peer review should lead to a definite conclusion that should be made publicly available. Possible conclusions include:

1. The validation of the QSAR was scientifically sound, and the model should be considered suitable for regulatory use within specified limits. This should either reiterate the predefined purpose and applicability of the QSAR, or could potentially redefine them, on the basis of the scientific evidence.
2. The validation of the QSAR was scientifically sound, but the model should be considered unsuitable for regulatory use for specified reasons. Such a recommendation could be accompanied by an indication of the further work that could be undertaken to obtain an improved QSAR that could be reconsidered for regulatory use. It is important to emphasize that such a recommendation should not be made simply because the peer reviewers fail to see how the validated QSAR would fit into a regulatory scheme, since this issue can be addressed at a later stage.
3. The validation of the QSAR was not scientifically sound, so conclusions cannot be made regarding the suitability of the model for regulatory purposes. Such a conclusion should be accompanied by a detailed description of the reasons why the validation was not sound, along with recommendations of how the validation could be reperformed in a scientifically sound manner.

An independent peer-review panel should make an official statement on the validity of the QSAR. In the EU, such a statement could be made, for example, by the Scientific Advisory Committee (ESAC) of the ECVAM (see Worth and Balls [2001] for information about ECVAM and the ESAC).

## F.  Progression toward Regulatory Acceptance

All QSARs should be associated with a clear scientific and regulatory purpose. A possible outcome of the peer-review step is the recommendation that the validated QSAR should be considered for inclusion into a regulatory framework. In such a case, the recommendation, and all of the supporting evidence, should be forwarded to the appropriate regulatory body (or bodies) for consideration. In the EU, if the QSAR is considered suitable for the assessment of chemicals, deliberations take place at a technical level by the National Coordinators for Testing Methods, and subsequently, a decision is taken at the policy level by representatives of the EU Competent Authorities for Directive 67/548/EEC.

In addition to the scientific evidence supporting the scientific validity of the QSAR, it would be helpful if the proposal for regulatory acceptance also included a proposal on how the validated

QSAR could be incorporated into a testing strategy for a particular regulatory endpoint; the proposal should address the known limitations of the QSAR model.

## III. TRANSPARENCY OF THE OVERALL PROCESS

To promote the use of scientifically valid QSARs, details regarding the development, selection, confirmation, and validation of QSARs, along with information regarding the independent peer review of the validation process, should be made publicly available, ideally via the Internet. Where possible, the complete training and test sets should also be made available.

## IV. AN IMAGINARY CASE STUDY: A QSAR FOR NONPOLAR NARCOSIS IN FISH

To illustrate how the general process described above could be performed, an imaginary case study is described in this section, using a simple QSAR for nonpolar narcosis to the fathead minnow (*Pimephales promelas*). This QSAR was chosen because it represents a simple, relatively well-understood endpoint for toxicity. Furthermore, the endpoint is included in OECD Test Guideline 203 on the Fish Acute Toxicity Test (Organization for Economic Cooperation and Development, 1992), and the QSAR is recommended by the EU TGD (European Economic Community, 1996). A further consideration was the availability of test data to form an external test set.

### A. Independent Selection of Candidate QSARs

It can be imagined that a QSAR for acute aquatic toxicity was selected by an independent body during a literature review of available QSARs, including the final report of a Directorate General (DG) Research Project (European Economic Community, 1995). In particular, the following QSAR for predicting the acute toxicity of organic chemicals to the fathead minnow (*Pimephales promelas*) was reported:

$$\text{Log } (LC_{50}) = -0.846 \log K_{ow} - 1.39 \tag{20.1}$$

where $LC_{50}$ is the concentration (in moles per liter) causing 50% lethality in *Pimephales promelas* after an exposure of 96 h, and $K_{ow}$ is the octanol-water partition coefficient. It should be noted that the inverse of toxicity (negative logarithm) was not reported in the original form of this equation.

The following statistics were reported for this QSAR: $n = 58$, $r^2 = 0.937$, $Q^2 = 0.932$ and s.e. = 0.361. The training set is given in Table 20.1.

The QSAR is a regression model, based on a single parameter. Linear regression analysis is considered to be one of the most transparent methods for the development of QSARs (Cronin and Schultz, 2003; Schultz and Cronin 2003), and the use of a single predictor variable makes the QSAR both simple and user friendly.

The QSAR was developed for chemicals considered to act by a single mechanism of toxic action, non-polar narcosis, as defined by Verhaar et al. (1992), and therefore has a clear mechanistic basis. Non-polar narcosis has been established experimentally by using the Fish Acute Toxicity Syndrome methodology (McKim et al., 1987). The QSAR is based on a descriptor for hydrophobicity (log $K_{ow}$), which is relevant to the mechanism of action (i.e., toxicity results from the accumulation of molecules in biological membranes). There are numerous regression models based on log $K_{ow}$ for this mechanism of action, so the development of such models has enabled interspecies comparisons (Dimitrov et al., 2003).

**Table 20.1    Training Set of Data Used to Develop a QSAR for Non-polar Narcosis**

| No. | Chemical | Log $K_{ow}$ | Log ($LC_{50}$) (mol/L) |
|---|---|---|---|
| 1 | 1-Butanol | 0.88 | −1.63 |
| 2 | 1-Decanol | 4.57 | −4.81 |
| 3 | 1-Dodecanol | 5.13 | −5.26 |
| 4 | 1-Hexanol | 2.03 | −3.02 |
| 5 | 1-Nonanol | 4.26 | −4.40 |
| 6 | 1-Octanol | 2.97 | −3.98 |
| 7 | 1-Undecanol | 4.52 | −5.21 |
| 8 | 1,1,2-Trichloroethane | 1.89 | −3.21 |
| 9 | 1,1,2,2-Tetrachloroethane | 2.39 | −3.91 |
| 10 | 1,2-Dichloroethane | 1.48 | −2.92 |
| 11 | 1,2,3,4-Tetrachlorobenzene | 4.63 | −5.29 |
| 12 | 1,2,4-Trichlorobenzene | 4.05 | −4.79 |
| 13 | 1,3-Dichlorobenzene | 3.52 | −4.27 |
| 14 | 1,4-Dichlorobenzene | 3.44 | −4.56 |
| 15 | 1,4-Dimethoxybenzene | 2.15 | −3.07 |
| 16 | 2-(2-Ethoxyethoxy)ethanol | −0.54 | −0.70 |
| 17 | 2-Butanone | 0.29 | −1.35 |
| 18 | 2-Decanone | 3.73 | −4.43 |
| 19 | 2-Hydroxy-4-methoxyacetophenone | 1.98 | −3.48 |
| 20 | 2-Methyl-1-propanol | 0.76 | −1.71 |
| 21 | 2-Methyl-2,4-pentanediol | −0.67 | −1.04 |
| 22 | 2-Octanone | 2.37 | −3.55 |
| 23 | 2-Phenoxyethanol | 1.16 | −2.60 |
| 24 | 2-Propanol | 0.05 | −0.76 |
| 25 | 2,2,2-Trichloroethanol | 1.42 | −2.69 |
| 26 | 2,3,4-Trichloroacetophenone | 3.57 | −5.04 |
| 27 | 2,3,4-Trimethoxyacetophenone | 1.12 | −3.08 |
| 28 | 2,4-Dichloroacetophenone | 2.84 | −4.20 |
| 29 | 2,6-Dimethoxytoluene | 2.64 | −3.87 |
| 30 | 3-Furanmethanol | 0.30 | −2.28 |
| 31 | 3-Methyl-2-butanone | 0.56 | −1.99 |
| 32 | 3-Pentanone | 0.79 | −1.74 |
| 33 | 3,3-Dimethyl-2-butanone | 0.96 | −3.06 |
| 34 | 3,4-Dichlorotoluene | 4.06 | −4.74 |
| 35 | 4-Methyl-2-pentanone | 1.31 | −2.29 |
| 36 | 5-Methyl-2-hexanone | 1.88 | −2.85 |
| 37 | 5-Nonanone | 2.90 | −3.66 |
| 38 | 6-Methyl-5-hepten-2-one | 1.70 | −3.16 |
| 39 | Acetone | −0.24 | −0.85 |
| 40 | Acetophenone | 1.58 | −2.87 |
| 41 | Benzophenone | 3.18 | −4.07 |
| 42 | Cyclohexanol | 1.23 | −2.15 |
| 43 | Cyclohexanone | 0.81 | −2.27 |
| 44 | Dibutyl ether | 3.21 | −3.60 |
| 45 | Diisopropyl ether | 1.52 | −3.04 |
| 46 | Dipentyl ether | 4.04 | −4.69 |
| 47 | Diphenyl ether | 4.21 | −4.62 |
| 48 | Ethanol | −0.31 | 0.51 |
| 49 | Furan | 1.34 | −3.04 |
| 50 | Hexachloroethane | 4.14 | −5.19 |
| 51 | Methanol | −0.77 | −0.057 |
| 52 | 4-Nitrophenyl phenylether | 4.28 | −4.90 |
| 53 | Pentachloroethane | 3.62 | −4.44 |
| 54 | (*Tert*)butylmethyl ether | 0.94 | −2.09 |
| 55 | Tetrachloroethene | 3.40 | −4.08 |
| 56 | Tetrahydrofuran | 0.46 | −1.52 |

**Table 20.1 (continued)  Training Set of Data Used to Develop a QSAR for Non-polar Narcosis**

| No. | Chemical | Log $K_{ow}$ | Log $(LC_{50})$ (mol/L) |
|-----|----------|--------------|-------------------------|
| 57 | Trichloroethene | 2.42 | −3.47 |
| 58 | Triethylene glycol | −1.24 | −0.33 |

Note: The data were taken from the final report of an EU project.

*Source:* European Commission, QSAR for Predicting Fate and Effects of Chemicals in the Environment, Final report of DG XII contract No. EV5V-CT92-0211, Brussels, 1995.

Furthermore, the QSAR could potentially fulfill a clear regulatory need, since the 96-h $LC_{50}$ in the fathead minnow is one of the endpoints referred to in OECD Test Guideline 203.

The domain of applicability of the QSAR was well defined by the model developer. The QSAR was stated to be applicable to chemicals having log $K_{ow}$ values in the range from −1.24 to 5.13, and operating by a non-polar narcosis mechanism of action. Such chemicals can be identified on a structural basis (Verhaar et al., 1992), or from physicochemical descriptors (Boxall et al., 1997).

## B.  Independent Confirmation of Candidate QSARs

The application of linear regression to the training set of data enabled the QSAR to be confirmed. However, it was decided to re-express the QSAR as follows:

$$\text{Log } (1/LC_{50}) = 0.846 \text{ log } K_{ow} + 1.39 \tag{20.2}$$

This is a convenient way of expressing models for the prediction of $LC_{50}$ values, and is easy to understand since a numerical increase in the response variable means an increase in toxicity (acute lethality in this case). Furthermore, it is apparent that the QSAR relates an increasing toxicity to an increasing partition coefficient.

Another decision that could be taken in this step would be to redevelop the QSAR by using a more homogeneous set of $K_{ow}$ data, (i.e., by obtaining new experimental data obtained by a single protocol), or by estimating the $K_{ow}$ by using a single computer algorithm/software package.

## C.  Independent Assessment of Data Quality and Development of Predictivity Criteria

The training set consists of the biological (96-h $LC_{50}$) data being modeled, and data for a single descriptor ($K_{ow}$) that is being used as a predictor variable. The biological data can be considered to be of very high quality, since they were obtained by applying a single protocol and measured in the same laboratory, possibly by the same worker.

The descriptor ($K_{ow}$) data are a mixture of experimental and calculated values. Generally, $K_{ow}$ is considered to be a high-quality physicochemical descriptor, and the range of log $K_{ow}$ values (Table 20.1) is well within that usually considered to provide adequate measured values. However, there is no certainty that the measurements of $K_{ow}$ were made by the same protocol, or in the same laboratory, so this could result in a small amount of variability. Furthermore, using a mixture of calculated and experimental values will also result in some variability.

For illustrative purposes, it can be assumed that the following predictivity criteria were established before undertaking the independent assessment of predictive capacity:

1. The QSAR should make predictions of an independent test set with an $r^2$ value $\geq 0.8$.
2. All predictions should lie within a 95% confidence interval of the model, meaning all estimates of $LC_{50}$ should fall within 0.704 log units of the regression line ($0.704 = 1.95 \times$ s.e.).

**Table 20.2    Test Set of Data Used to Validate the QSAR for Non-polar Narcosis**

| No. | Chemical | Log $K_{ow}$ | MW | $LC_{50}$ (mg/L) | Observed Log($LC_{50}$) (mol/L) | Predicted Log($LC_{50}$) (mol/L) | Residual (Observed − Predicted) |
|---|---|---|---|---|---|---|---|
| 1 | 1-Pentanol | 1.56 | 88.15 | 472 | −2.27 | −2.71 | 0.44 |
| 2 | N,N-Dimethyl-4-toluidine | 2.81 | 135.2 | 48.9 | −3.44 | −3.77 | 0.33 |
| 3 | 1-Heptanol | 2.72 | 116.2 | 34.5 | −3.53 | −3.69 | 0.16 |
| 4 | N,N-Dimethylaniline | 2.31 | 121.2 | 64.1 | −3.28 | −3.34 | 0.06 |
| 5 | 3-Bromobenzamide | 1.65 | 200.04 | 92.7 | −3.33 | −2.79 | −0.54 |
| 6 | 4-(Tert-butyl)benzamide | 2.51 | 177.25 | 31.9 | −3.74 | −3.51 | −0.23 |
| 7 | Urethane | −0.15 | 89.09 | 5240 | −1.23 | −1.26 | 0.03 |
| 8 | 4'-Aminopropiophenone | 1.43 | 149.19 | 146 | −3.01 | −2.60 | −0.41 |
| 9 | 2-Methyl-2-propanol | 0.35 | 74.12 | 6410 | −1.06 | −1.69 | 0.63 |
| 10 | 2-Ethylpyridine | 1.69 | 107.16 | 414 | −2.41 | −2.82 | 0.41 |
| 11 | 1-Bromopropane | 2.10 | 122.99 | 67.3 | −3.26 | −3.17 | −0.09 |
| 12 | 2-Heptanone | 1.98 | 114.19 | 131 | −2.94 | −3.07 | 0.13 |
| 13 | 1,4-Dichlorobutane | 2.24 | 127.01 | 51.6 | −3.39 | −3.29 | −0.10 |
| 14 | 2-Undecanone | 4.09 | 170.3 | 1.50 | −5.06 | −4.85 | −0.21 |
| 15 | 2-Dodecanone | 4.49 | 184.32 | 1.18 | −5.19 | −5.19 | 0.00 |
| 16 | 1,2,4-Trimethylbenzene | 3.78 | 120.2 | 7.72 | −4.19 | −4.59 | 0.40 |
| 17 | Ethylbenzene | 3.15 | 106.17 | 10.5 | −4.00 | −4.05 | 0.05 |
| 18 | Phenyl ether | 4.21 | 170.21 | 4 | −4.63 | −4.95 | 0.32 |
| 19 | Pyrrole | 0.75 | 67.09 | 210 | −2.50 | −2.02 | −0.48 |
| 20 | 1,3,5-Trioxane | −0.43 | 90.08 | 5950 | −1.18 | −1.03 | −0.15 |

## D.  Independent Assessment of Predictive Capacity

The amount of data needed to validate a QSAR model is a matter of debate. The better the selection of test data (i.e., the greater the coverage of the applicability domain), the fewer data that are needed. Another consideration is whether some of the test data should fall outside the predefined domain of the QSAR, in order to explore further the limits of applicability. In the absence of other considerations, a working suggestion is that a minimal amount of 30% of the total number of compounds in the training set should be included in the test set.

To illustrate the validation of the confirmed QSAR, physicochemical and toxicity data for 20 chemicals were taken from Russom et al. (1997). This is also the source of data used to develop the training set (Table 20.1), so the $LC_{50}$ and $K_{ow}$ data in the test set (Table 20.2) can be assumed to be of a similar quality.

The domain of the external test set falls within that of the training set. The range of log $K_{ow}$ values for the test set is from -0.43 to 4.50. The chemical structures represented by the test set are consistent with those representing non-polar narcosis (Verhaar et al., 1992), and are similar to those in the training set. The predicted toxicities of the test set chemicals are reported in Table 20.2. There is a statistically significant relationship between predicted and observed toxicity:

$$\text{Log } (1/LC_{50})\text{observed} = 0.963 \text{ log } K_{ow} - 0.08 \tag{20.3}$$

The following statistics were reported: $n = 20$, $r^2(adj) = 0.922$, $s = 0.329$, $F = 225$. This relationship is shown in Figure 20.1. Investigation of the residuals of predicted toxicity, as well as the standard errors from the above equation, suggests that the toxicity of chemicals may be predicted within a 95% confidence interval of ±0.64 log unit.

## E.  Independent Peer Review of the Validation Process

It can be imagined that the appropriate peer-review panel would take into account a report on the selection and validation of the QSAR before issuing a statement on the validity of the QSAR.

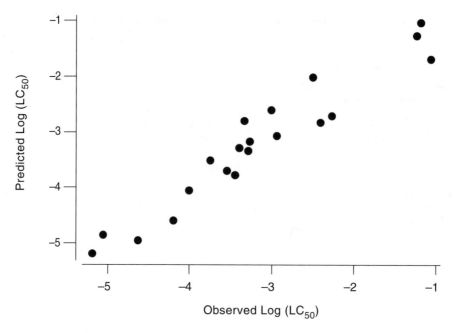

**Figure 20.1**  Plot of observed against predicted $LC_{50}$ values for the test set.

The statement would describe the background to the validation study and the principles of validation followed, and might include the following wording:

> The QSAR is acceptable for the prediction of the acute toxicity to *Pimephales promelas* of organic chemicals, considered to act by the non-polar narcosis mechanism of action. The range of acceptable log $K_{ow}$ values for which it can be applied is from –1.24 to 5.13. The coefficient of determination ($r^2$ value) of the model is 0.9, and its expected accuracy (95% prediction interval) is ±0.64 log unit

The statement would be signed by an appropriate person (e.g., the chairman of the working group), to testify that the wording was endorsed by the members of the peer review panel. An appendix to the statement would give details of the membership of the peer review panel.

## F.  Progression toward Regulatory Acceptance

This step is difficult to illustrate, because the process is unprecedented. However, it can be imagined that the validated and endorsed QSAR would be published in an official document, such as the EU TGD on risk assessment (European Economic Community, 1996), or possibly in an instrument of the chemicals legislation (in the EU, this might be Annex V to Directive 67/548/EEC).

## V. CONCLUSIONS AND DISCUSSION

The use of QSARs for the regulatory assessment of chemicals is very limited, mainly because there is widespread disagreement on the possible applications of QSARs and the extent to which QSAR predictions can be relied upon. To address the credibility problem of QSARs, it will be necessary to obtain international agreement on acceptability criteria for the development and validation of QSARs, and to apply the criteria in the context of a formal framework that guarantees independence in the selection, confirmation, and validation of QSARs.

With a view to promoting the use and regulatory acceptance of QSARs, an activity was agreed by the Heads of Delegation to the 34th OECD Joint Meeting of the Chemicals Committee and the Working Party on Chemicals, Pesticides and Biotechnology, during a special session on QSARs, held on November 6, 2002. It was also agreed that the first focus of the OECD activity should be to obtain international agreement on acceptability criteria for QSARs. To coordinate the new activity, the OECD has established an *Ad Hoc* Expert Group on QSARs.

In parallel to the OECD initiative, the European Commission has initiated a new activity on the Development, Validation, and Dissemination of QSARs, at the European Commission's Joint Research Center (JRC). The general role of the JRC is to provide independent scientific and technical advice to the policy-making Directorates-General of the Commission. The JRC is recognized to be independent of national and sectoral interests, and is therefore ideally suited to providing an independent framework for promoting the regulatory acceptance and widespread use of QSARs. The JRC framework will be developed according to the proposals presented in this chapter, taking into account future agreements on acceptability criteria. It is hoped that the new framework will lead to a more widespread use of QSARs, particularly for regulatory purposes under the future REACH system. The safety of chemicals can then be adequately assessed in a way that minimizes not only financial costs, but also the use of animal experimentation.

## REFERENCES

Balls, M., Blaauboer, B.J., Fentem, J.H., Bruner, L., Combes, R.D., Ekwall, B., Fielder, R.J., Guillouzo, A., Lewis, R.W., Lovell, D.P., Reinhardt, C.A., Repetto, G., Sladowski, D., Spielmann, H. and Zucco, F., Practical aspects of the validation of toxicity test procedures: the report and recommendations of ECVAM workshop 5, *Alternatives Lab. Anim.*, 23, 129–147, 1995.

Boxall, A.B.A., Watts, C.D., Dearden, J.C., Bresnen, G.M., and Scoffin, R., Predicting the toxic mode of action for environmental pollutants based on physico-chemical properties, in *Quantitative Structure-Activity Relationships in Environmental Sciences – VII*, Chen, F. and Schüürmann, G., Eds., SETAC Press, Pensacola, FL, 1997, pp. 263–275.

Cronin, M.T.D. and Schultz, T.W., Pitfalls in QSAR, *J. Mol. Struct. (Theochem)*, 622, 39–51, 2003.

Cronin, M.T.D., Jaworska, J.S., Walker, J.D., Comber, M.H.I, Watts, C.D., and Worth, A.P., Use of QSARs in international decision-making frameworks to predict health effects of chemical substances, *Environ. Health Perspect.*, 111, 1391–1401, 2003a.

Cronin, M.T.D., Walker, J.D., Jaworska, J., Comber, M.H.I., Watts, C.D., and Worth, A.P., Use of QSARs in international decision-making frameworks to predict ecologic effects and environmental fate of chemical substances, *Environ. Health Perspect.,* 111, 1376–1390, 2003b.

Dimitrov, S.D., Mekenyan, O.G., Sinks, G.D., and Schultz, T.W., Global modeling of narcotic chemicals: ciliate and fish toxicity, *J. Mol. Struct. (Theochem)*, 622, 63–70, 2003.

Eriksson, L., Jaworska, J., Worth, A.P., Cronin, M.T.D., McDowell, R.M., and Gramatica, P., Methods for reliability and uncertainty assessment and applicability evaluations of classification- regression-based and QSARs, *Environ. Health Perspect*, 111, 1361-1375, 2003.

European Commission, QSAR for Predicting Fate and Effects of Chemicals in the Environment, Final report of DG XII contract No. EV5V-CT92-0211, 1995.

European Economic Community, Technical Guidance Document in Support of Commission Directive 93/67/EEC on Risk Assessment for New Notified Substances and Commission Regulation (EC) No 1488/94 on Risk Assessment for Existing Substances, European Commission, Office for Official Publications of the European Communities, Luxembourg, 1996.

European Commission, *White Paper on a Strategy for a Future Chemicals Policy*, Commission of the European Communities, Brussels, europa.eu.int/comm/environment/chemicals/whitepaper.htm, 2001.

Jaworska, J.S., Comber, M., Auer, C. and Van Leeuwen, C.J., Summary of a workshop on regulatory acceptance of (Q)SARs for human health and environmental endpoints, *Environ. Health Perspect.*, 111, 1358–1360, 2003.

McKim, J.M., Schmieder, P.K., Carlson, R.W., Hunt, E.P., and Niemi, G.I., Use of respiratory-cardiovascular responses of rainbow-trout (*Salmo gairdneri*) in identifying acute toxicity syndromes in fish. 1. Pentachlorophenol, 2,4-dinitrophenol, tricaine methanesulfonate and 1-octanol, *Environ. Toxicol. Chem.,,* 6, 295–312, 1987.

Organization for Economic Cooperation and Development, OECD Guidelines for the Testing of Chemicals No 203: Fish Acute Toxicity Test, Organization for Economic Cooperation and Development, Paris, www.oecd.org/ehs/test/testlist.htm, 1992.

Russom, C.L., Bradbury, S.P., Broderius, S.J., Hammermeister, D.E., and Drummond, R.A., Predicting modes of toxic action from chemical structure: acute toxicity in the fathead minnow (*Pimephales promelas*), *Environ. Toxicol. Chem.,* 16, 948–967, 1997.

Schultz, T.W. and Cronin, M.T.D., Essential and desirable characteristics of ecotoxicity quantitative structure-activity relationships, *Environ. Toxicol. Chem.,* 22, 599–607, 2003.

Verhaar, H.J.M., van Leeuwen, C.J., and Hermens, J.L.M., Classifying environmental pollutants. 1. Structure-activity relationships for prediction of aquatic toxicity, *Chemosphere*, 25, 471–491, 1992.

Worth A.P., Barratt M.D., and Houston J.B., The validation of computational prediction techniques, *Alternatives Lab. Anim.*, 26, 241–247, 1998.

Worth, A.P. and Balls, M., The role of ECVAM in promoting the regulatory acceptance of alternative methods in the European Union, *Alternatives Lab. Anim.*, 29, 525–535, 2001.

Worth, A.P. and Cronin, M.T.D., Report of the workshop on the validation of (Q)SARs and other computational prediction models, *Proceedings of the Fourth World Congress on Alternatives to Animal Use in Life Sciences, Alternatives Lab. Anim.,* 2004, in press.

# Index